Birkhäuser

Trends in Mathematics

Trends in Mathematics is a series devoted to the publication of volumes arising from conferences and lecture series focusing on a particular topic from any area of mathematics. Its aim is to make current developments available to the community as rapidly as possible without compromise to quality and to archive these for reference.

Proposals for volumes can be submitted using the Online Book Project Submission Form at our website www.birkhauser-science.com.

Material submitted for publication must be screened and prepared as follows:

All contributions should undergo a reviewing process similar to that carried out by journals and be checked for correct use of language which, as a rule, is English. Articles without proofs, or which do not contain any significantly new results, should be rejected. High quality survey papers, however, are welcome.

We expect the organizers to deliver manuscripts in a form that is essentially ready for direct reproduction. Any version of TEX is acceptable, but the entire collection of files must be in one particular dialect of TEX and unified according to simple instructions available from Birkhäuser.

Furthermore, in order to guarantee the timely appearance of the proceedings it is essential that the final version of the entire material be submitted no later than one year after the conference.

Thomas Dedieu · Flaminio Flamini ·
Claudio Fontanari · Concettina Galati · Rita Pardini
Editors

The Art of Doing Algebraic Geometry

Editors
Thomas Dedieu
Institut de Mathématiques
de Toulouse
Université Paul Sabatier
Toulouse, France

Flaminio Flamini
Dipartimento di Matematica
Università di Roma "Tor Vergata"
Rome, Italy

Claudio Fontanari
Dipartimento di Matematica
Università degli Studi di Trento
Povo, Trento, Italy

Concettina Galati
Dipartimento di Matematica
e Informatica
Università della Calabria
Arcavacata di Rende, Cosenza, Italy

Rita Pardini
Dipartimento di Matematica
Università di Pisa
Pisa, Italy

ISSN 2297-0215 ISSN 2297-024X (electronic)
Trends in Mathematics
ISBN 978-3-031-11937-8 ISBN 978-3-031-11938-5 (eBook)
https://doi.org/10.1007/978-3-031-11938-5

Mathematics Subject Classification: 00B30, 14, 01

This book is published under the imprint Birkhäuser, www.birkhauser-science.com by the registered company Springer Nature Switzerland AG
The registered company address is: Gewerbestrasse 11, 6330 Cham, Switzerland

VIN**C**ENZO DI GENNARO *
LUCA CHIANTINI * R**I**CK
MIRANDA * FLAMINIO FLAMINI *
MA**R**GARIDA MENDES LOPES
* LUCA UGAGLIA * ENRIC**O**
ROGORA * DAVIDE FRANCO *
CLAUDIO PEDRINI * ANTONIO
LAFA**C**E * ALDO BRIGAGLIA
* CLAUDIO FONTANAR**I** * MARIA
CHIARA BRAMBILLA * ELISA
POSTINGHE**L** * RITA PARDINI *
FRANCESCO POLIZZI * PIETRO
SABAT**I**NO * ANGELO FELICE
LOPEZ * MIKHAIL ZAIDEN**B**ERG
* JOAQUIM ROÉ * THOMAS
DEDIEU * EDOARDO S**E**RNESI
* PAOLA SUPINO * GERARD
VAN DE**R** GEER * CONCETTINA
GALATI * OLIVIA DUMITRESCU
* FABRIZIO CA**T**ANESE * YURI
PROKHOROV * MASSIMILIANO
MELLA * GIORGIO **O**TTAVIANI

Preface

This book offers an overview of an important part of current research in algebraic geometry and related research in history of mathematics. It collects 4 surveys,12 research articles in algebraic geometry and 2 research articles in history of mathematics. All survey papers are about fundamental research topics: [G. van der Geer] is about Siegel modular forms and invariant theory; [A. F. Lopez] (with an appendix by T. Dedieu) is about the extendability of projective varieties; [M. Mendes Lopes and R. Pardini] is about the canonical map of projective surfaces of general type and related open questions; [F. Polizzi and P. Sabatino] discusses Kodaira fibrations and braid groups. The 12 research papers collected in this volume deal with very topical problems in algebraic geometry: the birational geometry of Mori dream spaces, Cremona orbits and interpolation theory, [M. C. Brambilla, O. Dumitrescu, E. Postinghel] and [O. Dumitrescu, R. Miranda]); Kummer surfaces, self-duality of hypersurfaces and existence of nodal surfaces [F. Catanese]; Severi varieties of nodal hypersurfaces and Hodge conjecture [V. Di Gennaro, D. Franco]; sextic plane projective curves and secant varieties of Veronese surfaces [L. Chiantini, G. Ottaviani]; deformation theory and extensions of weighted projective spaces [T. Dedieu, E. Sernesi]; Hilbert schemes of curves [F. Flamini, P. Supino], Seshadri constants of toric surfaces [A. Laface, L. Ugaglia], the Cremona equivalence problem [M. Mella], Hyperkähler varieties and motives [C. Pedrini]; the problem of flexibility of the affine cones over a projective variety and Fano-Mukai fourfolds [Y. Prokhorov and M. Zaidenberg]; ramified morphisms between smooth surfaces and singularities type of the ramification/branch locus [J. Roé]. Finally, this book contains two papers on peculiar topics of history of mathematics and in particular of geometry: [A. Brigaglia] analyzes the evolution of language in relation to the mathematical thinking through the study of the works of Apollonius and Simson; [E. Rogora] is dedicated to the course given by Corrado Segre in the a.y. 1920–21 on the geometric theory of differential equations, with emphasis on the geometric intuition present in the works of several Italian mathematicians in the period 1860–1940.

This volume is dedicated to Ciro Ciliberto on the occasion of his 70th birthday. The variety of topics covered in this volume reflects the versatility of interests of Ciro, master, colleague and friend.

Toulouse, France — Thomas Dedieu
Rome, Italy — Flaminio Flamini
Povo, Italy — Claudio Fontanari
Arcavacata di Rende, Italy — Concettina Galati
Pisa, Italy — Rita Pardini

Acknowledgements We are grateful to Francesca Bonadei, Springer, the Organizing and Scientific Committees of the conference *Riposte Armonie: Algebraic Geometry in Cetraro 2021, together with Ciro Ciliberto* (https://sites.google.com/site/agventotene2020/home) and all people who supported the idea of writing this book and who participated in this project.

The image on page v was created by Giulia Ciliberto. We warmly thank Giulia for her kind contribution.

Contents

Weyl Cycles on the Blow-Up of $\mathbb{P}^4$ at Eight Points

Maria Chiara Brambilla, Olivia Dumitrescu, and Elisa Postinghel

Abstract We define the *Weyl cycles* on X_s^n, the blown-up projective space $\mathbb{P}^n$ in s points in general position. In particular, we focus on the Mori Dream spaces X_7^3 and X_8^4, where we classify all the Weyl cycles of codimension two. We further introduce the *Weyl expected dimension* for the space of global sections of any effective divisor that generalizes the *linear expected dimension* of [2] and the *secant expected dimension* of [4].

1 Introduction

Let X_s^n be the blown-up projective space $\mathbb{P}^n$ in s points in general position. When the number of points s is small, the space X_s^n has an interpretation as certain moduli space, see, e.g., [1, 5]. Mori Dream Spaces of the form X_s^n were classified via the work of Mukai [19, 20] and techniques of birational geometry of moduli spaces. In previous work, in order to analyze properties of the pairs (X_s^n, D) with D a Cartier

---

Dedicated to Ciro Ciliberto, whose work inspired us throughout the years.

M. C. Brambilla
Università Politecnica delle Marche, via Brecce Bianche, 60131 Ancona, Italy
e-mail: brambilla@dipmat.univpm.it

O. Dumitrescu
University of North Carolina at Chapel Hill, 340 Phillips Hall CB 3250,
Chapel Hill, NC 27599-3250, USA
e-mail: dolivia@unc.edu

Simion Stoilow Institute of Mathematics Romanian Academy, 21 Calea Grivitei Street,
010702 Bucharest, Romania

E. Postinghel (✉)
Dipartimento di Matematica, Università degli Studi di Trento, via Sommarive 14,
38123 Povo di Trento, Italy
e-mail: elisa.postinghel@unitn.it

T. Dedieu et al. (eds.), *The Art of Doing Algebraic Geometry*, Trends in Mathematics,
https://doi.org/10.1007/978-3-031-11938-5_1

divisor, the authors of this article developed techniques of *polynomial interpolation theory* in [2] for $s = n + 2$ and in [4] for $s = n + 3$, respectively, via the study of the base loci.

An analogous approach, based on interpolation theory, is developed in this paper to define and to study the subvarieties determining the birational geometry of the Mori Dream Spaces X^3_7 and X^4_8. We will use this study as an opportunity to reveal the geometry hidden in the Weyl group action on fixed linear cycles of X^n_s and its consequences. For instance, we expect that for all Mori Dream Spaces of type X^n_s, Weyl cycles determine the birational geometry of such spaces, the cones of effective and movable divisors and their decomposition into Mori chambers.

In this article, we propose a definition of *Weyl cycles* on X^n_s as follows: (see Definition 1 for details).

(1) We call *Weyl divisor* any effective divisor D in $\mathrm{Pic}(X^n_s)$ in the Weyl orbit of an exceptional divisor $E_i \in \mathrm{Pic}(X^n_s)$.
(2) We call *Weyl cycle of codimension* i an element of the Chow group $A^i(X^n_s)$ that is an irreducible component of the intersection of Weyl divisors, which are pairwise orthogonal with respect to the Dolgachev-Mukai pairing on $\mathrm{Pic}(X^n_s)$.

For an arbitrary number s of points, Weyl divisors are always extremal rays of the cone of effective divisors of X^n_s. The correspondence between (-1)-curves of $\mathbb{P}^2$ and Weyl curves in X^2_s (i.e., Weyl divisors) was proved by Nagata [21], while giving a counterexample to the Hilbert 14-th problem. Moreover, in the case of $\mathbb{P}^2$, Weyl curves have been widely investigated since they are involved in the well-known Segre-Harbourne-Gimigliano-Hirschowitz conjecture, see, e.g., [6, 7] or, more widely, in the classification of algebraic surfaces (Castelnuovo's contraction theorem), the base of the minimal model program. The notion of *divisorial* (-1)*-classes* on X^n_s was introduced by Laface and Ugaglia in [16] and recently studied by the second author and Priddis in [14].

In the case of X^3_s, Laface and Ugaglia introduced the notion of elementary (-1)-curves and studied their properties in [17]. The case of Weyl cycles of X^4_8 has been studied in [11] with a different approach: indeed in such paper Weyl orbits in X^4_8 (as well as in X^3_7) of the strict transforms of linear cycles blown up along lines spanned by any two points are described. The classification of Weyl cycles obtained in [11] for the cases X^3_7, X^4_8 yields the same classification we determine here for Weyl curves in X^3_7 and for Weyl surfaces (see Eqs. (1)) in X^4_8. Therefore, we conclude that Definition 1 and the definitions used in [11] are equivalent for cycles of codimension 2 in X^3_7 and X^4_8. We believe that these two definitions are related in general, namely, for Weyl cycles in X^n_s for arbitrary n, s, and we will study their connection in forthcoming work.

In this article, we emphasize that basic methods of intersection theory, applied to pairs of orthogonal Weyl divisors, give an iterative method to compute the Weyl cycles of codimension 2 in X^3_7 and in X^4_8. Moreover, we show that every Weyl cycle is swept out by families of rational curves parametrized by Weyl cycles of larger codimension, see Proposition 3, Corollary 1 and Lemma 5. This allows us to give a formula for the multiplicity of containment of each Weyl cycle in the base locus of an effective divisor. We expect these formulas to give rise to the equations of the walls of the movable cone of divisors and of its decomposition in nef chambers.

Our main result, contained in Sect. 5.3, is a classification of all the Weyl surfaces of X^4_8. We compute the class of each such surface in the Chow ring of $X^4_{8,(1)}$, the blow-up of X^4_8 along the strict transforms of all lines through two base points and all rational normal quartic curves through seven base points. There are five such classes, up to index permutation, as listed in the following formula (see Sect. 5.2 for the precise notation):

$$\begin{aligned}
&S^1_{1,4,5} : h - e_1 - e_4 - e_5 - \sum_{i,j\in\{1,4,5\}} (e_{ij} - f_{ij})\\
&S^3_{1,\hat{8}} : 3h - 3e_1 - \sum_{i=2}^{7} e_i - (e_{C_{\hat{8}}} - f_{C_{\hat{8}}}) - \sum_{i=2}^{7} (e_{1i} - f_{1i})\\
&S^6_{6,7,8} : 6h - 3\sum_{i=1}^{5} e_i - \sum_{i=6}^{8} e_i - \sum_{i,j\in\{1,2,3,4,5\}, i\neq j} (e_{ij} - f_{ij}) - \sum_{k=6}^{8} (e_{C_{\hat{k}}} - f_{C_{\hat{k}}}) \qquad (1)\\
&S^{10}_{1,2} : 10h - 6e_1 - 6e_2 - \sum_{i=3}^{8} 3e_i - 3(e_{12} - f_{12}) - \sum_{i=1}^{2}\sum_{j=3}^{8} (e_{ij} - f_{ij}) - \sum_{k=3}^{8} (e_{C_{\hat{k}}} - f_{C_{\hat{k}}})\\
&S^{15}_8 : 15h - \sum_{i=1}^{7} 6e_i - 3e_8 - \sum_{1\leq i<j\leq 7} (e_{ij} - f_{ij}) - \sum_{i=1}^{7} (e_{C_{\hat{i}}} - f_{C_{\hat{i}}}) - 3(e_{C_{\hat{8}}} - f_{C_{\hat{8}}})
\end{aligned}$$

Recall that the birational geometry of X^4_8 has been investigated in [5, 20]. Casagrande, Codogni, and Fanelli studied in detail the relation between the geometry of X^2_8 and X^4_8 and in [5, Theorem 8.7] they described five types of surfaces in X^4_8 playing a special role in the Mori program. We emphasize that this list agrees with our classification of Weyl surfaces, (1). All the surfaces of table (1), except for the first one, are normal on $X^4_{8,(1)}$, but non-normal on $\mathbb{P}^4$. In particular, some of them have isolated singularities at the points p_i (when the coefficient of e_i is 2 or larger) and ordinary triple point singularities along lines L_{ij} or rational normal quartic curves $C_{\hat{i}}$ (when the coefficient of $(e_{ij} - f_{ij})$, or of $(e_{C_{\hat{i}}} - f_{C_{\hat{i}}})$ is 3, cf. $S^{10}_{1,2}$ and S^{15}_8). In classical language, the description of S^{15}_8 can be expressed as: there exists a surface of degree 15 in $\mathbb{P}^4$ passing through seven general points with multiplicity 6 and through another general point with multiplicity 3, containing lines L_{ij} and curves $C_{\hat{k}}$, where $1 \leq i, j, k \leq 7$ and triple at every point on the rational curve $C_{\hat{8}}$. Some of the conditions imposed by the curve containment could be redundant when the curve is already in the base locus forced by the points (as in the first case $S^1_{i,j,k}$), but perhaps not all of them. Indeed, since for instance the surface class $3h - 3e_1 - \sum_{i=2}^{7} e_i$ moves in a positive dimensional family, the curve $C_{\hat{8}}$ cannot be contained in all elements of such family (that contains also the union of 3 planes). Therefore we are imposing the containment of the curve in the surface class $S^3_{1,\hat{8}}$ which has a unique representative satisfying these conditions. In order to uniquely identify the surfaces, one needs to blow-up the curves.

Finally, we propose here a notion of expected dimension for a linear system which takes into account the contribution to the speciality given by the Weyl cycles contained in the base locus. In Definition 2, we introduce, for X^n_{n+4} and $n = 3, 4$, the *Weyl expected dimension* of a divisor D, as follows:

$$\mathrm{wdim(D)} := \chi(X, \mathcal{O}_X(D)) + \sum_{r=1}^{n-1}\sum_{A}(-1)^{r+1}\binom{n + k_A(D) - r - 1}{n},$$

where A ranges over the set of Weyl cycles of dimension r and $k_A(D)$ is the multiplicity of containment of the cycle A in the base locus of D. This notion extends the analogous definitions of *linear expected dimension* of [2] and *secant expected dimension* of [4]. We prove that any effective divisor D in X^3_7 satisfies $\mathrm{h}^0(X^3_7, \mathcal{O}_{X^3_7}(D)) = \mathrm{wdim}(D)$, see Theorem 1, and we conjecture that the same holds in X^4_8, see Conjecture 1.

The paper is organized as follows. In Sect. 2, we introduce the notation, recall basic facts on the blown-up of $\mathbb{P}^n$ at s general points, X^n_s, and on the action of standard Cremona transformations on $\mathrm{Pic}(X^n_s)$. In Sect. 3, we introduce the definition of Weyl cycles and we give some general results on Weyl curves in X^n_s. Section 4 is devoted to the preliminary case of X^3_7, where we classify Weyl divisors and Weyl curves and we describe their geometry. Section 5 concerns the case of X^4_8. The main result, i.e., the classification of the Weyl surfaces is contained in Sect. 5.3. In Sect. 5.4, we give the classification of Weyl divisors and their geometrical description. Section 6 is devoted to the dimensionality problem.

2 Preliminaries

We denote by X^n_s the blown-up of $\mathbb{P}^n$ at s general points $\mathcal{I} = \{p_1, \ldots, p_s\}$. The Picard group of X^n_s is $\mathrm{Pic}(X^n_s) = \langle H, E_1, \ldots, E_s\rangle$, where H is a general hyperplane class, and the E_i's are the exceptional divisors of the p_i's. For any subset $J \subseteq \{1, \ldots, s\}$ of cardinality $\leq n$, we denote by L_J the class, in the Chow ring of X^n_s, of the strict transform of the linear cycle spanned p_j, $j \in J$. If $|J| = n$, then $L_J = H - \sum_{i\in J} E_i \in \mathrm{Pic}(X^n_s)$ is the class of a fixed hyperplane.

The *Dolgachev-Mukai pairing* on $\mathrm{Pic}(X^n_s)$ is the bilinear form defined as follows (cf. [19]):

$$\langle H, H\rangle = n - 1, \quad \langle H, E_i\rangle = 0, \quad \langle E_i, E_j\rangle = -\delta_{i,j}.$$

The *standard Cremona transformation based on the coordinate points* on $\mathbb{P}^n$ is the birational transformation defined by the following rational map:

$$\mathrm{Cr} : (x_0 : \cdots : x_n) \to (x_0^{-1} : \cdots : x_n^{-1}),$$

see, e.g., [10, 14] for more details. Given any subset $I \subseteq \{1, \dots, s\}$ of cardinality $n+1$, we denote by Cr_I and call *standard Cremona transformation* the map obtained by precomposing Cr with a projective transformation which takes the points indexed by I to the coordinate points of $\mathbb{P}^n$. A standard Cremona transformation induces an automorphism of $\mathrm{Pic}(X^n_s)$, denoted again by Cr_I by abuse of notation, by sending a divisor

$$D = dH - \sum m_i E_i \tag{2}$$

to

$$\mathrm{Cr}_I(D) = (d-c)H - \sum_{i\in I}(m_i - c)E_i - \sum_{j\notin I}^{s} m_j E_j, \tag{3}$$

where $c = c_I := \sum_{i\in I} m_i - (n-1)d$. The canonical divisor $-(n+1)H + (n-1)\sum_{i=1}^{s} E_i$ is invariant under such an automorphism. The *Weyl group* $W_{n,s}$ acting on $\mathrm{Pic}(X^n_s)$ is the group generated by standard Cremona transformations, see [10]. We say that a divisor (2) is Cremona reduced if $c_I \leq 0$ for any I of cardinality $n+1$.

In [14, Theorem 3.2] the authors observed that the intersection pairing between divisors is preserved under Cremona transformations.

Lemma 1 *Let D, F be two divisors and let $\omega \in W_{n,s}$ be an element of the Weyl group. Then $\langle \omega(D), \omega(F)\rangle = \langle D, F\rangle$.*

Here we point out that the scheme-theoretic intersection of two divisors is in general not preserved under Cremona transformations. Let D, F be two divisors and let $\omega = \mathrm{Cr}_I$ be a standard Cremona transformation. Then

$$\omega(D \cap F) \cup \Lambda \subseteq \omega(D) \cap \omega(F)$$

where Λ is a union of linear cycles of the indeterminacy locus of ω. The following lemma provides an explicit recipe for Λ.

Lemma 2 *Let $I \subseteq \{1, \dots, s\}$ have cardinality $n+1$, and let $I = I_1 \cup I_2$, with $|I_1| = m+1$ and $|I_2| = n-m$. Let $D = dH - \sum m_i E_i$ be a divisor in X^n_s. If $(n-m-1)d - \sum_{i\in I_2} m_i = a \geq 1$, then the m-plane L_{I_1} is contained in $\mathrm{Cr}_I(D)$ exactly a times.*

Proof Set $c = \sum_{i\in I} m_i - (n-1)d$. By [2] and [13, Proposition 4.2], we can compute the multiplicity of containment of the m-plane L_{I_1} in $\mathrm{Cr}(D)$:

$$\sum_{i\in I_1}(m_i - c) - m(d-c) = \sum_{i\in I_1} m_i - md - c = (n-m-1)d - \sum_{i\in I_2} m_i = a,$$

concluding the proof. □

3 Weyl Cycles in $\mathbb{P}^n$ Blown Up at s Points

In [14, Definition 4.1] a smooth divisor D in $\mathrm{Pic}(X^n_s)$ is called (-1)-*class* (or (-1)-*divisorial cycle*) if D is effective, integral and it satisfies $\langle D, D\rangle = -1$ and $\langle D, -K_{X^n_s}\rangle = n-1$. In [14, Theorem 0.5], it is proved that D is a (-1)-class if and only if it is in the Weyl orbit of some exceptional divisor E_i. Notice that if $i \in I$, then $\mathrm{Cr}_I(E_i) = L_{I\setminus\{i\}}$ is a hyperplane through n base points.

Here we generalize the definition of (-1)-classes to cycles of higher codimension in X^n_s, as follows. We will say that two divisors D and F are *orthogonal* if $\langle D, F\rangle = 0$.

Definition 1 We introduce the following:

(1) A *Weyl divisor* is an effective divisor $D \in \mathrm{Pic}(X^n_s)$ which belongs to the Weyl orbit of an exceptional divisor E_i.
(2) A *Weyl cycle of codimension i* is a non-trivial effective cycle $C \in A^i(X^n_s)$ which is an irreducible component of the intersection of pairwise orthogonal Weyl divisors.

Remark 1 Let $s \geq n+1$ and $1 \leq m \leq n-1$. Any m-plane L spanned by $m+1$ points is a Weyl cycle. Indeed, it is easy to check that L is the intersection of $r = n-m$ pairwise orthogonal hyperplanes spanned by n points. By Lemma 1, any effective cycle C contained in the Weyl orbit of a m-plane L spanned by $m+1$ base points is a Weyl cycle. In particular, the Weyl planes and Weyl lines studied in [11] are always Weyl cycles, according to Definition 1.

We point out that two distinct non-orthogonal Weyl divisors intersect in a cycle which may not be a union of Weyl cycles according to our definition. For example, in X^3_5 the plane through p_1, p_2, p_3 and the plane through p_1, p_4, p_5 intersect in a line through p_1 which is not a Weyl cycle.

3.1 *Weyl Curves*

We collect here some results on Weyl cycles of codimension $n-1$ in X^n_s, which we call *Weyl curves*. The following examples show explicitly that the strict transforms of lines through two points and of the rational normal curves of degree n through $n+3$ points are Weyl curves in X^n_s, according to Definition 1.

Example 1 Let $L = L_{12}$ be the line through p_1 and p_2, then $L = D_1 \cap \cdots \cap D_{n-1}$ where $D_i = L_{I_i}$ and $I_i = \{1, 2, \ldots, n+1\} \setminus \{i+2\}$ for any $1 \leq i \leq n-1$.

Example 2 For any $i = 1, \ldots, n-1$, consider the pairwise orthogonal Weyl divisors $D_i = 2H - 2E_1 - \ldots - 2E_{n-1} - E_n - E_{n+1} - E_{n+2} - E_{n+3} + E_i$. One can easily check that $D_1 \cap \cdots \cap D_{n-1}$ is the union of $L_{1\ldots n-1}$ and the rational normal curve of degree n through $n+3$ points.

We recall that the Chow group of algebraic curves $A^{n-1}(X^n_s)$ is generated by h^1, e^1_i, the classes of a general line in X^n_s and of a general line on the exceptional divisor E_i, respectively. The following formula describes the action on curves of the standard Cremona transformation Cr_J, based on the set J if $C = \delta h^1 - \sum_{i=1}^{s} \mu_i e^1_i$, then [12] implies

$$\mathrm{Cr}_J(C) = (n\delta - (n-1)\sum_{j\in J} \mu_j)h^1 - \sum_{j\in J}(\delta - \sum_{i\in J\setminus\{j\}} \mu_i)e^1_j - \sum_{j\notin J} \mu_j e^1_j. \tag{4}$$

Remark 2 Given a divisor D in X^n_s and a line $L_{ij} = h^1 - e^1_i - e^1_j$, then the multiplicity of containment of the line L_{ij} in the base locus of D is exactly $\max\{0, -D \cdot L_{ij}\}$, where $\cdot$ denotes the intersection product in the Chow ring of X^n_s (cf. [13, Proposition 4.2]). If $n = 3, 4$, the same holds for any curve C in the Weyl orbit of the line L_{ij}, thanks to formulas (4).

4 $\mathbb{P}^3$ Blown Up in Seven Points

In this section, we consider Weyl cycles of X^3_7, the blow-up of $\mathbb{P}^3$ at seven points in general position. Recall that X^3_7 is a Mori Dream Space and that the cone of effective divisors is generated by the divisors of anticanonical degree $\frac{1}{2}\langle D, -K_{X^3_7}\rangle = 1$. These are exactly the Weyl divisors and they fit in five different types, modulo index permutation.

Proposition 1 *The Weyl divisors in X^3_7 are modulo index permutation:*

(1) E_i (exceptional divisor);
(2) $H - E_1 - E_2 - E_3$ (planes through three points);
(3) $2H - 2E_1 - E_2 - E_3 - E_4 - E_5 - E_6$ (pointed cone over the twisted cubic);
(4) $3H - 2(E_1 + E_2 + E_3 + E_4) - E_5 - E_6 - E_7$ (Cayley nodal cubic)*;*
(5) $4H - 3E_1 - 2(E_1 + E_2 + E_3 + E_4 + E_5 + E_6 + E_7)$.

Proof It is easy to compute the Weyl orbit of a plane through three points, by applying Formula (3). □

Proposition 2 *The Weyl curves in X^3_7 are the fixed lines $L_{ij} = h^1 - e^1_i - e^1_j$ and the fixed twisted cubics $C_{\hat{j}} = 3h^1 - \sum_{i=1}^{7} e^1_i + e^1_j$.*

Proof For every pair of orthogonal Weyl divisors as in Proposition 2, one can check that the intersection is always the union of fixed lines and twisted cubics.

We give here some details only in one example, that is, the case of a cubic Weyl divisor and a quartic one. Since the divisors are orthogonal, we can assume that

$$D_1 = 3H - 2E_1 - 2E_2 - 2E_3 - 2E_4 - E_5 - E_6 - E_7$$

and

$$D_2 = 4H - 3E_1 - 2(E_1 + E_2 + E_3 + E_4 + E_5 + E_6 + E_7)$$

and we easily see, by using Remark 2, that the intersection is

$$D_1 \cap D_2 = C_{\hat{5}} \cup C_{\hat{6}} \cup C_{\hat{7}} \cup L_{12} \cup L_{13} \cup L_{23}.$$

All the other cases can be analogously analyzed. □

From the previous result, we can conclude that our Definition 1 of Weyl curves in X^3_7 is equivalent to the definition of Weyl line of [11].

In the following result, we describe the intrinsic geometry of the Weyl divisors of X^3_7, showing that they are covered by pencils of rational curves parametrized by a Weyl curve.

Proposition 3 *Let D be Weyl divisor on X^3_7. If $C \subset D$ is a Weyl curve, then there is a pencil of rational curves $\{C_q : q \in C\}$ with $C_q \cdot D = 0$ sweeping out D.*

Proof We will consider the divisors (2)–(5) from Proposition 2. It is easy to check what Weyl curves are contained in D, using Remark 2. For each such containment $C \subseteq D$, we will find a suitable pencil of curves, parametrized by C, sweeping out D.

(2) Let us consider the fixed hyperplane $D = H - E_1 - E_2 - E_3$ and the Weyl line $L_{12} \subset D$. Such plane is swept out by the pencil of lines through p_3 and with a point $q \in L_{12}$: $\{C^1_3(q) : q \in L_{12}\}$. Since the cycle class of $C^1_3(q)$ is $h^1 - e^1_3$, then we obtain $C^1_3(q) \cdot D = 0$.

(3) The quadric surface $D = 2H - 2E_1 - E_2 - E_3 - E_4 - E_5 - E_6$ contains the fixed twisted cubic $C^3_{1,\dots,6} = 3h^1 - \sum_{i=1}^{6} e^1_i$. Since it is the strict transform of a pointed cone, it is swept out by the pencil of lines $\{C^1_1(q) : q \in C_{1,\dots,6}\}$. We have $C^1_1(q) \cdot D = 0$.

Notice also that D can be obtained from $H - E_1 - E_2 - E_3$ through the transformation $\mathrm{Cr}_{1,4,5,6}$ (cf. (3)). The latter preserves the line L_{12} and, for every $q \in L_{12}$, it sends the line $C^1_3(q)$ to the cubic curve $C^3_{1,3,4,5,6}(q)$, see Formula (4). Therefore, we see that D is also swept out by the pencil $\{C^3_{1,3,4,5,6}(q) : q \in L_{12}\}$. Moreover, since the general element of $C^1_3(q)$ is not contained in the indeterminacy locus of $\mathrm{Cr}_{1,4,5,6}$, the intersection number is preserved $0 = C^1_3(q) \cdot (H - E_1 - E_2 - E_3) = C^3_{1,3,4,5,6}(q) \cdot D$.

(4) This surface is obtained from (3) via the standard Cremona transformation Cr_{2347}. The image of the first pencil sweeping out (3) is the pencil of cubics $\{C^3_{1,\dots,4,7}(q) : q \in C_{1,\dots,6}\}$ and it sweeps out (4). The image of the second pencil sweeping out (3) is $\{C^5_{3,4}(q) : q \in L_{12}\}$, where $C^5_{3,4}(q)$ is a quintic curve with cycle class $5h - \sum_{i=1}^{7} e_i - e_3 - e_4$ and passing through $q \in L_{12}$: this pencil sweeps out (4). As before, we can argue that $C^3_{1,\dots,4,7}(q) \cdot D = 0$ and $C^5_{3,4}(q) \cdot D = 0$.

(5) This surface is obtained from (4) via $\mathrm{Cr}_{1,5,6,7}$. On the one hand, we obtain that (5) is swept out by the pencil of quintics $\{C^5_{1,7}(q) : q \in C_{1,\dots,6}\}$. On the other hand, the surface is covered by the pencil of septic curves $C^7_2(q)$ with class $7h - 2\sum_{i=1}^{7} e_i + e_2$

passing through $q \in L_{12}$: $\{C_2^7(q) : q \in L_{12}\}$. In both cases, the intersection product $\cdot$ is preserved under Cremona transformation because the general curve in the pencil is not contained in the indeterminacy locus. □

5 $\mathbb{P}^4$ Blown Up in Eight Points

5.1 *Curves in* X_8^4

Notation 1 We consider the following classes of moving curves in $A^3(X_8^4)$, each obtained from the previous via a standard Cremona transformation (see Formula (4) and a permutation of indices. They each live in a four-dimensional family.

- $h^1 - e_i^1$, for any $i \in \{1, \ldots, 8\}$;
- $4h^1 - \sum_{i\in J} e_i^1$, for any $J \subset \{1, \ldots, 8\}$ with $|J| = 6$;
- $7h^1 - \sum_{i\in J} 2e_i^1 - \sum_{i\notin J} e_i^1$, for any J with $|J| = 3$;
- $10h^1 - e_{i_1} - 3e_{i_2} - \sum_{i\neq i_1,i_2} 2e_i^1$, for any $i_1 \neq i_2$, $i_1, i_2 \in \{1, \ldots, 8\}$;
- $13h^1 - \sum_{i\in J} 2e_i^1 - \sum_{i\notin J} 3e_i^1$, for any J with $|J| = 3$;
- $16h^1 - \sum_{i\in J} 4e_i^1 - \sum_{i\notin J} 3e_i^1$, for any J with $|J| = 2$.

The families of curves in Notation 1 correspond to facets of the effective cone of divisors on X_8^4, see [5]. Here we include a proof via our geometrical approach.

Proposition 4 *Let $D = dH - \sum m_i E_i$ be a divisor in X_8^4. If D is effective, then we have*

- *$m_i \leq d$, for every $i \in \{1, \ldots, 8\}$;*
- *$\sum_{i\in J} m_i - 4d \leq 0$, for any $J \subset \{1, \ldots, 8\}$ with $|J| = 6$;*
- *$\sum_{i\in J} 2m_i + \sum_{i\notin J} m_i - 7d \leq 0$, for any J with $|J| = 3$;*
- *$m_{i_1} + 3m_{i_2} + \sum_{i\neq i_1,i_2} 2m_i - 10d \leq 0$, for any $i_1 \neq i_2$, $i_1, i_2 \in \{1, \ldots, 8\}$;*
- *$\sum_{i\in J} 2m_i + \sum_{i\notin J} 3m_i - 13d \leq 0$, for any J with $|J| = 3$;*
- *$\sum_{i\in J} 4m_i + \sum_{i\notin J} 3m_i - 16d \leq 0$, for any J with $|J| = 2$.*

The first two inequalities were also proved in [2, Lemma 2.2].

Proof Notice that each four-dimensional family of Notation 1 covers $X_8^4 \setminus \bigcup_i E_i$, indeed for each general point in $X_8^4 \setminus \bigcup_i E_i$ we find one curve of the family that passes through it. Now, if $D \cdot (h^1 - e_i^1) = d - m_i < 0$, then D contains each line in the family in its base locus, but this contradicts the assumption that D is effective. This proves the first inequality. The remaining inequalities are proved similarly. □

5.2 *Further Blow-up of* $\mathbb{P}^4$

For any $1 \leq i \leq 8$, we denote by $C_{\hat{i}}$ the rational normal quartic curve passing through seven base points and skipping the ith point. Consider now

$$X^4_{8,(1)} \xrightarrow{p} X^4_8,$$

the blow-up of X^4_8 along the 28 lines L_{ij} and the 8 curves $C_{\hat{i}}$. The strict transforms on X^n_s of a line passing through two points and that of the unique rational normal curve of degree n passing through $n+3$ points are (-1)-curves, i.e., rational curves with homogeneous normal bundle $\mathcal{O}(-1)^{\oplus(n-1)}$. Since these curves are rational, the projection on the first factor of their exceptional divisors is $\mathbb{P}^1$. Since their normal bundle is homogeneous, a twist by a line bundle will make it trivial, so the projection onto the second factor is $\mathbb{P}^2$.

The Picard group of $X^4_{8,(1)}$ is $\mathrm{Pic}(X^4_{8,(1)}) =< H, E_i, E_{ij}, E_{C_{\hat{i}}} >$, where, abusing notation, we denote again by E_i the pullback $p^*(E_i)$ and by H the pullback $p^*(H)$, while E_{ij} and $E_{C_{\hat{i}}}$ are the exceptional divisors of the curves. Notice that E_i is a $\mathbb{P}^3$ blown up in 14 points, coming from the intersection with 7 lines and 7 rational normal quartic curves, that lie on a configuration of twisted cubics, while $E_{ij} \cong \mathbb{P}^1 \times \mathbb{P}^2$ and $E_{C_{\hat{i}}} \cong C_{\hat{i}} \times \mathbb{P}^2$.

For any $D \in \mathrm{Pic}(X^4_8)$, of the form $D = dH - \sum_{i=1}^8 m_i E_i$, the strict transform $\widetilde{D}$ of D under p satisfies

$$\widetilde{D} := D - \sum k_{ij} E_{ij} - \sum k_{C_{\hat{i}}} E_{C_{\hat{i}}}, \tag{5}$$

where k_{ij} and $k_{C_{\hat{i}}}$ are defined in Remark 2.

Let us consider now the Chow group of 2-cycles of $X^4_{8,(1)}$:

$$A^2(X^4_{8,(1)}) = \langle h, e_i, e_{ij}, f_{ij}, e_{C_{\hat{i}}}, f_{C_{\hat{i}}} \rangle,$$

where h is the pullback of a general plane of $\mathbb{P}^4$, e_i is the pullback of a general plane contained in E_i, $f_{ij} \cong \mathbb{P}^2$ is the fiber over a point of the line and $e_{ij} \cong \mathbb{P}^1 \times \mathbb{P}^1$ is the transverse direction, $f_{C_{\hat{i}}}$ is the fiber over a point of the curve $C_{\hat{i}}$ and $e_{C_{\hat{i}}}$ is the transverse direction. In the Chow ring $A^*(X^4_{8,(1)})$, we have the following relations:

$$H^2 = h, \quad E_i^2 = -e_i, \quad HE_i = 0, \quad E_iE_j = 0 \tag{6}$$

$$HE_{ij} = E_iE_{ij} = f_{ij}, \quad E_iE_{jk} = 0 \tag{7}$$

$$E_{ij}^2 = -e_{ij} - f_{ij}, \quad E_{ij}E_{ik} = 0, \quad E_{ij}E_{kl} = 0 \tag{8}$$

$$H^4 = h^2 = 1, \quad E_i^4 = e_i^2 = -1, \quad f_{ij}e_{ij} = e_{ij}^2 = -1 \tag{9}$$

$$f_{ij}^2 = e_if_{ij} = 0, \quad he_i = hf_{ij} = 0, \quad e_ie_{ij} = he_{ij} = 0. \tag{10}$$

5.3 Classification of the Weyl Surfaces

The section contains one of the main results of this paper. We construct five Weyl surfaces in X^4_8 and we prove that they are the only such cycles, modulo index permutation. For any surface, we also give its exact multiplicity of containment in a given divisor and its class in the Chow ring of $X^4_{8,(1)}$.

Proposition 5 *Let $S^1 = S^1_{1,4,5}$ be the plane L_{145} through three points in $\mathbb{P}^4$.*

- *Given an effective divisor $D = dH - \sum m_i E_i$ in X^4_8, let*

$$k_{S^1}(D) = \max\{0, m_1 + m_4 + m_5 - 2d\}.$$

Then the surface S^1 is contained in the base locus of D exactly $k_{S^1}(D)$ times.

- *The class of the strict transform $\widetilde{S^1}$ of S^1 in the Chow group $A^2(X^4_{8,(1)})$ is*

$$h - e_1 - e_4 - e_5 - \sum_{i,j\in\{1,4,5\}} (e_{ij} - f_{ij}).$$

Proof The first part of the statement follows from [2] and [13, Proposition 4.2].

Consider the fixed hyperplanes $D_0 := H - E_1 - E_3 - E_4 - E_5$ and $F_0 := H - E_1 - E_2 - E_4 - E_5$. Let $\widetilde{D_0}$ and $\widetilde{F_0}$ be their strict transforms on $X^4_{8,(1)}$, see (5). Clearly we have $S^1 = D_0 \cap F_0$, and $\widetilde{S^1} = \widetilde{D_0} \cap \widetilde{F_0}$. By using Relations (6), (7), (8), we compute $\widetilde{D_0} \cap \widetilde{F_0} = h - e_1 - e_4 - e_5 - \sum_{i,j\in\{1,4,5\}}(e_{ij} - f_{ij})$. □

Using Lemma 2 we obtain the following.

Lemma 3 *Given a subset $I = \{i_1, \ldots, i_5\} \subseteq \{1, \ldots, 8\}$ and a divisor $D = dH - \sum m_i E_i$ in X^4_8. If*

$$d - m_{i_1} - m_{i_2} = a \geq 1,$$

then the 2-plane $L_{i_3 i_4 i_5}$ is contained in $\mathrm{Cr}_I(D)$ exactly a times.

Lemma 4 *Let $I = \{i_1, i_2, i_3, i_4, i_5\}$ and $J = \{i_1, i_2, i_6\}$ be two subsets of $\{1, \ldots, 8\}$, such that $|I \cap J| = 2$. If Cr_I is the standard Cremona transformation based on I, then the plane L_J is Cr_I-invariant, that is, $\mathrm{Cr}_I(L_J) = L_J$.*

Proof Consider the hyperplanes $D = L_{i_1 i_2 i_3 i_6}$ and $F = L_{i_1 i_2 i_4 i_6}$. We have $D \cap F = L_{i_1 i_2 i_6} = L_J$. Clearly $\mathrm{Cr}_I(D) = D$ and $\mathrm{Cr}_I(F) = F$ and hence also $\mathrm{Cr}_I(L_J) = \mathrm{Cr}(D \cap F) \subseteq \mathrm{Cr}_I(D) \cap \mathrm{Cr}_I(F) = D \cap F = L_J$. □

Proposition 6 *Let $J := \{1, 2, 3, 6, 7\}$ and consider the Cremona transformation Cr_J. Let $S^1 = L_{145}$. Then $S^3 := \mathrm{Cr}_J(S^1)$ is the strict transform of cubic pointed cone over the rational normal curve $C_{\hat{8}}$ and the point p_1.*

- *Given a divisor $D = dH - \sum m_i E_i$, let*

$$k_{S^3}(D) = \max\{0, 2m_1 + m_2 + m_3 + m_4 + m_5 + m_6 + m_7 - 5d\}.$$

Then the surface S_3 is contained in the base locus of D exactly $k_{S^3}(D)$ times.

- *The class of the strict transform $\widetilde{S}^3$ of S^3 in the Chow group $A^2(X^4_{8,(1)})$ is*

$$3h - 3e_1 - \sum_{i=2}^{7} e_i - (e_{C_{\bar{8}}} - f_{C_{\bar{8}}}) - \sum_{i=2}^{7}(e_{1i} - f_{1i})$$

Proof The plane $S^1 = L_{145}$ is swept out by the pencil of lines $\{C^1(q) : q \in L_{14}\}$, where the cycle class of $C^1(q)$ is $h^1 - e^1_5$ and it passes through the point $q \in L_{14}$. Using Formula (4) and the same idea as in the proof of Proposition 3, we compute the images of the line $L_{14} = h^1 - e^1_1 - e^1_4$ and of the pencil of lines $\{C^1(q) : q \in L_{14}\}$ of class $h^1 - e^1_5$ via the transformation Cr_J. We have $\mathrm{Cr}_J(L_{14}) = L_{14}$ and $\mathrm{Cr}_J(C^1(q)) = C^4(q)$ where $C^4(q)$ is a rational curve with class $4h^1 - e^1_1 - e^1_2 - e^1_3 - e^1_5 - e^1_6 - e^1_7$ and passing through q. Thus we get that the surface S_3 is swept out by the pencil $\{C^4(q) : q \in L_{14}\}$. Therefore, D contains any curve $C^4(q)$, and hence S_3, in its base locus at least $\max\{0, m_1 + m_2 + m_3 + m_5 + m_6 + m_7 + \max\{0, m_1 + m_4 - d\} - 4d\}$ times. Notice that we have $m_1 + m_2 + m_3 + m_5 + m_6 + m_7 - 4d \leq 0$, since D is effective, by Proposition 4. Hence the claim follows.

Now we prove the second statement. Given D_0 and F_0 defined in the previous proposition, recall that $D_0 \cap F_0 = S^1$. We consider now their images $D_1 = \mathrm{Cr}_J(D_0)$ and $F_1 = \mathrm{Cr}_J(F_0)$:

$$D_1 = 2H - 2E_1 - E_2 - 2E_3 - E_4 - E_5 - E_6 - E_7$$
$$F_1 = 2H - 2E_1 - 2E_2 - E_3 - E_4 - E_5 - E_6 - E_7.$$

Clearly $S^3 \subseteq D_1 \cap F_1$. By Proposition 5 we easily see that the only plane contained in $D_1 \cap F_1$ is L_{123}. Moreover, it is easy to check that the intersection $D_1 \cap F_1$ does not intersect the indeterminacy locus of the Cremona transformation Cr_J in any other two-dimensional component. Hence, $D_1 \cap F_1$ is the union of the plane L_{123} and an irreducible cubic surface with one triple point in p_1 and six simple points. We conclude that S^3 is exactly such cubic surface.

We now shall describe the class of S^3 in $A^2(X^4_{8,(1)})$. Let $\widetilde{D_1}$ and $\widetilde{F_1}$ be the corresponding strict transforms under the blow-up of lines and rational normal curves in $X^4_{8,(1)}$. By (5), we have

$$\widetilde{D_1} = 2H - 2E_1 - E_2 - 2E_3 - \sum_{i=4}^{7} E_i - 2E_{13} - \sum_{i\in\{1,3\},k\in\{2,4,5,6,7\}} E_{ik} - E_{C_{\bar{8}}}$$

$$\widetilde{F_1} = 2H - 2E_1 - 2E_2 - \sum_{i=3}^{7} E_i - 2E_{12} - \sum_{1\leq i\leq 2, 3\leq k\leq 7} E_{ik} - E_{C_{\bar{8}}}.$$

By using Relations (6), (7), (8), we compute the intersection:

$$\widetilde{D_1} \cap \widetilde{F_1} = (h - e_1 - e_2 - e_3 - \sum_{i,j \in \{1,2,3\}} (e_{ij} - f_{ij})) +$$
$$(3h - 3e_1 - \sum_{i=2}^{7} e_i - (e_{C_{\widehat{8}}} - f_{C_{\widehat{8}}}) - \sum_{i=2}^{7} (e_{1i} - f_{1i})).$$

Finally by Proposition 5, we can conclude. □

We will denote by $S^3_{i,\widehat{j}}$ the cubic surface with a triple point at p_i and multiplicity zero at p_j.

Proposition 7 *Let* $J := \{2, 3, 4, 5, 8\}$ *and consider the Cremona transformation* Cr_J. *Then* $S^6 := \mathrm{Cr}_J(S^3)$ *is a surface of degree* 6 *with five triple points.*

- *Given an effective divisor* $D = dH - \sum m_i E_i$, *let*

$$k_{S_6}(D) = \max\{0, 2(m_1 + m_2 + m_3 + m_4 + m_5) + m_6 + m_7 + m_8 - 8d\}.$$

Then the surface S_6 *is contained in the base locus of D exactly* $k_{S_6}(D)$ *times.*

- *The class in* $A^2(X^4_{8,(1)})$ *of strict transform* $\widetilde{S^6}$ *of* S^6 *in* $X^4_{8,(1)}$ *is*

$$6h - 3\sum_{i=1}^{5} e_i - \sum_{i=6}^{8} e_i - \sum_{i,j \in \{1,2,3,4,5\}, i \neq j} (e_{ij} - f_{ij}) - \sum_{k=6}^{8} (e_{C_{\widehat{k}}} - f_{C_{\widehat{k}}}).$$

Proof We know from the previous proposition that the surface S^3 is swept out by a pencil of rational normal quartic curves $\{C^4(q) : q \in L_{14}\}$. By (4), we obtain that the image of the pencil is $\{C^7(q) : q \in L_{14}\}$, where $C^7(q)$ is a rational septic curve with class $7h^1 - \sum_{i=1}^{8} e^1_i - e^1_2 - e^1_3 - e^1_5$ and passing through $q \in L_{14}$. Since the surface S^6 is swept out by this, we can say that D contains S^6 in its base locus at least $\max\{0, m_1 + 2m_2 + 2m_3 + m_4 + 2m_5 + m_6 + m_7 + m_8 + \max\{0, m_1 + m_4 - d\} - 7d\}$ times. Since D is effective, by Proposition 4, we have $m_1 + 2m_2 + 2m_3 + m_4 + 2m_5 + m_6 + m_7 + m_8 - 7d \leq 0$, hence the claim follows.

Now we prove the second statement. Given D_1 and F_1 defined in the previous proposition, recall that $D_1 \cap F_1 = L_{123} \cup S^3$. We consider now $D_2 = \mathrm{Cr}_J(D_1)$ and $F_2 = \mathrm{Cr}_J(F_1)$ to be their image under the Cremona transformation and we get

$$D_2 = 3H - 2E_1 - 2E_2 - 3E_3 - 2E_4 - 2E_5 - E_6 - E_7 - E_8$$
$$F_2 = 3H - 2E_1 - 3E_2 - 2E_3 - 2E_4 - 2E_5 - E_6 - E_7 - E_8.$$

We now analyze the intersection $D_2 \cap F_2$. Note that $\mathrm{Cr}_J(L_{123}) = L_{123}$, by Lemma 4. By Proposition 5 we see that the only planes contained in $D_2 \cap F_2$ are L_{123}, L_{234}, L_{235}. Finally, we check that the intersection of $D_2 \cap F_2$ with the indeterminacy locus of Cr_J does not contain any other two-dimensional component, besides the planes L_{234} and L_{235}. Hence, the intersection $D_2 \cap F_2$ splits into four components: the three planes L_{123}, L_{234}, L_{235} and a sextic surface with five triple points at p_1 and three simple points. Hence, we conclude that S^6 is exactly the sextic irreducible surface.

We now describe the class of S^6 in $X^4_{8,(1)}$. Let $\widetilde{D}_2$ and $\widetilde{F}_2$ be the corresponding strict transforms under the blow-up of lines and rational normal curves in $X^4_{8,(1)}$, see (5). We have

$$\widetilde{D}_2 = 3H - \sum_{i\in\{1,2,4,5\}} 2E_i - 3E_3 - \sum_{i=6}^{8} E_i$$

$$- \sum_{i\in\{1,2,4,5\}} 2E_{3i} - \sum_{i=6}^{8} E_{3i} - \sum_{i,j\in\{1,2,4,5\},i\neq j} E_{ij} - \sum_{i=6}^{8} E_{C_{\widehat{i}}}$$

$$\widetilde{F}_2 = 3H - \sum_{i\in\{1,3,4,5\}} 2E_i - 3E_2 - \sum_{i=6}^{8} E_i$$

$$- \sum_{i\in\{1,3,4,5\}} 2E_{2i} - \sum_{i=6}^{8} E_{2i} - \sum_{i,j\in\{1,3,4,5\},i\neq j} E_{ij} - \sum_{i=6}^{8} E_{C_{\widehat{i}}}.$$

Computing their complete intersection, we have

$$\widetilde{D}_2 \cap \widetilde{F}_2 = (h - e_1 - e_2 - e_3 - \sum_{i,j\in\{1,2,3\},i\neq j} (e_{ij} - f_{ij}))+$$

$$+ (h - e_2 - e_3 - e_4 - \sum_{i,j\in\{2,3,4\},i\neq j} (e_{ij} - f_{ij}))$$

$$+ (h - e_2 - e_3 - e_5 - \sum_{i,j\in\{2,3,5\}} (e_{ij} - f_{ij}))+$$

$$(6h - 3\sum_{i=1}^{5} e_i - \sum_{i=6}^{8} e_i - \sum_{i,j\in\{1,2,3,4,5\},i\neq j} (e_{ij} - f_{ij}) - \sum_{k=6}^{8}(e_{C_{\widehat{k}}} - f_{C_{\widehat{k}}})),$$

where we use relations (6), (7), (8), and we conclude. □

We will denote by $S^6_{i,j,k}$ the sextic surface with five triple points at $\{p_h\}$ for $h \neq i, j, k$.

Proposition 8 *Let $J := \{1, 2, 6, 7, 8\}$ and consider the Cremona transformation Cr_{IJ}. Then $S^{10} := \mathrm{Cr}_J(S^6)$ is a surface of degree 10 with two sextuple points and six triple points.*

- *Given an effective divisor $D = dH - \sum m_i E_i$, let*

$$k_{S^{10}}(D) = \max\{0, 3(m_1 + m_2) + 2(m_3 + m_4 + m_5 + m_6 + m_7 + m_8) - 11d\}.$$

Then the surface S^{10} is contained in the base locus of D exactly $k_{S^{10}}(D)$ times.

- *The class of the strict transform $\widetilde{S^{10}}$ of S^{10} in $A^2(X^4_{8,(1)})$ is*

$$10h - 6e_1 - 6e_2 - \sum_{i=3}^{8} 3e_i - 3(e_{12} - f_{12}) - \sum_{i=1}^{2}\sum_{j=3}^{8}(e_{ij} - f_{ij}) - \sum_{k=3}^{8}(e_{C_{\widehat{k}}} - f_{C_{\widehat{k}}}).$$

Proof We know from the previous proposition that the surface S^6 is swept out by the pencil of rational septic curves $\{C^7(q) : q \in L_{14}\}$. By (4), we obtain that the image of the pencil is $\{C^{10}(q) : q \in L_{14}\}$, where $C^{10}(q)$ is a rational curve with class $10h^1 - 2e^1_1 - 3e^1_2 - 2e^1_3 - e^1_4 - 2e^1_5 - 2e^1_6 - 2e^1_7 - 2e^1_8$ and passing through $q \in L_{14}$. Since the surface S^{10} is swept out by this pencil, we can say that D contains S^{10} in its base locus at least $\max\{0, 2m_1 + 3m_2 + 2m_3 + m_4 + 2m_5 + 2m_6 + 2m_7 + 2m_8 + \max\{0, m_1 + m_4 - d\} - 10d\}$ times. Since $2m_1 + 3m_2 + 2m_3 + m_4 + 2m_5 + 2m_6 + 2m_7 + 2m_8 - 10d \leq 0$, the claim follows by Proposition 4.

Now we prove the second statement. Given D_2 and F_2 defined in the previous proposition, recall that $D_2 \cap F_2 = S^6 \cup L_{123} \cup L_{234} \cup L_{235}$. We consider now $D_3 = \mathrm{Cr}_J(D_2)$ and $F_3 = \mathrm{Cr}_J(F_2)$ to be their image under the Cremona transformation and we get

$$D_3 = 5H - 4E_1 - 4E_2 - 3E_3 - 2E_4 - 2E_5 - 3E_6 - 3E_7 - 3E_8$$
$$F_3 = 4H - 3E_1 - 4E_2 - 2E_3 - 2E_4 - 2E_5 - 2E_6 - 2E_7 - 2E_8.$$

It is easy to check, by applying the previous propositions, that the intersection $D_3 \cap F_3$ contains the planes $L_{123}, L_{126}, L_{127}, L_{128}$ and the cubic surfaces $S^3_{2,\widehat{4}}$ and $S^3_{2,\widehat{5}}$. Notice that $\mathrm{Cr}_J(L_{123}) = L_{123}$ by Lemma 4, and $\mathrm{Cr}_J(L_{234}) = S^3_{2,\widehat{5}}, \mathrm{Cr}_J(L_{235}) = S^3_{2,\widehat{4}}$, by Proposition 6. By computing the intersection of $D_3 \cap F_3$ with the indeterminacy locus of Cr_J we see that there are no other two-dimensional components, besides the planes $L_{126}, L_{127}, L_{128}$. Hence we conclude that S^{10} is an irreducible surface with degree 10 and two sixtuple points at p_1 and p_2 and six triple points.

Finally, we describe the class of S^{10} in $X^4_{8,(1)}$. Let $\widetilde{D_3}$ and $\widetilde{F_3}$ be the corresponding strict transforms under the blow-up of lines and rational normal curves in $X^4_{8,(1)}$, see (5). Computing their complete intersection, as in the previous case we get our claim. □

Proposition 9 *Let $J := \{3, 4, 5, 6, 7\}$ and consider the Cremona transformation Cr_J. Then $S^{15} := \mathrm{Cr}_J(S^{10})$ is a surface of degree 15 with one triple point and seven sextuple points.*

- *Given an effective divisor $D = dH - \sum m_i E_i$, let*

$$k_{S^{15}}(D) = \max\{0, 3(m_1 + m_2 + m_3 + m_4 + m_5 + m_6 + m_7) - 2m_8 - 14d\}.$$

Then the surface S^{15} is contained in the base locus of D exactly $k_{S^{15}}(D)$ times.
- *The class of the strict transform $\widetilde{S^{15}}$ of S^{15} in $A^2(X^4_{8,(1)})$ is*

$$15h - \sum_{i=1}^{7} 6e_i - 3e_8 - \sum_{1\le i<j\le 7} (e_{ij} - f_{ij}) - \sum_{i=1}^{7}(e_{C_{\hat{i}}} - f_{C_{\hat{i}}}) - 3(e_{C_{\hat{8}}} - f_{C_{\hat{8}}}).$$

Proof We know from the previous proposition that the surface S^{10} is swept out by the pencil of rational septic curves $\{C^{10}(q) : q \in L_{14}\}$. By (4), we obtain that the image of the pencil is $\{C^{13}(q) : q \in L_{14}\}$, where $C^{13}(q)$ is a rational curve with class $13h^1 - 2e^1_1 - 3e^1_2 - 3e^1_3 - 2e^1_4 - 3e^1_5 - 3e^1_6 - 3e^1_7 - 2e^1_8$ and passing through $q \in L_{14}$. Since the surface S^{15} is swept out by this pencil, we can say that D contains S^{15} in its base locus at least $\max\{0, 2m_1 + 3m_2 + 3m_3 + 2m_4 + 3m_5 + 3m_6 + 3m_7 + 2m_8 + \max\{0, m_1 + m_4 - d\} - 13d\}$ times. Since $2m_1 + 3m_2 + 3m_3 + 2m_4 + 3m_5 + 3m_6 + 3m_7 + 2m_8 - 13d \le 0$ the claim follows by Proposition 4.

Now we prove the second statement. Given D_3 and F_3 defined in the previous proposition, recall that

$$D_3 \cap F_3 = S_{10} \cup L_{123} \cup L_{126} \cup L_{127} \cup L_{128} \cup S^3_{2,\hat{4}} \cup S^3_{2,\hat{5}}.$$

We consider now $D_4 = \mathrm{Cr}_J(D_3)$ and $F_4 = \mathrm{Cr}_J(F_3)$ to be their image under the Cremona transformation.

$$D_4 := 7H - 4E_1 - 4E_2 - 5E_3 - 4E_4 - 4E_5 - 5E_6 - 5E_7 - 3E_8$$
$$F_4 := 6H - 3E_1 - 4E_2 - 4E_3 - 4E_4 - 4E_5 - 4E_6 - 4E_7 - 2E_8.$$

Now the intersection $D_4 \cap F_4$ contains $S^3_{3,\hat{8}} = \mathrm{Cr}_J(L_{123})$ (by Proposition 6), $S^3_{6,\hat{8}} = \mathrm{Cr}_J(L_{126})$ (by Proposition 6), $S^6_{148} = \mathrm{Cr}_J(S^3_{2,\hat{4}})$ (by Proposition 7), $S^6_{158} = \mathrm{Cr}_J(S^3_{2,\hat{5}})$ (by Proposition 7). Moreover, we have the components $S^3_{7,\hat{8}}$, S^6_{128}, and it can be easily proved that $S^3_{7,\hat{8}} = \mathrm{Cr}_J(L_{127})$ and $S^6_{128} = \mathrm{Cr}_J(L_{128})$. Finally, we check that the intersection of $D_4 \cap F_4$ with the indeterminacy locus of Cr_J does not contain any two-dimensional component. Hence we conclude that S^{15} is an irreducible surface of degree 15 and with a triple point at p_8 and seven sextuple points.

Finally, as in the previous case, we compute the complete intersection of the strict transforms $\widetilde{D}_4$ and $\widetilde{F}_4$, and we get our statement. □

Remark 3 We point out that the five Weyl surfaces described above correspond to the same list computed by Casagrande, Codogni, and Fanelli in [5, Theorem 8.7].

Remark 4 Notice that the cone of effective surfaces of X^4_8 is not invariant under the Weyl action, as already observed by [8]. In particular, in [8, Theorem 4.4], the authors proved that the cone of effective 2-cycles of X^4_8 is linearly generated, namely, each effective cycle can be written as a sum of linear cycles. Indeed, for instance, the class of S^3 in the Chow ring of X^4_8 is $3h - 3e_1 - \sum_{i=2}^{7} e_i$, but so is the class of the union of the three planes L_{123}, L_{145} and L_{167}. However, the three planes do not contain the rational normal curve, whereas S^3 does. From this observation, it is clear that the cone of effective cycles of codimension 2 of $X^4_{8,(1)}$ will not be linearly generated. Therefore, in order to identify the irreducible surface S^3 we need to work in the Chow ring of $X^4_{8,(1)}$.

Remark 5 Notice that, in Propositions 5, 6, 7, 8, and 9, we used a specific sequence of Cremona transformations to obtain each Weyl surface of X^4_8 from the previous. This choice is clearly not unique, in fact, there are multiple paths going from one Weyl surface to another. Similarly, for each Weyl surface S we found a suitable pencil of curves over a Weyl curve $C \subseteq S$ that covers it. This description is also not unique, in particular, for every Weyl curve $C \subseteq S$, we can find one such pencil.

Proposition 10 *The five surfaces S^1, S^3, S^6, S^{10}, S^{15} are the only Weyl surfaces in X^4_8.*

Proof The statement can be proved by direct inspection. In Proposition 11, we classify all the Weyl divisors in X^4_8. Then we consider all the possible intersection of two orthogonal Weyl divisors and by using Propositions 5, 6, 7, 8, 9, and computing degrees and multiplicities, we have checked that all the irreducible components of the intersections are surfaces of type S^1, S^3, S^6, S^{10}, S^{15}. □

By the previous proposition we conclude that any Weyl surface of X^4_8 is contained in the orbit of a plane through three points. Hence, our Definition 1 of Weyl surface in this case coincides with the definition of Weyl plane given in [11].

From the proofs of Propositions 5, 6, 7, 8, and 9, we get the following consequence.

Corollary 1 *Every Weyl surface on X^4_8 is swept out by a pencil of rational curves $\{C(q) : q \in C\}$ over a Weyl curve C.*

5.4 Weyl Divisors

Recall that X^4_8 is a Mori Dream Space and, in particular, the cone of effective divisors is finitely generated by the divisors of anticanonical degree $\frac{1}{3}\langle D, -K_{X^4_8}\rangle = 1$. A simple application of Formula (3) gives the following classification of all the Weyl divisors in X^4_8; they are exactly the generators of the effective cone, see also [20].

Proposition 11 *The Weyl divisors in X^4_8 are modulo permutation of indices:*

(1) E_i *(the exceptional divisor);*
(2) $H - \sum_{i=1}^4 E_i$ *(hyperplane through four points);*
(3) $2H - 2E_1 - 2E_1 - \sum_{i=3}^7 E_i$ *(quadric cone, join of a rational normal quartic and a line);*
(4) $3H - \sum_{i=1}^7 2E_i$ *(the 2-secant variety to a rational normal quartic);*
(5) $3H - 3E_1 - \sum_{i=2}^5 2E_i - \sum_{i=6}^8 E_i$ *(cone on the Cayley surface of $\mathbb{P}^3$);*

(6) $4H - \sum_{i=1}^{4} 3E_i - \sum_{i=5}^{7} 2E_i - E_8$, *with* $|J| = 4$ *and* $j \notin J$;
(7) $4H - 4E_1 - 3E_2 - \sum_{i=3}^{8} 2E_i$ *(cone on a quartic surface of* $\mathbb{P}^3$*)*;
(8) $5H - 4E_1 - 4E_2 - \sum_{i=3}^{6} 3E_i - 2E_7 - 2E_8$;
(9) $6H - 5E_1 - \sum_{i=2}^{4} 4E_i - \sum_{i=5}^{8} 3E_i$;
(10) $6H - \sum_{i=1}^{6} 4E_i - 3E_7 - 2E_8$;
(11) $7H - \sum_{i=1}^{3} 5E_i - \sum_{i=4}^{7} 4E_i - 3E_8$;
(12) $7H - 6E_1 - \sum_{i=2}^{8} 4E_i$;
(13) $8H - 6E_1 - \sum_{i=2}^{6} 5E_i - 4E_7 - 4E_8$;
(14) $9H - \sum_{i=1}^{4} 6E_i - \sum_{i=5}^{8} 5E_i$;
(15) $10H - 7E_1 - \sum_{i=2}^{8} 6E_i$.

We conclude this section with the following geometrical descriptions of the Weyl divisors on X_8^4. As pencils of curves with cycle class as in Notation 1 sweep out Weyl surfaces of X_8^4, nets of such curves sweep out Weyl divisors.

Lemma 5 *Let D be a Weyl divisor on X_8^4 containing a Weyl surface S. Then there is a net of curves $\{C(q) : q \in S\}$ with $C(q) \cdot D = 0$ sweeping out D.*

Proof Notice that every divisor (2)–(15) of Proposition 11 satisfies the hypotheses. For one such divisor, let $S \subset D$. By Propositions 5, 6, 7, 8, and 9, we can find a sequence of standard Cremona transformations such that the image of S is a plane S_1. Applying the same sequence of transformations to D, we obtain a Weyl divisor, D_1, containing such plane. Modulo reordering the points, the possible outputs for the image of D are the divisors (2), (3), (5), (6), (7), (8), (9), (11) of Proposition 11. For each such output, we shall exhibit a sequence of Cremona transformations that preserve the plane S_1 and takes D to a hyperplane containing S_1. Without loss of generality, we will assume that S_1 is the class of the plane passing through the first three points. In the following tables, for every (i), on the left-hand side we will describe the class of the Weyl divisor and on the right-hand side the class of the curve $C(q)$ of the net:

(11)	7	5 5 5 3 4 4 4 4	19	4 4 4 3 4 4 4 4
(9)	6	5 4 4 3 4 3 3 3	16	4 3 3 3 2 3 3 3
(8)	5	4 4 3 2 3 3 3 2	13	3 3 2 2 3 3 3 2
(7)	4	4 3 2 2 2 2 2 2	10	3 2 1 2 2 2 2 2
(5)	3	3 2 2 2 2 1 1 1	7	2 1 1 2 2 1 1 1
(3)	2	2 2 1 1 1 1 1 0	4	1 1 0 1 1 1 1 0
(2)	1	1 1 1 1 0 0 0 0	1	0 0 0 1 0 0 0 0

This concludes the proof. □

6 Weyl Expected Dimension

Let $n = 3, 4$. For $r \in \{1, 2, 3\}$, let $L_{I(r)}$ be a linear cycle of dimension r spanned by $r + 1$ base points. Recall that $W_{n,n+4}$ denotes the Weyl group of X^n_{n+4}. Consider the following set of Weyl r-cycles: $W_n(r) := \{w(L_{I(r)}) : w \in W_{n,n+4}\}$, and let $k_A(D)$ denote the multiplicity of containment of the r-cycle A in the base locus of the divisor D.

By Remark 2 we know that for any Weyl curve $A \in W_n(1)$, then $k_C = \max\{0, -D \cdot A\}$. For every Weyl divisor $A \in W_n(n-1)$ (i.e., those listed in Propositions 2 and 11), we have that $k_A = -\max\{0, \langle D, A\rangle\}$, see [2, Proposition 2.3] and [13, Proposition 4.2] for details. Finally, for $n = 4$, by the results of Sect. 5.3 we know that $W_4(2)$ is the set of the Weyl surfaces (i.e., those listed in Eq. (1)) and the multiplicity of containment $k_A(D)$ of any Weyl surface $A \in W_4(2)$ in the base locus of an effective divisor D is computed in Propositions 5, 6, 7, 8, 9.

We introduce now the notion of *Weyl expected dimension*.

Definition 2 Let $n = 3, 4$ and D be an effective divisor on $X = X^n_{n+4}$. We say that D has *Weyl expected dimension* $\mathrm{wdim}(D)$, where

$$\mathrm{wdim(D)} := \chi(X, \mathcal{O}_X(D)) + \sum_{r=1}^{n-1} \sum_{A \in W_n(r)} (-1)^{r+1} \binom{n + k_A(D) - r - 1}{n}.$$

We now show that the Weyl expected dimension is invariant under the action of the Weyl group.

Proposition 12 *Let $n = 3, 4$ and D an effective divisor on X^n_{n+4}. The Weyl dimension of D is preserved under standard Cremona transformations.*

Proof Let $D{=}dH - \sum_{i=1}^{n+4} m_i E_i$. We need to prove that $\mathrm{wdim}(D){=}\mathrm{wdim}(\mathrm{Cr}_I(D))$ for Cr_I a standard Cremona transformation. Let $D' = dH - \sum_{i \in I} m_i E_i$ be the divisor obtained from D by forgetting three points. From [2], [Corollary 4.8, Theorem 5.3], we have that $\mathrm{wdim}(D') = \mathrm{wdim}(\mathrm{Cr}_I(D'))$, where the formula $\mathrm{wdim}(D')$ only takes into account the Weyl cycles of D based exclusively at the points parametrized by I that are therefore fixed linear subspaces through base points.

We claim that, for all the remaining Weyl cycles A of D, interpolating at least a point away from the indeterminacy locus and for which $k_A(D) \geq 1$, we have $k_A(D) = k_{\mathrm{Cr}(A)}(\mathrm{Cr}(D))$. If A is a curve, the claim is true because $k_A(D) = -A \cdot D = -\mathrm{Cr}(A) \cdot \mathrm{Cr}(D)$. If A is a divisor, the claim is true because $k_A(D) = -\langle A, D\rangle = -\langle \mathrm{Cr}(A), \mathrm{Cr}(D)\rangle$. It only remains to show the claim for $A = S$ a surface of X^4_8. It follows from the proofs of Propositions 5, 6, 7, 8, 9 and Remark 5 that for a Weyl curve $C \subseteq S$ such that S is swept out by a pencil $\{C(q) : q \in C\}$, then $k_S(D) = -C(q) \cdot D + k_C(D)$. Since D is effective, then $C(q) \cdot D \geq 0$ by Proposition 4, so $k_C(D) = -C \cdot D \geq 1$. Since $k_S(D) = -C(q) \cdot D - C \cdot D = -\mathrm{Cr}(C_q) \cdot \mathrm{Cr}(D) - \mathrm{Cr}(C) \cdot \mathrm{Cr}(D) = -\mathrm{Cr}(C(q)) \cdot D + k_{\mathrm{Cr}(C)}(\mathrm{Cr}(D))$ and $\mathrm{Cr}(D)$ is swept out by $\{\mathrm{Cr}(C(q)) : q \in \mathrm{Cr}(C)\}$, we conclude. □

This yields an explicit formula for the dimension of any linear system in X^3_7.

Theorem 1 *For any effective divisor $D \in Pic(X^3_7)$, we have*

$$h^0(X^3_7, \mathcal{O}_{X^3_7}(D)) = wdim(D).$$

Proof For the sake of simplicity, we will abbreviate $h^0(X^3_7, \mathcal{O}_{X^3_7}(D))$ with $h^0(D)$. Consider a sequence of standard Cremona transformations which takes D to a Cremona reduced divisor D': it is well known that $h^0(D) = h^0(D')$. By the previous proposition, we have that $\mathrm{wdim}(D) = \mathrm{wdim}(D')$. Since D' is Cremona reduced, by [9, Theorem 5.3], we know that D' is linearly non-special, i.e., its dimension equals its linear expected dimension introduced in [2]: $h^0(D') = \mathrm{ldim}(D') = \mathrm{wdim}(D')$, where the last equality is easy to check for Cremona reduced divisors in X^3_7. Hence we conclude that $h^0(D) = \mathrm{wdim}(D)$. □

For the case of X^4_8, we propose the following conjecture.

Conjecture 1 For any effective divisor $D \in \mathrm{Pic}(X^4_8)$, we have

$$h^0(X^4_8, \mathcal{O}_{X^4_8}(D)) = \mathrm{wdim}(D).$$

Solving Conjecture 1 would complete the analysis of the dimensionality problem for all the Mori Dream Spaces of the form X^n_s, which are X^n_{n+3}, X^2_8, X^3_7 and X^4_8. Indeed we recall that the case of $s \leq n+2$ was solved in [2] and the case of $s = n+3$ is studied in [15, 22]. It is clear that the notion of Weyl dimension extends both that of *linear expected dimension* of [2] and that of *secant expected dimension* of [4]. In fact, first of all, notice that linear cycles of dimension at most $n-1$ spanned by the collection of s points are *Weyl cycles*, according to our definition. This holds because hyperplanes passing through n base points are always Weyl divisors. We recall that for $s = n+2$, the only Weyl divisors are the exceptional divisors and the hyperplanes spanned by n base points. We conclude that for $s = n+2$ the linear cycles spanned by base points are the only Weyl cycles, hence we have that for divisors in X^n_{n+2}, the Weyl expected dimension equals the linear expected dimension, so the analogous of Conjecture (1) in X^n_{n+2} holds by [2]. Moreover, by [2, Corollary 4.8], we can say that the analogous of Conjecture 1 holds in arbitrary dimension for a small number of points. Secondly, in [4], the authors considered cycles $J(L_I, \sigma_t)$, joins over the t secant variety to the rational normal curve of degree n passing through $n+3$ points, and they gave a *secant expected dimension* for an effective divisor. It matches the Weyl expected dimension for $n = 4$. For X^4_7, these varieties are just the unique rational normal quartic curve through the seven points and the pointed cones over it, namely, cone over rational normal curve, labeled $S^3_{1,\hat{8}}$ as in notation (1). Therefore, we propose the following conjecture.

Conjecture 2 The varieties $J(L_I, \sigma_t)$ are the only Weyl cycles on X^n_{n+3}.

Acknowledgements We thank the referee for their useful comments and suggestions. We thank Cinzia Casagrande for many useful discussions and Luis J. Santana-Sánchez for several comments on a preliminary version of this article. The first and third authors are members of INdAM-GNSAGA. The second author is supported by the NSF grant DMS-1802082. The third author was partially supported by the EPSRC grant EP/S004130/1.

References

1. C. Araujo, C. Casagrande, On the Fano variety of linear spaces contained in two odd-dimensional quadrics. Geom. Topol. **21**:3009–3045 (2017)
2. M.C. Brambilla, O. Dumitrescu, E. Postinghel, On a notion of speciality of linear systems in $\mathbb{P}^n$. Trans. Am. Math. Soc. **367**, 5447–5473 (2015)
3. M.C. Brambilla, O. Dumitrescu, E. Postinghel, On linear systems of $\mathbb{P}^3$ with nine base points. Ann. Mat. Pura Appl. **195**, 1551–1574 (2016)
4. M.C. Brambilla, O. Dumitrescu, E. Postinghel, On the effective cone of $\mathbb{P}^n$ blown-up at $n+3$ points. Exp. Math. **25**(4), 452–465 (2016)
5. C. Casagrande, G. Codogni, A. Fanelli, The blow-up of $\mathbb{P}^4$ at 8 points and its Fano model, via vector bundles on a degree 1 del Pezzo surface. Revista Matematica Complutense **2**, 32 (2019)
6. C. Ciliberto, Geometrical aspects of polynomial interpolation in more variables and of Waring's problem, in *European Congress of Mathematics*, vol. I (Barcelona, 2000), pp. 289–316 (2001). Progr. Math., 201, Birkhäuser, Basel
7. C. Ciliberto, B. Harbourne, R. Miranda, J. Roé, Variations on Nagata's conjecture. Clay Math. Proc. **18**, 185–203 (2013). Am. Math. Soc
8. I. Coskun, J. Lesieutre, J. Ottem, Effective cones of cycles on blowups of projective space. Algebra Number Theory **10**(9), 1983–2014 (2016)
9. C. De Volder, A. Laface, On linear systems of $\mathbb{P}^3$ through multiple points. J. Algebra **310**(1), 207–217 (2007)
10. I. Dolgachev, Weyl groups and Cremona transformations, Singularities, Part 1 (Arcata, CA, 1981). Proc. Sympos. Pure Math. Am. Math. Soc. Providence, RI **40**, 283–294 (1983)
11. O. Dumitrescu, R. Miranda, Cremona orbits in $\mathbb{P}^4$ and applications. *The Art of Doing Algebraic Geometry*, this volume. arXiv:2103.08040
12. O. Dumitrescu, R. Miranda, *On (i)curves in P^r*. arXiv:2104.14141
13. O. Dumitrescu, E. Postinghel, Vanishing theorems for linearly obstructed divisors. J. Algebra **477**, 312–359 (2017)
14. O. Dumitrescu, N. Priddis, *On (-1) classes*. arXiv:1905.00074
15. A. Laface, E. Postinghel, L. J. Santana Sánchez, *On linear systems with multiple points on a rational normal curve*, *Linear Algebra Appl.* 657 (2023), 197–220
16. A. Laface, L. Ugaglia, Standard classes on the blow-up of $\mathbb{P}^n$ at points in very general position. Commun. Alg. **40**, 2115–2129 (2012)
17. A. Laface, L. Ugaglia, Elementary (-1) curves of $\mathbb{P}^3$. Commun. Algebra **35**, 313–324 (2007)
18. J. Lesieutre, J. Park, Log Fano structures and Cox rings of blowups of products of projective spaces. Proc. Am. Math. Soc. **145**(10), 4201– 4209 (2017)
19. S. Mukai, Geometric realization of T-Shaped root systems and counterexamples to Hilbert's fourteenth problem. Algebraic Trans. Groups Algebraic Variet. Encycl. Math. Sci. **132**, 123–129 (2004)
20. S. Mukai, *Finite Generation of the Nagata Invariant Rings in A-D-E Cases*, RIMS Preprint n. 1502, Kyoto (2005)
21. M. Nagata, On the fourteenth problem of Hilbert, in *Proceedings of the International Congress of Mathematicians* (Cambridge University Press, 1958), pp. 459–462
22. L.J. Santana Sánchez, *On Blow-ups of Projective Spaces at Points on a Rational Normal Curve*, Ph.D. thesis, Loughborough University, UK, 2021

Simson's Reconstruction of Apollonius' *Loci Plani*. Modern Ideas in Classical Language

Aldo Brigaglia

Abstract This paper deals with Simson's reconstruction of Apollonius' *Loci Plani*. In the first book we have a fixed point, A, and a point B varying on a given straight line or on a given circumference. The book is devoted to the study of the locus described by the end C of another segment AC making a given angle with AB and in a given relation with it. Therefore, when the relation is of proportionality, we have what we call homothety and when it is inverse proportionality, we have circular inversion. In the second book, Simson studies many loci connected with what we now call polar of a line (with respect to a circle) or radical axis. I think that it is interesting, from a historical point of view, to see how many important mathematical ideas evolved, following the general development of mathematics.

1 Introduction

During a recent, interesting talk, Ciro said (*non verbatim* quote) that "to understand what Italian geometers, we must learn their mathematical language". While this is true for algebraic geometry, it is even truer for ancient geometers: we should learn their language to detect their aims and their different approaches to mathematical problems. This could help us to understand that the development of mathematics depends not only on the development of technical tools, but also on the way in which we look at the various concepts. For example, we may look at the same object either as a locus or as a transformation, and this may radically change the way we develop our mathematical skills.

A. Brigaglia (✉)
Dipartimento di Matematica e Informatica, Palermo, Italy
e-mail: aldo.brigaglia@gmail.com

T. Dedieu et al. (eds.), *The Art of Doing Algebraic Geometry*, Trends in Mathematics,
https://doi.org/10.1007/978-3-031-11938-5_2

Simson's[1] book [18] represents, in my opinion, a good instance of what were, at the time, diverging styles in the study of geometrical problems: algebraic versus purely geometrical language.

At the beginning of the seventeenth century, the rediscovery of Greek treatises (Apollonius and Pappus above all) had led mathematicians to different readings: I refer above all to the ones of Desargues and Descartes. Analytical geometry became the first chapter of the emerging Calculus, and it was rightly considered prominent. Consequently, the geometrical approach was overlooked, to gain new interest only in the beginning of the nineteenth century with Poncelet and Steiner.

In particular, book 7 of Pappus' treatise[2] refers to many of the lost books of Euclid and Apollonius that formed the so-called *Dominion of Analysis*. I follow Jones[3] in dividing these treatises into *Problems* (Euclid's *Data* and *Porisms* and Apollonius' *Conics*—the only one surviving, of the first four books, in its Greek original—*Cutting off of a Ratio, Cutting off of an Area, Determined Section, Neuses* and *Tangencies*) and *Loci* (Aristeus' *Loci solidi*, Euclid's *Loci on Surfaces*, Apollonius' *Loci Plani*, Erathostenes' *On Means*).

Analysis is intended as a tool to solve problems. Starting from the problem as solved, we must go backwards until we get to a conclusion known to be true (or false). In the first case, by reversing the path followed, we obtain the synthesis of the solution; in the second case, we obtain the falsity of what was proposed.

This paper deals with Simson's reconstruction of Apollonius' *Loci Plani*. I have not the intention (nor, alas, the competency) to try to divine Apollonius' ideas in writing the *Loci Plani*. I will only try to show the way in which Simson, two thousand years after Apollonius, read it.

The treatise, following Apollonius, is divided into two books. In the first part of book I, two segments (AB and AC), issuing from a point and forming an angle (possibly of $0°$ or $180°$), are given and we must study the various loci described by C when B varies on a plane locus (i.e. a straight line or a circle). Apollonius considered only the cases in which also C varies in a plane locus. The resulting locus depends on the relation between AB and AC. This relation may be of proportionality (homothety) or inverse proportionality (inversion) composed of rotations (when the angle is different from $0°$). Thus we have, for the first time, a fairly detailed study of those relations, for instance the centres of homothety, the circumference described by a point inverse of one varying in a straight line, etc.

We could consider these loci as "transformations", and this is indeed the opinion (for example) of Hudson [10] or Coolidge [4]. I will use the word "transformation" only to indicate its use as a tool to simplify the study of a figure by "transforming" it in an equivalent one.

---

[1] Robert Simson (1687–1768) was a professor of mathematics in the University of Glasgow. He had restored not only the *Loci Plani*, but also Euclid's *Porisms* and Apollonius' *Determined Section*. On his biography I refer to [19].

[2] The first Latin translation of this text is [3]. I used the translation and the notes by Paul Ver Eecke [20] and Alexander Jones [13]. The citations in English are taken from the last one.

[3] Simson does the same division in his *Preface*, with the only exception of *Porisms*, which he poses among the Loci. In reality, *Porisms* have a mixed nature.

Regarding inversion, there are many different historical readings of its birth, but I think that this is a false problem. Inversion is strictly bound to inverse proportionality, so we can find everywhere "mutually inverse points" in ancient geometry. It is not a merely technical problem: as we will see, many of the properties of inversion in Simson's reconstruction (and much earlier, in Fermat [6], van Schooten [17], Newton [15]) were well known to any mathematician who had read those reconstructions. The difference lies in how we look at such properties. For example, we might immediately translate the proposition "a circumference passing through a point" simply as "a straight line", because we can transform the one into the other by inversion. But it would not be the same thing as saying, "the locus described by the inverse of points varying in a circumference passing through a point A is a straight line", and using this property in a problem.

The same thing may be said for the "polar" of a conic (in this case of a circumference, as we are dealing only with plane loci). I find it interesting how the same concept may change over time. I will return to this later: at the end of book II, in fact, there is an interesting treatment of the polars.

I will use Simson's (and therefore classical) notations: $r(ABC)$, to be read as rectangle ABC, is the rectangle of sides AB and BC and is equivalent to $AB \times BC$; $q(AB)$ is the square of side AB and is equivalent to AB^2. There are some simple relations, easy to prove, but bothersome when one has to verify them when it is necessary during a proof. Simson uses these relations often automatically. For instance: given three points A, B, C aligned in this order, we have $r(ACB) - r(ABC) = q(BC)$; $r(BAC) - r(ABC) = q(AB)$; $r(ACB) - r(BAC) = q(BC) - q(AB)$.

I will also use the symbol *(O, A)* for a circumference of centre O through a point A. Sometimes I will give the modern equivalent of the propositions. For now, I prefer to expose directly some of Simson's propositions and to postpone my comments and a part of his preface in the conclusions.

2 Harmonic Group, Tangent and Polar in Apollonius' Conics

Before moving to *Plane Loci*, I think it is useful to resume, without any ambition to completeness, Apollonius' use of harmonic ratio and his idea of polar of a point with respect to a conic section.[4] Apollonius uses the harmonic ratio many times. I will give a brief look at propositions I. 32, I. 34 and I. 37. Proposition I. 32 tells us how to draw a tangent to a given conic section in a point on the section: it is the parallel line (CE) to the diameter (GH) conjugate to the one passing by C (CI, see the next figure).

I.34 *Let there be a hyperbola* (remember that hyperbola, for Greek mathematicians, means only a branch of it), *or an ellipse or a circumference of circle whose diameter is AB, and let some point C be taken on the section, and from C let CD*

[4] I use the translation and the notes of [21].

be drawn as an ordinate, and let it be contrived that $BD : DA = BE : EA$, [i.e. (A, B, D, E) is a harmonic range] *and let* EC *be joined. I say that* EC *touches the section.*[5]

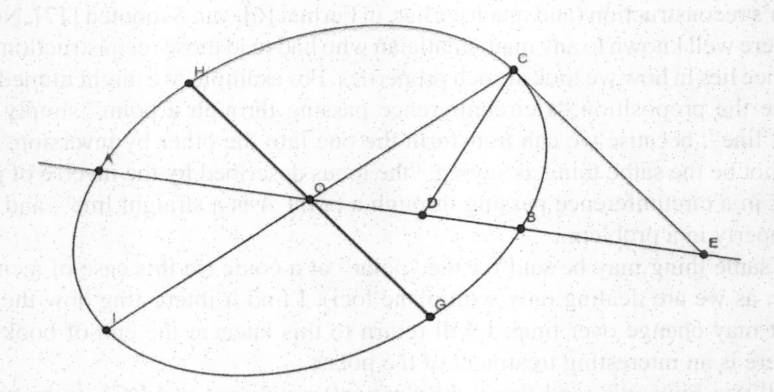

Proposition I. 36 is the converse of I. 34: *Let there be a hyperbola or an ellipse or the circumference of a circle whose diameter is* AB, *and let* CE *be tangent, and let* CD *be dropped as an ordinate. I say that as* BD *is to* D*A, so* BE *is to* EA. It is evident that now we are able to draw a tangent from any point E external to the conic section: Let EO be the diameter through E and (A, B, E, D) a harmonic range. Draw the parallel EC to the conjugate diameter to $AB.EC$ is the desired tangent.

For our needs, particularly interesting is the first part of proposition I. 37:

Let a line tangent to a hyperbola or an ellipse or the circumference of a circle in a point C *intersect a diameter* AB *in a point* E*. Let* CD *be dropped as an ordinate on the diameter* AB*, and let* Z *be the centre of the section. I say (1) that* $r(DZE)$ *is equal to the square of* ZB*.*

We already know from I. 34 that (A, B, D, E) is a harmonic range, and Apollonius obtains (1) only by manipulating the known relation, $BD : DA = BE : EA$. So we could restate (1) by the following (which does not involve any reference to the conic): (A, B, E, D) is a harmonic range if and only if $r(DOE) = q(OB)$, where O is the middle point between A and B. Here we have also an evident construction of the harmonic conjugate of the triple (A, B, D) using Euclid's construction of the third proportional of OE and OB (Eucl. VI. 11).

[5] The translation is in [16].

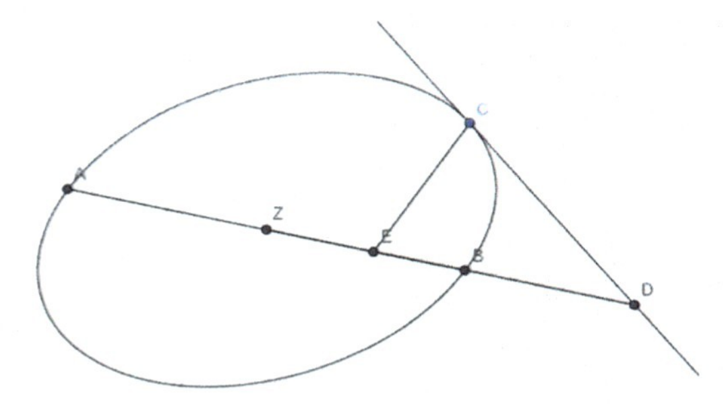

In modern language, this means the well-known fact that the harmonic conjugate of a triple (A, B, D) is the inverse of the point D with respect to the circle (O, B). This will be a very important relation in what follows, and we must bear in mind that a mathematician of the seventeenth century knew it very well. We note also that in Desargues' language D and E are points in involution.

The proposition III. 37 says that if we have two tangents, in A and B, to a conic section, intersecting in C, and we take a point Z in AB and draw ZC which intersects the section in D, E, then (D, E, C, Z) is a harmonic range. Therefore, we have a new property for the polar of a point C (outside the conic): not only the centre of the chord AB, but also any of its points is the harmonic conjugate of the triple $(E, D.C)$.

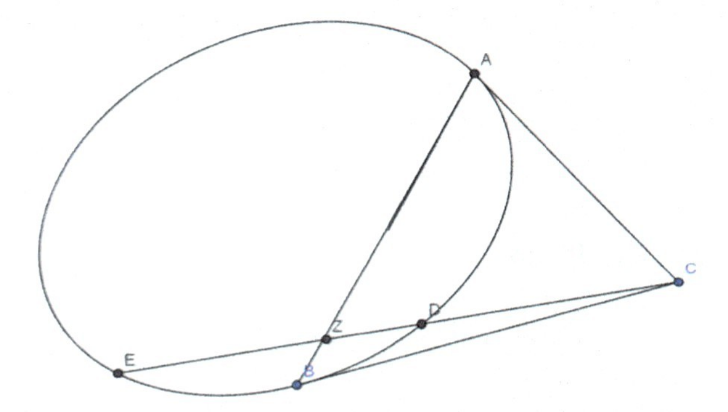

The Proposition III. 38 says that, if C is the intersecting point of two tangents (in A and B, with E middle point between them) to a section, and if, from a point O of the parallel through C to AB we draw the line AE, intersecting the section in D, Z, then (D, Z, O, E) is a harmonic range. Therefore, in modern words, the parallel to the polar of a point C external to a conic section is the polar of the middle point between the points of tangency. Therefore, we have a construction of the "polar" of a point E internal to the conic and a defining property.

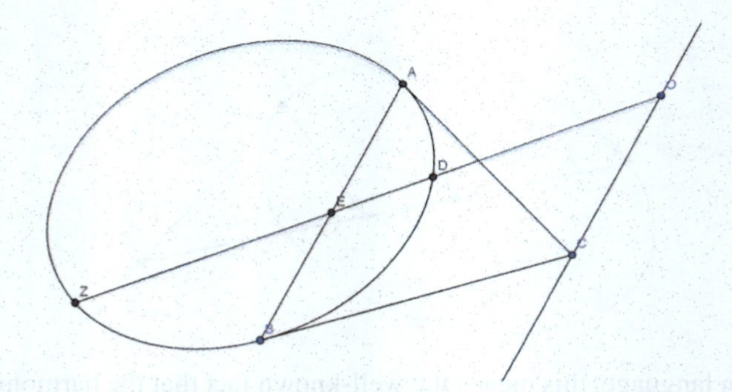

In 1588, Commandino published the Latin translation of Pappus' *Mathematicae Collectiones*. Even if some parts of this work were already known from Greek manuscripts, the appearance of this text, translated and printed, was a capital event for the development of modern mathematics. I will refer, here, exclusively to book 7. I will not attempt any philological examination or interpretation of Apollonius' intent, nor will I sketch any analysis of the status of mathematics when the text was edited (fourth century). My only intent is to discuss one of the most important readings of this work in the first half of eighteenth century.

Apollonius' *Loci Plani* were reconstructed (divined) in the seventeenth century by Fermat, van Schooten and Newton (in his *mathematical papers*) and, in the following century, by Simson. I will sketch the latter one, and only occasionally refer to the former three.

I start with a short summary of Pappus' account, using, with some freedom, a more or less modern mathematical language.

3 Book I of Loci Plani

Apollonius' book is divided in two books. According to Pappus, Book I consists of two parts: the first one consists in what *the ancients compiled,* and the second in what *people who came after them … added*. Here follows Jones' English translation of the first part: *If two straight lines are drawn either from one given point or from two and either in a straight line or parallel or containing a given angle, and either holding a ratio to one another or containing a given area, and the end of one touches a plane locus given in position, the end of the other will touch a plane locus given in position, sometimes of the same kind, sometimes of the other, and sometimes similarly situated respect to the straight line, sometimes oppositely.*

I will divide this part into various cases, following Simson, but not in the same order.

Case (1): straight lines issued from one point; case (2): straight lines issued from two points.

Case (1a): aligned ends of the segments; case (1b) segments containing a given angle.

For each case, we have two possibilities: the two segments may hold a given ratio or contain a given area (they may be directly or inversely proportional).

For each case, we have, again, two possibilities: one end may touch a straight line or a circumference.

Finally, for each case the end of the second segment may be on the same or on the opposite side with respect to the fixed point.

I will now examine a few cases.

3.1 A, B, C Aligned, a Given Ratio Between AB and AC (Central Similitude)

We have two straight lines AB and AC issued from the point A; A, B, C aligned; $AC : AB = k(cost.)$, B touching a straight line (BD); C touches a straight line (CF) parallel to BD (Proposition 4 of Simson).

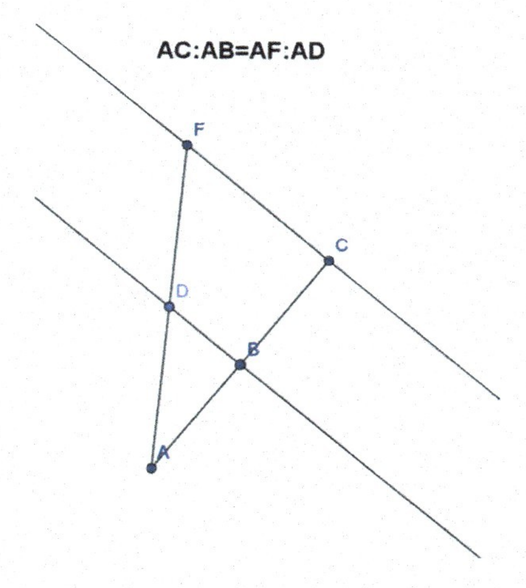

This is what we call a homothety (or a central similitude). If AB and AC are opposite, we have an inverse homothety; when, additionally, $k = 1$, we have the central symmetry.

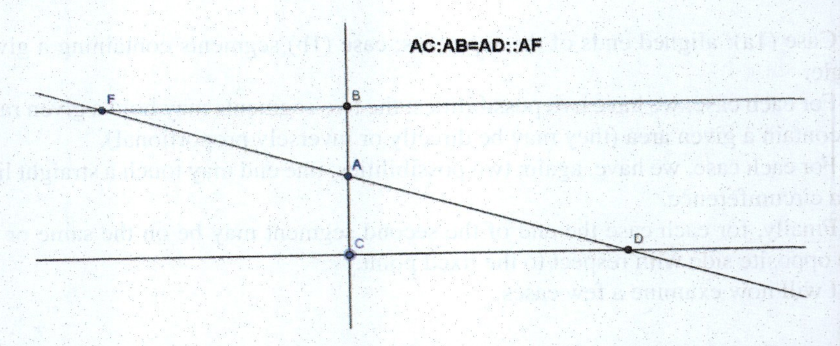

The proof of the above is very easy and I will omit it. We find the same proposition (with less details), with an almost identical proof, in Fermat and van Schooten (Fermat Proposition 1 and van Schooten Theorem 1, both contain also the following case, in which the point B touches a circumference).

Case 2. The same as case 1, but point B touches a circumference.

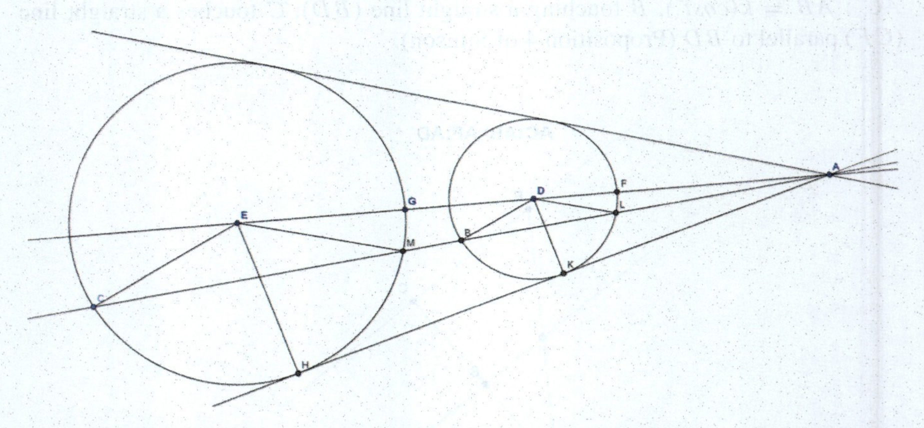

I will sketch the construction of Simson. (Proposition 5, pp. 4–7). Circle (D, F), with line DF prolonged passing through A, is given. Take E in DA so that $AD : AE = k$ (the given ratio; E, in one side or in the other in two cases). Call F one of the intersections of AD with the given circumference, and let G be such that $AF : AG = k$. The required circle is (E, G). The proof (I will give only the case illustrated by the first figure) goes in this way: we must prove that if B touches (D, F) and C touches (E, G) then $AB : AC = k$. Let AK be tangent to (D, F) and EH the radius of (E, G) parallel to AH. Using Euclid 6.32 we have that H, K, A are aligned; therefore also the triangles DBL and ECM are similar and the proposition has been proved. Following Simson's argument, we get also that A is the meeting point of DE and the common tangent AHK and, in the same way, of the other tangent. Looking at the proof given above, it is possible to note that here Simson also answers the problem of finding the centre of similitude of two given circles. Inverting Simson's reasoning we may indeed solve the problem in which the two circles are given and it is requested to find the point A. Therefore, we may think that how to find the two similitude centres of given circles was well known, and this construction is often used in solving various problems.

One important instance is Pappus' proposition 118, which refers to Apollonius' lost book "On Tangencies".[6] In Lemma I before Problem IX of [22], which is a reconstruction of Apollonius' "On Tangencies", Viète finds the centre of similitude in the same way as Pappus had done. Some years before him Adrien van Roomen, counting correctly the four common tangents to two circles, had also used freely this concept.[7]

3.2 A, B, C Aligned; AB and AC Inversely Proportional; B on a Given Line

We have two straight lines AB and $A\ C$ issued from the point A; A, B, C aligned; $r(CAB) = k(cost.)$, B touching a straight line (BD) : C touches a circumference passing through A.[8] I will sketch the construction of Simson (Proposition 8), which is a detailed construction of the inverse of a straight line. ([18], p. 11).

From the given point A, we draw the perpendicular AF to the given line. In AF (or in the opposite side), we take a point G such that $r(GAF) = k$; the required locus will be the circumference having AG as a diameter. The easy proof is to take any point (C) in the circumference and the point B, intersection of AC with the given line. The triangles AFB and ACG are similar, therefore $AC : AG = AF : AB$ and $r(CAB) = k$.

We find, in this way, that the "transformation" amounts to inversion,[9] or to inversion combined with central reflection.

---

[6] In an interesting paper, [1] Raymond Archibald reconstructs the history of the centres of similitudes. He says that M. Cantor attributed to Viète the first use of this concept, but he rightly vindicates this to Apollonius or perhaps to a more ancient Greek mathematician. He also shows—following Flauti—how from Pappus' 118, it is possible to reliably reconstruct Apollonius' construction of the problem of the circle tangent to three given circles. This reconstruction is interesting also because it involves Pappus' 117, a particular case of the famous Castillon problem.

[7] Van Roomen, too, has worked on the reconstruction of Apollonius' "Tangencies". He uses the angle between two common tangents to two circles, but he never says how to get the centre of similitude. Therefore, I think that this was something well known for contemporary mathematicians.

[8] Simson (proposition 9) does not consider the obvious case in which the given line passes through A.

[9] It is necessary to underline that Pappus' rendering of Apollonius' proposition speaks always of "the point C *touches* the locus" and not "the locus *is*". Therefore, there is no problem with the point A, which does not correspond to any of the points of the line. In the Greek original the word used is always ἅψεται [11, passim].

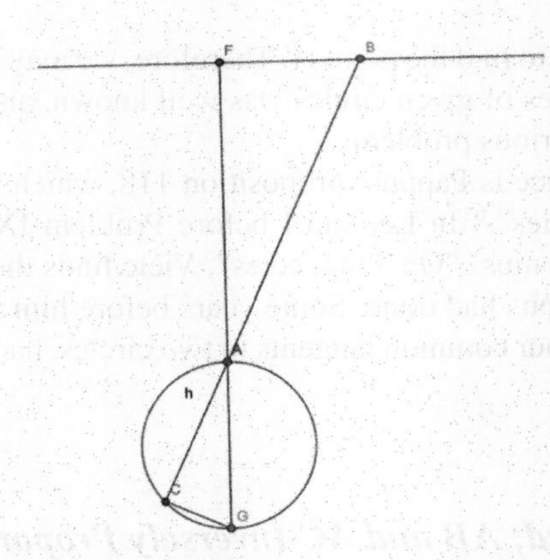

Fermat, van Schooten and Newton do almost the same, but with fewer details. I give Whiteside's translation of Newton's reconstruction: *If from a given point A to the straight line BD given in position there be drawn any straight line AB and in it be taken a point C with the stipulation that the rectangle $BA \times AC$ be given, then the point C is in a circle passing through A.* (Vol. IV, p. 233.)

3.3 A, B, C Aligned; AB and AC Inversely Proportional; B on a Given Circumference

Proposition 9 of Simson [18, pp. 12–15] states correctly (using the same notations as in the preceding case) that, (1) if the given circumference passes through A, the requested locus is a straight line and, (2) if not, the locus is another circumference.

Simson proves case (1) in a very short way, noting that proposition (1) is the reverse of 8. I will examine (2), which divides in many subcases depending on the position of A relative to the given circle, only in the case in which point A is external to the circle and $k > 1$.

I think that Simson's construction is very interesting. Let DE be a diameter of the given circle and let DE be aligned with A. Let $r(DAG)$ be the given constant area, and $k = AE : AG$. Use proposition 5 to construct the circle Γ when the ratio $AB : AC = k$. The same circle will be the required locus. The proof is choose in Γ a point C and let B and F be the points in which AC intersects the given circle; we have $AF : AC = k$ and $r(BAF) : r(BAC) = r(DAE) : r(DAG)$; but, $r(BAF) = r(DAE)$; it follows that $r(BAC) = r(DAG)$; therefore, any point in Γ is also in the required locus.

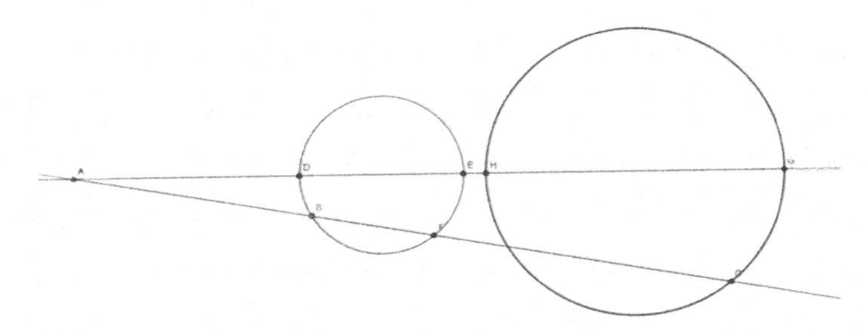

Viète in [22] (Lemma II for problem IX) proves almost exactly the same thing. Precisely, if A is the centre of similitude of two circles (as was found by Lemma I), then (I refer to the figure above; the notations will be—I hope—clear after looking at the figure), for every point B in the first circumference, we have: ($r(BAC)$ = $r(DAG)$. In modern terms: **given two circumferences with centre of similitude A, they are mutually inverse relatively to a circle of centre A and radius the proportional mean between AD and AG.**

Viète uses the two Lemmas in Problem IX (to find a circumference passing through one given point B and tangent to two given circles), showing that the required circumference has to pass also through the point C such that $r(CAB) = r(DAG)$.

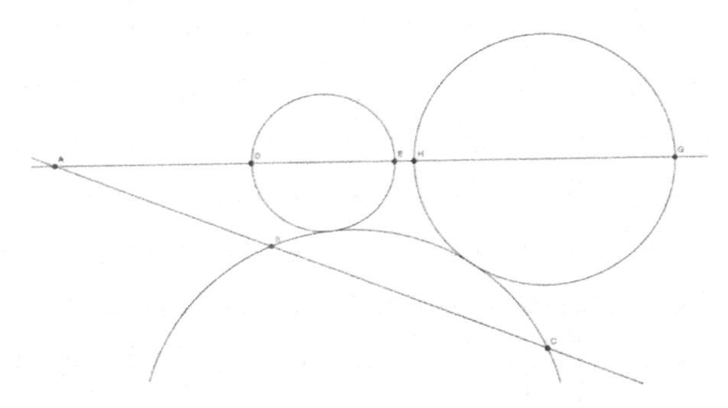

Fermat, Schooten and Newton have similar constructions, but they do not note explicitly that the centre of inversion coincides with the centre of similitude, and they do not explicitly distinguish the two cases of a circumference passing through A or not.

3.4 AB and AC Forming a Given Angle and Having a Given Ratio; B on a Given Line or on a Given Circumference. (Proposition 6)

This is obviously a homotopy composed with a rotation (with the same centre), so I will not go deep in the propositions concerning those cases. However, I would like to note that, when the given ratio is 1, this "transformation" is a rotation. I will only sketch Simson's construction when a circumference is given (Proposition 7 [18 pp. 9–11]). The construction is easy: Let A be the given point and D the centre of the given circle. Draw AD and let E be the intersection of DA and the given circumference (D, E). We take AF making the given angle with AD, and two points G, F such that $AD : AG = AE : AF = k$ (the given ratio). The circle (G, F) is the requested locus. The proof is a little cumbersome, showing how, sometimes, Euclidean notation is very heavy. We must prove that, if we take a point B on (D, E), and a line AC making with AB the given angle and meeting (G, F) in C, then $AB : AC = k$. We consider A internal to (D, E) and we need a Lemma.

Lemma: in the hypotheses of the theorem, if, from A, we draw AK and AM tangents to (D, E), and AL and AN tangents to (G, F), we have $K\widehat{A}L = M\widehat{A}N = D\widehat{A}N$. I omit the simple proof.

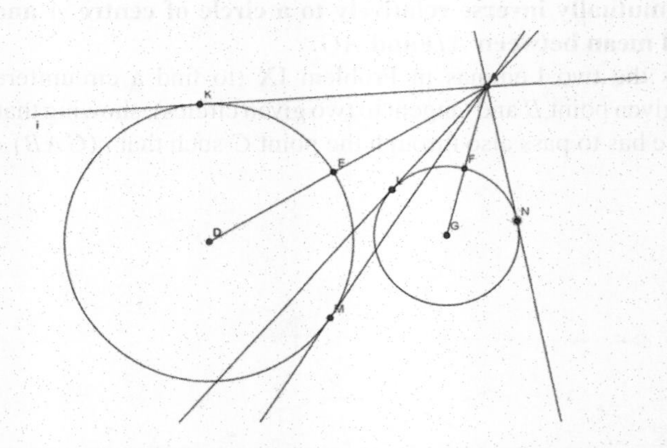

With the aid of the Lemma, the proof amounts to demonstrate that the triangles ABD and ACG are similar, and hence the thesis.[10]

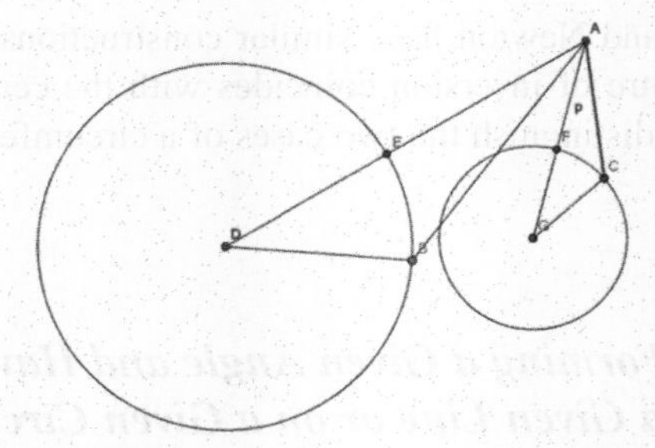

The case of inverse proportionality is not very different and I will omit it.

[10] I have preferred to avoid providing the details of the proof, but these could have been useful to show how Euclidean-style demonstrations are generally far from a naïf recourse to evidence from drawings. This, however, would be another story.

Now we may go to the case of two segments issued from different points. The segments may be (a) parallel; (b) forming a given angle. In any of the two cases, there are many different possibilities, as in the case of aligned segments. I will discuss only a few of them.

Two parallel straight lines AC and BD, A and B fixed; the ratio between AC and BD given; C touching a given line, CE: D touches a given line parallel to CE (Proposition 12 [18 pp. 19–20]). This is a homothety (or the composition of a homothety with a translation). When the given ratio is equal to 1, we have a translation alongAB.

Two parallel straight lines AC and BD, A and B fixed; the ratio between AC and BD given; C touching a given circumference, (E, C): D touches a given circumference (Proposition 13 [18 pp. 20–21]). It is not necessary to provide the easy proof.

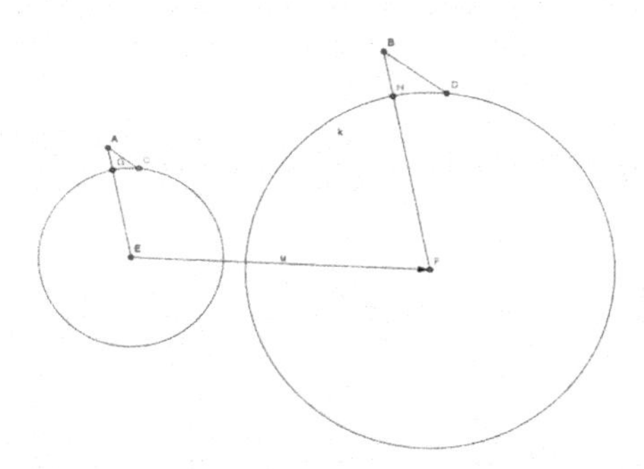

Two parallel straight lines AC and BD, A and B fixed; r(AC, BD) given; C touching a given line, CH: D touches a given circumference (Proposition 14 [18 pp. 21–22]).[11] Evidently, this is the composition of an inversion of centre A and a translation of vector $\overrightarrow{AB}$. It is interesting to note that, for the first time, Simson acknowledges this fact, and constructs the circle by using Proposition 8 from A, with the given product. Then, he uses Proposition 13 with the ratio 1 (i.e. translation). More precisely, after the construction of the circle (E, A) in which $r(GAC) = r(KAH) = k$(the given product), he says: draw BL parallel and equal to AK; the requested circle will be the one whose diameter is BL.

[11] It is evident from the drawing and the proof—though this is not explicitly stated—that the circumference passes through B.

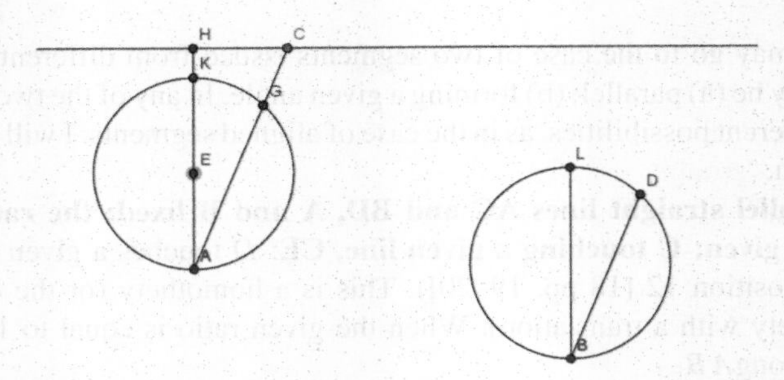

Two parallel straight lines AC and BD, A and B fixed; r(AC, BD) given; C touching a given circumference;

(a) if A is on the given circumference: D touches a straight line; (b) if A is not on the given circumference, D touches a given circumference (Proposition 15). In the first case, Simson says that this proposition is the inverse of 14 and its demonstration is too easy; in case (b), he uses Proposition 9 (case b) to obtain the inverse (O', H) of the given (O, E), and then he translates (O', H) to obtain the requested (O'', L). I omit the construction.

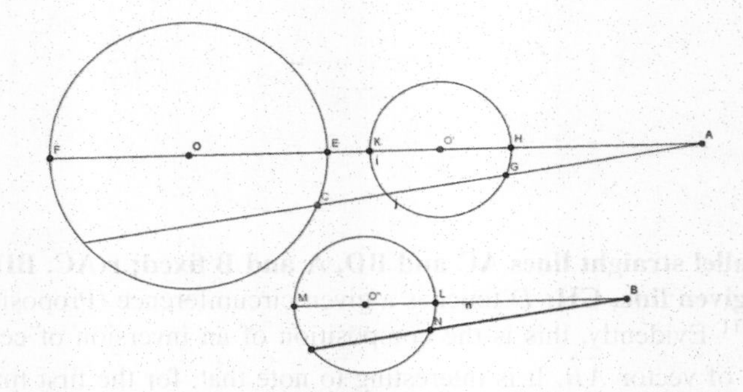

Following are four propositions (16–19) which examine the various cases in which the lines AB and AC contain a given angle and have a given ratio or form a given rectangle, and B touches a line or a circumference. They are evidently compositions of rotations with the cases previously examined. I omit to show them. It is worth noting that this is (besides being a homothety or inversion) a rotation of centre A with a translation of vector $\overrightarrow{AB}$ or a translation of vector $\overrightarrow{AB}$ with a rotation of centre B.

Starting with Proposition 20, there are many other propositions that, in Pappus' words, were not of Apollonius, but were added later. A mathematician known only by Pappus' citation, Chamandros, added the first three. They are all elementary and I will omit them. Four additional general propositions follow, each of a metrical content: I will give one case of each.

(1) *If one end* (A) *of a straight line given in magnitude* (AB) *and drawn parallel to some straight line given in position* (CD) *should touch a straight line given in*

position, the other end (B) *too will touch a straight line* (BF) *given in position.* This is too easy, and Simson rightly says (Proposition 20 [18 p. 33]) that it "manifesta est". This amounts to say that the translation given by the vector $\overrightarrow{AB}$ brings straight lines into straight lines. Simson also gives us (Proposition 21) a proposition added by Fermat, which states that the same transformation brings also circumferences into circumferences.

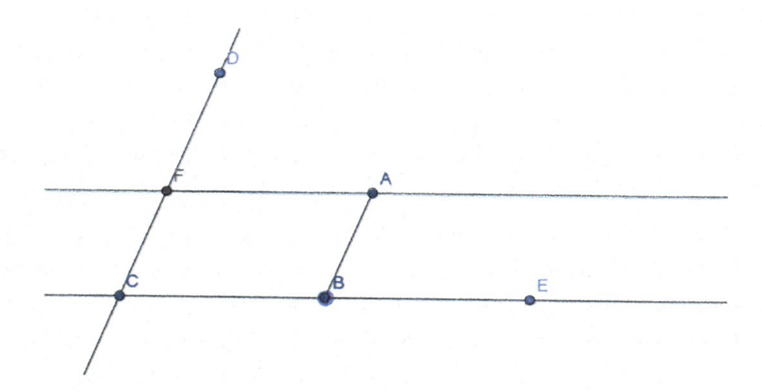

(2) *If from a point to two straight lines given in position, whether parallel or intersecting, straight lines are drawn at given angles, either having a given ratio to one another, or with one of them plus that to which the other has a given ratio being given, the point will touch a straight line given in position.*

There are obviously many different cases of the proposition. I will give a simple, but significant example: intersecting lines and segments having a given ratio.

Simson (Proposition 23 [18 pp. 38-41]) immediately recognized in this proposition the (at the time, already well known) construction of a given straight line from its equation read as a proportion. Following his knowledge of Cartesian geometry, he first studied the case in which the segments are drawn parallel to the given lines, or—which amounts to the same—they form an angle equal to the angle between the given lines (coordinates axes). The given ratio, in this case, is the ratio between AG and AH. We have immediately the construction of the locus requested (i.e. the line AF). Even if this proposition is very easy, one must pay attention to the fact that the mathematicians of the time did not consider negative numbers, so what Simson obtains here is not the line, but the half-line EA. There are, in fact, two points (one for either side of FH) which satisfy $HA : HF = k$, and we should consider one or the other depending on whether A belongs to the half-line AF or to its opposite. In other words, if the equation $y = ax$ means that the non-oriented segments AH and AG have the constant ratio a, it may indicate the half-line AH as well as the half-line $A'H$. It is useful to bear this in mind, even if it is obvious.

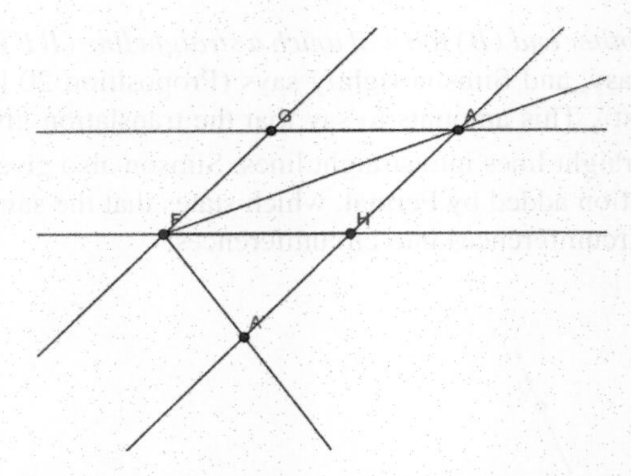

Simson's proposition 25 ([18 pp. 45–48] treats the case in which the requested locus is a generic line (not passing through the origin F). It is expressed in this way: *If from a point to two intersecting straight lines given in position, straight lines are drawn at given angles with one of them plus that to which the other has a given ratio being given, the point will touch a straight line given in position.* In order to better grasp the meaning of the above, I will give a short account of what was (I think) Simson's strategy. The idea is to use proposition 23 to obtain the line in the new "axes" FG (the given one) and MK (obtained from the given FH by a translation of vector $\overrightarrow{HK}$). From the figure, you may get (a) if HK is the given segment (*one of them plus that to which the other has a given ratio*); (b) if the segment AK has the given ratio to AG, then (using 23) A is on the line MN (because of the absence of oriented segments, its locus is only the segment MN) and AH is the *one of them* which added to AK (*that to which the other has a given ratio*) is equal to the given HK. In this way, it is possible to construct any straight line as a locus relative to any couple of intersecting straight lines.

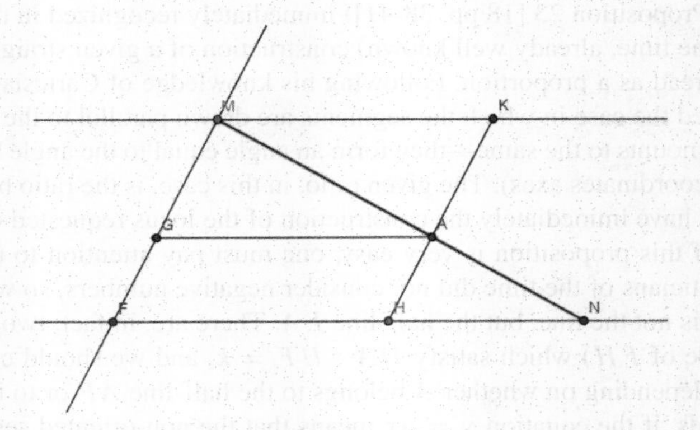

A short break may be useful to better understand the meaning of this construction: I will try to translate it in the algebraic language. After the construction of the line $y = ax$ (Proposition 23 in Simson) in which, evidently, a is the given ratio, we want to construct the non-linear equation: $y + ax = b$ (one of them, y, *plus that to*

which the other, x, *has a given ratio*, a, *is given*, b). Simson's choice is clear (at least I hope so). Simson already knew Cartesian geometry, therefore it is possible that, to interpret Apollonius' statement, he used coordinates; Fermat was looking for an algebraic expression of geometrical objects, and it is possible that he was inspired by Apollonius' proposition to get a clearer insight into the matter.

I will briefly describe also the proof of the two preceding propositions in Fermat's reconstruction, because it is strictly connected with his more famous *Isagoge*, in which they are explicitly used to obtain the equation of a straight line in a given frame.[12]

Fermat studies (2) in his proposition 6. He gives no construction but gives a proof in the style of Euclid's Data. He notes that, in the hypotheses of this proposition, the angles $\widehat{A}$, $\widehat{B}$, $\widehat{D}$ in the quadrilateral $ABCD$ are given, and therefore also $\widehat{C}$ is given; then, in the triangle BCD, the sides BC and CD in the given ratio and $\widehat{C}$ given, are given "in specie" (i.e. for any choice of C, satisfying the given conditions, the triangles are similar to each other); therefore, the triangle ABD is also given "in specie" and the ratio $AB : BD$ is given; but also $BD : BC$ is given; therefore $AB : BC$ is given and the line AC is given.

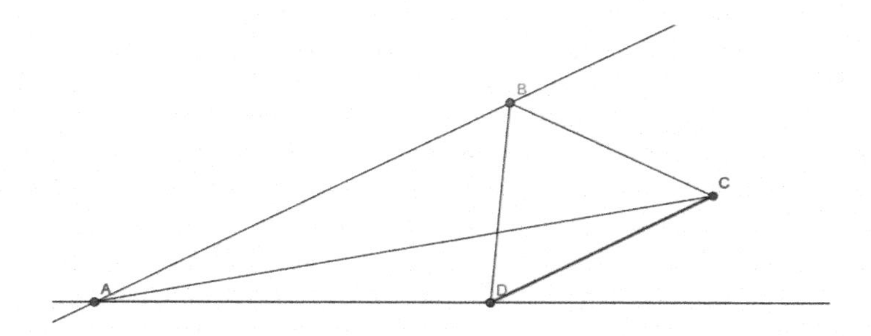

In the second part of the same proposition, Fermat studies the non-linear case. Here, as in Simson's proposition 23, two lines, intersecting in A, are given and it is required to inflect on them, at given angles, two segments from a point E, such that the coordinate EB plus a segment—whose ratio to the other coordinate ED is given—is a given segment.

[12] To get more information about the links between Fermat's reconstruction of *Loci Plani* and the *Isagoge*, I refer to his well-known biography by Mahoney [14] and to a Masters' dissertation by Irene Barbensi [2].

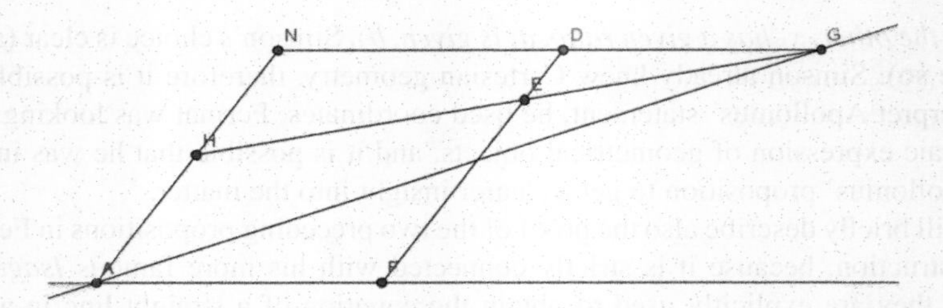

I will skip the many other "loci" that follow in book I. These are mainly related to problems similar to the preceding ones, involving more than two lines.

4 Book II of *Loci Plani*

Book II refers to loci obtained intersecting straight lines.

Generally, we have two (sometimes more) fixed points and a relation between the segments cut off by the point of intersection of the lines issuing from the points. Obviously, the simplest would be that the sum or the difference of the segments be given, but, as it is well known, this relation would bring to a conic section, a "solid locus", so it is not examined in this work. We may think of the loci examined in this part as a classification of the relations that give rise to lines or circles.

I can sketch this classification and the reader should bear in mind that the data below are not usually given in the form expressed by me, but in an equivalent form.

There are, in general, two or more segments issuing from given points. If between the segments it is given (1) their ratio: the locus is a circumference (Propositions 2, 3, 6 in the following); (2) the sum of their squares: the locus is a circumference (Propositions 4, 5); (3) the difference of their squares: the locus is a straight line (Propositions 1, 7). I will follow the order of propositions suggested by this classification: 2, 3, 6, 4, 5, 1, 7.

Pappus says: The second book contains these. *(1) If straight lines from two given points inflect and their squares differ by a given area* (Q), *the point will touch a straight line given in position.*[13] *(2) But if they be in a given ratio, (the point will touch) either a straight line or a circumference.*[14]

I will be concerned with Proposition (1) later.

Proposition 2 *Proposition* 2 [18 pp. 120-124] *obviously relates to the famous "Apollonius circle". Simson gives us many different constructions (his own or from other mathematicians), therefore I will give only Simson's second construction which is, in my opinion, the most interesting one.*

Let A and B be the given points and k the given ratio. In AB we take the point E such that $AE : EB = k$ and the point D such that $AD : DE = k$. The circle

[13] Jones' text is *given in area*. I think that it is a misprint. The Latin text reads *positione datas.*

[14] Jones' text is *arc*, while Commandino's is *circumferentiae.*

(D, E) is the requested one. In his proof, using the similarity between ADC and BDC, Simson proves the important and well-known fact that A and B are mutually inverse with respect to the Apollonius circle. Here we have the "circle of inversion".

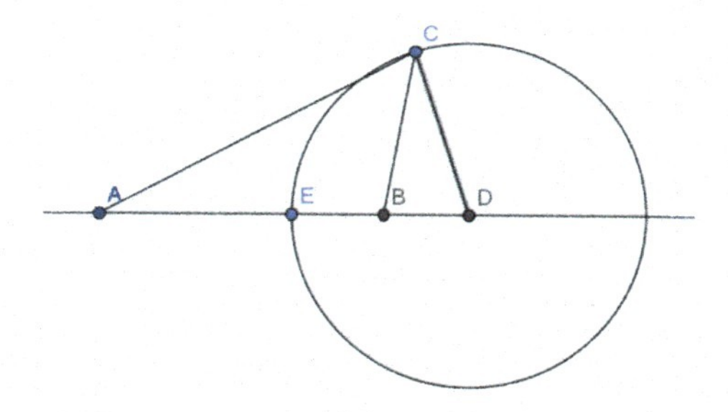

This construction was well known before Apollonius. Aristotle gives it in his Metereologica.[15] Simson himself speaks of a construction by Huygens [12], who uses it for optical problems (in an analogous way as Aristotle does) and by van Schooten. Newton has it as the first construction in his *Loci Plani* [15, p. 230].

Proposition 3. Pappus text reads[16]: *If a straight line be given in position, and a point be given in it, and from this some bounded (line) be drawn, and from the end a (straight line) be drawn at right angles to the (line) <given> in position, and the square of the (first line) drawn equals the (rectangle contained) by a given and what (the perpendicular) cuts off either as far as the given point or as far as another given point on the (line) given in position, the end of this (line) will touch an arc given in position.*

I will give only two of the many cases in Simson's reconstruction. The first one is a simple consequence of Euclid's theorem. It may be stated in this way: *if from a given point A on a given line, r, we take a segment AC such that—if we call D the point in which the perpendicular to r intersects r and E is a fixed point in r—we have (a) $q(CD) = r(ADE)$, then C is on the circumference having AE as a diameter (or, C describes this circumference varying the point D on r.).* In terms of analytical geometry, taking A as the origin, r as x-axis and the perpendicular to r in A as y-axis, we have that the equation: $y^2 = x(a - x)$ which is the equation of a circumference and a is AE, the given segment.

---

[15] There is a complete analysis of Aristotle's construction in [9].

[16] I always cite from Jones [13].

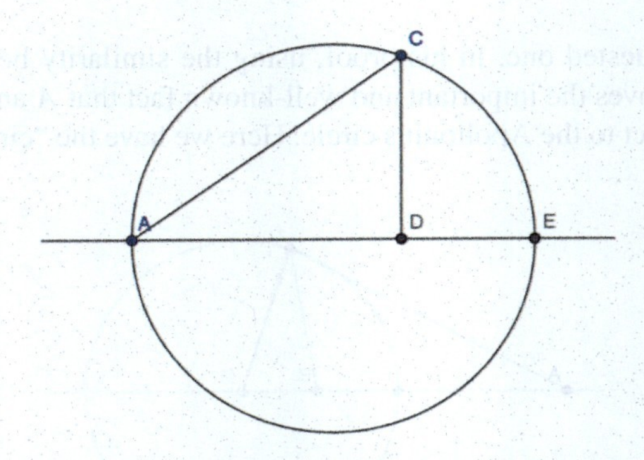

Simson gives a little variation to this proposition, taking CD not perpendicular to r, but parallel to a given direction, i.e. changing coordinates from orthogonal to oblique. In this case, the requested condition is $q(AC) = r(DAE)$.

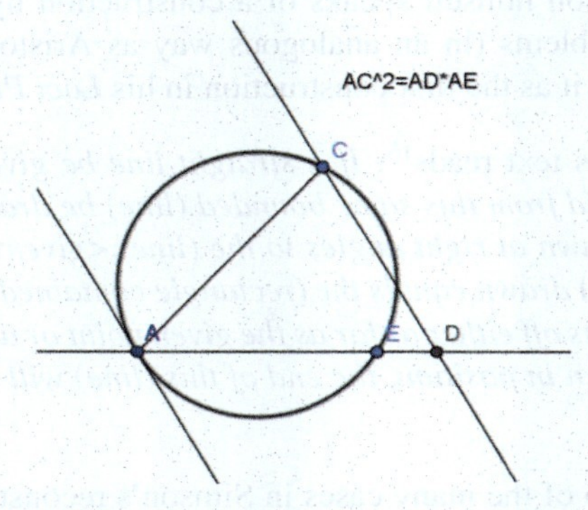

Proposition 6. *If straight lines from two given points inflect, and a straight line is drawn from the point parallel to (a line given) in position and cuts off (an abscissa) from a straight line given in position (extending) as far as a given point, and the shapes (constructed) on the inflecting (lines) equal the (rectangle contained) by a given and the abscissa, the point at the inflection will touch an arc given in position.*

Simson connects the different cases of this proposition to Proposition 3, and I prefer to skip them. The use of the word shapes, instead of squares, means that we may use any polygons different form squares, but they must be similar each other.

Proposition 4. *If straight lines from two given points inflect and the square of the one is greater than the square of the other by a given amount than in ratio, the point will touch an arc given in position.*

Simson corrects this statement adding an arc *or a straight line*. I will treat this case, which I call Proposition 4a—which appears when the given ratio is 1—later.

In other words, if we have two fixed points A and B; X is a given area; and k a given ratio, the points C such that $q(AC) - X = kq(AB)$ are on a given circumference, or the locus of the points C such that $q(AC) - kq(AB) = X$, is a circumference.

There are three different cases when $X >, <, = q(AB)$. I give Simson's solution when $X = q(AB)$. In this case, the construction is simply the following: Take the point E in the line AB such that $AE : EB = k$. The requested circle is (E, B).

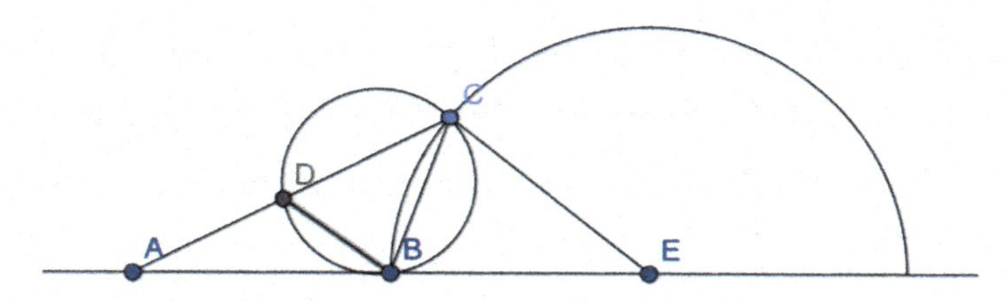

To prove this, for any point C in (E, B), he finds the point D in AC, such that $CAD = X = q(AB)$, using the fact that the circle through D, B, C is tangent in B to AB.

Proposition 5. If straight lines from any number of points inflect at one point, and the shapes (constructed) on all them equal a given area, the point will touch an arc given in position.

First, I will give the most elementary and well-known case. We have two points, A and B, and the shapes are all of squares, and the given area is $q(AB)$. In this way, we have if $q(AC)+q(BC) = q(AB)$ then C is on a circumference (a well-known fact). Proposition 5 generalizes this to a number whatsoever of points and to whatsoever shapes. I will give Simson's solution when we have still two points and the shapes involved are still squares and the proposition reads: if $q(AC) + q(BC) = Z$, the point C describes a circumference. (Here Z is a given area, in general different from $q(AB)$). The construction is: Let D be the middle point of AB, draw the line AB and in it, take a point E such that $r(BAD) + 2q(DE) = Z$. The circle (D, E) is the requested locus.

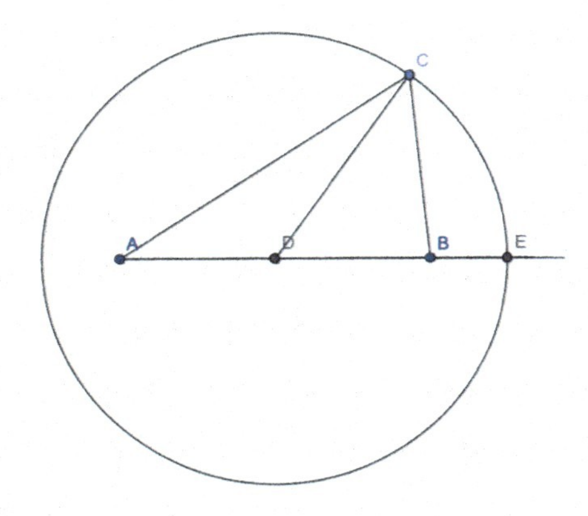

The proof of this proposition is based on Lemma 6, which is a generalization of Pythagoras' theorem: in any triangle ABC we have the following relation: $q(CA) + q(CB) = 2(q(CD) + q(DB))$ (here CD is the median of the triangle). The proof of this Lemma may be useful to compare Euclidean versus Algebraic methods. Here the given triangle is ABC and CE is one of its heights. Simson starts using Eucl. 2.9 which states (see next figure) that $q(AE) + q(BE) = 2(q(BD) + q(DE))$. If we call $AB = 2a$; $EB = b$ this proposition is algebraically equivalent to the identity.

$(2a - b)^2 + b^2 = 2(a^2 + (a - b)^2)$; we have also $2q(CE) + 2q(DC) = 2q(CD)$

$[2h^2 + 2(a - b)^2 = 2AD^2]$; but $q(CA) + q(CB) = q(AE) + q(EB) + 2q(CE) = 2(q(CD) + q(DB)$

$[q(CA) + q(CB) = (2a - b)^2 + b^2 + 2h^2 = 2(CD^2 + b^2)]$.

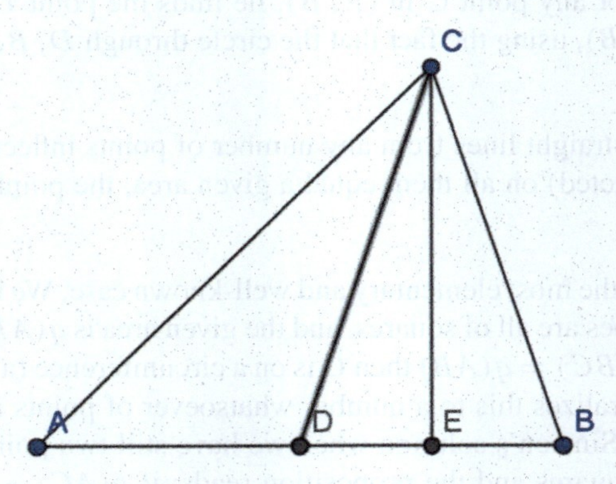

When we have three points A, B, C, and take E as middle point between A and B. Draw CE, the median (from C) of the triangle ABC, and take G in such a way that GC is double of GE (evidently G is centroid of the triangle). In the line CE, take the point H such that $r(B, A, E) + r(ECG) + q(GH) = Z(the\ given\ area)$. The requested locus is the circle (G, H). The proof depends on two of Pappus' lemmas and I will skip it.

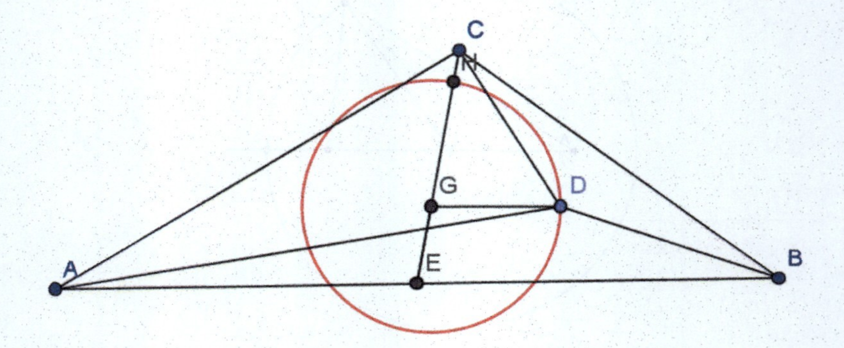

I find interesting the way that Simson uses to demonstrate the proposition for *any* number of points. He writes: *It is useful ... in the Propositions in which the number of given objects may grow without any end, to show the way to deduce the proposed Locus with a given number of points from the Locus in which their number is immediately inferior, as we are going to do* (...He finds the Locus when four points are given, deducing it from the preceding case of three points). *In the same way, if five points are given, the analysis and synthesis may be made by deducing it from the four-points locus, and the solution of six-points Locus can be deduced from five-points locus, in a similar uniform way for any number of given points.*[17]

Another interesting point in that proposition is the fact that, for any number of points having equal masses, the centre of the circle coincides with the barycentre. Simson knows that Huygens, in his *Horologium Oscillatorium,* had demonstrated this property in the following formulation: that for any circle drawn with its centre in the barycentre of a number whatsoever of points, the sum of the squares of the distances from the given points to the barycentre is constant. Simson proves, using his own proof of (5), the same thing, and concludes by saying that Huygens proved this proposition *calculo algebraico satis prolixo.*

To conclude with this important proposition, I only want to remind what Fermat states in a letter to Roberval (1937, February): *Je trouve assez de loisir pour vous envoyer encore la construction du lieu plan (...) que je tiens une des plus belles propositions de la Géométrie.* (Fermat, Oeuvres p. 100).

Proposition 1. *If straight lines from two given points inflect and their squares differ by a given area (Q), the point will touch a straight line given in position.*

I will follow Simson's steps. He begins with Lemma I (proposition 120 in Pappus[18]) of a clear Euclidean flavour. In any triangle ABC(we suppose $AC > BC$), if the height is CD (see the following figure), $q(AC)-q(BC) = 2r(AB, ED)$. From this equality, we see at once that this difference depends only on D and AB, and not on C. Therefore, we have immediately proposition (1): the requested locus is the perpendicular to AB in the point D with $2r(AB, ED) = Q$ (the given area).

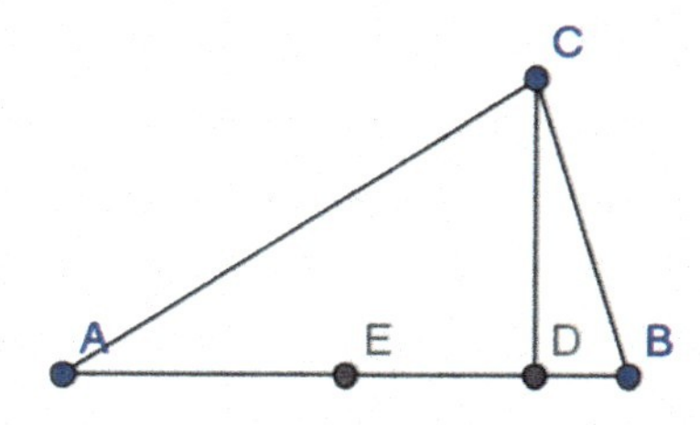

[17] I will not afford the discussion about explicit or implicit use of mathematical induction by Simson: there are too many terms there that would need to be defined in a clearer way.

[18] This proposition is clearly connected to Apollonius' first locus, but Pappus indicates it as a Lemma for the second locus. On the contrary, Pappus indicates his proposition 119 as useful for the first locus, while it is clearly connected with the second one. Simson notes "en passant" this contradiction. See also [13].

E is medium point of AB. We can so state an elementary Corollary to proposition 1. If r and r' are two given straight lines, perpendicular to each other in D, and if A and B are any two points in r and E is their middle point, then any point C of r' will satisfy $q(AC) - q(BC) = 2r(AB, ED)$.

Proposition 7 *If in a circle given in position some point is given, and through it is drawn some straight line, and some point is taken on it outside (the line) and the square of the (segment) as far as the point given inside equals the (rectangle contained) by the whole and the segment outside, either (the square) by itself or this and the (rectangle contained) by the two segments inside, the point outside will touch a straight line given in position.*

Case 1.

In this case are given: (a) a circle (A) whose centre is A; (b) a point B inside the circle. We need to find the locus of points E such that $q(EB) = r(CED)$ (C, D being the points in which EB intersects (A)). The construction is easy. We draw the line AB intersecting (A) in G, H. On it, we draw the point F such that $GB : BH = FB : FH$. Now, we draw the perpendicular EF in F to AB. This is the requested locus.

The proof is the following: take any point E on EF. We have $r(GB, FH) = r(FBH)$; $r(GB, FH) + r(BFH) = r(FBH) + r(BFH)$; $q(BF) = r(GFH)$; $q(BF) + q(EF) = r(GFH) + q(EF)$; $q(EB) = r(GFH) + q(EF)$; $q(BE) = r(GEK) = r(CED)$.

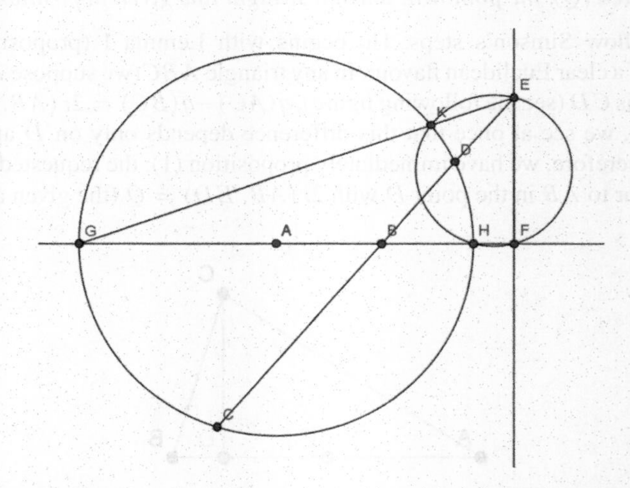

This proposition is identical to Pappus, 159,[19] which refers to Euclid's Porisms. There are, indeed, many coincidences between a group of Porismatic propositions and this proposition (7) of the Plane Loci.

[19] I refer to Ver Eecke's notes [20] for details about the calculations on proportions in the proof of the proposition.

Case 2.

In this case, with identical notations we must have $q(BE) + r(CBD) = r(CDE)$. The construction is the same as in the preceding case 1, but now F satisfies $r(ABF) = r(HBG)$. This is easily seen to be equivalent to $r(FAB) = q(AH)$, i.e. point F is the inverse of B or, if you prefer, (G, H, B, F) is a harmonic range and EF is the polar of B relatively to (A, H).

Simson continues with some generalizations of the preceding propositions, which are not in Pappus.

The connections with proposition (1) may be better understood by looking at Simson's demonstrations. In his (8), Simson specifies that: *The first part* (when the point B is inside the given circle) *may be considered as another way for the second part of Apollonius'* (7). Here the generalizations go in two ways: 1. The given point is not necessarily inside the given circle, but may be external. 2. The relation to be satisfied is $q(BE) + P = r(DEC)$, where P is a given but arbitrary surface. For the meaning of the points, I refer to the figure. In the case where $P = 0$, we have $q(BE) = r(DEC) = r(GEF); q(BE) + q(r) = r(GEF) + q(AG) = q(AE)$. Therefore, $q(AE) - q(BE) = q(r)$. We may now use (1) to have that the locus of points E is the perpendicular to AB in the point H, which satisfies $2r(AB, KH) = q(r)$ (K, as in the earlier case, is the medium point between A and B).

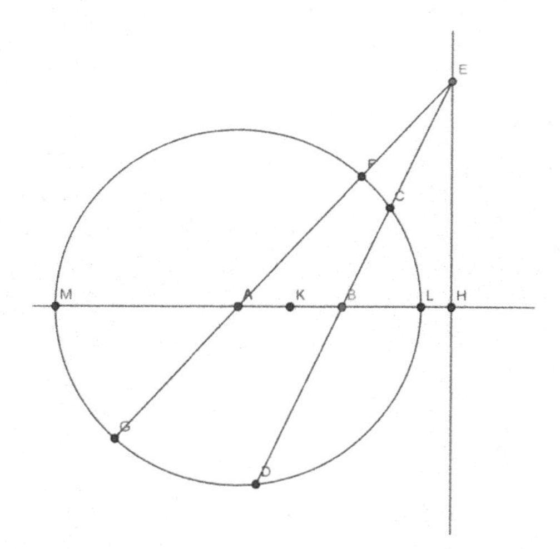

If $q(AE) - q(BE) = q(r) + r(MBL)$, the line HE is the polar of B by the given circle. It is only a matter of calculations with proportions to have that, in this case $EC : ED = BC : BD$, thus recovering the well-known result that (B, E, C, D) form a harmonic range.

If the given point is H, outside the circle, the equality must be changed because now $HE' > AE'$ and the relation is (a): $q(HE') - q(AE') = r(MHL) - q(r)$. If the point E' is inside the circle, this amounts, again, to saying that (b) the range (H, E', F', G') is harmonic. However, while (b) is valid only if E' is inside the circle,

(a) is valid for every point in the line. Therefore, the segment of $E'B$ inside the circle coincides with Apollonius' III.37 (or with Pappus' 154, related to Euclid's Porisms). As far as I know, this is the first definition of the polar of a point external to a circle as a geometrical locus extending to the whole line.

I think that the idea of shifting from the "locus of the harmonic conjugates" to the "locus of the points satisfying (a)" is a good example of "generalizing by a change in the definition".

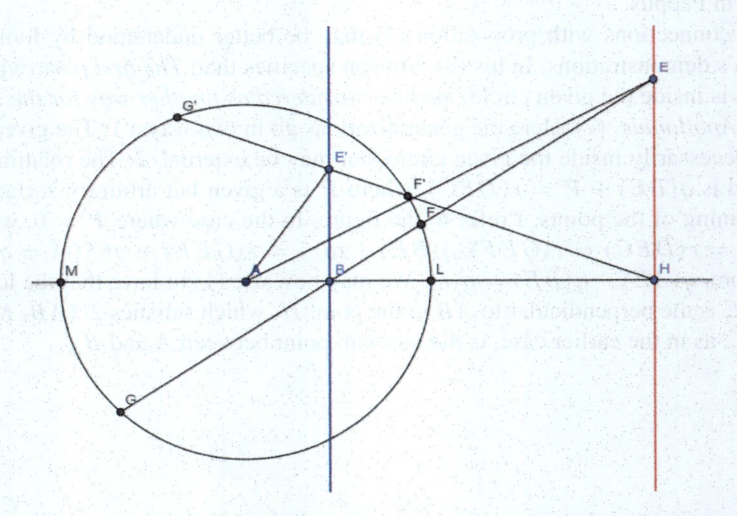

In the figure, the locus generated when the given point is B (inside the circle) is in red, and the generating point is E. The locus generated when the given point is H (outside the circle) is in blue, and the generating point is E'.

It is worth noting that this is probably the first time that we find the construction of the entire polar of a point external to a circle.

5 Simson's Preface

In his Preface, Simson shows the use of Loci in solving problems, a use that does not differ from our*: If, ex. gr. from a hypothesis of the Problem, it is possible to prove that a point demanded touches a Locus given in position, and from another condition has been showed that the same point touches another Locus, drawing those Loci the required point lays necessarily in their intersection and the problem has been solved.*

Therefore, *Loci Plani* are the main tool to solve a problem by ruler and compass.

After a general discussion about the method of geometrical analysis to solve problems, Simson makes an interesting comparison between algebraic versus geometrical methods. He observes that algebraic method gives no (or too little) information about the synthesis of a problem. We can—without going into details—show his observations about van Schooten's solution of a problem in his reconstruction of *Loci Plani*.[20] The problem asked to show that a point satisfying some given conditions touched a given straight line. After many calculations, van Schooten writes, for the required point (x, y) the formula.

$y = \frac{cdio+efko+ghio+abnx+dox-efox-ghox}{mzz+bnz-doz-foz+hoz}$, where the symbols different from x and y refer to known magnitudes. Posing, "brevitatis causa", $p = \frac{cdio+efko+ghio}{mzz+bnz-doz-foz+hoz}$ and $\frac{q}{r} = \frac{abn+cdo-efo-gho}{mzz+bnz-doz-foz+hoz}$, he gets the standard form for the equation of a line: $y = p + \frac{q}{r}x$. Now we may ask: is the problem solved? Van Schooten answers in the affirmative. He shows that the infinite (*innumera*) points satisfying the required conditions touch a given line. Simson, on the contrary, answers in the negative: van Schooten, he argues, has not explicitly constructed the required line. It is impossible not to see that the most difficult part of the "composition" of the problem was, precisely, to show how, from the given magnitudes, we can get p and $\frac{q}{r}$: *quis non videt multo difficilius fore invenire ipsas p, q, r geometrice, ut earum ope aequatio construatur, quam aequationem ipsam invenire.*

In 1656, when van Schooten published his *Exercitationes*, only seven years were elapsed from the appearance of his translation of Descartes' Geometry. Van Schooten reads Apollonius' work as a testing ground to prove the effectiveness of the new methods. In practice, however, what was going on was a completely new idea of the meaning of "solving a problem": it was not necessary anymore, in fact, to give the explicit geometrical construction of the requested object; it sufficed to give its equation. This episode is very similar to the attitude of Descartes himself in his famous correspondence with Elisabeth of Bohemia.[21] Studying Apollonius' problem of the Tangencies (to find the circles tangent to three given ones), assuming x, y, z respectively as the radius and the coordinates of the centre of the required circle, he had found a complicated relation between data and variables and, instead of trying to give their construction, he wrote: *On trouve une équation ou il n'y a que x et xx inconnue; de façon que le problème est plane, et il n'est plus besoin de passer outre. Car le reste ne sert point pour cultiver ou recréer l'esprit, mais seulement pour exercer la patience de quelque calculateur laborieux* [5, p. 42].

There is a deep difference between the two points of view. Simson's ideas had certainly been influenced by the ones of Newton, who, at the end of the 1690s, wrote: *A question ... is not solved before the construction's enunciation and its complete demonstration is ... with the equation now neglected, composed* [8, p. 76]. Newton's stand prevailed in the UK, but the emergence and development of differential calculus changed profoundly the debate.

In any case, Simson's ideas were not aimed at a historical or philological study of "ancient" heritage, but at finding new sources for a new mathematics. This is particularly clear in the many treatises that Newton left manuscript and unfinished.

---

[20] Simson indicates a reference to van Schooten's reconstruction of proposition 2.11 of *loci plani* at p. 210 of [17]. In reality, he refers to Problem (not proposition) XI at p. 246. In any case, Simson cites the exact van Schooten's formula.

[21] Descartes' letters to Elisabeth of Bohemia are in [5].

The ideas of Simson did not influence in any way the works on projective geometry of the beginning of the following century; in my opinion, however, they moved in the same direction: what was needed for them to bear fruit, was a deep shift in the attitude of mathematicians that was yet to come.

6 Conclusions

To conclude, I would like to summarize some of the principal new ideas contained in Simson's *Loci Plani*.[22]

In the first part of book I, there is the idea to classify all the loci obtained by moving a point in a plane locus and resulting in another plane locus. This classification is complete. We have, indeed, all the simple Moebius transformations (translation, rotation, homothety, reflection, inversion), as well as many compositions of the latter. May we now say that in Simson's work there is already a clear idea of transformations? I do not think so. The answer, however, as I said above, largely depends on what exactly we mean with the word "transformation". As Simson himself states, they are Loci, in our sense. It is not necessary to see the Conchoid of Nicomedes as the transformation of a straight line, to use it to trisect a given angle.

In the last part of book II, there is the study of the straight line's loci of the points where two segments issuing from two points A, B inflect in a point C with $q(AC) - q(BC) = Q$ given, particularly when A is the centre of a given circle. Propositions 1 and 7 are, in my opinion, deeply fascinating. Obtained with very simple means, they however bear a great quantity of consequences. When A is the centre of a circle of radius r, if $Q = r^2$, the line is the Locus of the centres of circles orthogonal to the given one through B; when $Q = r^2 + r(MBL)$, the Locus is the polar of B. When B is also the centre of a circle whose radius is $r_1 < r$, if $Q = r^2 - r_1^2$ we have the radical axis of the two circles.

Am I arguing that the radical axis was discovered by Apollonius two thousand years before Steiner and Poncelet? Not at all. What I am arguing is that all the technical tools needed for this interesting theory were already at the disposal of the mathematicians. The real change was in their minds and eyes. Prompted by the great advances of algebraic methods, pure geometers began to see geometric objects in a different way. For instance, after algebraic "imaginary" intersections of curves, they could be impressed by the simple observation that the "radical axis" was completely independent from whether the two circles did or did not intersect, and this, in turn, could lead to the completely new idea of "ideal chords". Old results and old mathematical objects can get a new life on the basis of new mathematical ideas.[23]

[22] As I already said, I do not discuss the adherence of Simson's reconstruction to Apollonius' original.

[23] In this regard, I found interesting Fried's comparison of the different treatments by Euclid and Steiner of what is now known as "power of a point". See [7].

I will now try to give an (hypothetical) solution of Problem VIII of Apollonius' Tangencies to show the possible use of this very important Locus. The Problem is: *To describe a circumference passing through two given points and tangent to a given circumference.*

Analysis: Let A, B be the given points and (O, F) the given circumference; consider the problem as already solved and let the circumference in red in the figure be tangent to (O, F) in D; let DM be the common tangent. If M is the point of intersection of DM and AB, it has obviously the same power (here I use a "modern" word for an old concept) with respect to the two circumferences. Therefore, we must have (taking H as the middle point of AB), $r(FMG) = r(AMB)$; $q(OM) - q(OF) = q(HM) - q(HA)$; $q(OM) - q(HM) = q(OF) - q(HA)$. But the second member of this equation is given; therefore, M is in the Locus of points that have a given difference of squares from two given points, and it is given from proposition 2.1 (the line ML perpendicular to OH). Therefore, M, intersection of two given lines, is also given and so is D. The composition, or synthesis, comes directly from Proposition II.1.

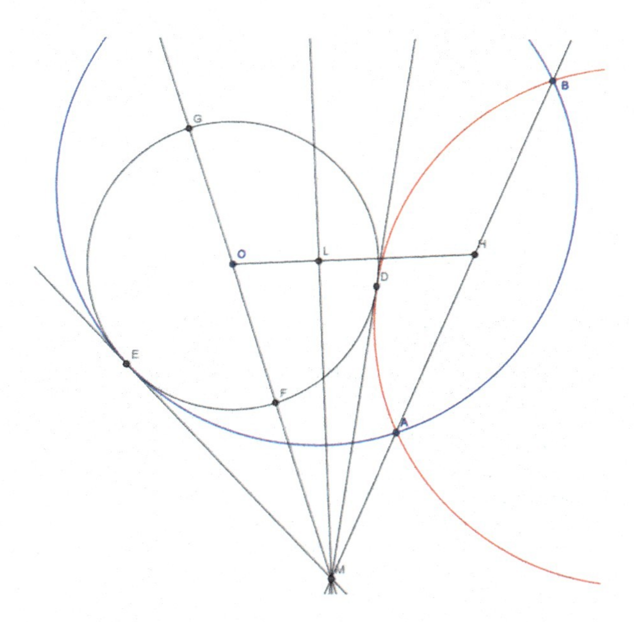

We may compare this solution with an algebraic one.[24] Descartes chooses the line OA and its perpendicular BE from B as orthogon axes. The coordinates of the given points will be: $A \equiv (d = AE, 0)$, $B \equiv (0, e = BE)$, $O \equiv (f = OE, 0)$: the unknowns are the radius r and the centre $D \equiv (x, y)$ of the required circle. Therefore, we have: $AD^2 = r^2 = y^2 + (x - d)^2 = y^2 + x^2 + d^2 - 2dx$; $BD^2 = r^2 = x^2 + (e - y)^2 = y^2 + x^2 + e^2 - 2ey$; $OD^2 = (r + c)^2 = y^2 + x^2 + f^2 + 2fx$.

[24] I have adapted Descartes' solution in [5, p. 42].

From these equations, we get the solution. We may see that it is not at all a mere change of the mathematical language used. The algebraic solution is not the translation of the geometric one in another language. The linguistic difference implies a radically different approach to the solution of the problem.

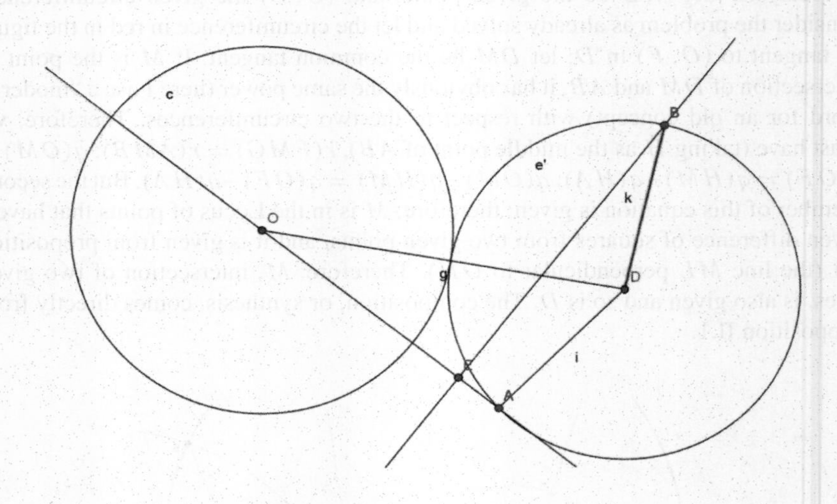

References

1. R. Archibald, Centres of similitude of circles and certain theorems attributed to Monge. Were they known to the Greeks?. Am. Math. Mon. **22**, 6–12 (1915)
2. I. Barbensi, Dai Loci Plani all'Isagoge: Fermat e l'invenzione di una nuova geometria, master thesis, 2016–2017, University of Pisa, https://etd.adm.unipi.it/theses/available/etd-01182018-114642/unrestricted/tesi_Irene_Barbensi.pdf
3. F. Commandino, *Pappi Alexandrini mathematicae Collectiones* (Pesaro, 1588)
4. J. Coolidge, *A History of Geometrical Methods* (Clarendon, Oxford, 1940)
5. R. Descartes, *Oeuvres,* ed. by C. Adam, P. Tannery, vol. 4 (Correspondance, Paris, 1901)
6. P. Fermat, Apollonii Pergaei Libri duo de Locis Planis restituti, in ed. by P. Fermat, *Œuvres*, par les soins de MM. Paul Tannery et Charles Henry (Gauthier-Villars, Paris, 1891) pp. 1–59
7. M. Freid, *Mathematics as the Science of Patterns* (Convergence, 2004). https://www.maa.org/press/periodicals/convergence/mathematics-as-the-science-of-patterns
8. N. Guicciardini, *Isaac Newton on Mathematical Certainty and Method* (The MIT Press, Cambridge (Mass.), 2009)
9. T. Heath, *A history of Greek Mathematics* (Clarendon, Oxford, 1921)
10. H. Hudson, *Cremona Transformations in Plane and Space* (Cambridge UP, Cambridge, 1927)
11. F. Hultsch (ed.), *Pappi Aleandrini Collectionis quae supersunt*, vol. II (Berlin, 1877)
12. C. Huygens, *Opuscula Postuma quae continent Dioptrica* (Leyde, 1703)
13. A. Jones (ed.), *Pappus of Alexandria. Book 7 of the Collections* (Springer, Heidelberg, 1985)
14. M. Mahoney, *The Mathematical Career of Pierre de Fermat*, 2nd edn. (Princeton UP, Princeton, 1994)
15. I. Newton, Simple locus problems (from Newton's Waste Book in the University of Cambridge Library), in *The Mathematical Papers of Isaac Newton*, ed. by D.T. Whiteside vol. IV (Cambridge UP, Cambridge, 1971) pp. 230–250

16. B. Rosenfeld, *Apollonius of Perga Conics. Books One – Seven*. http://www.personal.psu.edu/sxk37/Commentaries-new.pdf
17. F. Schooten, Apollonii Pergaei Loca Plana Restituta, in ed. by F. Schooten, *Exercitationum Mathematicarum Libri Quinque,* Leida, vol. 1657, pp. 203–292
18. R. Simson, *Apollonii Pergaei Locorum Planorum Libri II* (Glasgow, 1749)
19. W. Trail, *Account of the life and writings of Robert Simson* (Bath, 1812)
20. P. Ver Eecke (ed.), *Pappus d'Alexandrie. La Collèction Mathématique* (Paris, 1933)
21. P. Ver Eecke (ed.), *Les Coniques d'Apollonius de Perge*, vol. 2 (Blanchard, Paris, 1963)
22. F. Viète, *Apollonius Gallus* (Paris, 1600)

Kummer Quartic Surfaces, Strict Self-duality, and More

Fabrizio Catanese

Abstract In this paper, we first show that each Kummer quartic surface (a quartic surface X with 16 singular points) is, in canonical coordinates, equal to its dual surface and that the Gauss map induces a fixed-point free involution γ on the minimal resolution S of X. Then we study the corresponding Enriques surfaces S/γ. We also describe in detail the remarkable properties of the most symmetric Kummer quartic, which we call the Cefalú quartic. We also investigate the Kummer quartic surfaces whose associated Abelian surface is isogenous to a product of elliptic curves through an isogeny with kernel $(\mathbb{Z}/2)^2$, and show the existence of polarized nodal K3 surfaces X of any degree $d = 2k$ with the maximal number of nodes, such that X and its nodes are defined over $\mathbb{R}$. We take then as parameter space for Kummer quartics, an open set in $\mathbb{P}^3$, parametrizing nondegenerate $(16_6, 16_6)$-configurations, and compare with other parameter spaces. We also extend to positive characteristic some results which were previously known over $\mathbb{C}$. We end with a section devoted to remarks on normal cubic surfaces and providing some other examples of strictly self-dual hypersurfaces.

Keywords Kummer surfaces · Projective Self-duality · Theta functions · Automorphisms · K3 surfaces · Enriques surfaces · Projective Configurations · Gauss Maps · Quartic surfaces · Cefalu' surface · Cubic surfaces · Kummer–Enriques graph · Segre cubic Hypersurface

AMS Classification: 14J28, 14K25, 14E07, 14J10, 14J25, 14J50, 32J25

Dedicated to Ciro, the 'Prince of Stromboli', on the occasion of his 70-th birthday.

F. Catanese (✉)
Lehrstuhl Mathematik VIII, Mathematisches Institut der Universität Bayreuth, NW II
Universitätsstr. 30, 95447 Bayreuth, Germany

Korea Institute for Advanced Study, Hoegiro 87, Seoul 133–722, Korea
e-mail: Fabrizio.Catanese@uni-bayreuth.de

T. Dedieu et al. (eds.), *The Art of Doing Algebraic Geometry*, Trends in Mathematics,
https://doi.org/10.1007/978-3-031-11938-5_3

1 Introduction

This paper originated from some simple examples I gave in a course held in Bayreuth in the summer of 2018 and in a series of lectures held in November 2018 in Udine, on the topic of surfaces in $\mathbb{P}^3$ and their singularities.

Given an irreducible surface X of degree d in $\mathbb{P}^3$ which is normal, that is, with only finitely many singular points, one can give an explicit upper bound $\mu'(d)$ for the number of its singular points (for instance, over $\mathbb{C}$, one can use the birational map of X to its dual surface $X^\vee$ to obtain a crude estimate).

Letting $\mu(d)$ be the maximal number of singular points of a normal surface of degree d, the case of $d = 1, 2$ being trivial ($\mu(1) = 0$, $\mu(2) = 1$), the first interesting cases are for $d = 3, 4$: $\mu(3) = 4$, $\mu(4) = 16$ if char(K) $\neq 2$.[1]

For $d = 3$ (see Propositions 35 and 37), a normal cubic surface X can have at most four singular points, no three of them can be collinear, and if it does have four singular points, these are linearly independent; hence, X is projectively equivalent to the so-called Cayley cubic, first apparently found by Schläfli, see [10, 14, 53].

The Cayley cubic has the simple equation

$$X := \{x := (x_0, x_1, x_2, x_3) | \sigma_3(x) := \sum_i \frac{1}{x_i} x_0 x_1 x_2 x_3 = 0\}.$$

Here, σ_3 is the third elementary symmetric function (the four singular points are the four coordinate points).

A normal quartic surface can have, if char(K) $\neq 2$, at most 16 singular points: over $\mathbb{C}$, one can use in general the birational map of X to its dual surface $X^\vee$ to obtain the crude estimate that the dual surface $X^\vee$ of a normal surface X with ν singular points has degree $\leq d(d-1)^2 - 2\nu$. By biduality, the degree of $X^\vee$ is at least 3 if $d \geq 3$.

Hence, for $d = 3$, we get $\nu \leq 4$, as already mentioned, while for $d = 4$ $\nu \leq 16$, equality holding if and only if $X^\vee$ is also a quartic surface.

More generally, if char(K) $\neq 2$ and X is a normal quartic surface, by Proposition 37, it has at most 7 singular points if it has a triple point, else it suffices to project from a double point of the quartic to the plane and to use the bound for the number of singular points for a plane curve of degree 6, which equals 15, to establish $\nu \leq 16$.

Quartics with 16 singular points (char(K) $\neq 2$) have necessarily nodes as singularities, and they are the so-called Kummer surfaces [36] (the first examples were found by Fresnel, 1822).

There is a long history of research on Kummer quartic surfaces, for instance, it is well known that if $d = 4$, $\nu = 16$, then X is the quotient of a principally polarized Abelian surface A by the group $\{\pm 1\}$, embedded (for $K = \mathbb{C}$) by the theta functions of second order on A.

The simplest example of such a surface is the following

[1] In $char(K) = 2$, using also a remark by the referee on the discriminant of supersingular K3 surfaces, we can prove that $\mu(4) \leq 20$.

Main example: The Duke of Cefalú quartic surface is the 16-nodal quartic X with equation

$$s_1(z_i^2)^2 - 3s_2(z_i^2) := \left(\sum_1^4 z_i^2\right)^2 - 3\sum_1^4 z_i^4 = 0 \Leftrightarrow \sigma_2(z_i^2) - s_2(z_i^2) = 0.$$

Here, σ_i is the i-th elementary symmetric function, while s_i is the i-th Newton function (sum of i-th powers, in particular $s_2(z_i^2) = s_4(z_i)$).

The Duke of Cefalú K3 surface shall denote the minimal resolution S of X, and for simplicity we shall also call X the Cefalú quartic. Here are some remarkable properties of this surface.

Theorem 1 *The Cefalú quartic surface*

$$X := \{\sigma_2(z_i^2) - s_2(z_i^2) = 0\} = \{\left(\sum_1^4 z_i^2\right)^2 - 3\sum_1^4 z_i^4 = 0\} \subset \mathbb{P}^3_K$$

(K an algebraically closed field of characteristic $\neq 2, 3$) enjoys the following properties:

(i) The group $\mathcal{G}$ of projectivities which leave X invariant contains the semidirect product

$$G := (\mathbb{Z}/2)^3 \rtimes \mathfrak{S}_4,$$

central quotient of the canonical semidirect product G'

$$G := G'/Z(G'),\ G' := (\mathbb{Z}/2)^4 \rtimes \mathfrak{S}_4,$$

and $\mathcal{G} = G$ if the characteristic of K is $\neq 5$, else $|\mathcal{G}| = 10|G| = 1920$.

(ii) The singular points of X are 16, and they are the G-orbit $\mathcal{N}$ of the point $(1, 1, 1, 0)$, and also its $\mathcal{K}$-orbit, where $\mathcal{K} \cong (\mathbb{Z}/2)^4$ is the subgroup generated by double transpositions and by diagonal matrices with entries ± 1 and determinant 1.

(iii) X is strictly self-dual in the strong sense that $X^\vee = X$.

(iv) Via the standard scalar product the set $\mathcal{N}$ yields a set $\mathcal{N}'$ of 16 planes, such that $(\mathcal{N}, \mathcal{N}')$ form a $(16_6, 16_6)$ configuration.

(v) The Gauss map gives an automorphism γ of the minimal resolution $S := \tilde{X}$ of X, which has order 2 and which centralizes G.

(vi) The group $G \times \mathbb{Z}/2$ generated by G and γ contains a subgroup G^s of index 2 and order 192, $G^s \cong (\mathbb{Z}/2)^3 \rtimes \mathfrak{S}_4$, of automorphisms of S acting symplectically (acting trivially on the holomorphic two forms of S). The full group of symplectic automorphisms of S has order 384 and is a semidirect product $(\mathbb{Z}/2)^4 \rtimes \mathfrak{S}_4$.

(vii) γ acts without fixed points, hence S/γ is an Enriques surface, which contains a configuration of 16 (-2)-curves, which we describe via the called Kummer–

Enriques graph, and containing several sets of cardinality four of disjoint such curves, but none of cardinality 5 *(in particular* S/γ *is the minimal resolution of several* 4*-nodal Enriques surfaces).*

(viii) *The group of automorphisms of* S *is larger than* $G \times \mathbb{Z}/2$ *and indeed infinite.*

(ix) *The surface* X *is the Kummer surface of an Abelian surface isogenous to the product of two Fermat (equianharmonic) elliptic curves.*

(x) *The Cefalú quartic surface*

$$X := \{\sigma_2(z_i^2) - s_2(z_i^2) = 0\} \subset \mathbb{P}^3_{\mathbb{Z}}$$

defines a Kummer surface over $\mathcal{S} := Spec(\mathbb{Z}(\frac{1}{2}, \frac{1}{3}))$.

Do some of these properties carry over to all Kummer quartic surfaces? Yes, (ii), (iii), (iv), (v), (vii), and (viii) do, as we shall see in this paper; the other properties are very special of this surface or extend only in part or for special Kummer quartics.

There are several comments to be made.

- Concerning (i), we get a group of symplectic automorphisms of order 192, which is an index two subgroup of the group of symplectic automorphisms of the Fermat quartic surface (see Mukai's paper [37], item 5) on page 184, and line 8 on page 191).[2] We show in Proposition 22 that indeed the full group of symplectic automorphisms of the Cefalú K3 surface S is equal to the group of symplectic automorphisms of the Fermat quartic surface.
- Concerning (vii), the existence of (other) fixed-point free involutions on Kummer surfaces (thereby leading to new constructions of Enriques surfaces) was first shown by Hutchinson, who used Cremona involutions based at the so-called Göpel tetrads [25].
- Concerning (viii), for our assertion, leading to question 18, we use the Klein involution ι induced from projection from one node. Our argument allows us to prove Theorem 18 asserting that for each Kummer quartic surface the group of automorphisms of S is infinite, in characteristic $\neq 2$. Our result confirms the unpublished result of Jong Hae Keum (over $\mathbb{C}$), cited in [31], and based on the combination of the results of [29, 33].
  Over $\mathbb{C}$, since for the Cefalú surface S $Pic(S)$ has rank 20, the fact that $Aut(S)$ is infinite follows from the old result of Shioda and Inose [57].
- Also Hutchinson asserts that the group $Aut(S)$ of a Kummer surface is infinite (at least for general choice of the Kummer surface, compare the argument given on page 214 of [25]). Hutchinson uses the involutions mentioned above, but we show

---

[2] The vague question here is whether the Cefalú quartic and the Fermat quartic are somehow related, 'mirror' to each other? Motivation comes from the pencil containing them both and consisting precisely of the quartics having G-symmetry:

$$\lambda_1 s_1(z_i^2)^2 + \lambda_2 s_2(z_i^2) = 0 \Leftrightarrow \mu_1 \sigma_2(z_i^2) - \mu_2 s_2(z_i^2) = 0.$$

The Cefalú quartic is the only Kummer quartic in the pencil, which has five singular elements, for $\lambda_1 = 1, \lambda_2 = 0, -1, -2, -3, -4$.

that these do not exist as claimed by Hutchinson on the Cefalú surface, see Remark 23; more generally, they do not exist for Segre-type Kummer surfaces, because, if some Thetanull vanishes, see Remark 13, then the Göpel tetrads become linearly dependent.

- Concerning (iii), we observe that in all articles and textbooks one finds the weaker statement (inspired by the great article by Klein [32] treating quadratic line complexes) that a Kummer quartic is projectively equivalent to its dual surface (see, for instance, [22], page 784, [19] Corollary 4.27, page 95, [17] Theorem 10.3.19, and Remark 10.3.20). To clarify the issue, we give the following:

Definition 2 A projective variety $X \subset \mathbb{P}^n$ is strictly self-dual if there are coordinates such that $X^\vee = X$.

X is said to be weakly self-dual if X is projectively equivalent to $X^\vee$.

To focus on the difference between the two notions, it suffices to observe that X is strictly self-dual if and only if there is a projectivity sending X to $X^\vee$ whose matrix A is of the form ${}^tBB$, equivalently, such A is symmetric.

Up to now the main series of examples of self-dual varieties ([7, 47, 48]) yielded strongly self-dual varieties.

The question of whether there are weakly self-dual varieties which are not strongly self-dual remains (as far as we know) open, since for Kummer quartics we prove the analogue of (iii) and (vii) above:

Theorem 3 *(1) All Kummer quartics $X \subset \mathbb{P}^3_K$, K an algebraically closed field of characteristic $\neq 2, 3$, are strongly self-dual*[3] *and*

(2) the Gauss map produces an automorphism γ, of order 2, of the minimal resolution $S := \tilde{X}$ of X, which acts without fixed points so that S/γ is an Enriques surface.

We give two proofs for the above statement, a short transcendental proof over $\mathbb{C}$ using theta functions, and then an algebraic one (valid in characteristic $\neq 2$) which relies on earlier results, especially the ones by Gonzalez-Dorrego ([19]).

Observe that (2) provides, over $\mathbb{C}$, a simpler proof of an important result of Keum [29], that every Kummer surface is an unramified covering of an Enriques surface; and it extends the result also to fields K of positive characteristic.

Many explicit equations have been given for Kummer quartics (I refer for this to the beautiful survey by Dolgachev [18], which appeared just before this paper was finished), but here we use new free parameters for Kummer quartics.

These are based on the following theorem, essentially due to Gonzalez-Dorrego, extending (iv) above to all Kummer quartic surfaces over an algebraically closed field K of characteristic $\neq 2$.

We consider for this purpose the action on $\mathbb{P}^3$ of the (already mentioned) group

$$\mathcal{K} := (\mathbb{Z}/2)^4 = (\mathbb{Z}/2)^2 \oplus (\mathbb{Z}/2)^2$$

[3] Probably a similar computation shows the result also in $char = 3$.

such that the first summand acts via the group of double transpositions, the second summand through $z_i \mapsto \epsilon_i z_i$, for $\epsilon_i = \pm 1$, $\epsilon_1 = \epsilon_1\epsilon_2\epsilon_3\epsilon_4 = 1$, and we recall that a nondegenerate $(16_6, 16_6)$ configuration in $\mathbb{P}^3$ is a configuration $\mathcal{N}, \mathcal{N}'$ of 16 points and 16 planes such that each point in $\mathcal{N}$ belongs to six planes of the set $\mathcal{N}'$, and each plane in $\mathcal{N}'$ contains exactly six points of the set $\mathcal{N}$, and nondegenerate means that two planes of $\mathcal{N}'$ contain exactly two common points of $\mathcal{N}$.

Theorem 4 *Every nondegenerate $(16_6, 16_6)$ configuration in $\mathbb{P}^3_K$, K algebraically closed with $char(K) \neq 2$ is projectively equivalent to the configuration $(\mathcal{N}, \mathcal{N}')$ where $\mathcal{N}'$ is the set of planes orthogonal to the elements of $\mathcal{N}$, and $\mathcal{N}$ is the $\mathcal{K} = (\mathbb{Z}/2)^4$ orbit of a point $(a_1, a_2, a_3, a_4) \in \mathbb{P}^3$ such that:*
(I) no two coordinates are equal to zero, $\sum_i a_i^2 \neq 0$,

$$(II)\ a_1a_2 \pm a_3a_4 \neq 0,\ a_1a_3 \pm a_2a_4 \neq 0,\ a_1a_4 \pm a_2a_3 \neq 0,$$

$$(III)\ a_1^2 + a_2^2 \neq a_3^2 + a_4^2,\ a_1^2 + a_3^2 \neq a_2^2 + a_4^2,\ a_1^2 + a_4^2 \neq a_2^2 + a_3^2.$$

For each such configuration $\mathcal{N}$, which is strictly self-dual, there is a unique Kummer quartic X having $\mathcal{N}$ as a singular set, and all Kummer quartics arise in this way.

The equation of X has the Hudson normal form:

$$\alpha_0\left(\sum_i z_i^4\right) + 2\alpha_{01}(z_1^2z_2^2 + z_3^2z_4^2) + 2\alpha_{10}(z_1^2z_3^2 + z_2^2z_4^2)$$
$$+ 2\alpha_{11}(z_1^2z_4^2 + z_2^2z_3^2) + 4\beta z_1z_2z_3z_4 = 0,$$

where, setting $b_i := a_i^2$, $b := a_1a_2a_3a_4$, the vector of coefficients

$$v :=^t (\alpha_0, \alpha_{01}, \alpha_{10}, \alpha_{11}, \beta)$$

is given, for $b \neq 0$, by the solution of the system of linear equations

$$Bv = 0,$$

where B is the matrix

$$B := \begin{pmatrix} b_1^2 & b_1b_2 & b_1b_3 & b_1b_4 & b \\ b_2^2 & b_2b_1 & b_2b_4 & b_2b_3 & b \\ b_3^2 & b_3b_4 & b_3b_1 & b_3b_2 & b \\ b_4^2 & b_4b_3 & b_4b_2 & b_4b_1 & b \end{pmatrix}.$$

The case $b = 0$ is governed by other equations and, assuming without loss of generality that $a_1 = 0$, we obtain $\beta = 0$, and we get the coefficients:

$$\begin{aligned}\alpha_0 &= 2b_2b_3b_4\\ \alpha_{01} &= b_2(b_2^2 - b_3^2 - b_4^2)\\ \alpha_{10} &= b_3(b_3^2 - b_4^2 - b_2^2)\\ \alpha_{11} &= b_4(b_4^2 - b_2^2 - b_3^2)\end{aligned}$$

For $b \neq 0$, by the theorem of Rouché-Capelli, the coefficients of v are given by the determinants of the 4×4 minors of the matrix B, which we omit to spell out in detail.

The case $b = 0$ is quite interesting and indeed related to how we found the Cefalú surface as the most symmetric among the Segre-type Kummer quartics. These are the quartics whose equation is of the form $Q(z_1^2, z_2^2, z_3^2, z_4^2)$, where $Q(x) = 0$ is a smooth quadric and the four planes $x_i = 0$ are tangent to it at a point not lying on the edges of the tetrahedron $T := \{x_1x_2x_3x_4 = 0\}$.

Observe in fact that the dual quadric $Q^*(y) = 0$ has a matrix with diagonal entries equal to 0; hence, the most symmetric solution is $\sigma_2(y) = 0$, leading to the Cefalú quartic.

We devote a section to the geometry of these special Kummer quartics and show that these are the ones whose associated Abelian surface A is isogenous to a product of elliptic curves $A_1 \times A_2$ through an isogeny with kernel group $(\mathbb{Z}/2)^2$. And we use them in order to show a result which is used in [8] to determine the components of the variety of nodal K3 surfaces.

Theorem 5 *For each degree $d = 4m$, there is a nodal K3 surface X' of degree d with 16 nodes which is defined over $\mathbb{R}$, and such that all its singular points are defined over $\mathbb{R}$.*

Similarly, for each degree $d = 4m - 2$, there is a nodal K3 surface X'' of degree d with 15 nodes which is defined over $\mathbb{R}$, and such that all its singular points are defined over $\mathbb{R}$.

The same results hold replacing $\mathbb{R}$ by an algebraically closed field K of characteristic $\neq 2$.

Remark 6 As shown, for instance, in [8], if the degree d is not divisible by 4, then 15 is the maximal number of nodes.

Later on, we compare the equations provided by Theorem 4 with other representations of Kummer surfaces, for instance, the one given by Coble [12] and later van der Geer [58] through the Siegel quartic modular threefold, which is the dual of the 10-nodal Segre cubic threefold $\mathcal{S}$.

Projective duality is then shown to provide the relation between two different representations of Kummer surfaces,

(1) the first one as discriminants of the projection to $\mathbb{P}^3$ of the Segre cubic threefold $\mathcal{S}$ with centre a smooth point $x \in \mathcal{S}$, not lying in the 15 planes contained in $\mathcal{S}$,

(2) the second one as intersections of $\mathcal{S}^\vee$ with a tangent hyperplane.

In the end, we consider the Enriques surface $Z := S/\gamma$ obtained from a Kummer K3 surface, study the set of 16 (-2)-curves on it, and the associated graph (we call it the Enriques–Kummer graph), which turns out to be associated to a triangulation of the 2-torus. In this way, we are able to show that Z contracts in several ways to a 4-nodal Enriques surface.

In a final section, we describe simple illuminating examples, concerning self-dual hypersurfaces and simple proofs of results on normal cubic surfaces.[4]

A question we could not yet answer is whether the full group of birational automorphisms of the Cefalú quartic is generated (in characteristic $\neq 2, 3, 5$) by the subgroup G of projectivities, and by the involutions γ and ι (ι is induced by the projection from one node) (see [30, 34] for results in the case of generic Kummer surfaces, that is, with Picard number 17, and [31] for Kummer surfaces of the product of two elliptic curves); as already mentioned, we show at any rate in Proposition 22 that the group G^s of order 192 is smaller than the group $Aut(S)^s$ of symplectic automorphisms of S, which has order 384 and is isomorphic to the group of symplectic projectivities of the Fermat quartic surface.

Since the group $Aut(S)^s$ of symplectic automorphisms of S is a normal subgroup of $Aut(S)$, together with the involution γ, it generates a finite subgroup of $Aut(S)$ of cardinality $768 = 384 \cdot 2$ (and we can play the same game with ι). Keum and Kondo, in Theorem 5.6 of [31], exhibit, for the Kummer surface associated to the product of two Fermat curves, a finite group of order $1152 = 384 \cdot 3$. It seems interesting to relate automorphism groups of Kummer surfaces associated to isogenous Abelian surfaces.

1.1 Notation

For a point in projective space, we shall freely use two notations: the vector notation $(a_1, \dots, a_{n+1})$, and the notation $[a_1, \dots, a_{n+1}]$, which denotes the equivalence class of the above vector (the second notation is meant to point out that, in some cases, we do not have an equality of vectors, but only an equality of points of projective space).

2 Proof of Theorem 3 Over $\mathbb{C}$ via Theta Functions

As a preliminary observation, we observe that, as proven by Nikulin [42] (see also [8] for another argument), every quartic with exactly 16 singular points is the Kummer surface $K(A)$ associated to a principally polarized Abelian surface A with a smooth Theta divisor.

---

[4] These formed an important chapter in my lectures.

Hence, we consider these Kummer varieties $K(A)$ in general, under the assumption that (A, Θ) is not a product of polarized varieties

$$(A, \Theta) \neq (A_1, \Theta_1) \times (A_2, \Theta_2).$$

Let therefore A be a principally polarized complex Abelian variety

$$A = \mathbb{C}^g / \Gamma, \ \Gamma := \mathbb{Z}^g \oplus \tau \mathbb{Z}^g,$$

where τ is a $(g \times g)$ matrix in the Siegel upper halfspace

$$\mathcal{H}_g := \{\tau | \tau = {}^t\tau, Im(\tau) > 0\}.$$

The Riemann Theta function

$$\theta(z, \tau) := \sum_{m \in \mathbb{Z}^g} \mathbf{e}\left({}^t m z + \frac{1}{2}\, {}^t m \tau m\right), \ z \in \mathbb{C}^g, \tau \in \mathcal{H}_g, \ \mathbf{e}(x) := exp(2\pi i x)$$

defines a divisor Θ such that the line bundle $\mathcal{O}_A(\Theta)$ has a space of sections spanned by θ.

The space $H^0(\mathcal{O}_A(2\Theta))$ embeds (see [44]) the Kummer variety

$$K(A) := A / \pm 1$$

in the projective space $\mathbb{P}^{2^g - 1}$, and

Definition 7 The canonical basis of the space $H^0(\mathcal{O}_A(2\Theta))$ is given by the sections

$$\theta_\mu(z, \tau) := \theta[a, 0](2z, 2\tau),$$

for $a = \frac{1}{2}\mu$, $\mu \in \mathbb{Z}^g / 2\mathbb{Z}^g$, and where the theta function with rational characteristics a, b is defined as

$$\theta[a, b](w, \tau') := \sum_{p \in \mathbb{Z}^g} \mathbf{e}\left(\frac{1}{2}\, {}^t(p + a)\tau'(p + a) +^t (p + a)(w + b)\right).$$

The following is the main formula for self-duality of Kummer surfaces:

Proposition 8 *For each $u \in \mathbb{C}^g$, the product $\theta(z + u)\theta(z - u)$ defines a section $\psi_u \in H^0(\mathcal{O}_A(2\Theta))$ with*

$$\psi_u = \sum_{\mu \in \mathbb{Z}^g / 2\mathbb{Z}^g} \theta_\mu(u, \tau)\theta_\mu(z, \tau).$$

Proof

$$\psi_u := \theta(z+u)\theta(z-u) = \sum_{m,p\in\mathbb{Z}^g} \mathbf{e}\left({}^t m(z+u) + \frac{1}{2}\,{}^t m\tau m +{}^t p(z-u) + \frac{1}{2}\,{}^t p\tau p\right) =$$

$$= \sum_{m,p\in\mathbb{Z}^g} \mathbf{e}\left({}^t(m+p)(z) +{}^t (m-p)(u) + \frac{1}{2}\,{}^t(m+p)\tau(m+p) - {}^t m\tau p\right).$$

We set now:

$$m' := m+p =: 2\left(M+\frac{\mu}{2}\right),\ M\in\mathbb{Z}^g,\ \mu\in\mathbb{Z}^g/2\mathbb{Z}^g;$$

$$p' := (m-p) =: 2M'+\mu,\ M' := M-p.$$

Then we can rewrite ψ_u as

$$\sum_{\mu\in\mathbb{Z}^g/2\mathbb{Z}^g}\sum_{M\in\mathbb{Z}^g} \mathbf{e}\left(2\,{}^t(M+\frac{\mu}{2})z + {}^t(M+\frac{\mu}{2})\tau(M+\frac{\mu}{2})\right)\cdot$$

$$\cdot\sum_{p'\in\mathbb{Z}^g} \mathbf{e}\left({}^t p'u + {}^t(M+\frac{\mu}{2})\tau(M+\frac{\mu}{2}) -{}^t m\tau p\right) =$$

$$= \sum_{\mu}\theta_\mu(z,\tau)\sum_{p'\in\mathbb{Z}^g} \mathbf{e}\left(2\,{}^t(M'+\frac{\mu}{2})u + {}^t(M'+\frac{\mu}{2})\tau(M'+\frac{\mu}{2})\right) =$$

$$= \sum_{\mu\in\mathbb{Z}^g/2\mathbb{Z}^g} \theta_\mu(z,\tau)\theta_\mu(u,\tau).$$

□

Corollary 9 *Consider the Kummer variety $K(A)$ of a principally polarized Abelian variety with an irreducible Theta divisor, embedded in $\mathbb{P}^{2^g-1}$ by the canonical basis for $H^0(\mathcal{O}_A(2\Theta))$.*

Then $K(A)$ is contained in the dual variety of $K(A)$.

In particular, for dimension $g=2$, we have self-duality

$$K(A) = K(A)^\vee.$$

Proof By Proposition 8, we have that to each u is associated an element $\psi_u \in H^0(\mathcal{O}_A(2\Theta))$, to which there corresponds a Hyperplane section H_u of $K(A)$ whose pull-back is the divisor $(\Theta+u)+(\Theta-u)$.

Since the two divisors $(\Theta + u)$ and $(\Theta - u)$ intersect in some pair of points z and $-z$ (notice that all sections are even functions), we obtain that H_u is singular, and its singular points are, for general u, not points of 2-torsion in A.

For $g = 2$, we conclude since $K(A)^\vee$ has dimension at most 2 and is irreducible. □

Theorem 10 *Let $K(A)$ be the Kummer surface of a principally polarized Abelian surface ($g = 2$), with an irreducible Theta divisor, embedded in $\mathbb{P}^3$ by the canonical basis.*

Then the Gauss map $\gamma : K(A) \dashrightarrow K(A)$ induces a fixed-point free involution on the minimal resolution of $K(A)$.

γ blows up the node P_η of $K(A)$, image point of a 2-torsion point η, to the conic C_η image of $\Theta + \eta$, such that $2C_\eta$ is a plane section H_η of $K(A)$.

Proof In this case, Θ being an irreducible divisor with $\Theta^2 = 2$, $(\Theta + u) \cap (\Theta - u)$ consists either of two distinct points z and $-z$ or of a point z of 2-torsion counted with multiplicity 2, unless $(\Theta + u) = (\Theta - u) \Leftrightarrow (\Theta + 2u) = \Theta$.

The last case, since Θ is aperiodic, can only happen if $2u = 0$.

Let us argue first for $2u \neq 0$.

We have $\gamma(\pm z) = \pm u \in K(A)$, and if we had a fixed point on $K(A)$, then we would have $u \in (\Theta + u) \cap (\Theta - u)$, hence, $\theta(0) = 0$; since θ is even, it follows that θ vanishes of order 2 at 0, contradicting the fact that the divisor Θ is a smooth curve of genus $g = 2$.

If instead u is a 2-torsion point η, then there is a plane H_η which cuts the curve C_η with multiplicity two so that the Gauss map contracts the conic image of C_η to the point η, by Proposition 8.

Te rest of the proof is identical, since $\eta \in C_\eta \to 0 \in \Theta$ is again a contradiction.

γ is an involution ($\gamma^2 = Id$) since the Gauss map of the dual variety is the rational inverse of the original variety. □

Remark 11 Many authors use the weaker assertion that a Kummer quartic X is projectively equivalent to its dual surface in order to obtain an involution on the minimal resolution S, called a switch, which exchanges the (-2)-curves E_i which are the blow-up of the nodes with the 16 tropes E'_i, the strict transforms of the conics C_η (see, for instance, [32, 34]). They compose the Gauss map with a projectivity B giving an isomorphism of $X^\vee$ with X. Denote by f the birational automorphism thus obtained, then the square $f \circ f$ of f yields an automorphism of S leaving invariant the set of all the nodal curves and the set of tropes; since the pull back via f of H equals $3H - \sum_1^{16} E_i$, the pull back via $f \circ f$ of H equals

$$3\left(3H - \sum_1^{16} E_i\right) - \sum_1^{16} E'_i \equiv H,$$

because

$$2\sum_1^{16} E_i' \equiv 16\,H - 6\sum_1^{16} E_i \Rightarrow \sum_1^{16} E_i' \equiv 8\,H - 3\sum_1^{16} E_i.$$

Since $f \circ f$ sends $H \mapsto H$, the square $f \circ f$ is a projectivity of X. What is incorrect is the claim of several authors that this square is the identity.

In fact, we have shown for the Cefalú quartic that $\gamma^2 = 1$ and that γ centralizes the whole group G. If we take now as B a projectivity $g \in G$, we have that $\gamma g \gamma g = \gamma^2 g^2 = g^2 \neq Id$ if g does not have order 2 (and such a g exists for the Cefalú quartic).

We end this section observing that the image of a 2-torsion point η easily determines all others, essentially because ([26, 41]) $H^0(\mathcal{O}_A(2\Theta))$ is the Stone–von Neumann representation of the Heisenberg group $H(G)$, central extension by $\mathbb{C}^*$ of

$$G \times G^* \cong (\mathbb{Z}/2\mathbb{Z})^g \times (\tau\mathbb{Z}^g/2\tau\mathbb{Z}^g).$$

This representation (which is studied and used in [5]) is the space of functions on G, on which G acts by translations, while the dual group of characters $G^* = Hom(G, \mathbb{C}^*)$ acts by multiplication with the given character.

In simpler words (where we look at the associated action on projective space) recalling that

$$\theta_\mu(z,\tau) := \theta[a,0](2z,2\tau) = \sum_{p\in\mathbb{Z}^g} \mathbf{e}\left({}^t(p+a)\tau(p+a) +{}^t (p+a)(2z)\right),$$

where $a = \frac{1}{2}\mu$, and $\mu \in \mathbb{Z}^g/2\mathbb{Z}^g$, if we add to z a half-period $\frac{1}{2}(\epsilon + \tau\epsilon')$, $\epsilon, \epsilon' \in (\mathbb{Z}/2\mathbb{Z})^g$, we obtain

$$\theta_\mu\left(z + \frac{1}{2}\tau\epsilon', \tau\right) = \sum_{p\in\mathbb{Z}^g} \mathbf{e}\left({}^t(p+a)\tau(p+a) +{}^t (p+a)(2z+\tau\epsilon')\right) =$$

$$= \mathbf{e}\left(-\frac{1}{4}\,{}^t\epsilon'\tau\epsilon' -{}^t \epsilon' z\right) \cdot \sum_{p\in\mathbb{Z}^g} \mathbf{e}\left({}^t(p+\frac{1}{2}(\mu+\epsilon'))\tau(p+\frac{1}{2}(\mu+\epsilon')) +{}^t (p+\frac{1}{2}(\mu+\epsilon'))(2z)\right) =$$

$$= \mathbf{e}\left(-\frac{1}{4}\,{}^t\epsilon'\tau\epsilon' -{}^t \epsilon' z\right) \cdot \theta_{\mu+\epsilon'}(z,\tau),$$

whereas

$$\theta_\mu\left(z + \frac{1}{2}\epsilon, \tau\right) = \sum_{p\in\mathbb{Z}^g} \mathbf{e}\left({}^t(p+a)\tau(p+a) +{}^t (p+a)(2z+\epsilon)\right) =$$

$$= \mathbf{e}\left(\frac{1}{2}\,{}^t\epsilon\mu\right)\theta_\mu(z,\tau) = \pm\theta_\mu(z,\tau).$$

Corollary 12 *The image points P_η of the 2-torsion points η of A, which are the singular points of $K(A)$, are obtained from the point $P_0 = [\theta_\mu(0, \tau)]$ (whose coordinates are called the Thetanullwerte), via the projective action of the groups $(\mathbb{Z}/2\mathbb{Z})^g$ acting by diagonal multiplication via $\mathbf{e}(\frac{1}{2}{}^t\epsilon\mu)$, and $(\tau\mathbb{Z}^g/2\tau\mathbb{Z}^g)$, sending the point $[\theta_\mu(z, \tau)]$ to $[\theta_{\mu+\epsilon'}(z, \tau)]$. In particular, for $g = 2$, letting $P_0 = [a_{00}, a_{10}, a_{01}, a_{11}]$, the 16 nodes of $K(a)$ are the orbit of P_0 for the action of the group $\mathcal{K} \cong (\mathbb{Z}/2)^4$ which acts via double transpositions and via multiplication with diagonal matrices having entries in $\{\pm 1\}$ and determinant $= 1$.*

Remark 13 (I) Observe that the formulae given by Weber on page 352 of [60] for the double points do not exhibit such a symmetry.

(II) Observe that the orbit of P_0 for the subgroup of diagonal matrices having entries in $\{\pm 1\}$ and determinant $= 1$ consists of four linearly dependent points if $a_{00}a_{10}a_{01}a_{11} = 0$, that is, one of the Thetanullwerte vanishes (this implies, as we shall see, that the Abelian surface A is isogenous to a product of elliptic curves).

3 Segre's Construction and Special Kummer Quartics

Beniamino Segre [54] introduced a method in order to construct surfaces with many nodes: take polynomials of the form

$$F(z_1, z_2, z_3, z_4) := \Phi(z_1^2, z_2^2, z_3^2, z_4^2).$$

Since $\frac{\partial F}{\partial z_i} = 2z_i \frac{\partial \Phi}{\partial y_i}$, the singular points are the inverse images of the singular points of $Y := \{y | \Phi(y) = 0\}$ not lying on the coordinate tetrahedron $T := \{y | y_1 y_2 y_3 y_4 = 0\}$, or of the points where $y_j = 0$ exactly for $j \in J$, and $\frac{\partial F}{\partial y_i} = 0$ for $i \notin J$ (geometrically, these are the vertices of T lying in Y, or the points where faces, respectively, the edges of T are tangent to Y.

The first interesting case is the one where Y is a smooth quadric Q, and we shall assume that the four faces are tangent in points which do not lie on any edge. Hence, in this case, $F = 0$ defines a quartic X with 16 different singular points, which are nodes.

The matrix $\mathcal{A}$ of the dual quadric $Q^\vee$ has therefore diagonal entries equal to 0, and the condition that no edge is tangent is equivalent to the condition that all nondiagonal entries are nonzero.

The branch locus of $X \to Q \cong (\mathbb{P}^1 \times \mathbb{P}^1)$ consists of $T \cap Q$, and in $\mathbb{P}^1 \times \mathbb{P}^1$ it gives four horizontal plus four vertical lines. Since we know that double transpositions do not change the cross-ratio of four points, we can assume that the quadric Q is invariant for the Klein group $\mathcal{K}'$ in $\mathbb{P}^3$ generated by double transpositions. In short, we may assume that the matrix $\mathcal{A}$ has the form

$$\mathcal{A} := \begin{pmatrix} 0 & b_2 & b_3 & b_4 \\ b_2 & 0 & b_4 & b_3 \\ b_3 & b_4 & 0 & b_2 \\ b_4 & b_3 & b_2 & 0 \end{pmatrix},$$

where $b_i \neq 0 \ \forall i$ and, using the $\mathbb{C}^*$-action (K^*-action if we work more generally with an algebraically closed field K of characteristic $\neq 2$), we may also assume that $b_2 = 1$.

Remark 14 Set $b_i = a_i^2$, $i = 2, 3, 4$. Then the 16 nodes of X are the points whose coordinates are the square roots of the coordinates of a column of $\mathcal{A}$.

That is, the four points $[0, \pm a_2, \pm a_3, \pm a_4]$ and their orbit under the Klein group $\mathcal{K}' := (\mathbb{Z}/2)^2$.

We get a set $\mathcal{N}$ such that, if we define $\mathcal{N}'$ to be the set of planes orthogonal to some point of $\mathcal{N}$, then $\mathcal{N}, \mathcal{N}'$ form a configuration of type $(16_6, 16_6)$.

In fact, for instance, the plane $a_2 z_2 + a_3 z_3 + a_4 z_4$ contains exactly the six points of $\mathcal{N}$ where $z_j = 0$ for some $j = 2, 3, 4$, $z_1 = a_j$, and, if $\{j, h, k\} = \{2, 3, 4\}$, then $z_k = \epsilon a_h$, $z_h = -\epsilon a_k$, where $\epsilon = \pm 1$.

To calculate the equation of X it suffices here to find the matrix of Q, which is, up to a multiple, the inverse of $\mathcal{A}$. And since Q is $\mathcal{K}'$-symmetric, it suffices to compute its first column.

The calculation of the matrix of Q can be done by hand, since

$$\mathcal{A}(e_1 + e_3) = b_3(e_1 + e_3) + (b_2 + b_4)(e_2 + e_4)$$
$$\mathcal{A}(e_2 + e_4) = (b_2 + b_4)(e_1 + e_3) + b_3(e_2 + e_4)$$
$$\mathcal{A}(e_1 - e_3) = -b_3(e_1 - e_3) + (b_2 - b_4)(e_2 - e_4)$$
$$\mathcal{A}(e_2 - e_4) = (b_2 - b_4)(e_1 - e_3) - b_3(e_2 - e_4)$$

and moreover the inverse of

$$C_1 := \begin{pmatrix} b_3 & b_2 + b_4 \\ b_2 + b_4 & b_3 \end{pmatrix},$$

is $(b_3^2 - (b_2 + b_4)^2)^{-1} \cdot D_1$, where:

$$D_1 := \begin{pmatrix} b_3 & -(b_2 + b_4) \\ -(b_2 + b_4) & b_3 \end{pmatrix},$$

while the inverse of

$$C_2 := \begin{pmatrix} -b_3 & b_2 - b_4 \\ b_2 - b_4 & -b_3 \end{pmatrix},$$

equals $(b_3^2 - (b_2 - b_4)^2)^{-1} \cdot D_2$, where:

$$D_2 := \begin{pmatrix} -b_3 & b_4 - b_2 \\ b_4 - b_2 & -b_3 \end{pmatrix}.$$

Now, a straightforward calculation yields the following coefficients for the first column of the matrix of Q:

$$\begin{aligned} \alpha_0 &= 2b_2b_3b_4 \\ \alpha_{01} &= b_2(b_2^2 - b_3^2 - b_4^2) \\ \alpha_{10} &= b_3(b_3^2 - b_4^2 - b_2^2) \\ \alpha_{11} &= b_4(b_4^2 - b_2^2 - b_3^2), \end{aligned}$$

hence X has the following equation in Hudson normal form :

$$\alpha_0 \left(\sum_i z_i^4 \right) + 2\alpha_{01}(z_1^2z_2^2 + z_3^2z_4^2) + 2\alpha_{10}(z_1^2z_3^2 + z_2^2z_4^2) + 2\alpha_{11}(z_1^2z_4^2 + z_2^2z_3^2) = 0.$$

Observe here that the Cefalú quartic corresponds to the special case, where $Q^\vee$ is $\sigma_2(y) = 0$, hence, the case where all $b_2, b_3, b_4 = 1$,

$$Q = \{y | \sigma_2(y) - s_2(y) = 0\}, \ \ X = \{z | \sigma_2(z_i^2) - s_4(z) = 0\}.$$

We want to explain now the geometrical meaning of the family of Segre-type Kummer quartics, as Kummer quartics whose associated Abelian surface is isogenous to a product of two elliptic curves.

Proposition 15 *The Segre-type (special) Kummer quartics are exactly the Kummer surfaces whose associated principally polarized Abelian surface A is isogenous to a product $A_1 \times A_2$ of elliptic curves,*

$$A = (A_1 \times A_2)/H, \ \ H \cong (\mathbb{Z}/2)^2,$$

and where $H \to A_i$ is injective for $i = 1, 2$.

Proof Choose an isomorphism of the quadric Q with $\mathbb{P}^1 \times \mathbb{P}^1$, with coordinates $(u_0, u_1), (v_0, v_1)$.

Then the linear forms x_i, since the plane section $\{x_i = 0\}$ is tangent to Q, splits as $x_i = L_i(u)M_i(v)$.

The Galois cover $X \to Q$ with group $(\mathbb{Z}/2)^3$ yields sections z_i such that $z_i^2 = x_i = L_i(u)M_i(v)$, hence the divisor of z_i is reducible, and we can write $z_i = \lambda_i\mu_i$, where $\lambda_i^2 = L_i(u)$ and $\mu_i^2 = M_i(v)$.

We define A_1 as the $(\mathbb{Z}/2)^3$-Galois cover of $\mathbb{P}^1$ branched on $L_1L_2L_3L_4(u) = 0$, and similarly A_2 as the $(\mathbb{Z}/2)^3$-Galois cover of $\mathbb{P}^1$ branched on $M_1M_2M_3M_4(v) = 0$.

The sections λ_i, μ_i are defined on $A_1 \times A_2$, hence X is the quotient of $A_1 \times A_2$ by the action of $(\mathbb{Z}/2)^3$ diagonally embedded in $(\mathbb{Z}/2)^3 \oplus (\mathbb{Z}/2)^3$.

Clearly, $A_i \to \mathbb{P}^1$ is the quotient by the group of automorphisms of the form $z \mapsto \pm z + \eta$, with $\eta \in A_i[2]$ (a 2-torsion point). Hence, $A = (A_1 \times A_2)/H$, where $H \subset A_1[2] \times A_2[2]$ is diagonally embedded for some isomorphism between $A_1[2]$ and $A_2[2]$.

The embedding of X into $\mathbb{P}^3$ is given by the sections z_i, which on $A_1 \times A_2$ can be written as $z_i = \lambda_i \mu_i$, hence they are sections of a line bundle of bidegree $(4, 4)$. The principal polarization then corresponds to a line bundle of bidegree $(2, 2)$.

A can also be viewed as an étale $(\mathbb{Z}/2)^2$-Galois covering of

$$A_1 \times A_2 \cong (A_1/A_1[2]) \times (A_2/A_2[2]).$$

□

Remark 16 We can now prove (ix) of Theorem 1.

Choosing an isomorphism of the smooth quadric Q with $\mathbb{P}^1 \times \mathbb{P}^1$, the $\mathfrak{S}_4$ action shows that the union of the four vertical lines and the four horizontal lines are $\mathfrak{S}_4$-invariant. Hence, the four first coordinates are fixed by a group of order 12, and they form an equianharmonic cross-ratio. Therefore, A_1 admits such group of order 12 permuting the set of torsion points $A_1[2]$, hence A_1 is the equianharmonic = Fermat elliptic curve. Same argument for A_2.

3.1 Proof of Theorem 5

Proof Take a Segre-type Kummer quartic corresponding to the choice of $a_2, a_3, a_4 \in \mathbb{R}$, so that X is defined over $\mathbb{R}$ and its singular points are real points.

Observe that $b_i > 0$, and that the determinant of $\mathcal{A}$ is a polynomial in b_4 with leading term b_4^4.

Hence, for $b_4 >> 0, det(\mathcal{A}) > 0$, as well as the determinant of the inverse. Since $\mathcal{A}$ is not definite, its signature is of type $(2, 2)$, the same holds for Q, so that Q is of hyperbolic type and we may choose an isomorphism of the quadric Q with $\mathbb{P}^1 \times \mathbb{P}^1$, with coordinates $(u_0, u_1), (v_0, v_1)$, defined over $\mathbb{R}$. And to the real singular points correspond real points in $\mathbb{P}^1 \times \mathbb{P}^1$.

We have seen that the Galois cover $X \to Q$ with group $(\mathbb{Z}/2)^3$ yields sections z_i such that $z_i^2 = x_i = L_i(u)M_i(v)$, the divisor of z_i is reducible, $z_i = \lambda_i \mu_i$, where $\lambda_i^2 = L_i(u)$ and $\mu_i^2 = M_i(v)$. Since z_i is real, and the two terms of its irreducible decomposition are not exchanged by complex conjugation, it follows that also λ_i, μ_i are real.

X is embedded in $\mathbb{P}^3$ by the divisor H_X whose double is the pull back of the hyperplane divisor H_Q of Q.

$Q \cong \mathbb{P}^1 \times \mathbb{P}^1$ is embedded in $\mathbb{P}^3$ by the homogeneous polynomials of bidegree $(1, 1)$, in other words the divisor $H_Q = H_1 + H_2$, where the two effective divisors H_1, H_2 are the respective pull backs of the divisors of degree 1 on $\mathbb{P}^1$. Pulling back to X, we see that $H_X = H_1' + H_2'$, as we argued in the proof of Proposition 15. The previous observation on the reality of the λ_i, μ_i's shows that both H_1', H_2' are real.

Q can be embedded in $\mathbb{P}^{2(m+1)-1}$ by the homogeneous polynomials of bidegree $(1, m)$, as a variety of degree 2 m. This embedding is given by a vector of rational functions $\Psi(y_0, y_1, y_2, y_3)$, ad corresponds to the linear system $|H_1 + mH_2|$.

X is a finite cover of Q, hence the pull back of the hyperplane H_m in $\mathbb{P}^{2(m+1)-1}$ yields the double of a polarization of degree 4 m, namely the divisor $(H_1' + mH_2')$. Hence we have shown that $H_1' + mH_2'$ is an ample divisor.

To show that $(H_1' + mH_2')$ is very ample we first observe that since $|H_1' + mH_2'| \supset |H_1' + H_2'| + (m-1)H_2'$ the linear system yields a birational map. Then we can invoke, for instance, Theorem 6.1 of Saint Donat's article [52] on linear systems on K3 surfaces.

The image of X under the linear system $|H_1' + mH_2'|$, which is associated to a real divisor, is the desired surface X' for $d = 4m$, and all the nodes are defined over $\mathbb{R}$, as well as X'.

Projecting X' from one node, and using the theorem of Saint Donat [52], as in [8], we get the desired surface X'' for degree $d = 4\text{m} - 2$.

The proof for the case where $\mathbb{R}$ is replaced by an algebraically closed field of characteristic $\neq 2$ is identical, we can find a_2, a_3, a_4 such that $det(\mathcal{A}) \neq 0$ because the determinant is a polynomial in the a_i's with leading term a_4^8. □

4 Proof of Theorem 1

In this section, we want to show how most of the statements, even if proven in other sections of the paper in greater generality, can be proven by direct and elementary computation.

First of all, we observe the necessity of the condition $char(K) \neq 2, 3$: because in both cases the equation F is a square. In the other direction, the determinant of the quadric dual to Q, which is $\{\sigma_2(x) = 0\}$, equals -3, hence Q is smooth if the characteristic is $\neq 2, 3$.

(ii), after the above remarks, is a straightforward computation (done in the section on special Kummer quartics in greater generality) based, if $F(z) = Q(z_i^2)$, on the formula $\frac{\partial F}{\partial z_i} = 2z_i \frac{\partial Q}{\partial x_i}(z_i^2)$ which implies that some $z_i = 0$.

(iii) is based on the fact that X admits G-symmetry, and the plane $H_1 := \{z_2 + z_3 + z_4 = 0\}$ intersects X in an irreducible conic $C_1 := \{-z_1^2 + z_2^2 + z_3^2 + z_2 z_3 = 0\}$ counted with multiplicity 2.

This can also be found by direct computation, however, it follows, since $H_1 \cap X$ contains six singular points, no three of which are collinear; hence, this intersection yields a conic counted with multiplicity 2.

Therefore, we have 16 curves mapping to 16 singular points of the dual, which are exactly the same set $\mathcal{N}$ of nodes of X. Since $X^\vee$ is also G-symmetric, it must equal X.

(iv) is straightforward using G-symmetry.

(v) γ has order 2 since the Gauss map of $X^\vee$ is the inverse of the Gauss map of X, and γ is a morphism on the minimal model S of X.

If A is the matrix associated to a projective automorphism of X, and we treat γ as a rational self-map of X, ${}^tA\gamma A = \gamma$, hence $\gamma A \gamma = {}^t A^{-1}$ and hence γ commutes with any orthogonal transformation A, in particular γ commutes with G.

(vii) since γ blows up the point P to the curve C_P which is the set-theoretical intersection $H_P \cap X$, where H_P is the orthogonal plane to P, and P does not belong to H, follows that any fixed point of γ should correspond to a smooth point of X not lying in any hyperplane H_P.

A fixed point would satisfy $[z_i] = [z_i \frac{\partial Q}{\partial x_i}(z_i^2)]$. If all $z_i \neq 0$, then we would have that the point $[1, 1, 1, 1]$ lies in the dual quadric of Q, a contradiction since this is the quadric $\sigma_2(x) = 0$.

No coordinate point lies on X; hence, exactly one or two coordinates z_i must be equal to 0. In the former case, we may assume by symmetry that $z_4 = 0$ and $z_i \neq 0$ for $i = 1, 2, 3$.

This leads to the conclusion that $(1, 1, 1)$ must be in the image of the matrix

$$\begin{pmatrix} -2 & 1 & 1 \\ 1 & -2 & 1 \\ 1 & 1 & -2 \end{pmatrix},$$

absurd since the sum of the rows equals zero.

If instead $z_3 = z_4 = 0$, $z_1, z_2 \neq 0$, then we get $x_3 = x_4 = 0$, $x_1, x_2 \neq 0$.

Here must be that $(1, 1)$ must be in the image of the matrix

$$\begin{pmatrix} -2 & 1 \\ 1 & -2 \end{pmatrix},$$

hence that $-2x_1 + x_2 = x_1 - 2x_2 \Leftrightarrow x_1 = x_2 = 1$.

But $[1, 1, 0, 0]$ is not a point of the quadric Q.

(ix) was proven in the section on special Kummer quartics, while (x) follows from the calculations we already made.

We turn now to the assertions (i), (vi), and (viii) concerning automorphisms.

(i): that G is contained in the group $\mathcal{G}$ of projective automorphisms of X is obvious, the question is whether equality $G = \mathcal{G}$ holds.

Remark 17 $\mathcal{G}$ induces a permutation of the singular set $\mathcal{N}$, and of the set $\mathcal{N}'$ of the 16 tropes, the planes dual to the points of $\mathcal{N}$. Given a projectivity $h : X \to X$, acting with the subgroup G, we can assume that it fixes the point $P_1 = [1, 1, 1, 0]$ and since the stabilizer of P_1 in G acts transitively on the six tropes passing through P_1, we may further assume that h stabilizes the plane $\pi : \{z_1 - z_2 + z_4 = 0\}$.

To determine these automorphisms h (among these, an element of G of order (2), we need more computations, so we shall finish the proof of (i) at the end of the section.

(vi) Observe that since γ acts freely on S, the quotient $Z := S/\gamma$ is an Enriques surface; therefore, γ acts as multiplication by -1 on a nowhere vanishing regular 2-form ω. The action of G on ω factors through the Abelianization of G, which is the Abelianization $\cong \mathbb{Z}/2$ of $\mathfrak{S}_4$, since in the semidirect product, for $\epsilon \in (\mathbb{Z}/2)^3$, $\sigma \in \mathfrak{S}_4$, $\sigma\epsilon\sigma^{-1} = \sigma(\epsilon)$, and there are no $\mathfrak{S}_4$-coinvariants.

The conclusion is that, for each $g \in G$, there is a unique element in the coset $g\{1, \gamma\}$ which acts trivially on ω, hence $G^s \cong G$.

The last assertion is proven in Proposition 22.

(viii) The group of birational automorphisms of X coincides with the group of biregular automorphisms of S. Among the former are the involutions determined by projection from a node P. Since, however, the group G acts transitively on the set $\mathcal{N}$ of nodes, two such are conjugate by an element of G, hence it suffices to consider just one such involution ι.

We shall show that ι does not lie in the group generated by γ and the projectivities just looking at its action on the Picard group.

Consider $P_1 = [1, 1, 1, 0]$, so that projection from P_1 is given by $w_2 := z_2 - z_1$, $w_3 := z_1 - z_3$, $w_4 = z_4$.

We can write the equation $F(z)$ as

$$z_1^2\phi(w) + 2z_1\psi(w) + f(w),$$

where

$$\phi(w) = 4(w_2 - w_3)^2 + 6s_2(w) - 18(w_2^2 + w_3^2) = -2(2w_2^2 + 2w_3^2 + 2(w_2 + w_3)^2 - 3w_4^2)$$

$$\psi(w) = 2(w_2 - w_3)s_2(w) - 6(w_2^3 + w_3^3)$$

$$f(w) = s_2(w)^4 - 3s_4(w),$$

hence the branch curve (the union of the six lines corresponding to the tropes through P_1) has an equation

$$\psi(w)^2 - \phi(w)f(w) = (w_4^2 - w_2^2)(w_4^2 - w_3^2)(w_4^2 - (w_2 + w_3)^2) =$$

$$= (w_4^2 - w_2^2)(w_4^2 - w_3^2)(w_4^2 - w_1^2) = 0,$$

where we define w_1 so that $w_1 + w_2 + w_3 = 0$.

And

$$\iota(w_2, w_3, w_4, z_1) = (\phi(w)z_1w_2, \phi(w)z_1w_3, \phi(w)z_1w_4, f(w)) =$$

$$= (\phi(w)w_2, \phi(w)w_3, \phi(w)w_4, -z_1\phi(w) - 2\psi(w)).$$

The inverse image of the conic $\{\phi(w)=0\}$, which is everywhere tangent to the branch locus, splits in S as the exceptional curve E_1 plus a curve $C_1 := \{\phi(w) = 0,\ 2z_1\psi(w) + f(w) = 0\}$ such that $\iota(C_1) = E_1$.

Since the projection is given by the system $2H - 2E_1$ and $C_1 + E_1$ belongs to this system, the class of C_1 equals $2H - 3E_1$.

Then $\iota(H) = 3H - 4E_1, \iota(E_1) = 2H - 3E_1$.

On the other hand, each projectivity and γ leave invariant the two blocks given by $\{E_i | P_i \in \mathcal{N}\}$, and $\{D_i | P_i \in \mathcal{N}\}$, so the group generated by G, γ, ι is larger than the group $G \times \mathbb{Z}/2$ generated by G, γ.

We show now that the group generated by G, ι is infinite. Indeed, this result holds for all Kummer quartics, namely we have the following:

Theorem 18 *Let X be a Kummer quartic surface, S its minimal resolution, and assume that $char(K) \neq 2$. Then $Aut(S)$ is infinite, and actually, the subgroup generated by the group G of projective automorphisms of X, and by an involution ι obtained via projection from one node, is infinite.*

Proof Observe in fact that ι leaves the other nodes fixed, because, if P_2 is another node, then the line $\overline{P_1P_2}$ intersects X only in P_1, P_2.

Hence, ι acts on the Picard group as follows:

$$\iota(H) = 3\,H - 4E_1, \iota(E_1) = 2H - 3E_1, \iota(E_j) = E_j,\ \ j \neq 1.$$

Let $g \in G$ be a projectivity such that $g(P_1) = P_2, g(P_2) = P_1$.

Then $\phi := g \circ \iota$ acts on the Picard group like this:

$$\phi(H) = 3\,H - 4E_2,\ \phi(E_1) = 2\,H - 3E_2,\ \phi(E_2) = E_1,$$

and leaves the subgroup generated by the E_j, for $j \neq 1, 2$ invariant.

The subgroup generated by H, E_1, E_2 is also invariant and on it ϕ acts with the matrix:

$$M := \begin{pmatrix} 3 & 2 & 0 \\ 0 & 0 & 1 \\ -4 & -3 & 0 \end{pmatrix}.$$

An easy calculation shows that the characteristic polynomial of M is

$$P_M(\lambda) = (\lambda - 1)^3;$$

moreover, $rank(M - Id) = 2$. Therefore, the Jordan normal form of M is a single 3×3 Jordan block with eigenvalue 1, and it follows that M and ϕ have infinite order. □

The proof of (viii) is then concluded.

End of the proof of (i).
Giving an automorphism of X fixing P_1 and the plane $z_4 - z_2 + z_1 = 0$, which projects to the line $w_4 - w_2 = 0$, is equivalent to giving a projectivity of the plane with coordinates (w_2, w_3, w_4) leaving invariant the line $w_4 - w_2 = 0$, and permuting the other five lines of the branch locus. Since all these lines are tangent to the conic $\mathcal{C} := \{\phi(w) = 0\}$, it is equivalent to give an automorphism of $\mathcal{C}$ fixing the intersection point $P' := \mathcal{C} \cap \{w_4 - w_2 = 0\}$, and permuting the other five points.

Projecting $\mathcal{C}$ with centre P', we see that it is equivalent to require an affine automorphism permuting these five points.

We note that one such is the permutation

$$\tau : w_2 \mapsto w_2, w_4 \mapsto w_4, w_3 \mapsto w_1 = -w_2 - w_3.$$

So, we calculate:

$$P' = (-2, 1, -2) \in \{(-2, 1, \pm 2), (1, 1, \pm 2), (1, -2, \pm 2)\}.$$

The projection with centre P' is given by $(w_4 - w_2, w_4 + 2w_3)$, maps P' to ∞, and the other five points to

$$\{1, 4, 0, -2, 2\}.$$

This set is affinely equivalent to

$$\mathcal{M} := \{-3, -1, 0, 1, 3\},$$

and the above permutation τ corresponds to multiplication by -1.

If $char(K) = 5$, then $\mathcal{M} = \mathbb{Z}/5 =: \mathbb{F}_5 \subset K$, and obviously, we have then the whole affine group $Aff(1, \mathbb{F}_5)$, with cardinality 20, acting on $\mathcal{M}$.

If instead $char(K) \neq 5$, we observe that the barycentre of $\mathcal{M}$ is the origin 0. Assume now that α is an affine isomorphism leaving $\mathcal{M}$ invariant. Then necessarily α fixes the barycentre; hence, α is a linear map, and $0 \in \mathcal{M}$ is a fixed point. Hence, α is either of order 4 or of order 2, and we must exclude the first case. But this is easy, since then $\alpha(1) \in \{-1, 3, -3\}$ should be a square root of -1, and this is only possible (recall here that $2, 3 \neq 0$) if $5 = 0$, a contradiction which ends the proof.

Question 19 *Is there a larger finite subgroup of $Aut(S)$ (S the Cefalú K3 surface, the minimal model of the Cefalú quartic) containing the group isomorphic to $G \times (\mathbb{Z}/2)$ which is generated by G, γ?*

In order to see whether there are larger subgroups than $G^s \cong G$ acting symplectically on the Cefalú K3 surface, we use first the following:

Lemma 20 *The group G cannot be a subgroup of the group $AL(2, \mathbb{F}_4)$ of special affine transformations of the plane $(\mathbb{F}_4)^2$, which is a semidirect product*

$$AL(2, \mathbb{F}_4) \cong (\mathbb{F}_4)^2 \rtimes \mathfrak{A}_5 \cong (\mathbb{Z}/2)^4 \rtimes \mathfrak{A}_5.$$

Proof Since the order of $G \cong (\mathbb{Z}/2)^3 \rtimes \mathfrak{S}_4$ is 192, and the order of $\Gamma := AL(2, \mathbb{F}_4)$ is $920 = 5 \cdot 192$, the existence of a monomorphism $G \to \Gamma$ would imply that the two groups have isomorphic 2-Sylow subgroups H_G, H_Γ. But we claim now that, for the respective Abelianizations, we have

$$H_G^{ab} \cong (\mathbb{Z}/2)^3, \quad H_\Gamma^{ab} \cong (\mathbb{Z}/2)^4.$$

Indeed, $H_G = (\mathbb{Z}/2)^3 \rtimes H_{\mathfrak{S}_4}$, where $H_{\mathfrak{S}_4} \cong D_4$ is generated by a four-cycle $(1, 2, 3, 4)$ and the double transposition $(1, 2)(3, 4)$. The four-cycle $(1, 2, 3, 4)$ acts conjugating all the generators for $(\mathbb{Z}/2)^3$; hence, it follows easily that the image of $(\mathbb{Z}/2)^3$ in H_G^{ab} is $\mathbb{Z}/2$, while the Abelianization of $H_{\mathfrak{S}_4} \cong D_4$ is $(\mathbb{Z}/2)^2$. Hence, the first assertion.

We have that $H_\Gamma = (\mathbb{Z}/2)^4 \rtimes H_{\mathfrak{A}_5}$, and $H_{\mathfrak{A}_5} \cong (\mathbb{Z}/2)^2$ is a Klein group of double transposition fixing one element in $\{1, 2, 3, 4, 5\} \cong \mathbb{P}^1(\mathbb{F}_4)$.

Hence, $H_{\mathfrak{A}_5}$ is the group of projectivities $(x_1, x_2) \mapsto (x_1, x_2 + a x_1)$, for $a \in \mathbb{F}_4$. Hence, the image of $(\mathbb{Z}/2)^4$ in H_Γ^{ab} is $\mathbb{F}_4 \cong (\mathbb{Z}/2)^2$, and $H_\Gamma^{ab} \cong (\mathbb{Z}/2)^4$. □

Remark 21 (1) By using Mukai's theorem 0.6 [37], it follows that either G^s is a maximal subgroup acting symplectically, or it is an index 2 subgroup of Γ which acts symplectically, where Γ is the group of symplectic projectivities of the Fermat quartic,

$$\Gamma \cong (\mathbb{Z}/4)^2 \rtimes \mathfrak{S}_4 \cong (\mathbb{Z}/2)^4 \rtimes \mathfrak{S}_4,$$

(see [37] page 191 for the last isomorphism).

(2) The Cefalú K3 surface S (as all the other Segre-type Kummer surfaces) possesses two elliptic pencils (induced by the $(\mathbb{Z}/2)^3$ covering $X \to Q = \mathbb{P}^1 \times \mathbb{P}^1$) and these are exchanged by the transpositions in $\mathfrak{S}_4$, since, for instance, the plane $z_4 = 0$ intersects X in two conics:

$$X \cap \{z_4 = 0\} = \{(z_1^2 + \omega z_2^2 + \omega^2 z_3^2)(z_1^2 + \omega^2 z_2^2 + \omega z_3^2) = z_4 = 0\}$$

which are exchanged by the transposition $(2, 3)$ (ω is here a primitive third root of unity).

It follows then easily that there is a symplectic automorphism exchanging the two pencils, and it suffices to study the subgroup G^0 of those symplectic automorphisms which do not exchange the two pencils. By Mukai's theorem and since $G^0 \cap G^s$ has order 96, the cardinality of G^0 is either 96 or 192.

(3) Next, we divide S by the group $(\mathbb{Z}/2)^3 \rtimes \mathfrak{A}_4$, obtaining

$$Q/\mathfrak{A}_4 = (\mathbb{P}^1 \times \mathbb{P}^1)/\mathfrak{A}_4,$$

and set S' to be the quotient by the symplectic index two subgroups $(\mathbb{Z}/2)^2 \rtimes \mathfrak{A}_4$.

Letting S'' be the Kummer surface double cover of Q branched on the four plane sections (S'' is the Kummer surface of $\mathcal{E} \times \mathcal{E}$, where $\mathcal{E}$ is the Fermat elliptic curve), $S' = S''/\mathfrak{A}_4$ and is a double cover of $Q/\mathfrak{A}_4$.

The second projection of Q onto $\mathbb{P}^1$ yields an elliptic pencil on S', corresponding to the projection $\mathbb{P}^1 \to \mathbb{P}^1/\mathfrak{A}_4 \cong \mathbb{P}^1$ (and the fibration $Q \to \mathbb{P}^1/\mathfrak{A}_4$ has just three singular fibres, with multiplicities (2, 3, 3)). $G^0/((\mathbb{Z}/2)^2 \rtimes \mathfrak{A}_4)$ is a group G'' of cardinality 2 or 4, hence $(\mathbb{Z}/2)$ or $(\mathbb{Z}/2)^2$.

Let G''' be the subgroup of G'' preserving the fibration and acting as the identity on the base, hence sending each fibre to itself.

Because it is a group of exponent two acting symplectically, it must act on the general fibre via translations of order 2, from this follows that $|G'''| \leq 4$. But we shall see next that $|G'''| = 4$, hence $G'' = G'''$ has cardinality 4.

Proposition 22 *The group of symplectic automorphisms of the Cefalú K3 surface S has order 384 and is isomorphic to the group Γ of symplectic projectivities of the Fermat quartic,*

$$\Gamma \cong (\mathbb{Z}/4)^2 \rtimes \mathfrak{S}_4.$$

Proof In view of Mukai's theorem, see Remark 21, (1), (2) and (3), we just need to show that the Kummer surface S'' of $\mathcal{E} \times \mathcal{E}$, where $\mathcal{E}$ is the Fermat elliptic curve, admits a group of automorphisms preserving the two fibrations induced by the two product projections of $\mathcal{E} \times \mathcal{E}$ onto $\mathcal{E}$, acting sympletically, and of order 48, since then $|G^0| = 192$.

For this purpose, it suffices to take the group of automorphisms of $\mathcal{E} \times \mathcal{E}$ generated by translations by points of order 2, and by the automorphism of order 3 which acts on the uniformizing parameters (t_1, t_2) by

$$(t_1, t_2) \mapsto (\omega t_1, \omega^2 t_2), \ \omega^3 = 1.$$

This group descends faithfully to a group of automorphisms of $S'' = Km(\mathcal{E} \times \mathcal{E})$. And the translations on the first factor act trivially on the Picard group, hence they extend to the étale covering $A \to \mathcal{E} \times \mathcal{E}$, where A is the ppav such that $S = Km(A)$; therefore, they lift to automorphisms of S, thereby showing that $|G'''| = 4$, as required. □

Remark 23 We discuss now some claim by Hutchinson, that if P_1, P_2, P_3, P_4 are nodes of X such that the faces $w_i = 0$, $(i = 1, \dots, 4)$ of the Tetrahedron T having P_1, P_2, P_3, P_4 as vertices are not tangent to X, then the Cremona transformation $(w_i) \mapsto (w_i^{-1})$ leaves X invariant.

This is false, let in fact

$$e := \sum_1^4 e_i = (1, 1, 1, 1), \ P_i := e - e_i,$$

so that these nodes are an orbit for the group $\mathfrak{S}_4$.

The faces of T are the orbit of $w_1 := 2z_1 - z_2 - z_3 - z_4 = 0$.

Since the point $(2, -1, -1, -1) \notin X$, as $7^2 - 3(19) \neq 0$, the faces are not tangent to X (we use here $X = X^\vee$).

Consider the node $(0, 1, 1, -1)$: the Cremona transformation $(w_i) \mapsto (w_i^{-1})$ sends this node to the point

$$\left(-1, \frac{1}{2}, \frac{1}{2}, -\frac{1}{4}\right),$$

which is not a node of X.

Hutchinson involutions do of course exist, for instance, if we have the Hudson normal form with $\alpha_0 = 0$,

$$\alpha_{01}(z_1^2 z_2^2 + z_3^2 z_4^2) + \alpha_{10}(z_1^2 z_3^2 + z_2^2 z_4^2) + \alpha_{11}(z_1^2 z_4^2 + z_2^2 z_3^2) + 2\beta z_1 z_2 z_3 z_4 = 0,$$

the surface is invariant for the Cremona transformation $(z_i) \mapsto (z_i^{-1})$.

More generally, to have a Hutchinson involution, one must take a Göpel tetrad, that is, the four nodes should correspond to an isotropic affine space in the Abelian variety A, see [38].

But we saw in Remark 13 that a Göpel tetrad does not consist of four independent points if $a_{00}a_{10}a_{01}a_{11} = 0$, that is, in classical terminology, if some Thetanull is vanishing.

5 Algebraic Proofs for Theorems 4 and 3

We gave short self-contained proofs over the complex numbers, but indeed the theorems hold also for an algebraically closed field of characteristic $p \neq 2$. And the results follow easily from existing literature.

The first result, which I found and explained to Maria Gonzalez-Dorrego, inspired by the talk she gave in Pisa, and is contained in [19], is that there are exactly three isomorphism classes of abstract nondegenerate $(16_6, 16_6)$ configurations, but only one can be realized in $\mathbb{P}^3$, because of the theorem of Desargues.

The second important result of [19] is that all these configurations in the allowed combinatorial isomorphism class are projectively equivalent to the ones obtained as the $\mathcal{K}$ orbit $\mathcal{N}$ of a point $[a_1, a_2, a_3, a_4]$ satisfying the inequalities (I), (II), (III), and by the set $\mathcal{N}'$ of planes orthogonal to points of $\mathcal{N}$. These inequalities are necessary and sufficient to ensure that the cardinality of $\mathcal{N}$ is exactly 16.

After that, it is known that two Kummer quartics with the same singular points are the same (one can either see this via projection from a node, since there is only one conic in the plane tangent to the six projected lines, or another proof can be found in [19], Lemma 2.19, page 67).

Hence, it suffices to find explicitly such a Kummer quartic passing through these 16 points, and this can be done by finding such a surface in the family of quartics in Hudson's normal form: the result follows via the solution of a system of linear equations (mentioned in the statement of Theorem 4).

After these calculations, the proof of the first assertion of Theorem 3 follows right away since the singular points of $X^\vee$ are exactly the ones of X, hence, by the previous observation, $X = X^\vee$.

To prove that the Gauss map induces a fixed-point free involution γ on the minimal resolution S, instead of carrying out a complicated calculation, we use the easy computations used in the case of the Cefalú quartic, which show the assertion for one surface in the family (hence, the assertion holds on a Zariski open subset of the parameter space, which is contained in $\mathbb{P}^3$).

We show now the result for all Kummer quartics.

In fact, since γ is an involution, it acts on the nowhere vanishing 2-form ω by multiplication by ± 1. Since, for a surface of the family, it acts multiplying by -1, the same holds for the whole family since the base is connected. This implies that γ can never have an isolated fixed point x, because then for local parameters u_1, u_2 at x we would have $u_i \mapsto -u_i$ and $\omega = (du_1 \wedge du_2 +$ higher order terms$) \mapsto \omega$, a contradiction.

Exactly as shown in the case of the Cefalú quartic, there are no fixed points on the union of the nodal curves E_i and of the tropes D_i.

Hence, an irreducible curve C of fixed points (therefore smooth) necessarily does not intersect the curves E_i, nor the tropes D_i. Hence, $C \cdot E_i = 0$ and $C \cdot D_i = 0$ for all i; since $2D_i \equiv H - \sum_{P_j \in P_i^\perp} E_j$, we infer that $C \cdot H = 0$. This implies that C is one of the E_i's, a contradiction.

Remark 24 An argument similar to the one we just gave, put together with Theorem 5, should allow to recover the full result of [29] also for polarized Kummer surfaces of degree $d = 4m > 4$ in positive characteristic; the only point to verify is that the involution γ, which is defined on the subfamily corresponding to Segre-type Kummer surfaces, extends to the whole three-dimensional family of polarized Kummer surfaces.

6 The Role of the Segre Cubic Hypersurface

An important role in the theory of Kummer quartic surfaces is played by the 10-nodal Segre cubic hypersurface in $\mathbb{P}^4$ of equations (in $\mathbb{P}^5$)

$$s_1(x) := s_1(x_0, \dots, x_5) = 0, \quad s_3(x) = 0.$$

It has a manifest $\mathfrak{S}_6$-symmetry, it contains 10 singular points, and 15 linear subspaces of dimension 2, as we shall now see.

Since surfaces of degree $d = 4$ with 16 nodes contain a weakly even set of cardinality 6 (see [4, 8]), they can be obtained as the discriminant of the projection of a 10-nodal cubic hypersurface $X \subset \mathbb{P}^4$ from a smooth point.

It is known (see [23]) that the maximal number of nodes that a complex cubic hypersurface $X \subset \mathbb{P}^m(\mathbb{C})$ can have equals $\gamma(m)$, and $\gamma(3) = 4, \gamma(4) = 10, \gamma(5) = 15, \gamma(6) = 35$ (we have in general $\gamma(m) = \binom{m+1}{[m/2]}$).

Equality is attained, for $m = 2h$ by the Segre cubic

$$\Sigma(m-1) := \{x \in \mathbb{P}^{m+1} | \sigma_1(x) = \sigma_3(x) = 0\} = \{x \in \mathbb{P}^{m+1} | s_1(x) = s_3(x) = 0\},$$

where as usual σ_i is the i-th elementary symmetric function, and $s_i = \sum_j x_j^i$ is the i-th Newton function.

Whereas, for $m = 2h + 1$ odd, Goryunov produces

$$T(m-1) := \{(x_0, \dots, x_m, z) | \sigma_1(x) = 0, \sigma_3(x) + z\sigma_2(x) + \frac{1}{12}h(h+1)(h+2)z^3 = 0\} =$$

$$= \{(x_0, \dots, x_m, z) | s_1(x) = 0, 2s_3(x) - 3z\sigma_2(x) + \frac{1}{2}h(h+1)(h+2)z^3 = 0\}.$$

The nodes of the Segre cubic are easily seen to be the $\mathfrak{S}_{2h+2}$-orbit of the point x satisfying $x_i = 1,\ i = 0, \dots, h,\ x_j = -1,\ j = h+1, \dots, 2h+2$.

Whereas the linear subspaces of maximal dimension contained in the Segre cubic hypersurface are the $\mathfrak{S}_{2h+2}$-orbit of the subspace $x_i + x_{i+h+1} = 0,\ i = 0, 1, \dots, h$.

These are, in the case $h = 2$, exactly 15 $\mathbb{P}^2$'s.

We have the following result, which is due to Corrado Segre [56], and we give a simple argument here based on an easy result of [8].

Theorem 25 *Any nodal maximizing cubic hypersurface X in $\mathbb{P}^4$ is projectively equivalent to the Segre cubic.*

Proof Let P be a node of $X \subset \mathbb{P}^4$, and consider the Taylor development of its equation at the point $P = (0, 0, 0, 0, 1)$, $F(x, z) = zQ(x) + G(x)$.

Then by corollary 89 of [8], the section on cubic fourfolds with many nodes, the curve in $\mathbb{P}^3$ given by $Q(x) = G(x) = 0$ is a nodal curve contained in a smooth quadric and has $\gamma - 1$ nodes, where γ is the number of nodes of X. But a curve of bidegree $(3, 3)$ on $Q = \mathbb{P}^1 \times \mathbb{P}^1$ has at most 9 nodes, equality holding if and only if it consists of three vertical and three horizontal lines.

Hence, $\gamma \leq 9 + 1 = 10$, and the equality case is projectively unique. □

It is an open question whether a similar result holds for all Segre cubic hypersurfaces (Coughlan and Frapporti in [13] proved the weaker result that the small equisingular deformations of the Segre cubic are projectively equivalent to it).

It follows that all Kummer quartic surfaces X are obtained as discriminants of the projection of the Segre cubic hypersurface $\Sigma \subset \mathbb{P}^4$ from a smooth point P, which does not lie in any of the 15 subspaces $L \subset \Sigma,\ L \cong \mathbb{P}^2$.

Of the 16 nodes of X, 10 are images of the 10 nodes of Σ, while six nodes correspond to the complete intersection of the linear, quadratic, and cubic terms of the Taylor expansion at P of the equation of Σ (see [8], Sect. 9 on discriminants of cubic hypersurfaces).

We point out here an interesting connection to the work of Coble, ([12], page 141) and of van der Geer [58], who proved that a compactification of the Siegel modular threefold, the moduli space for principally polarized Abelian surfaces with a level 2 structure, is the dual variety $\mathcal{S}$ of the Segre cubic Σ,

$$\Sigma^\vee = \mathcal{S} := \{x \in \mathbb{P}^5 | s_1(x) = s_2(x)^2 - 4s_4(x) = 0\}$$

The singular set of $\mathcal{S}$, also called the Igusa quartic, or Castelnuovo-Richmond quartic, consists exactly of 15 lines, dual to the subspaces $L \subset \Sigma,\ L \cong \mathbb{P}^2$.

Coble and Van der Geer show that to a general point P' of $\mathcal{S}$ corresponds the Kummer surface X' obtained by intersecting $\mathcal{S}$ with the tangent hyperplane to $\mathcal{S}$ at P'.

Observe that the Hudson equation

$$\alpha_{00}^3 - \alpha_{00}(\alpha_{10}^2 + \alpha_{01}^2 + \alpha_{11}^2 - \beta^2) + 2\alpha_{10}\alpha_{01}\alpha_{11} = 0$$

defines a cubic hypersurface with 10 double points as singularities, hence projectively equivalent to the Segre cubic ([16, 17, 24, 58]). And to a point of this hypersurface corresponds the Kummer surface in Hudson's equation above, intersection of $\mathcal{S}$ with the tangent hyperplane to $\mathcal{S}$ at the dual point.

Proposition 26 *Let $P \in \Sigma$ be a smooth point of the Segre cubic which does not lie in any of the* 15 *planes contained in Σ.*

Then the Kummer quartic X_P obtained as the discriminant for the projection $\pi_P : \Sigma \to \mathbb{P}^3$ is the dual of the Kummer quartic

$$X'_P := \mathcal{S} \cap T_{P'}\mathcal{S} \subset T_{P'}\mathcal{S} \cong \mathbb{P}^3,$$

where $P' \in \mathcal{S}$ is the dual point of P.

Proof By biduality, the points y of X'_P correspond to the hyperplanes tangent to Σ and passing through P. If $z \in \Sigma$ is the tangency point of the hyperplane y, then the line $P * z$ is tangent to Σ, hence z maps to a point $x \in X_P$. To the hyperplane y corresponds a hyperplane w in $\mathbb{P}^3$ which contains x.

Hence, we have to show that, if $x \in X_P$ is general, then the inverse image of the tangent plane w to X_P at x is the tangent hyperplane y at the point $z \in \Sigma$ where the line $P * z$ is tangent to Σ.

To this purpose, take coordinates $(u, x_0, x_1, x_2, x_3) =: (u, x)$, so that $P = (1, 0)$. Then we may write the equation F of Σ as

$$L(x)u^2 + 2Q(x)u + G(x) = 0.$$

P being a smooth point means that $L(x)$ is not identically zero.

The equation of $X := X_P$ is $f(x) := L(x)G(x) - Q(x)^2 = 0$. If $x \in X$, and $L(x)$ does not vanish at x, there is a unique point (u, x) such that $\frac{\partial F}{\partial u} = 2(uL(x) + Q(x)) = 0$. In this point, the other partial derivatives are proportional to the partials of f, since

$$\frac{\partial F}{\partial x_i} = u^2 \frac{\partial L}{\partial x_i} + 2u \frac{\partial Q}{\partial x_i} + \frac{\partial G}{\partial x_i},$$

which multiplied by L^2 and evaluated at $x \in X$ yields

$$u^2 L^2 \frac{\partial L}{\partial x_i} + 2uL^2 \frac{\partial Q}{\partial x_i} + L^2 \frac{\partial G}{\partial x_i} = LG \frac{\partial L}{\partial x_i} - 2QL \frac{\partial Q}{\partial x_i} + L^2 \frac{\partial G}{\partial x_i} = L \frac{\partial f}{\partial x_i},$$

q.e.d. for the main assertion.

Let us now prove the first assertion; first of all, let us explain why we must take P outside of these 15 planes. Because, if $P \in \Lambda$, then there are coordinates x_2, x_3 such that the equation F of Σ lies in the ideal (x_2, x_3); then the equation f of X lies in the square of the ideal (x_2, x_3), hence X has a singular line.

If P is a smooth point, it cannot be collinear with two nodes of Σ unless it lies in one of the 15 planes Λ contained in Σ. In fact, using the $\mathfrak{S}_6$-symmetry, we see that any pair of nodes is in the orbit of the pair formed by $(1, 1, 1, -1, -1, -1)$ and $(1, -1, -1, 1, 1, -1)$, which is contained in $\Lambda := \{z | z_i + z_{i+3} = 0\}$.

Hence, the 10 nodes of Λ project to distinct nodes of X (cf. [8]), where the forms L, Q, G do not simultaneously vanish.

There is now a last condition required in order that $\{L(x) = Q(x) = G(x) = 0\}$ consists of six distinct points in $\mathbb{P}^3$ (which are then nodes for X).

In fact, $\{L(x) = Q(x) = G(x) = 0\}$ is the equation of the lines passing through P. The Fano scheme $F_1(\Sigma)$ has dimension 2 (see for instance [8]), hence if

$$\mathcal{U} \subset F_1(\Sigma) \times \Sigma \subset F_1(\Sigma) \times \mathbb{P}^4$$

is the universal family of lines contained in Σ, the last condition is that P is not in the set $\mathcal{B}$ of critical values of the projection $\mathcal{U} \to \Sigma$, which contains the 15 planes $\Lambda \subset \Sigma$ ($\mathcal{B}$ is the set where the fibre is not smooth of dimension 0).

Indeed, it turns out that $\mathcal{B}$ is the union of the 15 planes $\Lambda \subset \Sigma$, see [55, 56]. A slick proof was given in [16] page 184: since the tangent hyperplane at a point of Σ yields a quartic which can be put in Hudson normal form, hence has $\mathcal{K} \cong (\mathbb{Z}/2)^4$-invariance, the same holds for the dual. And since there are already 10 distinct nodes, it follows that X has exactly 16 nodes, hence $\{L(x) = Q(x) = G(x) = 0\}$ consists of six distinct points in $\mathbb{P}^3$. □

Remark 27 It is interesting to study the modular meaning of the family with parameters (a_i). Gonzalez-Dorrego [19] showed that the moduli space of such nondegenerate $(16_6, 16_6)$ configurations is the quotient of the above open set in $\mathbb{P}^3$ for the action of a subgroup H of $\mathbb{P}GL(4, K)$, which is a semidirect product

$$(\mathbb{Z}/2)^4 \rtimes \mathfrak{S}_6,$$

and which is the normalizer of $\mathcal{K} \cong (\mathbb{Z}/2)^4$. The group $\mathfrak{S}_6$ appearing in this and in the other representations (the one of [58] for instance), is the group of permutations of the six Weierstrass points on a curve of genus 2, and it is known (see [34] page 590) that $\mathfrak{S}_6 \cong Sp(4, \mathbb{Z}/2)$, so that the extension is classified by the action of $\mathfrak{S}_6$ on the group of 2-torsion points of the Jacobian of the curve.

7 Enriques Surfaces Étale Quotients of Kummer K3 Surfaces

We have seen that the Gauss map of a quartic Kummer surface (in canonical coordinates) yields a fixed-point free involution $\gamma : S \to S$ on the K3 surface S which is the minimal resolution of X.

S contains the 16 disjoint exceptional curves E_i, for $P_i \in \mathcal{N}$, such that $E_i^2 = -2$, ($\mathcal{N}$ is the set of nodes) and 16 disjoint curves D_i, for $i \in \mathcal{N}$, such that $D_i^2 = -2$, corresponding to the tropes, that is, the planes orthogonal to P_i, such that

$$H \equiv 2D_i + \sum_{E_j \cdot D_i = 1} E_j.$$

And $E_j \cdot D_i = 1$ if and only if the point P_j belongs to the trope $P_i^{\perp}$, equivalently, if the scalar product $P_i \cdot P_j = 0$.

γ sends E_i to D_i, hence, in the quotient Enriques surface $Z := S/\gamma$, we obtain 16 curves E_i' with $(E_i')^2 = -2$ (E_i' is the image of E_i) and they have the following intersection pattern:

$$E_i' \cdot E_j' = 1 \Leftrightarrow P_i \cdot P_j = 0, \text{else } E_i' \cdot E_j' = 0.$$

Observe that, since the inverse image of E_i' equals $E_i + D_i$, there is no point of Z where three curves E_i' pass, because the curves E_i are disjoint, likewise the curves D_i.

Definition 28 We define the **Kummer–Enriques graph** the graph Γ_Z whose vertices are the points P_i, and an edge connects P_i and P_j if and only if $P_i \cdot P_j = 0$.

It is the dual graph associated to the configuration of curves E_i' in the Enriques surface Z.

Without loss of generality, to study Γ_Z it suffices to consider the special case where $\mathcal{N}$ is the set of nodes of the Cefalú quartic, the orbit of $[0, \pm 1, \pm 1, \pm 1]$ under the action of the Klein group $\mathcal{K}'$ acting via double transpositions. Because $\mathcal{N}$ is the orbit of the group $\mathcal{K} \cong (\mathbb{Z}/2)^2 \oplus (\mathbb{Z}/2)^2$, the graph Γ_Z is a regular graph. The advantage of seeing $\mathcal{N}$ as the set of nodes of the Cefalú quartic X is that we can use the group

G of projectivities leaving X invariant as group of symmetries of the graph, since G acts through orthogonal transformations.

Proposition 29 *The Kummer–Enriques graph Γ_Z contains 16 vertices, 48 edges, each vertex has exactly six vertices at distance 1, six vertices at distance 2, three vertices at distance 3.*

It contains 32 triangles and 48 edges, so that, adding a cell for each triangle, we get a triangulation of the real two-dimensional torus.

In particular, the maximal number of pairwise nonneighbouring vertices is 4.

Proof There are 16 vertices, and for each vertex, there are exactly six vertices at distance one. Hence, each vertex belongs to 6 edges and there are exactly 48 edges.

$G = C_2^3 \rtimes \mathfrak{S}_4$, and the set of vertices consists of four blocks of cardinality 4 (where one fixed coordinate is 0), permuted by $\mathfrak{S}_4$, while C_2^3 acts transitively on each block.

Hence, noticing that elements of the same block are not orthogonal, we see that G acts transitively on the set of oriented edges, with stabilizer of order 2.

We show that each edge belongs to exactly two triangles, hence the number of triangles is $\frac{1}{3}(2 \cdot 48) = 32$.

By transitivity, consider two neighbouring vertices, namely $P := [1, 1, 1, 0]$ and $P' := [0, 1, -1, 1]$. If P'' is orthogonal to both, then necessarily either the second or the third coordinate equals to 0 and we get the two solutions $P'' = [1, 0, -1, -1]$ or $P'' = [1, -1, 0, -1]$, which are not orthogonal, thereby proving that each edge is a side of exactly two triangles.

Adding a cell for each triangle, we get a triangulation of the real two-dimensional torus, since we get a 2-manifold whose Euler number is $16 - 48 + 32 = 0$.

Follows in particular that for each vertex P there are exactly six triangles with vertex P.

To count the diameter of the graph, start from $P_1 := [1, 1, 1, 0]$ and observe that P_1 is stabilized by $C_2 \times \mathfrak{S}_3$ and its 6 neighbours are the $\mathfrak{S}_3$-orbit of $[0, 1, -1, 1]$, where $\mathfrak{S}_3$ permutes the first three coordinates.

Each neighbour produces three more neighbours, but in total at distance 2 we have six vertices, the remaining ones with exception of the three vertices which form the $\mathfrak{S}_3$-orbit of $[1, 1, 0, 1]$, namely $[1, 1, 0, 1]$, $[1, 0, 1, 1]$, $[0, 1, 1, 1]$.

At distance 2 from P_1, we have the $\mathfrak{S}_3$-orbit $\mathcal{S}_1$ of $[1, 1, -1, 0]$ (three elements) and the $\mathfrak{S}_3$-orbit $\mathcal{S}_2$ of $[1, 1, 0, -1]$ (three elements).

To show that $[1, 1, 0, 1]$ is at distance 3 from $P_1 = [1, 1, 1, 0]$ we give the path through $P_2 = [0, 1, -1, 1]$ and $P_3 = [-1, 1, 1, 0]$.

Hence, our assertion is proven.

We try now to determine each maximal set $\mathcal{M}$ of pairwise non-neighbouring vertices.

We can immediately see two solutions:

$$\mathcal{M}_1 := \{[1, 1, 1, 0], [1, 1, 0, 1], [1, 0, 1, 1], [0, 1, 1, 1]\}$$

which consists of four vertices which are pairwise at distance 3.

Or, the elements of the same block, such as

$$\mathcal{M}_2 := \{[1, 1, 1, 0], [1, 1, -1, 0], [1, -1, 1, 0], [-1, 1, 1, 0]\},$$

which consists of vertices which are pairwise at distance 2.

We want to see whether there are other solutions, up to G-symmetries. Now, if there are two vertices at distance 3 we may assume that these are $P_1 = [1, 1, 1, 0]$, $P_4 = [0, 1, 1, 1]$. $\mathcal{M}$ cannot contain the six neighbours of P_1 and the six neighbours of P_4, hence $\mathcal{M} = \mathcal{M}_1$.

If all the vertices are at distance 2, then we may assume that $\mathcal{M}$ contains P_1, and then some of the six vertices at distance 2. We observe that these form two $\mathfrak{S}_3$-orbits $\mathcal{S}_1$ and $\mathcal{S}_2$, and these have the property that two vertices of the same orbit are not neighbouring, while two vertices of different orbits are neighbouring.

Hence, we find just another solution

$$\mathcal{M}_3 := \{[1, 1, 1, 0], [1, 1, 0, -1], [1, 0, 1, -1], [0, 1, 1, -1]\},$$

and we conclude that all the solutions are in the respective G-orbits of $\mathcal{M}_1$, $\mathcal{M}_2$, $\mathcal{M}_3$. □

Corollary 30 *The Enriques surface $Z = S/\gamma$ contains sets of four disjoint (-2) curves, hence Z is in several ways the minimal resolution of a 4-nodal Enriques surface.*

Remark 31 (i) One may ask, in view of Proposition 29, whether Z cannot be the minimal resolution of a 5-nodal Enriques surface: a set of five disjoint (-2) curves on Z would produce on the K3 double cover S at least two more (-2)-curves exchanged by the involution γ.

(ii) We can also consider the 32 vectors orbit of $(1, 1, 1, 0)$ for the action of G. Then we can define another graph, with 32 vertices and 96 edges; this is an unramified double covering of the graph Γ_Z, and is again associated to a triangulation of the two-dimensional real torus.

8 Remarks on Normal Cubic Surfaces and on Strict Selfduality

In this section, we are mainly concerned with the case of the complex ground field $\mathbb{C}$, and we make some topological considerations.

It is worthwhile to observe that, already for degree $d = 3$, the maximal number $\mu(d)$ of singular points can be smaller, for surfaces with isolated singularities, then the total sum of the Milnor numbers of the singularities. In fact, if a normal cubic surface has four singular points, it is projectively equivalent to the Cayley cubic of equation $\sigma_3(x) = 0$, where σ_3 is the third elementary symmetric function, and its singularities are just four nodes.

But there is the cubic surface $xyz = w^3$ which possesses three singular points of type A_2, hence realizing a sum of the Milnor numbers equal to 6.

Proposition 32 *The cubic surface $X := \{xyz = w^3\}$ is strictly self-dual.*

Proof We can change coordinates so that $X := \{xyz = \lambda w^3\}$, where we can choose $\lambda \neq 0$ arbitrarily.

The dual map is given by

$$\psi(x, y, z, w) = (yz, xz, xy, -3\lambda w^2) =: (a, b, c, d).$$

We want $\lambda d^3 = abc$, this requires that

$$-27\lambda^4 w^6 = (xyz)^2 \Leftrightarrow -27\lambda^4 w^6 = \lambda^2 w^6 \Leftarrow -27\lambda^2 = 1.$$

□

In [11], the authors study the limits of dual surfaces of smooth quartic surfaces: and they show that the above cubic surface, as well as the Kummer surfaces, appear as limits.

As we discuss in the following example, this normal cubic is also the one with the largest (finite) fundamental group of the smooth part among the normal cubic surfaces which are not cones over smooth plane cubic curves (this is easy to show since the degree of a normal Del Pezzo surface is at most 9).

Example 33 (A cubic with maximal Milnor number) This is the quotient $\mathbb{P}^2/(\mathbb{Z}/3)$ for the action such that

$$(u_0, u_1, u_2) \mapsto (u_0, \epsilon u_1, \epsilon^2 u_2),$$

where ϵ is a primitive third root of unity.

The quotient is embedded by $x_i := u_i^3,\ i = 0, 1, 2$ and by $x_3 := u_0 u_1 u_2$ so that

$$\mathbb{P}^2/(\mathbb{Z}/3) \cong Y := \{x_0 x_1 x_2 = x_3^3\}.$$

In this case, the triple covering is only ramified in the three singular points $x_3 = x_i = x_j = 0$ (for $0 \leq i < j \leq 2$), hence we conclude for $Y^* := Y \setminus Sing(Y)$, that (since $\mathbb{P}^2$ minus three points is simply connected)

$$\pi_1(Y^*) \cong \mathbb{Z}/3.$$

This cubic surface has three singular points with Milnor number 2 (locally isomorphic to the singularity $x_1 x_2 = x_3^3$), and $3 \cdot 2 = 6$ is bigger than the maximum number of singular points that a normal cubic can have, which equals to 4. The surface realizes the maximum for the total sum of the Milnor numbers of the singularities.

In fact, for each normal cubic surface Y, which is not the cone over a smooth cubic curve, its singularities are rational double points, hence the sum m of their Milnor numbers of the singular points equals the number of the exceptional (-2)-curves appearing in the resolution $\tilde{Y}$; since $\tilde{Y}$ is the blow-up of the plane in six points, the rank of $H^2(\tilde{Y}, \mathbb{Z})$ is at most 7, hence (these (-2)-curves being numerically independent, see [1]) $m \leq 6$.

We give now a very short proof of the following known theorem (see [15] for other approaches)

Theorem 34 *If Y is the Cayley cubic, and Y^* its smooth part, then*

$$\pi_1(Y^*) \cong \mathbb{Z}/2.$$

Proof The key point is that Y admits a double covering ramified only in the four singular points (this can also immediately be seen representing Y is the determinant of a symmetric matrix of linear forms), which is isomorphic to the blow-up of $\mathbb{P}^2$ in three points.

Indeed, consider the birational action of $\mathbb{Z}/2$ on $\mathbb{P}^2$ given by the Cremona involution

$$c : (u_i) \mapsto \left(\frac{1}{u_i}\right) = \left(\frac{u_0 u_1 u_2}{u_i}\right).$$

The involution becomes biregular on the blow-up Z of $\mathbb{P}^2$ at the three points of indeterminacy, the three coordinate points; Z is the so-called Del Pezzo surface of degree 3, and the fixed points of the involution are exactly four, the points $(\epsilon_1, \epsilon_2, 1)$, where $\epsilon_i \in \{1, -1\}$.

We have that the quotient Z/c is exactly a cubic Y with four singular points, hence isomorphic to the Cayley cubic.

Indeed, the quotient map ψ is given by the system of c-invariant cubics through the coordinate points e_i, hence

$$\psi(x) = (x_1 x_2 x_3, x_1(x_2^2 + x_3^2), x_2(x_3^2 + x_1^2), x_3(x_1^2 + x_2^2)),$$

and the image Y has degree $3 = \frac{1}{2}(3^2 - 3)$.

$Z \to Y$ is only ramified at the four fixed points, whose image points are the four singular points $(\epsilon_1\epsilon_2, 2\epsilon_1, 2\epsilon_2, 2)$, whereas ψ is otherwise injective on Z/c: hence, Y is a cubic with four singular points.

Since Z minus a finite number of points is simply connected, follows that $\pi_1(Y^*) \cong \mathbb{Z}/2$. □

Proposition 35 *If Y is a normal cubic surface, then it cannot contain three collinear singular points.*

If it contains at least four singular points, then these are linearly independent and Y is projectively equivalent to the 4-nodal Cayley cubic.

If Y is a nodal cubic with $\nu \leq 3$ nodes, then Y^* is simply connected.

Proof There cannot be three collinear singular points of Y, since otherwise we may take coordinates such that these are the points $e_0, e_1, e_0 + e_1$ and the equation of the cubic is then of the form $F = x_2 A(x) + x_3 B(x)$, since the line $\{x | x_2 = x_3 = 0\}$ is contained in Y. Vanishing of the partial derivatives $\frac{\partial F}{\partial x_2}, \frac{\partial F}{\partial x_3}$ in the three points imply that A, B also vanish in the three points, hence A, B belong to the ideal (x_2, x_3) and Y is singular along the line $x_2, x_3 = 0$.

Similarly, if e_0, e_1, e_2 are singular points of Y, then the equation F of Y contains only monomials x^m with $m_0, m_1, m_2 \leq 1$, hence we can write $F = c x_0 x_1 x_2 + x_3 B(x)$. If also $e_0 + e_1 + e_2$ is a singular point of Y, then $c = 0$ and Y is reducible. Hence, if there are four singular points, then we may assume them to be e_0, e_1, e_2, e_3 and then the equation F of Y contains only monomials x^m with $m_0, m_1, m_2, m_3 \leq 1$, hence $F = \sum_i \frac{a_i}{x_i} x_0 x_1 x_2 x_3$, and we may replace x_i by $\lambda_i x_i$ obtaining that $a_i = 1$.

Therefore, if Y is nodal, the ν nodes of Y are linearly independent, and it follows that the nodal cubics with ν nodes are parametrized by a Zariski open set in a linear space, in particular, for ν fixed, Y^* has always the same topological type.

We finish observing that a cubic surface with $\nu \leq 3$ nodes is the blow-up of the plane in six points, of which ν triples are collinear (the case of six points on a smooth conic, with $\nu = 1$, reduces to the previous case through a standard Cremona transformation based at 3 of the six points). In particular, the fundamental group $\pi_1(Y^*)$ is the quotient of the fundamental group of the complement in $\mathbb{P}^2$ of at most three general lines.

But $\mathbb{P}^2$ minus three general lines is $(\mathbb{C}^*)^2$, and $\pi_1((\mathbb{C}^*)^2) = (\mathbb{Z})^2$. Hence $\pi_1(Y^*)$ is Abelian, hence equal to $H_1(Y^*, \mathbb{Z})$, which is a trivial binary code for $\nu \leq 4$ (see [8]). □

Francesco Russo [51] informed us of other examples of strictly self-dual hypersurfaces, the simplest one being:

Example 36 (Francesco Russo' s generalization of the Perazzo cubic [45]) Consider in $\mathbb{P}^{2n+1}$ the hypersurface of equation

$$\{x_0 \dots x_n - y_0 \dots y_n = 0\}.$$

The dual map is given by

$$\psi(x, y) = (u, v), \ u_i = \frac{1}{x_i} x_0 \dots x_n, \ v_i = -\frac{1}{y_i} y_0 \dots y_n.$$

For n odd, we have $u_0 \dots u_n - v_0 \dots v_n = 0$, so X equals its dual variety.
For n even, it suffices to take

$$X = \{x_0 \dots x_n + \lambda y_0 \dots y_n = 0\}, \lambda^2 = -1.$$

The other examples constructed by Russo in [51] use determinants, symmetric determinants or Pfaffians, and the Perazzo trick.

9 Appendix on Monoids

We consider now a normal quartic surface $X = \{F = 0\} \subset \mathbb{P}^3_K$, where K is an algebraically closed field.

If X has a point of multiplicity 4, then this is the only singular point, while if X contains a triple point, we can write the equation

$$F(x_1, x_2, x_3, z) = zG(x) + B(x),$$

and, setting $G_i := \frac{\partial G}{\partial x_i}$, $B_i := \frac{\partial B}{\partial x_i}$, we have

$$Sing(X) = \{G(x) = B(x) = G_i z + B_i = 0, \ i = 1, 2, 3\} .$$

If $(x, z) \in Sing(X)$ and $x \in \{G(x) = B(x) = 0\}$, then $x \notin Sing(\{G = 0\})$, else $x \in Sing(\{G = 0\}) \Rightarrow x \in Sing(\{B = 0\})$ and the whole line $(\lambda_0 z, \lambda_1 x) \subset Sing(X)$. Hence, $\nabla(G)(x) \neq 0$ and there exists a unique singular point of X in the above line. Since the two curves $\{G(x) = 0\}$, $\{B(x) = 0\}$ have the same tangent at x the intersection multiplicity at x is at least 2, and we conclude:

Proposition 37 *Let X be quartic surface $X = \{F = 0\} \subset \mathbb{P}^3_K$, where K is an algebraically closed field, and suppose that $Sing(X)$ is a finite set. If X has a triple point then $|Sing(X)| \leq 7$.*

More generally, if X is a normal monoid, that is, a degree d normal surface $X = \{F = 0\} \subset \mathbb{P}^3_K$, where K is an algebraically closed field, possessing a point of multiplicity $d - 1$, then $|Sing(X)| \leq 1 + \frac{d(d-1)}{2}$.

Proof The second assertion follows by observing that in proving the first we never used the degree d, except for concluding that the total intersection number (with multiplicity) of G, B equals $d(d - 1)$. □

Finally, I would like to thank Thomas Dedieu, Igor Dolgachev, Francesco Russo, for useful comments on the first version of the paper; and Shigeyuki Kondo and especially the referee for very stimulating and useful remarks and queries.

Acknowledgements The author acknowledges the support of the ERC 2013 Advanced Research Grant—340258—TADMICAMT

References

1. Michael Artin, On isolated rational singularities of surfaces. Am. J. Math. **88**, 129–136 (1966)
2. Arnaud Beauville, Le théorème de Torelli pour les surfaces K3: fin de la démonstration. *Geometry of K3 Surfaces: Moduli and Periods (Palaiseau, 1981/1982)*. Astérisque No. 126, pp. 111–121 (1985)

3. D. Burns, J.M. Wahl, Local contributions to global deformations of surfaces. Inventiones Mathematicae **26**(1), 67–88 (1974)
4. F. Catanese, Babbage's conjecture, contact of surfaces, symmetric determinantal varieties and applications. Invent. Math. **63**(3), 433–465 (1981)
5. F. Catanese, C. Ciliberto, Symmetric products of elliptic curves and surfaces of general type with $p_g = q = 1$. J. Algebr. Geom. **2**(3), 389–411 (1993)
6. F. Catanese, J. Hae, K. Oguiso, Some remarks on the universal cover of an open K3 surface. Math. Ann. **325**(2), 279–286 (2003)
7. F. Catanese, Cayley forms and self-dual varieties. Proc. Edinb. Math. Soc., II. Ser. **57**(1), 89–109 (2014)
8. F. Catanese et al., Varieties of nodal surfaces, coding theory and discriminants of cubic hypersurfaces. *Preliminary Version*, 135 pages (2020)
9. F. Catanese, The number of singular points of quartic surfaces (char =2). Preprint, May 2021
10. A. Cayley, A memoir on cubic surfaces. Trans. London CLIX, 231–326 (1869)
11. C. Ciliberto, T. Dedieu, Limits of pluri-tangent planes to quartic surfaces, in *Algebraic and complex geometry. In honour of Klaus Hulek?s 60th birthday. Based on the conference on algebraic and complex geometry*, Hannover, Germany, September 10–14, 2012, vol. 71, ed. by A. Frühbis-Krüger et al. Springer Proceedings in Mathematics and Statistics (Springer, Cham, 2014), pp. 123–199
12. A.B. Coble, *Algebraic Geometry and Theta Functions*, vol. 10 (American Mathematical Society, New York, Colloquium Publications, 1929), VII + 282 p
13. S. Coughlan , D. Frapporti, *Segre Cubic Hypersurfaces*. Preliminary version, 14 pages (2020)
14. L. Cremona, Mémoire de géométrie pure sur les surfaces du troisième ordre. J. Reine Angew. Math. **68**, 1–133 (1868)
15. M. Dettweiler, M. Friedman, M. Teicher, M. Topol-Amram, The fundamental group for the complement of Cayley's singularities. Beiträge Algebra Geom. **50**(2), 469–482 (2009)
16. I. Dolgachev, D. Ortland, Point sets in projective spaces and theta functions. *Astérisque, 165. Paris: Société Mathématique de France; Centre National de la Recherche Scientifique*. 210 p. (1988)
17. I. Dolgachev, Classical algebraic geometry. *A Modern View* (Cambridge University Press, Cambridge, 2012), xii+639 pp
18. I. Dolgachev, Kummer Surfaces: 200 years of study. Not. A.M.S. **67**(10), 1527–1533 (2020)
19. M.R. Gonzalez-Dorrego, (16,6) configurations and geometry of Kummer surfaces in $\mathbb{P}^3$. Mem. Am. Math. Soc. **107**(512), vi+101 pp (1994)
20. R. Maria, Gonzalez-Dorrego, Construction of (16, 6) configurations from a group of order 16. J. Algebra **162**(2), 471–481 (1993)
21. Theoriae transcendentium Abelianarum primi ordinis adumbratio. J. Reine Angew. Math. **35**, 277–312 (1847)
22. P. Griffiths, J. Harris, Principles of algebraic geometry. *Pure and Applied Mathematics* (A Wiley-Interscience Publication. Wiley, New York etc., 1978), XII, 813 p
23. V.V. Goryunov, Symmetric quartics with many nodes. Singularities and bifurcations. Adv. Soviet Math. Am. Math. Soc. Providence, RI **21**, 147–161 (1994)
24. R.W.H.T. Hudson, Kummer's quartic surface. *Revised Reprint of the 1905 Original, with a Foreword by W. Barth. Cambridge Mathematical Library* (Cambridge University Press, Cambridge, 1990), xxiv+222 pp
25. J.I. Hutchinson, On some birational transformations of the Kummer surface into itself. Am. M. S. Bull. **2**(7), 211–217 (1900)
26. J.-I. Igusa, On the graded ring of Theta-constants. Am. J. Math. **86**(1), 219–246 (1964)
27. J.-I. Igusa, Theta functions. *Die Grundlehren der mathematischen Wissenschaften, Band*, vol. 194 (Springer, New York-Heidelberg, 1972), x+232 pp
28. C.M. Jessop, *Quartic Surfaces with Singular Points* (Cambridge University Press, 1916)
29. J.H. Keum, Every algebraic Kummer surface is the K3-cover of an Enriques surface. Nagoya Math. J. **118**, 99–110 (1990)

30. Jong Hae Keum, Automorphisms of Jacobian Kummer surfaces. Compositio Math. **107**(3), 269–288 (1997)
31. Jong Hae Keum, Shigeyuki Kondo, The automorphism groups of Kummer surfaces associated with the product of two elliptic curves. Trans. Am. Math. Soc. **353**(4), 1469–1487 (2001)
32. F. Klein, U. Configurationen, welche den Kummer'schen Flächen zugleich ein- und umgeschrieben sind. Math. Ann. XXVI **I**, 106–142 (1886)
33. S. Kondo, Enriques surfaces with finite automorphism groups. Jpn. J. Math. New Ser. **12**, 191–282 (1986)
34. S. Kondo, The automorphism group of a generic Jacobian Kummer surface. J. Algebr. Geom. **7**(3), 589–609 (1998)
35. S. Kondo, Niemeier lattices, Mathieu groups, and finite groups of symplectic automorphisms of K3 surfaces. With an appendix by Shigeru Mukai. Duke Math. J. **92**(3), 593–603 (1998)
36. E.E. Kummer, On surfaces of degree four containing sixteen singular points. (Über die Flächen vierten Grades mit sechzehn singulären Punkten.) Berl. Monatsber. 246–260 (1864)
37. S. Mukai, Finite groups of automorphisms of K 3 surfaces and the Mathieu group. Invent. Math. **94**, 183–221 (1988)
38. Shigeru Mukai, Hisanori Ohashi. Enriques surfaces of Hutchinson-Göpel type and Mathieu automorphisms. *Laza, Radu (ed.) et al., Arithmetic and geometry of K3 surfaces and Calabi-Yau threefolds. Proceedings of the workshop, Toronto, Canada, August 16–25, 2011. New York, NY: Springer Fields Institute Communications 67, 429–454 (2013).*
39. S. Mukai, H. Ohashi, The automorphism group of Enriques surfaces covered by symmetric quartic surfaces (English), in *Recent Advances in Algebraic Geometry. A Volume in Honor of Rob Lazarsfeld?s 60th birthday. Based on the Conference*, Ann Arbor, MI, USA, 16–19 May 2013, vol. 417, ed. by D.H. Christopher et al. London Mathematical Society Lecture Note Series (Cambridge University Press, Cambridge, 2014), pp. 307–320
40. D. Mumford, Prym varieties. I, *Contributions to Analysis (a collection of papers dedicated to Lipman Bers)* (Academic Press, New York, 1974), pp. 325–350
41. D. Mumford. Tata lectures on theta. I. With the assistance of C. Musili, M. Nori, E. Previato, M. Stillman, *Progress in Mathematics*, vol. 28 (Birkhäuser Boston, Inc., Boston, MA, 1983), xiii+235 pp
42. V.V. Nikulin. Kummer surfaces. Izv. Akad. Nauk SSSR Ser. Mat. **39**(2), 278–293, 471 (1975)
43. V.V. Nikulin, Integral symmetric bilinear forms and some of their geometric applications. Izv. Akad. Nauk SSSR Ser. Mat. **43**(1), 111–177, 238 (1979)
44. A. Ohbuchi, Some remarks on ample line bundles on abelian varieties. Manuscr. Math. **57**, 225–238 (1987)
45. U. Perazzo, Sopra una forma cubica con 9 rette doppie dello spazio a cinque dimensioni, e i correspondenti complessi cubici di rette nello spazio ordinario. Torino Atti **36**, 891–895 (1901)
46. I. Ilya, Pjateckii-Shapiro, Igor Rotislav Shafarevic, Torelli's theorem for algebraic surfaces of type K3. Izv. Akad. Nauk SSSR Ser. Mat. **35**, 530–572 (1971)
47. Vladimir L. Popov, Self-dual algebraic varieties and nilpotent orbits, in *Proceedings of the international colloquium on algebra, arithmetic and geometry, Mumbai, India, January 4–12, 2000. Part I and II*, vol. 16, ed. by Parimala, R. (Narosa Publishing House, New Delhi, Published for the Tata Institute of Fundamental Research, Bombay. Stud. Math., Tata Inst. Fundam. Res., 2002), pp. 509–533
48. V.L. Popov, E.A. Tevelev, Self-dual projective algebraic varieties associated with symmetric spaces, in *Algebraic Transformation Groups and Algebraic Varieties. Proceedings of the Conference on Interesting Algebraic Varieties Arising in Algebraic Transformation Group Theory*, Vienna, Austria, 22–26 Oct. 2001, ed. by V.L. Popov. Encyclopaedia of Mathematical Sciences 132. Invariant Theory and Algebraic Transformation Groups 3 (Springer, Berlin, 2004), pp. 131–167
49. K. Rohn, *Die Flächen vierter Ordnung hinsichtlich ihrer Knotenpunkte und ihrer Gestaltung*, vol. 9 (S. Hirzel, 1886)
50. K. Rohn, Die Flächen vierter Ordnung hinsichtlich ihrer Knotenpunkte und ihrer Gestaltung. Mathematische Annalen **29**(1), 81–96 (1887)

51. F. Russo, Selfdual hypersurfaces. *Manuscript* (2021)
52. B.S. Donat, Projective models of K3 surfaces. Am. J. Math. **96**, 602–639 (1974)
53. L. Schläfli, On the distribution of surfaces of the third order into species, in reference to the absence or presense of singular points, and the reality of their lines. Phil. Trans. of Roy. Soc. London **6**, 201–241 (1863)
54. B. Segre, Sul massimo numero di nodi delle superficie algebriche. Atti Acc. Ligure **10**(1), 15–2 (1952)
55. C. Segre, Sulla varietá cubica con dieci punti doppi dello spazio a quattro dimensioni . *Atti della Reale Accademia delle Scienze di Torino* vol. XXII, pp. 791–801 (1886–1887)
56. C. Segre, Sulle varietá cubiche dello spazio a quattro dimensioni e su certi sistemi di rette e certe superficie dello spazio ordinario. *Memorie della Reale Accademia delle Scienze di Torino, (2) tomo XXXIX*, pp. 3–48 (1888)
57. T. Shioda, H. Inose, On singular K3 surfaces. Complex Anal. Algebr. Geom. Collect. Pap. dedic. K. Kodaira 119–136 (1977)
58. G. van der Geer, On the geometry of a Siegel modular threefold. Math. Ann. **260**(3), 317–350 (1982)
59. D. van Straten, A quintic hypersurface in $\mathbb{P}^4$ with 130 nodes. Topology **32**(4), 857–864 (1993)
60. H.M. Weber, On the Kummer surface of order four with sixteen nodes and its relation to the theta function with two variables (Ueber die Kummer'sche Fläche vierter Ordnung mit sechszehn Knotenpunkten und ihre Beziehung zu den Thetafunctionen mit zwei Veränderlichen). Borchardt J. **84**, 332–355 (1877)

A Footnote to a Footnote to a Paper of B. Segre

Luca Chiantini and Giorgio Ottaviani

Si canimus silvas, silvae sint consule dignae.

(Vergilius)

Abstract The paper is devoted to a detailed study of sextics in three variables having a decomposition as a sum of nine powers of linear forms. This is the unique case of a Veronese image of the plane which, in the terminology introduced by Ciliberto and the first author in [12], is weakly defective, and non-identifiable. The title originates from a paper of 1981, where Arbarello and Cornalba state and prove a result on plane curves with preassigned singularities, which is relevant to extend the studies of B. Segre on special linear series on curves. We explore the apolar ideal of a sextic F and the associated catalecticant maps, in order to determine the minimal decompositions. A particular attention is played to the postulation of the decompositions. Starting with forms with a decomposition A of length 9, the postulation of A determines several loci in the 9-secant of the 6-Veronese image of $\mathbb{P}^2$, which include the lower secant varieties, and the ramification locus, where the decomposition is unique. We prove that equations of all these loci, including the 8th and the 7th secant varieties, are provided by minors of the catalecticant maps and by the invariant H_{27} that we describe in Sect. 4.

Dedicated to Ciro Ciliberto, for his 70th birthday.
Paper written while the authors were members of INdAM-GNSAGA.

L. Chiantini (✉)
Dipartimento di Ingegneria dell'Informazione e Scienze Matematiche, Università di Siena, Siena, Italy
e-mail: luca.chiantini@unisi.it

G. Ottaviani
Dipartimento di Matematica e Informatica 'Ulisse Dini', Università di Firenze, Florence, Italy
e-mail: giorgio.ottaviani@unifi.it

T. Dedieu et al. (eds.), *The Art of Doing Algebraic Geometry*, Trends in Mathematics,
https://doi.org/10.1007/978-3-031-11938-5_4

Keywords Symmetric tensors · Waring decomposition · Plane sextics

2000 Mathematics Subject Classification 14N07

1 Introduction

The paper is devoted to a detailed study of sextics in three variables having a decomposition as a sum of nine powers of linear forms. This is indeed the unique case of a Veronese image X of the plane which, in the terminology introduced by Ciliberto and the first author in [12], is *weakly defective*, and non-identifiable: a general sextic of the 9-secant variety of X has two minimal decompositions.

The title originates from a famous paper of 1981, [4], where Arbarello and Cornalba state and prove a result on plane curves with preassigned singularities, which is relevant to extend the studies of B. Segre on special linear series on curves. The result (Theorem 3.2) says that the linear system of sextics with 9 general nodes in $\mathbb{P}^2$ is the unique non-superabundant system of plane curves with general nodes whose (unique) member is non-reduced.

The result is of course relevant also in the theory of interpolation, and in the study of secant varieties to Veronese varieties, with consequences for the Waring decomposition of forms. The matter turns out to be strictly related to the uniqueness of a minimal Waring decomposition of a general form. From this point of view Theorem 3.2 of [4] and Theorem 2.9 of [13] imply that the unique case of general ternary forms of fixed (Waring) rank with a finite number greater than one of minimal decomposition holds for degree 6 and rank 9. Since the unique sextic with 9 general nodes is twice an elliptic curve, it turns out by Proposition 5.2 of [13] that a general ternary sextic of rank 9 has exactly two minimal decompositions. A complete list of cases, in any number of variables, in which the previous phenomenon appears is contained in [15].

Remaining in the case of ternary sextics, already in [4] the authors observe that if the 9 nodes are in special position (e.g., when they are complete intersection of two cubics), then the linear system of nodal curves has also reduced members. Our analysis starts here, and aims to describe the decompositions of specific ternary sextics, with respect to the postulation of the corresponding sets of projective points in $\mathbb{P}^2$.

We consider a fixed sextic form F in the polynomial ring $R = \mathbb{C}[x_0, x_1, x_2]$ in three variables. In order to effectively produce decompositions of F, the first natural step is to consider the *apolar* ideal $F^\perp$ of F.

In general, the apolar ideal $F^\perp$ of a form F of degree d is defined via the natural action of the dual ring $R^\vee$ on R. $F^\perp$ is the ideal of elements $D \in R^\vee$ that kill F. A classical theorem by Sylvester says that a set of points is a decompositions of F if and only if the corresponding ideal in $R^\vee$ is contained in $F^\perp$. Indeed, in order to find properties of the decompositions of a specific form F, it is often sufficient to look at homogeneous pieces of the ideal $F^{\perp}$, which correspond to the kernels of the (catalecticant) maps $(R^{\vee})_k \to R_{d-k}$ induced by F, and their projective versions $\mathcal{C}_F^k$. In [22], as well as in [27], it is explained that if the image of $\mathcal{C}_F^k$ has the expected dimension and cuts the corresponding Veronese variety in $\mathbb{P}(R_{d-k})$ in a finite set A, then the set A determines a decomposition of F, and many minimal decompositions can be found in this way.

Consider, in particular, the case of a sextic F of rank 8. In this situation, the image of the catalecticant map of order 3 determines a linear space of codimension 2 in $\mathbb{P}(R_3)$, which cuts the 3-Veronese of $\mathbb{P}^2$ in 9 points. The set Z of 9 points is the image of a complete intersection of two cubics in $\mathbb{P}^2$, via the Veronese map. Then the 8 points of a minimal decomposition of F are among the points of Z. It is quite easy to find the coefficients of F with respect to the points of Z, and then determine the 8 points that give a minimal decomposition. The procedure shows that there are sextic forms which lie in between forms of rank 8 and forms of rank 9. These are forms of rank 9, with a minimal decomposition Z coming from a complete intersection of two plane cubics. Forms of this sort fill a subvariety W of the secant variety $S^9(X)$ (X being the 6-Veronese image of $\mathbb{P}^2$) which contains $S^8(X)$.

We show that a general sextic in W has a unique decomposition with 9 powers of linear forms. From a certain point of view, this is quite surprising. A general sextic F' of rank 9 has two minimal decompositions of length 9, coming from the existence of an elliptic normal curve in X which spans F'. When F' sits in W, then there is a pencil of elliptic curves in X which span F', so one may expect infinitely many minimal decompositions, one for each elliptic curve.

The phenomenon can be investigated in terms of the map s^9 from the abstract secant variety $AS^9(X)$ to $S^9(X)$, which is generically 2:1.

The component of the ramification locus $\mathcal{R}$ is strictly connected with the theory of Terracini loci, introduced in [6]. Roughly speaking, Terracini loci contain finite subsets of X such that the tangent spaces to the points are not independent. A subset $A \subset X$ consisting of 9 points belongs to the Terracini locus $\mathbb{T}_9(X)$ exactly when the linear span of $v_6(A)$ is contained in $\mathcal{R}$. Thus, when a sextic form F has a decomposition A complete intersection of two cubics, which obviously lie in the Terracini locus, then in a neighborhood of F the secant map has degree 2, so it has degree 2 also in F, but the fiber is non-reduced.

The subvariety W of $S^9(X)$ lies in between $S^9(X)$ and $S^8(X)$, and it is easy to detect because it is easy to compute the rank and the kernel of the catalecticant map $\mathcal{C}_F^3$.

There are other relevant varieties of $S^9(X)$, which can be studied in terms of the geometry of the decomposition. Given a general set A of 9 points in $\mathbb{P}^2$, and a general sextic F in the span of $v_6(A)$, a second decomposition of F comes out by taking a set of 9 points B linked to A by a cubic and a sextic. In this case, as explained in [2], the sum $I_A + I_B$ of the homogeneous ideals of A and B determines a subspace of R_6, whose orthogonal direction gives coefficients for the (unique) form F in the intersection $\langle v_6(A)\rangle \cap \langle v_6(B)\rangle$. In general, the cubic is uniquely determined by A but the sextic moves and then the set of forms defined by $(I_A + I_B)_6$ dominates $\langle v_6(A)\rangle$. On the contrary, when A is complete intersection, then also B is complete intersection. We obtain that starting from a complete intersection A the linked sets B determine only a subvariety of forms in $\langle v_6(A)\rangle$, each having a 2-dimensional set of decompositions. When A varies in the Hilbert stratum of complete intersections of type (3, 3) in $\mathbb{P}^2$, we obtain a subvariety $W' \subset W$ of sextics in $S^9(X)$ with a 2-dimensional set of decomposition, which can be easily described because for such forms F the kernel of $\mathcal{C}^3_F$ has dimension 3, so that W' does not contain $S^8(X)$, but properly contains $S^7(X)$.

Let us notice that the picture is completed by another subvariety $\mathcal{R}'$ of the ramification locus, given by forms F with two different and linked decompositions A, B, both sets of nodes of irreducible sextics. The variety $\mathcal{R}'$ parametrizes forms with a 1-dimensional family of decompositions, and its intersection with W contains W'. The variety $\mathcal{R}'$ is the most mysterious object in the picture, for it cannot be completely described in terms of the catalecticant map $\mathcal{C}^3_F$ and its kernel.

While the third section is devoted to the geometry (and, somehow, parametric) description of loci in $S^9(X)$, in the last section we provide for most of them a set of equations. Equations for $S^9(X)$, W, W', can be easily described generically by the vanishing of appropriate minors of the catalecticant map $\mathcal{C}^3_F$. Equations for $S^8(X)$ and $S^7(X)$ require the determinant of a 27×27 matrix A_f which represents a flattening, introduced in [24].

It is remarkable that this invariant $H_{27} = \det A_f$, joint with the catalecticant matrix $\mathcal{C}^3$, allows to describe the equations of all the secant varieties to $X = v_6(\mathbb{P}^2)$. The ramification locus turns out to be described also in terms of $\det(A_f)$ (see Theorem 4.4). We wonder if a description of equations for $\mathcal{R}'$ can be achieved in terms of minors of A_f.

2 Preliminaries

2.1 Notation

In this section, as in most of the paper, we write $\mathbb{P}^n$ for the projective space of linear forms in $n+1$ variables, with complex coefficients.

In the sequel, by abuse, we will identify a form F with the point in the projective space $\mathbb{P}(\mathrm{Sym}^d(\mathbb{C}^{n+1}))$ *associated to F, and also with the hypersurface of equation* $F = 0$.

Fix $N = -1 + \binom{n+d}{n}$ and identify $\mathbb{P}^N$ with $\mathbb{P}(\mathrm{Sym}^d(\mathbb{C}^{n+1}))$, the projective space of degree d forms in $n+1$ variables.

If we denote with R the polynomial ring $R = \mathbb{C}[x_0, \dots, x_n]$, then $\mathbb{P}^N$ is the projective space over the degree d piece R_d, while $\mathbb{P}^n$ is the projective space over R_1.

We denote with $v_d : \mathbb{P}^n \to \mathbb{P}^N$ the dth Veronese map of $\mathbb{P}^n$ which, in the previous notation, maps $L \in \mathbb{P}^n$ to L^d.

A (Waring) expression of length r of F is an equality

$$F = a_1 L_1^d + \cdots + a_r L_r^d$$

where the $L_i's$ are linear forms and the a_i's are complex coefficients. Each L_i represents a point in $\mathbb{P}^n$. The expression is *non-redundant* if the L_i^d are linearly independent in $\mathrm{Sym}^d(\mathbb{C}^{n+1})$, and all the coefficients are nonzero. Non-redundant means that one cannot find a proper subset $A' \subset A$ such that also A' is a decomposition of F.

We are aware that, since we are working over the algebraically closed field $\mathbb{C}$, we could get rid of the coefficients a_i's in a Waring expression of F. It is convenient for us to maintain the coefficients, because we will compare, below, Waring expressions in which the linear forms L_i's are fixed and the coefficients move.

Given a Waring expression of F, we call (Waring) *decomposition* of F the finite set $A = \{L_1, \ldots, L_r\} \subset \mathbb{P}^n$. The length of the decomposition A is the cardinality of the finite set A, often denoted with $\ell(A)$.

From the geometric point of view, A is a decomposition of F if and only if F belongs to the linear span $\langle v_d(A) \rangle$ of $v_d(A)$, and A is non-redundant if one cannot find a proper subset $A' \subset A$ such that also A' is a decomposition of F.

The *rank* of F is the minimal r for which F has a Waring expression of length r. The *border rank* of F is the minimal r' such that F is limit of forms of rank r'.

2.2 Catalecticant Maps

The *projective catalecticant map* associated to a form F, as suggested in [22], is defined as follows.

Denote with $R^\vee = \mathbb{C}[\partial_0, \ldots, \partial_n]$ the ring of linear operators. There is a natural contraction map $\mathcal{C} : R^\vee \times R \to R$ defined by linearity and by

$$\mathcal{C}(\partial_{i_1} \cdots \partial_{i_k}, F) \mapsto \frac{\partial^k F}{\partial_{x_{i_1}} \cdots \partial_{x_{i_k}}}.$$

Notice that $\mathcal{C}$ maps $(R^\vee)_k \times R_d$ bilinearly to R_{d-k}.

For fixed $F \in R_d$ and $k \le d$, define the *catalecticant map of order* k associated to F as

$$\mathcal{C}_F^k : (R^\vee)_k \to R_{d-k} \qquad D \mapsto \mathcal{C}(D, F).$$

Call k*-polar space of* F the subspace $Im(\mathcal{C}_F^k)$ of R_{d-k}, and call k*-polar projective space of* F its projectification $P_F^k = \mathbb{P}(Im(\mathcal{C}_F^k))$.

We will use in the sequel the correspondence between R_d and $R_d^\vee$ induced by the monomial basis.

Properties of catalecticant maps in connection with decompositions of a form F are deeply studied in [22]. We recall some basic facts.

Theorem 2.1 *If $A = \{L_1, \dots, L_r\} \subset \mathbb{P}^n$ is a decomposition of F, then the projective polar space P_F^k is contained in the linear span of $L_1^{d-k}, \dots, L_r^{d-k}$.*

Thus, the projective dimension of P_F^k is at most equal to $\mathrm{rank}(F)$.

Comparing dimensions, one obtains.

Corollary 2.2 *In the setting of Theorem 2.1, if moreover P_F^k has (projective) dimension $r-1$, then the points $L_1^{d-k}, \dots, L_r^{d-k}$ are linearly independent, and*

$$P_F^k = \langle L_1^{d-k}, \dots, L_r^{d-k} \rangle.$$

In particular, if $\dim(P_F^k) = \mathrm{rank}(F)$, *then any minimal decomposition A of F satisfies*

$$v_{d-k}(A) \subset P_F^k \cap v_{d-k}(\mathbb{P}^n).$$

We notice that it is easy to construct the matrix of C_F^k with respect to the monomial basis, and then check the dimension of the image of C_F^k.

The Apolarity Theorem (see [22] Sect. 4) determines a fundamental link between catalecticant maps and decompositions. We will use it in the form below

Theorem 2.3 *The direct sum of the kernels of the catalecticant maps C_F^k is an (artinian, homogeneous) ideal in $R^\vee$, called the* apolar ideal $F^\perp$ *of F. If the homogeneous ideal $I_{A^\vee}$ of a finite set $A^\vee$, in the projective space $(\mathbb{P}^n)^\vee$ associated to $R^\vee$, is contained in $F^\perp$, then the corresponding set $A \subset \mathbb{P}^n$ determines a decomposition of F.*

Notice that when $I_{A^\vee} \subset F^\perp$, then the k-polar space of F is contained in the span of $v_{d-k}(A)$ for all k. It is well known that the converse fails. Thus in general one cannot hope to find a decompositions of F by looking only to one catalecticant map. Yet, we will see that there are cases in which just one catalecticant map is sufficient.

Remark 2.4 Consider an element D of the kernel of C_F^k, and an element D' of the kernel of $C_F^{k'}$. We can consider D, D' as polynomials in $R^\vee$, with degrees k, k' resp. Assume that D, D' have no common factors.

By the Apolarity Theorem 2.3 the complete intersection of D, D' determines a decomposition Z of F, of length kk'.

Since $F^\perp$ is strictly bigger than the ideal generated by D, D', for $F^\perp$ is artinian, one can have decompositions Z' of F which are different, even disjoint, from Z.

2.3 Hilbert Functions

Let Z be a finite, reduced subset of length r in $\mathbb{P}^n$. Fix a set of representatives for the coordinates points of Z. The evaluation of polynomials at the chose representatives provides for any degree d a linear map $\rho_d : R_d \to \mathbb{C}^r$ whose image has dimension which depends only on Z and not on the choice of the representatives. The Hilbert function of Z is the map

$$h_Z : \mathbb{Z} \to \mathbb{Z} \qquad h_Z(d) = \dim Im(\rho_d).$$

We will often use the difference $Dh_Z(d) = h_Z(d) - h_Z(d-1)$, and the h-vector of Z, which is the t-uple of nonzero values of Dh_Z.

Clearly $h_Z(i) = Dh_Z(i) = 0$ when i is negative. Also $h_Z(0) = Dh_Z(0) = 1$. Moreover $h_Z(d)$ is non decreasing, and $h_Z(d) = r$ for all large d. Consequently Dh_Z is non-negative and $Dh_Z(d) = 0$ if d is sufficiently large.

Other standard properties of h_Z and Dh_Z are known in the literature, and are collected, e.g., in [11]. We list the most relevant of them.

(a1) for all $d \gg 0$, $\sum_{i=0}^{d} Dh_Z(i) = h_Z(d) = r$;
(a2) if $Dh_Z(i) = 0$ for some $i > 0$, then $Dh_Z(j) = 0$ for all $j \geq i$;
(a3) if $Z' \subset Z$ then for all i, $h_{Z'}(i) \leq h_Z(i)$ and $Dh_{Z'}(i) \leq Dh_Z(i)$.
(a4) if Z is contained in a plane curve of degree q, then $Dh_Z(i) \leq q$ for all i.

The application of Hilbert functions to tensor analysis is based on the trivial observation that $h_Z(1) < \min\{n+1, r\}$, i.e., the evaluation map in degree 1 is not of maximal rank, if and only if Z is linearly dependent and fails to span $\mathbb{P}^n$. This implies that when $\binom{n+d}{n} \geq r$, then $h_Z(d) < r$ if and only if $v_d(Z)$ is linearly dependent.

It follows that if A, B are two different, non-redundant decompositions of the same form $F \in R_d$, and we take Z to be the union $Z = A \cup B$, then $h_Z(d) < r$, so that

(b0) $Dh_Z(d+1) > 0$.

The next properties of the union $Z = A \cup B$, which will be used in the sequel, follow from results in the theory of finite subsets in projective space, as the Cayley–Bacharach theorem, or the Macaulay Maximal Growth principle. For an account, we refer to [11] or [2].

Proposition 2.5 *Let A, B be two different, non-redundant decomposition of a form F of degree d in $n+1$ variables. Put $Z = A \cup B$ and assume $A \cap B = \emptyset$.*

(b1) For all $j \leq d+1$ one has

$$\sum_{i=0}^{j} Dh_Z(i) \leq \sum_{i=d+1-j}^{d+1} Dh_Z(i).$$

(b2) The projective dimension of the intersection $\langle v_d(A)\rangle \cap \langle v_d(B)\rangle$ is equal to $(\sum_{i>d} Dh_Z(i)) - 1$. In particular, if $Dh_Z(d+1) = 1$ and $Dh_Z(d+2) = 0$, then the intersection contains only F.

(b3) Assume $n = 2$, i.e., F is a ternary form. If for some $0 < i < j$ one has $Dh_Z(i) > Dh_Z(j) > 0$ then $Dh_Z(j+1) < Dh_Z(j)$.

2.4 Ramification and the Terracini Locus

The study of decompositions of tensors is naturally linked to the geometric theory of secant varieties. We refer to the book [21], or to Section III of the book [8] for an introduction to secant and abstract secant varieties.

In this setting, following [6], one defines the Terracini loci of a variety X as follows.

Definition 2.6 The Terracini locus $\mathbb{T}_r(X)$ of a variety $X \subset \mathbb{P}^N$ is the closure, in the symmetric product $X^{(r)}$, of the locus of subsets A of cardinality r, contained in the regular part X_{reg}, such that the span of the tangent spaces to X at the points of A has dimension smaller than the expected value $r(n+1)-1$.

The locus $\mathbb{T}_r(X)$ is connected to the structure of the natural map (projection) from the abstract secant variety $AS^r(X)$ to $S^r(X)$,

$$s^r : AS^r(X) \to S^r(X).$$

Given a linearly independent finite set $Y \subset X_{reg}$, by Terracini's Lemma and its proof (see [31] or, in a modern setting, Sect. 5.3 of [23] and Proposition 2.4 of [14]), it follows that the span of the tangent spaces to X at the points of Y correspond to the image of the differential of s^r. Since for a general $P \in \langle Y \rangle$, $AS^r(X)$ is smooth of dimension $r(n+1)-1$ at $(P, Y) \in AS^r(X)$, then Y *belongs to* $\mathbb{T}_r(X)$ *if and only if the linear span* $\langle Y \rangle$ *is contained in the ramification locus of the map* s^r.

3 Plane Sextics

In this section we will analyze in details the situation of forms of degree 6 in three variables, i.e., sextic plane curves. We will see how the study of the catalecticant map determines several loci in the secant varieties of the surface $X = v_6(\mathbb{P}^2)$.

Remark 3.1 Sextics in three variables are parameterized by $\mathbb{P}(\mathrm{Sym}^6(\mathbb{C}^3)) = \mathbb{P}^{27}$.

By [1], the (Waring) rank of a general ternary sextic is 10.

Sextics of (border) rank 9 determine a hypersurface in the space of sextics: the 9-secant variety to the 6-Veronese of $\mathbb{P}^2$. Moreover, by [4, 13], a general form of rank 9 has exactly two minimal decompositions.

Sextics of rank 8 determine an irreducible subvariety of dimension 23 in $\mathbb{P}^{27}$, i.e., the 8-secant variety to $v_6(\mathbb{P}^2)$. By [15], a general sextic of rank 8 has a unique decomposition of length 8.

Following [27], let us look what happens to the catalecticant map of order three applied to a sextic F of rank < 10. We will focus mainly on the cases in which F has rank either 8 or 9.

We start by recalling the following, elementary fact.

Remark 3.2 Let C be a form in the kernel of $\mathcal{C}_F^k$. Then C corresponds by duality to a form in R (that we continue to denote as C), and its coefficients correspond to the orthogonal to a hyperplane in $\mathbb{P}(R_k^\vee)$ that contains P_F^k. Since the Veronese map and the duality are both given in terms of the monomial basis, then the intersection of P_F^k with $v_k(\mathbb{P}^2)$ corresponds to a divisor that defines v_k, i.e., to a curve of degree k in $\mathbb{P}^2$, which is exactly C. Thus $P_F^k \cap v_k(\mathbb{P}^2) = v_k(C)$.

Proposition 3.3 *If F has rank $r < 10$ then $\dim(P_F^3) \leq r - 1$. If $\dim(P_F^3) = 8$ and the kernel of $\mathcal{C}_F^3$ is generated by an irreducible form C, then F has border rank at most 9, and rank ≥ 9.*

Proof If we take a minimal decomposition A of F, then the ideal of A certainly contains $10 - r$ independent cubics. Thus from Theorem 2.1 we know that the kernel of $\mathcal{C}_F^3$ has dimension at least $10 - r$, so that $\dim(P_F^3) \leq r - 1$.

Now assume that the dimension of P_F^3 is exactly equal to 8, so that F cannot have rank smaller than 9. P_F^3 is then a hyperplane in $\mathbb{P}(\mathrm{Sym}^3(\mathbb{C}^3)) = \mathbb{P}^9$ which intersects $v_3(\mathbb{P}^2)$ in $v_3(C)$. Since $F^\perp$ certainly contains some form C' independent from C, and by the Apolarity Theorem $C \cap C'$ is a decomposition of F, hence F lies in the span of $v_6(C)$, which is the Veronese image of an irreducible plane cubic of arithmetic rank 1. Since irreducible curves are never defective, then F has border rank at most 9 with respect to $v_6(C)$, hence also with respect to $v_6(\mathbb{P}^2)$. □

Notice that *not every* general set of 9 points in C determines a decomposition of F. So, it is not sufficient that the span of $v_3(A)$ contains P_F^3 for A to be a decomposition of F.

Example 3.4 Let F be a sextic of rank 8. Since 8 points always lie in a pencil of cubic curves, then $\dim(P_F^3) \leq 7$. Assume that P_F^3 has dimension exactly 7 and look at the intersection of P_F^3 with the 3-Veronese surface $S = v_3(\mathbb{P}^2)$.

Since $\deg(S) = 9$, it is clear that P_F^3 intersects S either in a curve Γ, or in a scheme of length 9.

The former case cannot happen. Namely the curve Γ would be the image in v_3 of a plane curve whose ideal in degree 3 has linear dimension 1, but no such plane curves exist.

Thus $Z = P_F^3 \cap S$ is a subscheme of length 9. It is easy to see (from Sect. 4 of [27]) that for a general choice of the form F the scheme Z is reduced, hence it consists of 9 distinct points.

It follows that any decomposition of length 8 *of F is contained in Z*, for any set of length 8 sits in two independent cubics, and by the Apolarity Theorem these two cubics are in the kernel of $\mathcal{C}_F^3$.

When Z is reduced, it consists of a minimal decomposition, plus one extra point.

Proposition 3.5 *Let F be a sextic ternary form such that $\dim(P_F^3) = 7$ and $Y = P_F^3 \cap S$ is reduced. Assume that F has rank* 8*. Then there exists only one decomposition of F of length* 8*, i.e., F is identifiable, in the sense of [13].*

Proof Put $Z = v_3^{-1}(Y)$. By Example 3.4 every decomposition of length 8 of F is a subset of Z. Since Z is complete intersection of two cubics, then Z imposes independent conditions to curves of degree $\geq 3+3-2=4$. In particular, $v_6(Z)$ is linearly independent, thus every subset of $v_6(Z)$ is linearly independent.

Assume that there are two different decompositions A, A' of length 8. Since $A, A' \subset Z$ by Example 3.4, then $B = A \cap A'$ has length 7. Since $v_6(A)$ and $v_6(A')$ are linearly independent, then the spaces $\langle v_6(A)\rangle$ and $\langle v_6(A')\rangle$, both containing F, meet exactly in $\langle v_6(B)\rangle$. Then $F \in \langle v_6(B)\rangle$, which means that $\mathrm{rank}(F) \leq 7$, a contradiction. □

In conclusion, we get

Corollary 3.6 *Let F be a sextic ternary form such that $\dim(P_F^3) = 7$ and $Y = P_F^3 \cap S$ is reduced. Then $Z = v_3^{-1}(Y) \subset \mathbb{P}^2$ is a complete intersection of two cubics, and either*

- *F has rank 8 and a unique decomposition of length 8, contained in Z; or*
- *F has rank 9, and Z is a minimal decomposition of F, i.e., there exists a minimal decomposition of F which is a complete intersection of 2 cubics.*

An algorithm that can guarantee that a ternary form F of degree 6 has rank 8, and find a decomposition, is the following (see [27]).

Compute $\dim(P_F^3)$. If $\dim(P_F^3) > 7$, then F cannot have rank 8. Assume that $\dim(P_F^3) = 7$ (this will be true for general forms of rank 8).

Compute two generators D, D' of the kernel of C_F^3 and compute the coordinates of the points of the complete intersection Z of D and D' (this can require approximation). Assume that Z has length 9 (this will be true for general forms of rank 8).

Then $Z = \{L_1, \ldots, L_9\}$. If we identify each L_i with a linear form in R, then there exists a linear combination $F = a_1L_1^6 + \cdots + a_9L_9^6$ and we know that the linear combination is unique, for $v_6(Z)$ is linearly independent. Compute the a_i's.

By Proposition 3.5, F has rank 8 if and only if one coefficient a_i is 0. In this case, by dropping the summand $a_iL_i^6$, we also obtain the unique decomposition of length 8 of F.

It is known that since a general point in $\mathbb{P}^{17}$ has exactly two different decompositions with respect to an elliptic normal curve, then a general sextic of rank 9 has exactly two minimal general decomposition.

A geometric way to find the second decomposition, once one of them is known, can be described in terms of liaison.

We need first a series of preparatory results on the interaction between Hilbert functions and decompositions.

Lemma 3.7 (Intersection Lemma) *Let $F \in R_d$ be a form with two non-redun-dant decompositions A, B, of length $\ell(A) \geq \ell(B)$. Assume $A \cap B \neq \emptyset$. Then there exists a form $F' \in R_d$ with two non-redundant decompositions $A' \subset A$, $B' \subset B$ such that $A' \cap B' = \emptyset$ and $\ell(A') \geq \ell(B')$, and moreover either $\ell(A') > \ell(B')$ or $A' \neq A$.*

Proof Let $A = \{L_1, \dots, L_r\}$ and $B = \{L_1, \dots, L_j, M_{j+1}, \dots, M_{r'}\}$ with $L'_i \notin A$ for $i = j+1, \dots, r'$ ($r = \ell(A), r' = \ell(B)$). Thus $A \cap B = \{L_1, \dots, L_j\}$ and $j > 0$. Write

$$F = a_1 L_1^d + \cdots + a_j L_j^d + a_{j+1} L_{j+1}^d + \cdots + a_r L_r^d$$
$$F = b_1 L_1^d + \cdots + b_j L_j^d + b_{j+1} M_{j+1}^d + \cdots + b_{r'} M_{r'}^d.$$

Since A, B are non-redundant, then $v_d(A)$, $v_d(B)$ are both linearly independent, and the a_i's and b_i's are all nonzero. Consider the form F' obtained by subtracting from F the first j summands of the second Waring expression above

$$F' = (a_1 - b_1) L_1^d + \cdots + (a_j - b_j) L_j^d + a_{j+1} L_{j+1}^d + \cdots + a_r L_r^d$$
$$F' = \qquad b_{j+1} M_{j+1}^d + \cdots + b_{r'} M_{r'}^d.$$

Then $B' = \{M_{j+1}, \dots, M_{r'}\}$ is a decomposition of F', of length $r' - j < r'$, which is non-redundant since $b_{j+1}, \dots, b_{r'} \neq 0$ and $v_d(B')$ is linearly independent. Let $A' = A \setminus \{L_i : i \leq j \text{ and } a_i = b_i\}$. Then also A' is a decomposition of F', which is non-redundant since $v_d(A')$ is linearly independent and $a_{j+1}, \dots, a_r \neq 0$. A' and B' are clearly disjoint, and $\ell(A') \geq r - j \geq r' - j = \ell(B')$.

Finally, if $A' = A$ then $\ell(A') = r > r' - j$. □

The next result tells us where we must look, geometrically, in order to find minimal decompositions of a sextic form.

We give a long proof in which we analyze one by one the several cases, just because we want to stress the use of the Hilbert function for tensor analysis. Indeed, some steps could be shortened by using [5] or Sect. 3 of [2]. See also [26] for an approach to the problem.

Proposition 3.8 *Let F be a ternary sextic with two different non-redundant decompositions A, B, with length $\ell(A) \leq 9$ and $\ell(B) \leq \ell(A)$. Assume that no 4 points of A are aligned, and that A does not lie in a conic.*

Then $\ell(B) = \ell(A) = 9$ and $A \cap B = \emptyset$.

Furthermore $A \cup B$ lies in a cubic curve C, and it is complete intersection of C and a sextic curve.

Proof We set $r = \ell(A)$, $r' = \ell(B)$ and we make induction on r, r'. Put $Z = A \cup B$ and notice that $Dh_Z(7) > 0$, by (b0).

Since A lies in no conics by assumption, then $r \geq 6$.

Assume $r = 6$, so that $\ell(Z) \leq 12$. Since A is not aligned, then $Dh_A(1) = 2$. If $Dh_A(2) \leq 1$, then A sits in two independent conics, which, by Bezout, is possible only if A has 5 aligned points, a contradiction. Thus $Dh_A(2) \geq 2$, so that the h-vector of A is either $(1, 2, 3)$, or $(1, 2, 2, 1)$. In the former case, arguing as in cases $r = 4, 5$, we get $Dh_Z(3) = 0$, a contradiction. In the latter case, we have $\sum_{i=0}^{3} Dh_Z(i) \geq 6$, thus by (b1) also $\sum_{i=4}^{7} Dh_Z(i) \geq 6$. Since $\ell(Z) \leq 12$, we must have $\sum_{i=0}^{3} Dh_Z(i) =$

6 which, by (a3), implies $Dh_Z(i) = Dh_A(i)$ for $i = 0, \dots, 3$. Then $Dh_Z(3) = 1$, so that, by (b3), $Dh_Z(4) = 0$, which is not consistent with (b0).

Assume $r = 7, 8$. Since A lies in no conics, then $Dh_A(2) = 3$. If $Dh_Z(3) \leq 2$, then by applying (b3) for $i = 4, 5$ we see that $Dh_Z(5) = 0$, which contradicts (b0). Thus $Dh_Z(3) \geq 3$, so that $\sum_{i=0}^{3} Dh_Z(i) \geq 9$. Then by (b1) also $\sum_{i=4}^{7} Dh_Z(i) \geq 9$, which is not consistent with $\ell(Z) \leq 2r$.

So, we are left with the case $r = 9$. As above $Dh_Z(2) = 3$, and $Dh_Z(3) \leq 2$ yields a contradiction. Thus the h-vector of Z starts with $(1, 2, 3, q, \dots)$ with $q = Dh_Z(3) \geq 3$. Then $\sum_{i=0}^{3} Dh_Z(i) = 6 + q$, so that by (b1) $\sum_{i=4}^{7} Dh_Z(i) \geq 6 + q$. Since $\ell(Z) \leq 18$, it follows $q = 3$. This means that $\ell(Z) = 18$, i.e., $\ell(B) = 9$, and moreover $h_Z(3) = 9$, so Z lies in a cubic curve.

By Proposition 2.5, one computes that the unique possibility for the h-vector of Z is $(1, 2, 3, 3, 3, 3, 2, 1)$. By [3] Lemma 5.3 we know that Z has the Cayley–Bacharach property. Since the h-vector of Z is the same than the h-vector of a complete intersection of type $(3, 6)$, by [17] Z is complete intersection of a cubic and a sextic curve. Then A and B are linked by a cubic and a sextic curve. Finally, the case where $A \cap B \neq \emptyset$ implies, by Lemma 3.7, the existence of a sextic F' with non-redundant decompositions $A' \subset A$ and B', such $\ell(B') \leq \ell(A')$, and either $\ell(A') \leq 8$ or $\ell(B') \leq 8$. The existence of F' is excluded by the previous argument. □

Given a general sextic F of rank 9, and a decomposition A of F, we can find a second decomposition B of F with a procedure introduced, e.g., in [2], Sect. 4.

Remark 3.9 Let F be a sextic of rank 9, with a decomposition A of length 9.

From the kernel of $\mathcal{C}_F^3$, which has dimension 1 if F is general, we find a cubic C containing A. A resolution of the homogeneous ideal of A is given by

$$0 \to R(-5)^3 \xrightarrow{M} R(-4)^3 \oplus R(-3) \to I_A \to 0.$$

M is the Hilbert–Burch matrix of A, whose minors provide generators for I_A. The degrees of M are

$$\begin{pmatrix} 1 & 1 & 1 \\ 1 & 1 & 1 \\ 1 & 1 & 1 \\ 2 & 2 & 2 \end{pmatrix}.$$

We know that $A \cup B$ is complete intersection of C with a sextic. The homogeneous ideal of the residue B of A in the complete intersection is obtained by erasing the bottom row of M and adding a column of three quadrics. The maximal minors of the matrix M' that we obtain generate the homogeneous ideal of a scheme B linked to A in a complete intersection $(3, 6)$.

The previous procedure produces a set B which decomposes some form $F' \in v_6(A)$. In other words, the intersection $\langle v_6(A) \rangle \cap \langle v_6(B) \rangle$ is a form F' in $v_6(A)$ which depends on the choice of the three quadrics in M'.

One can find F' by taking the sum $I_A + I_B$, which determines in R_6 a linear subspace H of codimension 1. The coefficients of F' form a vector orthogonal to H.

Thus, in order to find the second decomposition of the given F, one needs to choose the three quadrics Q_1, Q_2, Q_3 in M' so that the orthogonal to H matches with the coefficient of F. In practice, one needs to solve a linear system whose matrix has entries linear in the coefficients of the Q_i's. A convenient way to write this linear system into M2 [19] is to ask that the minors of M' are apolar to F.

With a procedure similar to the proof of Proposition 3.8, we can prove the following property of sextics of rank 9 with a decomposition A complete intersection of two cubics.

Proposition 3.10 *Let F be a sextic ternary form such that* $\dim(P^3_F) = 7$ *and* $Y = P^3_F \cap S$ *is reduced. Assume that F has rank 9. Then* $A = v_3^{-1}(Y)$ *is the unique decomposition of F of length 9, i.e., F is identifiable, in the sense of [13].*

Proof From Proposition 3.6 we know that A is a complete intersection of two cubics. In particular, no 4 points of A are aligned and A does not lie in a conic. The h-vector of A is $(1, 2, 3, 2, 1)$.

Assume that there exists another decomposition B of length 9. Then by Proposition 3.8 we know that $A \cap B = \emptyset$ and $Z = A \cup B$ lies in a cubic, so that the h-vector of Z starts with $(1, 2, 3, 3, \dots)$. Since Z lies in a plane curve of degree 3, by (a4) $Dh_Z(i) \leq 3$ for all i.

Next, we claim that $Dh_Z(4) = 3$. Indeed if $Dh_Z(4) \leq 2$ then by (b3) $Dh_Z(5) \leq 1$ and $Dh_Z(6) = 0$, which contradicts (b0). If $Dh_Z(4) \geq 4$, then $\sum_{i=5}^{7} Dh_Z(i) < 6$, which contradicts (b1). The same argument proves that $Dh_Z(5) = 3$, and $Dh_Z(6) \leq 2$. Then $Dh_Z(6) + Dh_Z(7) = 3$. Since by (b3) $Dh_Z(6) > Dh_Z(7) \geq 1$, then the h-vector of Z is $(1, 2, 3, 3, 3, 3, 2, 1)$.

By [3] Lemma 5.3 we know that Z has the Cayley–Bacharach property $CB(6)$. Since moreover the h-vector of Z is the same than the h-vector of a complete intersection of type $(3, 6)$, by [17] Z is complete intersection of a cubic and a sextic curve. Then A and B are linked by a cubic and a sextic curve. This implies, by [25] that the h-vector of B is $(1, 2, 3, 2, 1)$. Thus B lies in the base locus of a pencil of cubic curves. Since B is a decomposition of F, and $\dim(P^3_F) = 7$, there is only one pencil of cubic curves that contain B, and the base locus of the pencil is A. Comparing the degrees, we get $A = B$. □

From a certain point of view, Proposition 3.10 is rather surprising. Namely, as explained in Theorem 3.2 of [4], the 6-Veronese surface $v_6(\mathbb{P}^2)$ is 9-weakly defective, in the sense of [12]: a general hyperplane which is tangent to $v_6(\mathbb{P}^2)$ at 9 points is tangent along an elliptic normal curve C. Thus, by Theorem 2.9 and Proposition 5.2 of [13], F has two minimal decompositions of length 9, both lying in C. Notice that, in general, none of the two decompositions lies in two independent cubic curves.

If the form F has a decomposition Z of length 9 which is a complete intersections of two cubics, then there are infinitely many elliptic normal curves in $v_6(\mathbb{P}^2)$ containing $v_6(Z)$, so one may expect the existence of infinitely many decompositions of length 9 of F, for any elliptic normal curves $v_6(Z)$ could provide a second decomposition $Z' \neq Z$ of F. On the contrary, in this case the construction collapses and any elliptic normal curves provides only one decomposition for F, namely A alone.

A motivation for the peculiar behavior of sextics with a decomposition complete intersection of two cubics can be explained in terms of the ramification of the map s^9 from the abstract secant variety $AS^9(v_6(\mathbb{P}^2))$ to $S^9(v_6(\mathbb{P}^2))$, and the Terracini locus of $v_6(\mathbb{P}^2)$.

Remark 3.11 For $X = v_6(\mathbb{P}^2)$, by Definition 2.6 a linearly independent set $v_6(A)$ of length 9 belongs to $\mathbb{T}_9(X)$ if the 9 tangent planes at the points of $v_6(A)$ span a subspace of dimension ≤ 25. When A is complete intersection of two cubics Γ_1, Γ_2, then certainly $v_6(A)$ belongs to the Terracini locus. Indeed, there is a 2-dimensional family of sextic curves in $\mathbb{P}^2$ which are singular at the points of A: the sextics formed by the union of two cubics in the pencil generated by Γ_1, Γ_2. Any sextic like that corresponds to the intersection of X with a hyperplane which contains the tangent planes to X at the points of $v_6(A)$. Thus, the span of the tangent planes at the points of $v_6(A)$ has codimension at least 2 in $\mathbb{P}^N = \mathbb{P}^{27}$.

By Remark 3.1, the map $s^9 : AS^9(X) \to \mathbb{P}^{27}$ is generically 2 : 1, when A is a general set of 9 points in the plane. By Proposition 3.10 we see that general points $(F, v_6(A))$ such that A is complete intersection of two cubics are in the ramification locus of s^9.

The last observation is consistent with the fact that the differential of s^9 degenerates at points $(F, v_6(A))$ such that A is complete intersection of two cubics.

Definition 3.12 We denote with $\mathcal{R}$ the closure of the set of sextics with a decomposition formed by a set A of nine points which are nodes of an irreducible sextic. By [20], $\mathcal{R}$ is irreducible.

We denote by W the closure of the set of sextics with a decomposition formed by a set A of nine points, complete intersection of two cubics.

Remark 3.13 The previous discussion proves that, in the new notation, W properly contains $S^8(X)$, and both W and $\mathcal{R}$ are contained in the Terracini locus $\mathbb{T}_9(X)$, which is the closure of the image of the locus of smooth points in $AS^9(X)$ in which the differential of the map s^9 drops rank.

Indeed, we will show in the next section that $\mathcal{R}$ is the unique component of $\mathbb{T}_9(X)$ which intersect the smooth locus of $S^9(X)$, while W, whose codimension is bigger than 1, sits into the singular locus of $S^9(X)$.

We will also prove in the next section that W is not contained in $\mathcal{R}$, which means that the Terracini locus has at least two components.

Remark 3.14 For the 6-Veronese variety X of $\mathbb{P}^2$, the Terracini locus $\mathbb{T}_9(X)$ corresponds to sets A of 9 points which are singular in a pencil of sextics. General sets in $\mathbb{T}_9(X)$ are easy to describe (see Example 5.1 in [6]): just take C to be a reduced elliptic plane sextic whose singularities are nodes. C has a set A of 9 nodes which also sit in a cubic curve D. All sextics in the pencil generated by C and D^2 are singular at A. It follows that the 9 tangent planes to X at the points of $v_6(A)$ span a linear space of dimension 25 in $\mathbb{P}^{27}$, the linear space spanned by X.

Notice that a $\mathbb{T}_9(X)$ does not contain a general set of 9 points in the plane, and a general $A \in \mathbb{T}_9(X)$ is not complete intersection of two cubics.

General forms contained in the variety $\mathcal{R}$ have only one decomposition with 9 summands. On the other hand, both $\mathcal{R}$ and W contain subvarieties whose elements have many decompositions of length 9.

Namely, if A lies in the Terracini locus $\mathbb{T}_9(X)$, then the liaison procedure introduced in Remark 3.9 determine a second decomposition for some forms in $\langle v_6(A)\rangle$.

Lemma 3.15 *Let C be a general (smooth) cubic plane curve and let A be a reduced divisor of degree 9 on C such that $2A \in |\mathcal{O}_C(6)|$, but $A \notin |\mathcal{O}_C(3)|$. Then there exists a sextic plane curve G which is singular at the points of A and irreducible.*

Proof Call $P_1, \dots, P_9$ the points of A. For each i call ϵ_i the double structure on P_i contained in C. For each P_i choose a scheme μ_i of length 2 in $\mathbb{P}^2$, supported at P_i such that $\mu_i \notin C$. By assumption there exists a sextic G' such that G' does not contain C and $\epsilon_i \in G'$ for all i. Thus, the linear system $\mathcal{M}$ of sextics in $\mathbb{P}^2$ which contain $\{\epsilon_1, \dots, \epsilon_9\}$ has (projective) dimension 10. Note that $\mathcal{M}$ contains the double curve $2C$. Since $\epsilon_i \cup \mu_i$ has length 3, then every μ_i imposes at most one condition to curves in $\mathcal{M}$. Then there exists a sextic G'' which contains $\epsilon_i \cup \mu_i$ for all i and G'' is different from $2C$. Since G'' contains $\epsilon_i \cup \mu_i$, then G'' is singular at the points of A. G'' cannot contain C, because $A \notin |\mathcal{O}_C(3)|$. Thus a general curve G in the pencil generated by $2C$ and G'' is irreducible, by Bertini. □

Proposition 3.16 *Let A be a set of 9 points in $\mathbb{P}^2$, which are nodes of a general reduced elliptic sextic curve G. The generality of G implies that there exists a unique cubic curve C containing A, and moreover C is smooth. Let B be the residue of A in the intersection of C with a general sextic S that passes through A. Then also B is the set of nodes of a reduced sextic G'.*

Proof The first claim is classical. Just to see an argument, observe that for a general set D of 8 points in a general cubic curve C, the divisor $2D$ on C lies in some divisor D' of the linear series $|\mathcal{O}_C(6)|$. $D' - D$ consists of 2 points, that determine a linear series $\mathcal{L} = g^1_2$. Choose a Weierstrass point P_9 of $\mathcal{L}$. Since $\mathcal{L}$ cannot have 8 Weierstrass points, then by monodromy for a general choice of D we can assume that $P_9 \notin D$. Then $2D + 2P_9 \in |\mathcal{O}_C(6)|$ and then $D \cup \{P_9\}$ is contained in the singular locus of a sextic curve, by Lemma 3.15. Since the variety that parametrizes sextic plane curves with 9 singular points is irreducible, we see that on a general cubic C we can find a set of 9 nodes of an irreducible sextic.

For the second claim, observe that on C the divisor $2A$ belongs to $|\mathcal{O}_C(6)|$, and also the divisor $A + B$ belongs to $|\mathcal{O}_C(6)|$. Thus $2A + 2B$ belongs to $|\mathcal{O}_C(12)|$. Then also $2B \in |\mathcal{O}_C(6)|$, and the claim follows from the generality of G, C, S and from Lemma 3.15. □

In other words, the previous proposition says that some forms F of rank 9 with a decomposition A which lies in the Terracini locus $T_9(X)$, have a second decomposition B of length 9 which also sits in the Terracini locus $T_9(X)$. Thus the subscheme of the abstract 9-secant variety which maps to F is supported at two points, but it has length 4.

The closure $\mathcal{R}'$ of the locus of forms F as above provides a proper subvariety of $\mathcal{R}$, whose geometry has non-trivial aspects.

Proposition 3.17 *A general sextic in $\mathcal{R}'$ has infinitely many decompositions of length 9. Thus $\mathcal{R}'$ is the locus in $\mathcal{R}$ in which the dimension of fibers of the map $s^9 : AS^9(X) \to S^9(X)$ jumps.*

Proof Fix a general cubic curve C and a general set $A \subset C$ of length 9, such that there is an irreducible sextic S singular at the points of A. For a general sextic S' through A the intersection $S' \cap C$ is a set $A \cup B$ of 18 points, with B disjoint from A. As explained in Proposition 3.8, the linear spans $\langle v_6(A) \rangle$ and $\langle v_6(B) \rangle$ meet in a point $F \in \mathbb{P}^{27}$ which represents a point in $\mathcal{R}'$. Since any element of $\mathcal{R}' \cap \langle v_6(A) \rangle$ arises in this way, we get a surjective map form the projective space $\mathbb{P}$ over $(I_A)_6$ mod multiples of C, to $\mathcal{R}' \cap \langle v_6(A) \rangle$. Since A is separated by cubics, then $\mathbb{P}$ has dimension $27 - 19 = 8$. Since $\langle v_6(A) \rangle$ also has dimension 8, and $\mathcal{R}' \cap \langle v_6(A) \rangle$ is a proper subvariety, because $\mathcal{R} \neq \mathcal{R}'$, the claim follows. □

The same procedure of liaison produces a special subvariety of W.

Example 3.18 Let A be a set of 9 points, complete intersection of two cubics C_1, C_2 Fix a general cubic curve C_3 and consider the set $B = C_1 \cap C_3$. Then $A \cap B = \emptyset$, and $Z = A \cup B$ is a complete intersection of a cubic and a sextic curve. Thus the h-vector of Z is $(1, 2, 3, 3, 3, 3, 2, 1)$, By Proposition 2.19 of [2], it follows that the spans $\langle v_6(A) \rangle$ and $\langle v_6(B) \rangle$ meet in exactly one point, corresponding to a sextic form F.

Since F has a decomposition given by $C_1 \cap C_2$, then F corresponds to a point in the 9-secant variety of X spanned by a set in the Terracini locus. Moreover, from the Apolarity Theorem, C_1, C_2 lie in the apolar ideal of F. For the same reason, looking at the decomposition $C_1 \cap C_3$ of F, one obtains that also C_3 lies in $F^\perp$. Thus we have an example of a form F whose apolar ideal contains 3 independent cubics C_1, C_2, C_3.

Notice that, by the generality of C_3, the intersection $C_1 \cap C_2 \cap C_3$ is empty. Thus the image of C^3_F, which is a $\mathbb{P}^6$, will not cut the 3-Veronese surface in $\mathbb{P}^9$. In particular, F has not rank 7.

By the Apolarity Theorem, a decomposition of F can be found by taking any general pair of cubics in the linear system spanned by C_1, C_2, C_3. Thus F has a 2-dimensional family of decompositions.

Definition 3.19 We denote with W' the closure of the locus of sextic forms F such that the catalecticant map C^3_F has a kernel of dimension 3.

A general choice of three cubics C_1, C_2, C_3 determines $F \in W'$, by taking the intersection of the spans of $v_6(C_1 \cap C_2)$ and $v_6(C_1 \cap C_3)$ (any general choice of two pairs will be suitable and determine the same F). Thus W' has the dimension of the Grassmannian of planes in $\mathbb{P}^9$, i.e., 21.

We stress that there is no way to determine if F belongs to W' only by looking at the 9 projective points of a decomposition of F. Once one knows that F has a decomposition A which is complete intersection, and one fixes representatives $\{L_1, \dots, L_9\}$ for the points of A, then the membership of F in W' depends on the coefficients of a linear combination of powers $L_1^6, \dots, L_9^6$ that determine F in $\langle v_6(A) \rangle$.

Observe that the locus $S^7(X)$ lies in the intersection of $S^8(X)$ with W'. We will see in the next section that, at least set-theoretically, $S^7(X) = S^8(X) \cap W'$.

Example 3.20 Many forms F in the subvariety W' defined above can be computed following the procedure introduced in Sect. 4 of [2], as explained in Remark 3.9.

Just to see an example of such a form, consider

$$F = (x_0 x_1 x_2)^2.$$

One easily sees that $F^\perp$ contains three independent cubics, which are x_0^3, x_1^3, x_2^3.

We have the explicit decomposition of rank 9 [9]

$$810(x_0 x_1 x_2)^2 = \sum_{p,q=0}^{2} e^{2\pi i (p+q)/3} \left(x_0 + e^{2\pi i p/3} x_1 + e^{2\pi i q/3} x_2 \right)^6.$$

In the next section we will compute equations for the 8th secant variety $S^8(X)$. By applying these equations to F, one realizes that F does not belong to $S^8(X)$. Thus F has rank (and border rank) 9. This proves that W' is not contained in $S^8(X)$.

The relations among the subvarieties $\mathcal{R}$, $\mathcal{R}'$, W, W' of $S^9(X)$ reflect the rich geometry of secant varieties, as soon as the genus approaches the generic value.

The most complicate object to describe remains $\mathcal{R}'$, for which we do not have a set of equations. We hope that the geometric structure of $\mathcal{R}'$ will be clarified, in a future footnote, maybe.

4 Equations for Loci in S^9

We introduce a Young flattening (see [24]) to get a more refined study of plane sextics. Let T be the tangent bundle of $\mathbb{P}^2$, we consider the rank three bundle $E = \mathrm{Sym}^2 T$. The space of sections $H^0(\mathrm{Sym}^2 T)$ is 27-dimensional and can be identified with the $SL(3)$-module $\Gamma^{4,2}\mathbb{C}^3$, corresponding to the Young diagram

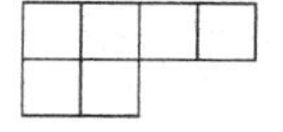

A presentation of $\mathrm{Sym}^2 T$ can be described as follows. First we recall that T is presented by the following exact sequence

$$0 \to \mathcal{O} \to \mathbb{C}^3 \otimes \mathcal{O}(1) \to T \to 0$$

which dualizes to

$$0 \to T^\vee \to \mathbb{C}^3 \otimes \mathcal{O}(-1) \to \mathcal{O} \to 0$$

and since $T^\vee = T(-3)$ we get

$$0 \to T \to \mathbb{C}^3 \otimes \mathcal{O}(2) \to \mathcal{O}(3) \to 0$$

The first and third sequence fit together into the presentation of T

$$\begin{array}{ccc} \mathbb{C}^3 \otimes \mathcal{O}(1) & \xrightarrow{f} & \mathbb{C}^3 \otimes \mathcal{O}(2) \\ & \searrow \quad \nearrow & \\ & T & \end{array}$$

where the horizontal skew-symmetric map $f(v) = v \wedge x$ is given by the matrix

$$\begin{pmatrix} 0 & x_2 & -x_1 \\ -x_2 & 0 & x_0 \\ x_1 & -x_0 & 0 \end{pmatrix}$$

The second symmetric power $\mathrm{Sym}^2 T = E$ appears in the exact sequence

$$0 \to \mathbb{C}^3 \otimes \mathcal{O}(1) \to \mathrm{Sym}^2\mathbb{C}^3 \otimes \mathcal{O}(2) \to \mathrm{Sym}^2 T \to 0$$

and repeating the above argument, thanks to the identification $\mathrm{Sym}^2 T^\vee = \mathrm{Sym}^2 T(-6)$ we get the presentation

$$\begin{array}{ccc} \mathrm{Sym}^2\mathbb{C}^3 \otimes \mathcal{O}(2) & \xrightarrow{\mathrm{Sym}^2 f} & \mathrm{Sym}^2\mathbb{C}^3 \otimes \mathcal{O}(4) \\ & \searrow \quad \nearrow & \\ & \mathrm{Sym}^2 T & \end{array}$$

where the horizontal symmetric map $\mathrm{Sym}^2 f(v^2) = (v \wedge x)^2$ is given by the matrix

$$B = \begin{pmatrix} 0 & 0 & 0 & x_2^2 & -2x_1x_2 & x_1^2 \\ 0 & -2x_2^2 & 2x_1x_2 & 0 & 2x_0x_2 & -2x_0x_1 \\ 0 & 2x_1x_2 & -2x_1^2 & -2x_0x_2 & 2x_0x_1 & 0 \\ x_2^2 & 0 & -2x_0x_2 & 0 & 0 & x_0^2 \\ -2x_1x_2 & 2x_0x_2 & 2x_0x_1 & 0 & -2x_0^2 & 0 \\ x_1^2 & -2x_0x_1 & 0 & x_0^2 & 0 & 0 \end{pmatrix} \tag{1}$$

Note that for $L = \mathcal{O}(6)$ we have $E = E^\vee \otimes L$ hence, as in [24] we have a contraction map

$$\begin{array}{ccc} End(\mathrm{Sym}^2\mathbb{C}^3) & \xrightarrow{P_f} & End(\mathrm{Sym}^2\mathbb{C}^3) \\ \downarrow & & \uparrow \\ H^0(E) & \xrightarrow{A_f} & H^0(E^\vee \otimes L)^\vee \end{array} \tag{2}$$

where the horizontal map P_f for $f = v^6 \in v_6(\mathbb{P}^2)$ is defined as

$$P_{v^6}(M^2)(w^2) = (M(v) \wedge v \wedge w)^2 v^2 \quad \forall M \in End(\mathbb{C}^3),\ w^2 \in \mathrm{Sym}^2\mathbb{C}^3$$

where $M^2 \in End(\mathrm{Sym}^2\mathbb{C}^3)$ is defined by $M^2(v^2) = (M(v))^2$ and then extended by linearity to any $N \in End(\mathrm{Sym}^2\mathbb{C}^3)$ and to any plane sextic f. Comparing with [28, Sect. 2] we see that the invariant det A_f of degree 27 of plane sextics has a construction similar to the Aronhold invariant of degree 4 of plane cubics.

The coordinate description of P_f is the following. Differentiate the 6×6 catalecticant $C(f)$ (given by differentiating on rows and columns by monomials of degree 2) by B in (1) . The output is a 36×36 matrix obtained by replacing any entry of B by a 6×6 catalecticant block.

The M2 [19] commands to get P_f are the following, after the matrix B has been defined as in (1)

```
R=QQ[x_0..x_2]
f=x_0^3+x_1^3+x_2^3+5*x_0*x_1*x_2---any cubic polynomial
P=diff(transpose basis(2,R),diff(basis(2,R),diff(B,f)))
```

The map A_f is symmetric and could be obtained by a convenient submatrix of P_f. An alternative way to get a coordinate description is to employ the M2 package "PieriMaps" by Steven Sam with the M2 commands [19]

```
loadPackage "PieriMaps"
pieri({6,4,2},{1,1,2,2,3,3},QQ[x_0..x_2]);
```

although the symmetry is not transparent from the coordinates chosen by the system.

The contraction A_f in (2) can be pictorially described (compare with [28]) by

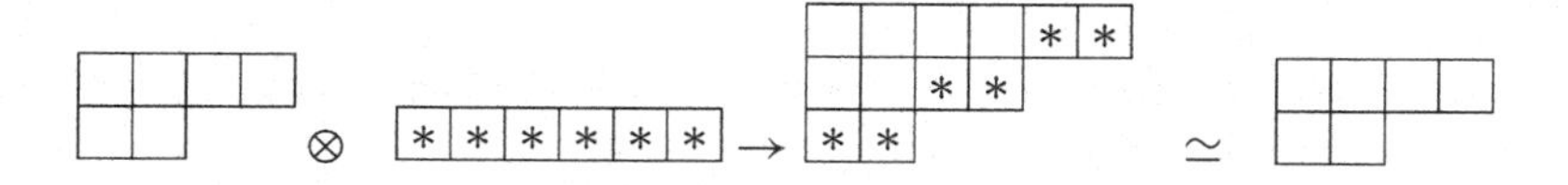

We have the decomposition

$$End(\mathrm{Sym}^2\mathbb{C}^3) = \Gamma^{4,2}\mathbb{C}^3 \oplus \Gamma^{2,1}V \oplus \wedge^3\mathbb{C}^3$$

corresponding to

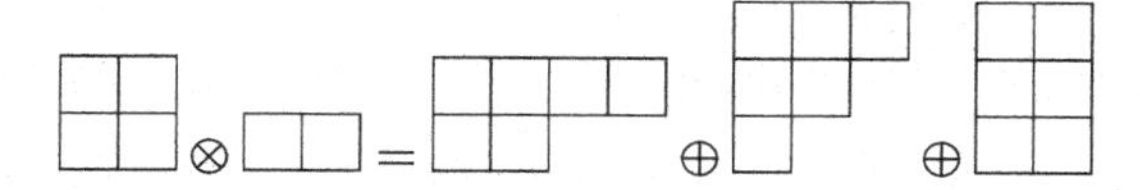

Note that the last two summands correspond to $End(\mathbb{C}^3)$.

If $f \in v_6(\mathbb{P}^2)$ then $\mathrm{rk}(A_f) = \mathrm{rk}(E) = 3$, so that if $f \in S^8(v_6(\mathbb{P}^2))$ then $rk(A_f) \leq 3 \cdot 8 = 24$ and in particular $\det A_f = 0$. This was observed already in [24, Theorem 4.2.9].

Proposition 4.1 *(1) The variety W of sextics in $S^9(v_6(\mathbb{P}^2))$ such that their Waring decomposition comes from a complete intersection of two cubics is cut by the 9-minors of C^3. It has codimension* 3 *and degree* 165 *and it coincides with the singular locus of $S^9(v_6(\mathbb{P}^2))$.*

(2) The secant variety $S^8(v_6(\mathbb{P}^2))$ coincides with the reduced structure on $W \cap V(\det A_f)$, it has codimension 4 *and degree* $1485 = 165 \cdot 9$.

Proof First recall that the variety of symmetric 10×10 matrices of rank ≤ 8 is irreducible of codimension 3 and degree 165 by Segre formula [30]. Since the variety C^3_9 given by the 9-minors of C^3 has again codimension 3, being a linear section of the previous one, it has the same degree 165. This variety C^3_9 corresponds to the row with $s = 8$ of Table 3.1 in [22], indeed with the notations of [22], it is the union of irreducible strata $Gor(T)$ for several Hilbert functions T, and one may check using Conca–Valla formula ([16] or [22, Theorem 4.26]) that only one strata has the maximum dimension. It follows that C^3_9 is irreducible. We know that W is irreducible and contained in the variety given by the 9-minors of C^3, since the two varieties have the same dimension, the equality follows, proving (i). The claim about the singular locus follows because by direct computation on a random linear subspace the singular locus of $S^9(v_6(\mathbb{P}^2))$, which obviously contains W, has codimension 3 and degree 165.

In order to prove (ii), recall that if $f \in S^8(v_6(\mathbb{P}^2))$ then $\det(A_f) = 0$. Pick 9 general points p_i for $i = 0, \ldots, 8$ and consider the linear span $f = \sum_{i=0}^{8} k_i p_i^6$ with coefficients k_i, in other words we restrict f to a general 9-secant space. The expression $\det A_f$ is symmetric in k_i, has total degree 27 and vanishes when $\prod_{i=0}^{8} k_i$ vanishes. An explicit computation with the coordinate expression found by the package "PieriMaps" as described above shows that $\det(A_f)$ coincides, up to scalar multiples, with $\left(\prod_{i=0}^{8} k_i\right)^3$. We get that the condition that $\det(A_f)$ vanishes on a general 9-secant space is equivalent to f being of rank 8. This expression is unchanged if $f \in W$, namely if f belongs to a 9-secant space corresponding to a complete intersection of two general cubics. This computation shows that $V(\det A_f)$ meets W with multiplicity 3 at a general point of the intersection, so that the reduced structure on $W \cap V(\det A_f)$ has degree $1485 = 165 \cdot 9$. Note the rank of A_f drops by 3 on W. The degree of $S^8(v_6(\mathbb{P}^2))$ can be computed with Numerical Algebraic Geometry. The M2 package *NumericalImplicitization* by Cho and Kileel shows indeed that its degree is 1485, which can be certified by the trace test. Since $S^8(v_6(\mathbb{P}^2)) \subseteq \left(W \cap V(\det A_f)\right)_{red}$, this proves (ii) □

Remark 4.2 (ii) of Proposition 4.1 strengthens [24, Theorem 4.2.9] where the 25-minors of A_f were considered.

We wish now to find the ramification locus of the $2:1$ map $s^9 : AS^9(v_6(\mathbb{P}^2)) \to S^9(v_6(\mathbb{P}^2))$. Let $p_1, \ldots, p_{10}$ be points in $\mathbb{P}^2$. Let $C(p_1, \ldots, p_{10})$ be a multihomogeneous polynomial of degree 3 in the coordinates of the points $p_1, \ldots, p_{10}$, skew-symmetric with respect to them and that vanishes if and only if $p_1, \ldots, p_{10}$ are on a cubic. A coordinate expression is given by the determinant of the 10×10 matrix such that its ith row is the evaluation on p_i of the monomial basis of $\mathrm{Sym}^3\mathbb{C}^3$.

We construct now the Terracini matrix of the points $p_1, \ldots, p_9$ which defines the tangent space at $S^9(v_6(\mathbb{P}^2))$ at any point in $\langle p_1^6, \ldots p_9^6\rangle$. Consider the 3×28 Jacobian matrix J of the monomial basis of $\mathrm{Sym}^6\mathbb{C}^3$. Let $J(p_i)$ be the evaluation of J at p_i and let $T(p_1, \ldots, p_9)$ be the 27×28 Terracini matrix obtained by stacking $J(p_1), \ldots, J(p_9)$. Let $R(p_1, \ldots, p_9; p_{10}) = \det \begin{pmatrix} & T(p_1, \ldots, p_9) & \\ \hline (p_{10})_0^6 & \cdots & (p_{10})_2^6 \end{pmatrix}$ be the determinant of the 28×28 matrix obtained by stacking $T(p_1, \ldots, p_9)$ and the monomial basis of $\mathrm{Sym}^6\mathbb{C}^3$ evaluated at p_{10}.

$R(p_1, \ldots, p_9; p_{10})$ is a multihomogeneous polynomial of degree 15 in the coordinates of the points $p_1, \ldots, p_9$, of degree 6 in the coordinates of p_{10}, skew-symmetric with respect to $p_1 \ldots, p_9$ and that vanishes if and only if p_{10} lies on a sextic singular at $p_1, \ldots, p_9$.

Since $C(p_1, \ldots, p_{10})^2$ is a (non-reduced) sextic singular at $p_1, \ldots, p_{10}$ we have the factorization

$$R(p_1, \ldots, p_9; p_{10}) = C(p_1, \ldots, p_{10})^2 N(p_1, \ldots, p_9)$$

where $N(p_1, \ldots, p_9)$ is a multihomogeneous polynomial of degree 9 in the coordinates of the points $p_1, \ldots, p_9$, symmetric with respect to $p_1 \ldots, p_9$ and that vanishes if and only if there is a reduced sextic singular at $p_1, \ldots, p_9$. Note that the result in [4] that the unique sextic singular at $p_1, \ldots, p_9$ is the non-reduced curve $C(p_1, \ldots, p_9, p)^2$ in p is equivalent to the fact that T has maximal rank for a general choice of $p_1, \ldots, p_9$, which can be checked by a random choice. This proves that $N(p_1, \ldots, p_9)$ is a nonzero polynomial, which defines the codimension 1 condition that the space of sextics singular at $p_1, \ldots, p_9$ has dimension ≥ 2. It can be checked that at general $p_1, \ldots, p_9$ on the hypersurface $N(p_1, \ldots, p_9) = 0$ then T has corank 1, while if $p_1, \ldots, p_9$ are distinct points defined by a general complete intersection then both $R(p_1, \ldots, p_9; p_{10})$ and $C(p_1, \ldots, p_{10})$ vanish $\forall p_{10}$ and $N(p_1, \ldots, p_9) \neq 0$.

We note also that the same argument given in the proof of [29, Theorem 45] shows that $N(p_1, \ldots, p_9)$ is irreducible, since there are no nonzero symmetric cubic invariants of 9 points in $\mathbb{P}^2$.

Remark 4.3 For general $p_1, \ldots, p_8$, the locus of ninth point p_9 such that there is a reduced sextic singular at $p_1, \ldots, p_8, p_9$ has two irreducible components, namely the nonic $N(p_1, \ldots, p_9) = 0$ and the point p_9 in the base locus of the linear system of cubics through $p_1, \ldots, p_8$. The general sextic arising from the first (resp. second) component is irreducible (resp. reducible) and lies in $\mathcal{R}$ (resp. in W), see Definition 3.12. More precisely, $\mathcal{R} \subset S^9(v_6(\mathbb{P}^2))$ is the closure of $\cup\langle p_1^6, \ldots, p_9^6\rangle$, where the union is taken for distinct $p_1, \ldots, p_9$ satisfying $N(p_1, \ldots, p_9) = 0$. As noted in

Remark 3.13, $\mathcal{R}$ and W are two irreducible components of the ramification locus of the $2:1$ map $AS^9(v_6(\mathbb{P}^2)) \to S^9(v_6(\mathbb{P}^2))$. They are distinct since H_{27} vanishes on $\mathcal{R}$ and not on W. When $N(p_1,\ldots,p_9)=0$ then $Z=\{p_1,\ldots,p_9\}$ is self-linked on the cubic $C(p_1,\ldots,p_9,p)$ with respect to any irreducible sextic singular at $p_1,\ldots,p_9$, which meets the cubic in $2p_1+\ldots+2p_9$.

Theorem 4.4 *Let $f=\sum_{i=1}^{9} p_i^6$ be a rank 9 sextic, in affine notation. Then*

$$\det A_f = \lambda N(p_1,\ldots p_9)^2$$

for a nonzero scalar $\lambda \in \mathbb{C}^*$.

Proof Recall we denoted $E=\mathrm{Sym}^2 T$ and denote $Z=\{p_1,\ldots,p_9\}$. Assume that $\det A_f$ vanishes, then $\ker A_f$ is a nonzero subspace of $H^0(E)$. Consider the restriction $H^0(E)\xrightarrow{j} H^0(E_{|Z})$, note both sides are 27-dimensional. Assume j is injective. Then by [24, Lemma 5.4.1] we have that $H^0(I_Z\otimes E)=\ker A_f$ is nonzero, which is a contradiction since a section vanishing on Z is in the kernel of j. It follows that j is not injective and there is a section $s\in H^0(E)$ vanishing at $Z=\{p_1,\ldots,p_9\}$. Again by [24, Lemma 5.4.1] the section s belongs to $\ker A_f$. Then the contraction

$$\mathrm{Sym}^2 H^0(I_Z\otimes E)\to H^0(I_Z\otimes E)\otimes H^0(I_Z\otimes E^\vee\otimes\mathcal{O}(6))\to H^0(I_{Z^2}\otimes\mathcal{O}(6))$$

as in [24, Theorem 5.4.3] takes s^2 to a sextic singular at Z. There are two cases, Z is a complete intersection or $N(p_1,\ldots,p_9)=0$. The first case has codimension 3 by Proposition 4.1 and can be excluded in proving a polynomial equality, like in our statement. It follows that $\det A_f=0$ implies generically $N(p_1,\ldots,p_9)=0$, hence a power of $N(p_1,\ldots p_9)$ is divided by $\det A_f$. The multihomogeneous polynomial $\det A_f$ is symmetric in $p_1,\ldots,p_9$ and has total degree $27\cdot 6$, hence must have degree 18 in each p_i. The result follows since $N(p_1,\ldots,p_9)$ is irreducible. □

Corollary 4.5 *The variety $\mathcal{R}$ is the complete intersection of the two hypersurfaces* $\det A_f$ *and* $S^9(v_6(\mathbb{P}^2))$ *and has degree* $10\cdot 27=270$.

Tables About the Loci we Have Studied

In the next table we list some informations about the loci we have studied in $\mathbb{P}^{27}=\mathbb{P}(\mathrm{Sym}^6\mathbb{C}^3)$. The main theme is that $(\mathcal{C})_3$ and the invariant $H_{27}=\det A_f$ are enough to describe all k-secant varieties to $v_6(\mathbb{P}^2)$.

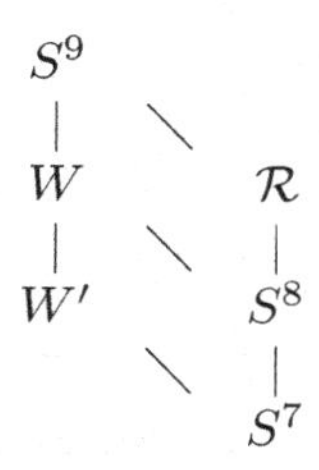

Fig. 1 Containment of loci in $\mathbb{P}^{27}$

	dim	*deg*	equations	
S^9	26	10	$\det C^3$	9-secant
$\mathcal{R}$	25	270	$\det C^3$, H_{27}	comp. of ramif. locus of $AS^9 \to S^9$
W	24	165	$(C^3)_9$	Sing S^9, span of compl. inters. 9-ples
S^8	23	$1485 = (165 \cdot 27)/3$	$(C^3)_9$, H_{27}	8-secant
W'	21	2640	$(C^3)_8$	sextics with three apolar cubics
S^7	20	$11880 = (2640 \cdot 27)/6$	$(C^3)_8$, H_{27}	7-secant

The fact that Sing $S^9 = W$ has been checked computationally, since the singular locus of S^9 has codimension 3 and degree 165, and contains the variety W given by the 9-minors of C^3, hence the equality follows. The following Fig. 1 (resp. 2) show the containments among the several (resp. dual) loci.

The right column is obtained by the left column after cut with the invariant hypersurface H_{27}. The three diagonal links correspond to cut with the invariant H_{27}. The matrix A_f drops rank by 1 in the first link, by 3 in the second one, and by 6 in the third one. Hence a check on the degrees shows that the intersection with H_{27} does not contain other components in all the three cases.

The following table lists the properties related to Waring decomposition of general member of the loci (brk is the border rank)

	rk C^3	brk	rk	#{ minimal Waring decompositions}
S^9	9	9	9	2
$\mathcal{R}$	9	9	9	1
W	8	9	9	1 (Proposition 3.10)
S^8	8	8	8	1
W'	7	9	9	∞
S^7	7	7	7	1

$$
\begin{array}{ccc}
(S^7)^\vee & & \\
| & & \\
(S^8)^\vee & & (W')^\vee \\
| & \diagdown & | \\
\mathcal{R}^\vee & & W^\vee \\
& \diagdown & | \\
& & (S^9)^\vee
\end{array}
$$

Fig. 2 Containment of dual loci in $\mathbb{P}^{27}$

Tables About the Dual Loci

	dim	deg	
$(S^7)^\vee$	20	34435125 [10]	irred. sextics with 7 singular pts
$(W')^\vee$	26	83200 [7]	sums of three squares
$(S^8)^\vee$	19	58444767 [10]	irred. sextics with 8 singular pts
$W^\vee$	18	$\frac{1}{2}\binom{18}{9} = 24310$	reducible in 2 cubics, or sums of two squares
$\mathcal{R}^\vee$	18	57435240 [10, 18]	irred. sextics with 9 singular pts
$(S^9)^\vee$	9	$2^9 = 512$	square of a cubic

Note that the last diagram (Fig. 2) about dual loci is not dual to the previous one (Fig. 1), since for dimensional reasons $(W')^\vee$ is not contained in $(S^7)^\vee$. Nevertheless we have the inclusion $S^7 \subset W'$ (Fig. 1).

Acknowledgements We are honored to have the chance of contributing to the volume that celebrates the 70th birthday of Ciro Ciliberto. A master, a guide, and a friend for us.

References

1. J. Alexander, A. Hirschowitz, Polynomial interpolation in several variables. J. Algebraic Geom. **4**, 201–222 (1995)
2. E. Angelini, L. Chiantini, On the identifiability of ternary forms. Lin. Alg. Applic. **599**, 36–65 (2020)
3. E. Angelini, L. Chiantini, A. Mazzon, Identifiability for a class of symmetric tensors. Mediterr. J. Math. **16**, 97 (2019)
4. E. Arbarello, M. Cornalba, Footnotes to a paper of B. Segre. Math. Ann. **256**, 341–362 (1981)
5. E. Ballico, An effective criterion for the additive decompositions of forms. Rend. Ist. Matem. Trieste **51**, 1–12 (2019)
6. E. Ballico, L. Chiantini, On the Terracini locus of projective varieties. Milan J. Math. **89**, 1–17 (2021)
7. G. Blekherman, J. Hauenstein, J.C. Otten, B. Ranestad, K. Sturmfels, Algebraic boundaries of Hilbert's SOS cones. Compos. Math. **148**, 1717–1735 (2012)
8. C. Bocci, L. Chiantini, *An Introduction to Algebraic Statistics with Tensors*, Unitext (Springer, Berlin, New York, 2019)

9. W. Buczyńska, J. Buczyński, Z. Teitler, Waring decompositions of monomials. J. Algebra **378**, 45–57 (2013)
10. L. Caporaso, J. Harris, Counting plane curves of any genus. Inventiones Math. **131**, 345–392 (1998)
11. L. Chiantini, *Hilbert Functions and Tensor Analysis, Quantum Physics and Geometry*, Lecture Notes of the Unione Matematica Italiana, vol. 25. (Springer, Berlin, New York, 2019), pp. 125–151
12. L. Chiantini, C. Ciliberto, Weakly defective varieties. Trans. Amer. Math. Soc. **354**, 151–178 (2002)
13. L. Chiantini, C. Ciliberto, On the concept of k-secant order of a variety. J. Lond. Math. Soc. **73**, 436–454 (2006)
14. L. Chiantini, G. Ottaviani, On generic identifiability of 3-tensors of small rank. SIAM J. Matrix Anal. Appl. **33**, 1018–1037 (2012)
15. L. Chiantini, G. Ottaviani, N. Vannieuwenhoven, On generic identifiability of symmetric tensors of subgeneric rank. Trans. Amer. Math. Soc. **369**, 4021–4042 (2017)
16. A. Conca, G. Valla, Hilbert function of powers of ideals of low codimension. Math. Zeit. **230**, 753–784 (1999)
17. E. Davis, Hilbert functions and complete intersections. Rend. Seminario Mat. Univ. Politecnico Torino **42**, 333–353 (1984)
18. E. Getzler, Intersection theory on $\bar{M}_{1,4}$ and elliptic Gromov-Witten invariants. J. Amer. Math. Soc. **10**, 973–978 (1997)
19. D. Grayson, M. Stillman, Macaulay 2, a software system for research in algebraic geometry.http://www.math.uiuc.edu/Macaulay2/
20. J. Harris, On the Severi problem. Invent. Math. **84**, 445–461 (1986)
21. J. Harris, *Algebraic Geometry, a First Course*. Graduate Texts in Math (Springer, Berlin, New York, 1992)
22. A. Iarrobino, V. Kanev, *Power Sums, Gorenstein Algebras, and Determinantal Loci*, Lecture Notes in Mathematics, vol. 1721. (Springer, Berlin, New York, 1999)
23. J.M. Landsberg, *Tensors: Geometry and applications, Graduate Studies in Mathematics*, vol. 128 (American Mathematical Society, Providence RI, 2012)
24. J.M. Landsberg, G. Ottaviani, Equations for secant varieties of Veronese and other varieties. Ann. Mat. Pura Appl. **192**, 569–606 (2013)
25. J. Migliore, *Introduction to Liaison Theory and Deficiency Modules*, Progress in Mathematics, vol. 165 (Birkäuser, Basel, Boston, 1998)
26. B. Mourrain, A. Oneto, On minimal decompositions of low rank symmetric tensors. Lin. Alg. Applic. **607**, 347–377 (2020)
27. L. Oeding, G. Ottaviani, Eigenvectors of tensors and algorithms for Waring decomposition. J. Symbolic Comput. **54**, 9–35 (2013)
28. G. Ottaviani, An invariant regarding Waring's problem for cubic polynomials. Nagoya Math. J. **193**, 95–110 (2009)
29. G. Ottaviani, E. Sernesi, On the hypersurface of Lüroth quartics. Michigan Math. J. **59**, 365–394 (2010)
30. C. Segre, Gli ordini delle varietá che annullano i determinanti dei diversi gradi estratti da una data matrice. Atti Accad. Lincei Classe Sci. **9**, 253–260 (1900)
31. A. Terracini, Sulle V_k per cui la varietà degli S_h (h+1)-seganti ha dimensione minore dell'ordinario. Rend. Circolo Mat. Palermo **31**, 392–396 (1911)

Deformations and Extensions of Gorenstein Weighted Projective Spaces

Thomas Dedieu and Edoardo Sernesi

Abstract We study the existence of deformations of all 14 Gorenstein weighted projective spaces **P** of dimension 3 by computing the number of times their general anticanonical divisors are extendable. In favorable cases (8 out of 14), we find that **P** deforms to a 3-dimensional extension of a general non-primitively polarized $K3$ surface. On our way, we show that each such **P** in its anticanonical model satisfies property N_2, i.e., its homogeneous ideal is generated by quadrics, and the first syzygies are generated by linear syzygies, and we compute the deformation space of the cone over **P**. This gives as a byproduct the exact number of times **P** is extendable.

Keywords Weighted projective space · Canonical curve · K3 surface · Fano variety

1 Introduction

Some topics related to this paper have been discussed and worked out with our friend and colleague Ciro Ciliberto. It is a great pleasure for us to dedicate this work to him.

It is well known that there are precisely 14 Gorenstein weighted projective spaces of dimension 3 (see [29]; we give the list in Table 1). In this paper, we introduce a method in the study of their deformations, consisting in studying simultaneously

Dedicated to Ciro Ciliberto on the occasion of his 70th birthday.

T. Dedieu
Institut de Mathématiques de Toulouse; UMR5219. Université de Toulouse; CNRS. UPS IMT, F-31062 Toulouse Cedex 9, France
e-mail: thomas.dedieu@math.univ-toulouse.fr

E. Sernesi (✉)
Dipartimento di Matematica e Fisica, Università Roma Tre, L.go S.L. Murialdo, 1-00146 Roma, Italy
e-mail: sernesi@gmail.com

T. Dedieu et al. (eds.), *The Art of Doing Algebraic Geometry*, Trends in Mathematics,
https://doi.org/10.1007/978-3-031-11938-5_5

the deformations and the extendability of their general anticanonical divisors. The underlying philosophy goes back to Pinkham [28], and then Wahl [35] who showed the close connection between the existence of extensions of a projective variety $X \subset \mathbf{P}^r$ and the deformation theory of its affine cone $CX \subset \mathbf{A}^{r+1}$. We discuss and recall this connection in Sect. 4. Wahl's interest was focused on canonical curves, aiming at a characterization of those curves that are hyperplane sections of a $K3$ surface ("$K3$ curves") by means of the behavior of their Gaussian map, thereafter called "Wahl map". His program was carried out in [2] and applied in [8] to the extendability of canonical curves, $K3$ surfaces and Fano varieties. The present work follows the same direction, as surface (resp. curve) linear sections of Gorenstein weighted projective 3-spaces in their anticanonical embeddings are $K3$ surfaces with at worst canonical singularities (resp. canonical curves).

Our starting point is the observation that most polarized general anticanonical divisors $(S, -K_{\mathbf{P}}|_S)$ of the weighted projective spaces $\mathbf{P}$ of Table 1 are non-primitively polarized (and singular). This suggests considering a 1-parameter smoothing $(\mathcal{S}, \mathcal{L})$ of $(S, -K_{\mathbf{P}}|_S)$ and exploiting the fact that the extendability of non-primitively polarized $K3$ surfaces is well understood, thanks to work of Ciliberto–Lopez–Miranda [9], Knutsen [18], and Ciliberto–Dedieu [5, 6] (see Sect. 3 for details). This plan works fine when the extendability of $(S, -K_{\mathbf{P}}|_S)$ coincides with that of the general non-primitively polarized $K3$ surface, i.e., the invariant

$$\alpha(S_t, L_t) = h^0(S_t, N_{S_t/\mathbf{P}^g} \otimes L_t^{-1}) - g - 1$$

introduced in Theorem 3.3 takes the same value for all fibers of $(\mathcal{S}, \mathcal{L})$. What we get in this case is that $\mathbf{P}$ deforms to a threefold extension of the general member of $(\mathcal{S}, \mathcal{L})$. The final output (see Sect. 7) is an understanding of the deformation properties of Cases $\#i$, $i \in \{1, \dots, 7, 9\}$, in the notation of Tables 1 and 3. A finer analysis is required for the other cases, which we don't try to carry out here, although we make a couple of observations at the end of Sect. 7.

Our strategy involves various substantial technical verifications. The main point is controlling the deformation theory of the affine and projective cones over possibly singular $K3$ surfaces. In the nonsingular case, this is a well-known chapter of deformation theory, due to Schlessinger [34]: we extend it to the singular case in Sect. 4. It is a non-trivial task to compute the relevant deformation spaces in our examples, and for this purpose we took advantage of the computational power of Macaulay2 [23]. Still this leaves some obstacles to the human user (see the proof of Proposition 6.2 and the comments thereafter), which we have found are best coped with by considering more generally deformations of cones over arithmetically Cohen–Macaulay surfaces or arithmetically Gorenstein curves.

As a further reward of this computation, we obtain the exact number of times each Gorenstein weighted projective space of dimension 3 is extendable (Corollary 6.4 and Table 3). We have not been able however to identify the maximal extension in all cases.

Another condition we had to verify in order to apply Wahl's criterion (Theorem 4.8) is that the projective schemes X involved satisfy condition N_2 (Definition 4.7), so that "each first order ribbon over X is integrable to at most one extension of X". We carry this out again with the computer and Macaulay2 (see Proposition 6.1), by explicitly computing the homogeneous ideals of all Gorenstein weighted projective spaces in their anticanonical embeddings, as well as their first syzygy modules.

Some of our end results about Gorenstein Fano threefolds in Sect. 7 can also be obtained by direct calculations, using computational tricks on weighted projective spaces similar to those employed by Hacking in [15, Sect. 11], and showed to us by the referee. We find it nice that the observations we made indirectly using deformation theory may be confirmed by direct computations of a different nature. Let us also mention the article [24] (which has been continued in [15]), in which degenerations of the projective plane to various weighted projective planes are exhibited: this is similar in spirit to what we do in Sect. 7.

The organization of the article is as follows. In Sect. 2, we gather some elementary facts about weighted projective geometry and give the list of all 14 Gorenstein weighted projective spaces of dimension 3. In Sect. 3, we give a synthetic account of the extension theory of non-primitive polarized $K3$ surfaces along the lines of [8] and taking advantage of [18]. Section 4 is the technical heart of the paper and is devoted to the deformation theory of cones and its application to extensions. This leads to our main technical result Theorem 5.2 in the following Sect. 5. In Sect. 6, we carry out the explicit computations required for our application of Theorem 5.2 to Gorenstein weighted projective spaces, and in the final Sect. 7, we give the explicit output of this application.

2 Gorenstein Weighted Projective Spaces

We will consider some weighted projective spaces (WPS for short) of dimension 3. In this section, we collect some preliminary definitions and basic facts. The authoritative reference is [11]; we will also rely on [3, 13].

2.1 Consider a weighted projective 3-space of the form $\mathbf{P} := \mathbf{P}(a_0, a_1, a_2, a_3)$, where the a_i's are relatively prime positive integers.

It is not restrictive to further assume, and we will do it, that any three of the a_i's are relatively prime, in which case one says that $\mathbf{P}$ is *well formed*. Let:

$$m := \operatorname{lcm}(a_0, a_1, a_2, a_3), \quad s := a_0 + a_1 + a_2 + a_3.$$

The following holds:

(1) For all $d \in \mathbf{Z}$ the sheaf $\mathcal{O}(d)$ is reflexive of rank 1, and it is invertible if and only if $d = km$ for some $k \in \mathbf{Z}$ [3, Sect. 4].
(2) $\operatorname{Pic}(\mathbf{P}) = \mathbf{Z} \cdot [\mathcal{O}(m)]$ [3, Theorem 7.1, p. 152].
(3) $\mathbf{P}$ is Cohen–Macaulay and its dualizing sheaf is $\omega_{\mathbf{P}} = \mathcal{O}(-s)$. Therefore $\mathbf{P}$ is Gorenstein if and only if $m|s$ [3, Corollary 6B.10, p. 151]. In this case, $\mathbf{P}$ has

canonical singularities because it is a Gorenstein orbifold. This follows for example from [32, Proposition 1.7].

(4) The intersection product in $\mathbf{P}$ is determined by (see [19, p. 240]):

$$\mathcal{O}(1)^3 = \frac{1}{a_0 a_1 a_2 a_3}.$$

Lemma 2.2 *Let $S \subset \mathbf{P}(a_0, a_1, a_2, a_3)$ be a general hypersurface of degree d, such that $\mathcal{O}(d)$ is locally free on $\mathbf{P}$. For all $k \in \mathbf{Z}$, the restriction to S of $\mathcal{O}(k)$ is locally free if and only if*

$$\forall i \neq j : \quad \gcd(a_i, a_j) | k.$$

Proof The local freeness needs only to be checked at the singular points. As S is general it may be singular only along the singular locus of $\mathbf{P}$, hence only along the lines joining two coordinate points $P_i = (0 : \ldots : 1 : \ldots : 0)$. Moreover, S avoids all coordinate points themselves thanks to our assumption on the degree.

Let P be a point on the line $P_i P_j$, off P_i and P_j. In the local ring $\mathcal{O}_P$ we have the invertible monomials $x_i^{n_i} x_j^{n_j}$ whose degrees sweep out

$$a_i \mathbf{Z} + a_j \mathbf{Z} = \gcd(a_i, a_j)\mathbf{Z}.$$

This shows that if $\gcd(a_i, a_j)$ divides k then $\mathcal{O}(k)$ is invertible at all points of $P_i P_j$ but P_i and P_j themselves, hence the "if" part of the statement. The "only if" part follows in the same way. □

2.3 It follows from 2.1 that there are exactly 14 distinct 3-dimensional weighted projective spaces which are Gorenstein, see [29]. We list them in Table 1, together with the following information. For each $\mathbf{P}$ in the list, we denote by S a general anticanonical divisor: it is a $K3$ surface with ADE singularities [33]. We also use the following notation:

- m is the lcm of the weights, so that $\mathcal{O}(m)$ generates Pic($\mathbf{P}$);
- s is the sum of the weights, so that $\omega_{\mathbf{P}} = \mathcal{O}(-s)$;
- i_S denotes the divisibility of $K_{\mathbf{P}}|_S$ in Pic(S), which is readily computed with Lemma 2.2 above;
- g_1 is the genus of the primitively polarized $K3$ surface (S, L_1), where $L_1 = -\frac{1}{i_S} K_{\mathbf{P}}|_S$; we reserve the symbol g to the common genus of the Fano variety $\mathbf{P}$ and the polarized $K3$ surface $(S, -K_{\mathbf{P}}|_S)$, i.e., $2g - 2 = -K_{\mathbf{P}}^3$.

The rows are ordered according to g_1, then i_S (decreasing), then the weights. We also indicate the singularities of S, which may be found following [13].

Table 1 Gorenstein 3-dimensional weighted projective spaces

#	Weights	$-K_{\mathbf{P}}^3$	m	s	i_S	g_1	Sings(S)
1	(1, 1, 1, 3)	72	3	6	6	2	Smooth
2	(1, 1, 4, 6)	72	12	12	6	2	A_1
3	(1, 2, 2, 5)	50	10	10	5	2	$5A_1$
4	(1, 1, 1, 1)	64	1	4	4	3	Smooth
5	(1, 1, 2, 4)	64	4	8	4	3	$2A_1$
6	(1, 3, 4, 4)	36	12	12	3	3	$3A_3$
7	(1, 1, 2, 2)	54	2	6	3	4	$3A_1$
8	(1, 2, 6, 9)	54	18	18	3	4	$3A_1$, A_2
9	(2, 3, 3, 4)	24	12	12	2	4	$3A_1, 4A_2$
10	(1, 4, 5, 10)	40	20	20	2	6	$A_1, 2A_4$
11	(1, 2, 3, 6)	48	6	12	2	7	$2A_1, 2A_2$
12	(1, 3, 8, 12)	48	24	24	2	7	$2A_2$, A_3
13	(2, 3, 10, 15)	30	30	30	1	16	$3A_1, 2A_2$, A_4
14	(1, 6, 14, 21)	42	42	42	1	22	A_1, A_2 , A_6

3 Extendability of Non-primitive Polarized $K3$ Surfaces

Let us first recall the following.

Definition 3.1 A projective variety $X \subset \mathbf{P}^r$ of dimension d is called *n-extendable* for some $n \geq 1$ if there exists a projective variety $\widetilde{X} \subset \mathbf{P}^{r+n}$ of dimension $d+n$, not a cone, such that $X = \widetilde{X} \cap \mathbf{P}^r$ for some linear embedding $\mathbf{P}^r \subset \mathbf{P}^{r+n}$. The variety $\widetilde{X}$ is called an *n-extension* of X. If $n = 1$ we call X *extendable* and $\widetilde{X}$ an *extension* of Y.

We refer to [21] for a beautiful tour on this subject.

If X is a Fano 3-fold of genus $g = -\frac{1}{2}K_X^3 + 1$, then a general $S \in |-K_X|$ is a $K3$ surface naturally endowed with the ample divisor $-K_X|_S$ which makes $(S, -K_X|_S)$ an extendable polarized $K3$ surface of genus g. Suppose that X has index $i_X > 1$, i.e., $-K_X = i_X H$ for an ample divisor H, indivisible in Pic(X). Then $(S, -K_X|_S)$ is non-primitive because $-K_X|_S = i_X H|_S$ is at least i_X-divisible. Therefore, by considering Fano 3-folds of index > 1, we naturally land in the world of extendable non-primitively polarized $K3$ surfaces.

Notation 3.2 We denote by $\mathcal{K}_g^k$ the moduli stack of polarized $K3$ surfaces of genus g and index k, i.e., pairs (S, L) such that S is a $K3$ surface, possibly with ADE singularities, and L is an ample and globally generated line bundle on S with $L^2 = 2g-2$, such that $L = kL_1$ with L_1 a primitive line bundle on S; note that (S, L_1) belongs to $\mathcal{K}_{g_1}^1$, which we usually denote by $\mathcal{K}_{g_1}^{\mathrm{prim}}$, where $2g_1 - 2 = L_1^2$ and $g = 1 + k^2(g_1 - 1)$.

We have the following necessary condition for the extendability of a projective variety:

Theorem 3.3 ([22]) *Let $X \subset \mathbf{P}^n$ be a smooth, projective, irreducible, non-degenerate variety, not a quadric, and write $L = \mathcal{O}_X(1)$. Set*

$$\alpha(X, L) = h^0(N_{X/\mathbf{P}^n} \otimes L^{-1}) - n - 1.$$

If $\alpha(X, L) < n$, then X is at most $\alpha(X, L)$-extendable.

When the polarization of X is clear from the context, we write $\alpha(X)$ instead of $\alpha(X, L)$. Note that if X is a smooth $K3$ surface or Fano variety (resp. a canonical curve, hence $L = K_X$) then

$$\alpha(X) = H^1(X, T_X \otimes L^{-1}) \quad (\text{resp. cork}(\Phi_{\omega_X}),$$

with Φ_{ω_X} the Gauss–Wahl map of X, see for instance [8, Sect. 3].

For $K3$ surfaces and canonical curves, the converse to Theorem 3.3 also holds, under some conditions. Precisely we have:

Theorem 3.4 ([8], Theorems 2.1 and 2.17) *Let (X, L) be a smooth polarized $K3$ surface (resp. $(X, L) = (C, K_C)$ a canonical curve) of genus g. Assume that $g \geq 11$ and* Cliff$(S, L) \geq 3$. *Then (X, L) is $\alpha(X, L)$-extendable.*

More precisely, every non-zero $e \in H^1(S, T_S \otimes L^{-1})$ (resp. $e \in \ker({}^{\mathsf{T}}\Phi_{\omega_C})$) defines an extension X_e of X which is unique up to projective automorphisms of $\mathbf{P}^{g+1}$ (resp. $\mathbf{P}^g$) fixing every point of $\mathbf{P}^g$ (resp. $\mathbf{P}^{g-1}$), and there exists a universal extension $\tilde{X} \subset \mathbf{P}^{g+\alpha(X,L)}$ *(resp. $\mathbf{P}^{g-1+\alpha(X,L)}$) of X having each X_e as a linear section containing X.*

We denoted by Cliff(S, L) the *Clifford index* of any nonsingular curve $C \in |L|$; by [12, 14, 31], this does not depend on the choice of C. Note that in case (X, L) is a $K3$ surface the extension X_e in the theorem is an arithmetically Gorenstein Fano variety of dimension 3 with canonical singularities.

Unfortunately $H^1(S, T_S \otimes L^{-1})$ is not easy to compute in general, but in the non-primitive setting, we can reduce to a more amenable case.

Lemma 3.5 *Let $S \subset \mathbf{P}^g$ be a smooth $K3$ surface. Then:*

$$H^1(S, T_S \otimes L^{-j}) = \begin{cases} \text{coker}\big[H^0(S, L)^\vee \to H^0(S, N_{S/\mathbf{P}^g}(-1))\big], & \text{if } j = 1 \\ H^0(S, N_{S/\mathbf{P}^g} \otimes L^{-j}), & \text{if } j \geq 2. \end{cases}$$

Proof See [9] (2.8). □

This lemma, applied to a smooth $(S, L_1) \in \mathcal{K}_{g_1}$ with L_1 very ample, tells us that $(S, L) = (S, jL_1)$ with $j \geq 2$ is extendable if and only if $H^0(S, N_{S/\mathbf{P}^{g_1}} \otimes L_1^{-j}) \neq 0$. The possibilities for the pair (S, L_1) and j are then very limited. In particular,

$H^0(S, N_{S/\mathbf{P}^{g_1}} \otimes L_1^{-j}) = 0$ for all $j \geq 2$ as soon as S satisfies the property N_2 (see Definition 4.7), see [18, Lemma 1.1] and the references therein. In fact the possibilities have been completely classified by Knutsen [18], see also [6, 9]. The result is the following:

Theorem 3.6 ([18]) *Let $(S, L_1) \in \mathcal{K}_{g_1}^{\text{prim}}$ with S smooth and L_1 very ample. Then $H^1(S, T_S \otimes L_1^{-j}) = 0$ for all $j \geq 2$ except in the following cases:*

g_1	j	$g(L_1^j)$	$h^1(S, T_S \otimes L_1^{-j})$	Notes
3	2	9	10	any (S, L_1)
3	3	19	4	any (S, L_1)
3	4	33	1	any (S, L_1)
4	2	13	5	any (S, L_1)
4	3	28	1	any (S, L_1)
5	2	17	3	any (S, L_1)
6	2	21	1	any (S, L_1)
7	2	25	1	(1)
8	2	29	1	(2)
9	2	33	1	(3)
10	2	37	1	(4)

where

(1) S is one of the following:

   (I) a divisor in the linear system $|3H - 3F|$ on the quintic rational normal scroll $T \subset \mathbf{P}^7$ of type $(3, 1, 1)$, with H a hyperplane section and F a fiber of the scroll.
   (II) a quadratic section of the sextic Del Pezzo threefold $\mathbf{P}^1 \times \mathbf{P}^1 \times \mathbf{P}^1 \subset \mathbf{P}^7$ embedded by Segre.
   (III) the section of $\mathbf{P}^2 \times \mathbf{P}^2 \subset \mathbf{P}^8$, embedded by Segre, with a hyperplane and a quadric.

(2) S is an anticanonical divisor in a septic Del Pezzo 3-fold (the blow-up of $\mathbf{P}^3$ at a point).
(3) S is one of the following:

   (I) the 2-Veronese embedding of a quartic of $\mathbf{P}^3$; equivalently a quadratic section of the Veronese variety $v_2(\mathbf{P}^3) \subset \mathbf{P}^9$.
   (II) a quadratic section of the cone over the anticanonical embedding of the Hirzebruch surface $\mathbf{F}_1 \subset \mathbf{P}^8$.

(4) S is a quadratic section of the cone over the Veronese surface $v_3(\mathbf{P}^2) \subset \mathbf{P}^9$.

In all cases, except $g_1 = 3$ and $j = 2$, we have $\text{Cliff}(S, L^j) \geq 3$.

Proof Using the identification given by Lemma 3.5, the proof reduces to list all possible cases described by Proposition 1.4 of [18]. The final statement is an easy calculation. □

The case $g_1 = 3$ and $j = 2$ is not liable to Theorem 3.4 as the Clifford index in this case is too small, but it has been studied by hand in [5, 6].

Theorem 3.6 does not cover the cases of (S, L_1) hyperelliptic. We shall only consider the case $g(L_1) = 2$, which will be sufficient for our purposes. The following result has been obtained in [5] by geometric means; we give here a cohomological proof.

Lemma 3.7 *Let $(S, L_1) \in \mathcal{K}_2^{\mathrm{prim}}$. Then the dimension of $H^1(S, T_S \otimes L_1^{-j})$ takes the values given by the following table:*

j	$g(L_1^j)$	$\mathrm{Cliff}(L_1^j)$	$h^1(S, T_S \otimes L_1^{-j})$
1	2	0	18
2	5	0	15
3	10	2	10
4	17	>2	6
5	26	>2	3
6	37	>2	1
≥7		>2	0

Recalling Theorem 3.4, this table implies that (S, L_1^j) is extendable for $4 \le j \le 6$, since $\mathrm{Cliff}(S, L_1^j) \ge 3$. In particular $(S, L_1^6) \in \mathcal{K}_{37}$ and is precisely 1-extendable; in fact it is hyperplane section of $\mathbf{P}(1, 1, 1, 3)$. Lemma 3.7 also tells us that the surfaces (S, L_1^4) and (S, L_1^5) are 6-extendable and 3-extendable, respectively, in agreement with the results in [5], (4.8). There, also the situation in case $j = 3$ (to which Theorem 3.4 does not apply), is completely described.

Proof The elementary computation of $\mathrm{Cliff}(L_1^j)$ is left to the reader. The surface S is a double plane $\pi : S \longrightarrow \mathbf{P}^2$ branched along a sextic Γ and $L_1 = \pi^* \mathcal{O}_{\mathbf{P}^2}(1)$. Denote by R the ramification curve of π. We have $\mathcal{O}_S(R) = L_1^3$. The cotangent sequence of π is

$$0 \to \pi^* \Omega^1_{\mathbf{P}^2} \longrightarrow \Omega^1_S \longrightarrow \Omega^1_{S/\mathbf{P}^2} \to 0$$

where $\Omega^1_{S/\mathbf{P}^2} = \mathcal{O}_R(-R) = L_1^{-3} \otimes \mathcal{O}_R = \omega_R^{-1}$. Therefore, for every j we have the following diagram, where the vertical sequence is the twisted Euler sequence restricted to S:

$$\begin{array}{ccccccc}
 & & 0 & & & & \\
 & & \downarrow & & & & \\
0 & \longrightarrow & \pi^*\Omega^1_{\mathbf{P}^2} \otimes L_1^j & \longrightarrow & \Omega^1_S \otimes L_1^j & \longrightarrow & L_1^{j-3} \otimes \mathcal{O}_R \to 0 \\
 & & \downarrow & & & & \\
 & & L_1^{j-1} \otimes H^0(S, L_1) & & & & \\
 & & \downarrow & & & & \\
 & & L_1^j & & & & \\
 & & \downarrow & & & & \\
 & & 0 & & & &
\end{array}$$

For $j \geq 7$, this diagram gives $H^1(S, \Omega^1_S \otimes L_1^j) = 0$. If $j = 1$ we get the following exact sequence:

$$\begin{array}{ccccccc}
0 \to H^1(S, \Omega^1_S \otimes L_1) & \longrightarrow & H^1(R, L_1^{-2} \otimes \mathcal{O}_R) & \longrightarrow & H^2(S, \pi^*\Omega^1_{\mathbf{P}^2} \otimes L_1) & \longrightarrow & 0 \\
 & & & & \| & & \\
 & & & & H^2(S, \mathcal{O}_S) \otimes H^0(S, L_1) & &
\end{array}$$

which gives $h^1(S, \Omega^1_S \otimes L_1) = h^1(R, L_1^{-2} \otimes \mathcal{O}_R) - 3 = 18$.

If $j = 2$ we have

$$h^1(S, \pi^*\Omega^1_{\mathbf{P}^2} \otimes L_1^2) = \operatorname{corank}\left[\operatorname{Sym}^2 H^0(S, L_1) \to H^0(S, L_1^2)\right] = 0$$

and the following exact sequence:

$$0 \to H^1(S, \pi^*\Omega^1_{\mathbf{P}^2} \otimes L_1^2) \longrightarrow H^1(S, \Omega^1_S \otimes L_1^2) \longrightarrow H^1(R, L_1^{-1} \otimes \mathcal{O}_R) \to 0$$

which gives $h^1(S, \Omega^1_S \otimes L_1^2) = 0 + h^1(R, L_1^{-1} \otimes \mathcal{O}_R) = 15$.

If $3 \leq j \leq 6$ then

$$h^1(S, \pi^*\Omega^1_{\mathbf{P}^2} \otimes L_1^j) = 0$$

thus $H^1(S, \Omega^1_S \otimes L_1^j) \cong H^1(R, L_1^{j-3} \otimes \mathcal{O}_R)$, and the conclusion is clear. □

4 Extendability and Graded Deformations of Cones

Consider a projective scheme $X \subset \mathbf{P}^r$ and let $A = R/I_X$ be its homogeneous coordinate ring, where $R = \mathbf{C}[X_0, \ldots, X_r]$ and I_X is the saturated homogeneous ideal of X in $\mathbf{P}^r$. The *affine cone* over X is

$$CX := \mathrm{Spec}(A) \subset \mathbf{A}^{r+1}$$

and the *projective cone* over X is

$$\overline{CX} := \mathrm{Proj}(A[t]) \subset \mathbf{P}^{r+1}.$$

Recall the following standard definitions. The scheme X is *projectively normal*, resp. *arithmetically Cohen–Macaulay*, resp. *arithmetically Gorenstein* if the local ring of CX at the vertex is integrally closed, resp. Cohen–Macaulay, resp. Gorenstein. Also recall that if X is normal and arithmetically Cohen–Macaulay then it is projectively normal.

The deformation theory of CX is controlled by the cotangent modules T^1_{CX} and T^2_{CX}, which are graded because of the $\mathbf{C}^*$-action on A. We will only need the explicit description of the first one.

Proposition 4.1 *Let $X \subset \mathbf{P}^r$ be a non-degenerate scheme of pure dimension $d \geq 1$. Consider the following conditions:*

(a) *X is arithmetically Cohen–Macaulay (aCM for short).*
(b) *X is projectively normal.*

If either (a) or (b) holds then we have an exact sequence of graded modules:

$$\bigoplus_{k\in\mathbf{Z}} H^0(X, T_{\mathbf{P}^r}|_X(k)) \longrightarrow \bigoplus_{k\in\mathbf{Z}} H^0(X, N_{X/\mathbf{P}^r}(k)) \longrightarrow T^1_{CX} \to 0. \tag{1}$$

Proof Let $v \in CX$ be the vertex, $W = CX \setminus \{v\}$ and let $\pi : W \longrightarrow X$ be the projection. By definition we have an exact sequence:

$$H^0(CX, T_{\mathbf{A}^{r+1}}|_{CX}) \longrightarrow H^0(CX, N_{CX/\mathbf{A}^{r+1}}) \longrightarrow T^1_{CX} \to 0. \tag{2}$$

We assume that (a) or (b) holds. Then CX verifies Serre's condition S_2 at the vertex. The two sheaves F respectively involved in the two first terms of (2) are reflexive, each being the dual of a coherent sheaf, hence they have depth ≥ 2 at v as well by [16, Proposition 1.3] (for the implication we use, it is enough that the X from the notation of ibid. be S_2, as the proof given there shows). Therefore

$$H^0(CX, F) \cong H^0(W, F|_W).$$

Thus (2) induces an exact sequence

$$H^0(W, T_{\mathbf{A}^{r+1}}|_W) \longrightarrow H^0(W, N_{W/\mathbf{A}^{r+1}}) \longrightarrow T^1_{CX} \to 0.$$

As in the proof of [34, Lemma 1], one sees that

$$H^0(W, T_{\mathbf{A}^{r+1}}|_W) = \bigoplus_{k\in\mathbf{Z}} H^0(X, \mathcal{O}_X(k+1))^{r+1}$$

and

$$H^0(W, N_{W/\mathbf{A}^{r+1}}) = \bigoplus_{k\in\mathbf{Z}} H^0(X, N_{X/\mathbf{P}^r}(k)).$$

Then we have a commutative diagram

$$\begin{array}{ccc} H^0(W, T_{\mathbf{A}^{r+1}}|_W) & \longrightarrow & H^0(W, N_{W/\mathbf{A}^{r+1}}) \\ \| & & \| \\ \bigoplus_{k\in\mathbf{Z}} H^0(X, \mathcal{O}_X(k+1))^{r+1} \xrightarrow{\phi} \bigoplus_{k\in\mathbf{Z}} H^0(X, T_{\mathbf{P}^r}|_X(k)) & \longrightarrow & \bigoplus_{k\in\mathbf{Z}} H^0(X, N_{X/\mathbf{P}^r}(k)) \end{array}$$

(in which the map ϕ comes from the Euler exact sequence), and (1) is proved. □

Considering the degree -1 pieces of the exact sequences (1), we get:

Corollary 4.2 *In the notation of Proposition 4.1, if (a) or (b) holds, then there is an exact sequence:*

$$H^0(X, T_{\mathbf{P}^r}|_X(-1)) \longrightarrow H^0(X, N_{X/\mathbf{P}^r}(-1)) \longrightarrow T^1_{CX,-1} \to 0. \quad (3)$$

The following corollary will be important in our applications. It applies in the cases under consideration in this article, because embedded $K3$ surfaces, Gorenstein weighted projective spaces of dimension 3 in their anticanonical embedding, and their linear curve sections are arithmetically Gorenstein.

Corollary 4.3 *Let $X \subset \mathbf{P}^r$ be either aCM of pure dimension ≥ 2 or arithmetically Gorenstein of pure dimension 1 and positive arithmetic genus. Then $\alpha(X) \leq \dim(T^1_{CX,-1})$.*

(See Theorem 3.3 for the definition of α).

Proof The twisted Euler exact sequence

$$0 \to \mathcal{O}_X(-1) \to H^0(X, \mathcal{O}(1)) \otimes \mathcal{O}_X \to T_{\mathbf{P}^r}|_X(-1) \to 0$$

induces the exact sequence

$$0 \to H^0(X, \mathcal{O}(1)) \to H^0(T_{\mathbf{P}^r}|_X(-1)) \\ \to H^1(\mathcal{O}_X(-1)) \to H^0(\mathcal{O}_X(1)) \otimes H^1(\mathcal{O}_X). \quad (4)$$

When $\dim(X) > 1$, since X is arithmetically Cohen–Macaulay we have that $H^1(\mathcal{O}_X(-1)) = 0$, and therefore by (4),

$$H^0(T_{\mathbf{P}^r}|_X(-1)) \cong H^0(X, \mathcal{O}(1)).$$

Then (3) gives a presentation

$$H^0(X, \mathcal{O}(1)) \longrightarrow H^0(X, N_{X/\mathbf{P}^r}(-1)) \longrightarrow T^1_{CX,-1} \to 0,$$

from which the desired inequality follows at once.

When $\dim(X) = 1$, (4) gives the following exact sequence of vector spaces,

$$0 \to H^0(X, \mathcal{O}(1)) \to H^0(T_{\mathbf{P}^r}|_X(-1)) \to \ker({}^{\mathsf{T}}\mu) \to 0,$$

where μ is the multiplication map

$$\mu : H^0(\mathcal{O}_X(1)) \otimes H^0(\omega_X) \to H^0(\omega_X(1)).$$

If X is arithmetically Gorenstein of positive genus we have $\omega_X = \mathcal{O}_X(\nu)$ for some $\nu \geq 0$, hence μ is the multiplication map

$$H^0(\mathcal{O}_X(1)) \otimes H^0(\mathcal{O}_X(\nu)) \to H^0(\mathcal{O}_X(\nu + 1))$$

which is surjective, and we conclude as before. □

If X is smooth, in most cases the leftmost map of (3) is injective, so that in fact $\alpha(X) = \dim(T^1_{CX,-1})$. The same holds in the cases under investigation in this article:

Corollary 4.4 *Let $X \subset \mathbf{P}^r$ be either a $K3$ surface with at most canonical singularities, or a Gorenstein weighted projective 3-space in its canonical embedding. Then $\alpha(X) = \dim(T^1_{CX,-1})$.*

Proof The kernel of the leftmost map of (3) is contained in $H^0(X, T_X(-1))$. If X is a $K3$ surface then $H^0(X, T_X) = 0$, hence $H^0(X, T_X(-1)) = 0$ as well. If X is a Gorenstein weighted projective space, then $H^0(X, T_X(-1)) = H^3(X, \Omega^1_X)^\vee$ by Serre duality, hence it is zero in this case as well. It follows that the leftmost map of (3) is injective.

On the other hand, it follows from the Euler exact sequence and the vanishing of $H^1(X, \mathcal{O}_X(-1))$ that $H^0(X, T_{\mathbf{P}^r}|_X(-1)) \cong H^0(X, \mathcal{O}_X(1))^\vee$ which has dimension $r + 1$, hence the result. □

The following will also be of fundamental importance for us.

Proposition 4.5 *Let $X \subset \mathbf{P}^g$ be a $K3$ surface with at worst canonical singularities. Then* $T^1_{CX,-1} = \mathrm{Ext}^1_X(\Omega^1_X, \mathcal{O}_X(-1))$.

Proof Taking $\mathrm{Hom}(\ .\ , \mathcal{O}_X(-1))$ of the conormal exact sequence of X in $\mathbf{P}^g$, and using the fact that the conormal sheaf of X in $\mathbf{P}^g$ and $\Omega^1_{\mathbf{P}^g}|_X$ are locally free, we obtain the exact sequence

$$H^0(X, T_{\mathbf{P}^g}|_X(-1)) \longrightarrow H^0(X, N_{X/\mathbf{P}^g}(-1)) \longrightarrow \mathrm{Ext}^1_X(\Omega^1_X, \mathcal{O}_X(-1)) \longrightarrow H^1(X, T_{\mathbf{P}^g}|_X(-1)). \qquad (5)$$

From the restricted and twisted Euler sequence

$$0 \longrightarrow \mathcal{O}_X(-1) \longrightarrow H^0(X, \mathcal{O}_X(1))^\vee \otimes \mathcal{O}_X \longrightarrow T_{\mathbf{P}^g}|_X(-1) \longrightarrow 0$$

we deduce that $H^1(X, T_{\mathbf{P}^g}|_X(-1)) = 0$. Therefore comparing the two exact sequences (5) and (3) gives the assertion. □

4.6 Consider now an extension $\widetilde{X}$ of a projectively normal $X \subset \mathbf{P}^r$. In such a situation, we let $e_{X/\widetilde{X}} \in \mathrm{Ext}^1(\Omega^1_X, \mathcal{O}_X(-1))$ be the class of the conormal exact sequence

$$0 \longrightarrow \mathcal{O}_X(-1) \longrightarrow \Omega^1_{\widetilde{X}}|_X \longrightarrow \Omega^1_X \longrightarrow 0.$$

If the extension $\widetilde{X}$ is non-trivial, i.e., it is not a cone over X, then we can also associate to it a family of deformations of $\overline{CX}$, the projective cone over X, as follows. Let $X = \widetilde{X} \cap H$, where $H \cong \mathbf{P}^r \subset \mathbf{P}^{r+1}$ is a hyperplane. Consider in $\mathbf{P}^{r+2}$ the projective cone $\overline{C\widetilde{X}}$ and the pencil of hyperplanes H_t with center H. Let H_o be the hyperplane containing the vertex v of $\overline{C\widetilde{X}}$. Then $H_o \cap \overline{C\widetilde{X}} = \overline{CX}$, while $H_t \cap \overline{C\widetilde{X}} \cong \widetilde{X}$ for all $t \neq o$. After blowing up X we obtain a family

$$f : \mathrm{Bl}_X(\overline{C\widetilde{X}}) \longrightarrow \mathbf{P}^1$$

which is flat because $\widetilde{X}$ is projectively normal, with $f^{-1}(t) = H_t \cap \overline{C\widetilde{X}}$. By restriction we get a deformation of the affine cone CX. If $\widetilde{X}$ is smooth then this deformation is a smoothing of $\overline{CX} = f^{-1}(o)$. This is a classical construction called *sweeping out the cone* (see, e.g., [28, (7.6) (iii)]). Algebraically, the above construction has the following description. Let $\widetilde{X} = \mathrm{Proj}(\mathcal{A})$, where $\mathcal{A} = \mathbf{C}[X_0, \ldots, X_r, t]/J$. Then

$$A = \mathcal{A}/t\mathcal{A} = \mathbf{C}[X_0, \ldots, X_r]/I$$

where $I = J/tJ$. Consider $C\widetilde{X} = \mathrm{Spec}(\mathcal{A}) \subset \mathbf{A}^{r+2}$. The pencil of parallel hyperplanes $V(t) \subset \mathbf{A}^{r+2}$ has as projective closure the pencil $\{H_t\}$ considered before. Therefore the morphism

$$\phi : \mathrm{Spec}(\mathcal{A}) \longrightarrow \mathrm{Spec}(\mathbf{C}[t])$$

is the corresponding family of deformations of CX. It is clear that if $e_{X/\widetilde{X}} \in T^1_{CX,-1}$ (e.g., X is nonsingular or is a singular $K3$ surface) then the first-order deformation of X associated to ϕ is $e_{X/\widetilde{X}}$. Note that, by construction, $e_{X/\widetilde{X}}$ *is unobstructed both as a first-order deformation of* CX *and of* $\overline{CX}$.

The upshot of the above construction is that the datum of an extension $\tilde{X}$ of X gives a deformation of the cone over X. In fact the two objects correspond to the same ring $\mathcal{A}$: the former is $\mathrm{Proj}(\mathcal{A})$ and the latter is $\mathrm{Spec}(\mathcal{A})$. We shall now state a result of Wahl which will be crucial in what follows. It may be considered as a reverse sweeping out the cone, in that it produces an extension of X from a first-order deformation of the cone over X. We first need the following standard definition.

Definition 4.7 Let $X \subset \mathbf{P}^r$ be a non-degenerate projectively normal scheme of pure dimension ≥ 1. we say that X *has the property* N_2 or *satisfies* N_2 if its homogeneous coordinate ring $A = R/I_X$ has a minimal graded presentation over $R := \mathbf{C}[X_0, \dots, X_r]$ of the form:

$$R(-3)^a \xrightarrow{\psi} R(-2)^b \xrightarrow{\phi} R \longrightarrow A \to 0. \tag{6}$$

Theorem 4.8 ([35], Proof of Theorem 7.1 and Remark 7.2) *Let* $X \subset \mathbf{P}^r$ *be a non-degenerate projectively normal scheme of pure dimension* ≥ 1 *and let A be its homogeneous coordinate ring. Consider the following two conditions:*

(a) X *has the property* N_2*;*
(b) $T^2_{A,k} = 0$ *for all* $k \leq -2$.

If (a) holds then any first-order deformation of CX *of degree* -1 *lifts to at most one graded deformation* $\mathcal{A}$ *over* $\mathbf{C}[t]$, *with* $\deg(t) = 1$. *Moreover* $Y := \mathrm{Proj}(\mathcal{A}) \subset \mathbf{P}^{r+1} = \mathrm{Proj}[t, X_0, \dots, X_r]$ *is an extension of* $X := \mathrm{Proj}(A) = \mathrm{Proj}(\mathcal{A}) \cap \{t = 0\} \subset \mathbf{P}^r$ *which is unique up to projective automorphisms of* $\mathbf{P}^{r+1}$ *fixing every point of* $\mathbf{P}^r = \{t = 0\}$.

If both (a) and (b) hold then every first-order deformation of CX *of degree* -1 *lifts to a graded deformation* $\mathcal{A}$ *as above.*

It is one of the main results of [2] that condition (b) above holds when X is a canonical curve. We shall use this and Theorem 3.4 to prove that the same holds when X is a Gorenstein weighted projective space of dimension 3, see Corollary 6.4.

For more details on the unicity statement, we refer to [8, Remark 4.8]. Note that assumption (a) implies that $H^0(N_{X/\mathbf{P}^r}(-k)) = 0$ for all $k \geq 2$, as we have already observed.

5 The Deformation Argument

We now come to our main technical result, and its application to deformations of weighted projective spaces.

5.1 Let (S, L) be a polarized $K3$ surface with canonical singularities and $g = h^0(L) - 1$. A *smoothing* of (S, L) is a pair $\big(p : \mathcal{S} \to (\Delta, o), \mathcal{L}\big)$, where p is a smoothing of S over an affine nonsingular pointed curve (Δ, o) and $\mathcal{L}$ extends L, i.e., $L = \mathcal{L}(o) := \mathcal{L}|_{p^{-1}(o)}$. There is a flat family of surfaces in $\mathbf{P}^g$ associated to such a smoothing:

$$\begin{array}{ccc} \mathcal{S} & \overset{j}{\hookrightarrow} & \mathbf{P}^g \times \Delta \\ & {\scriptstyle p}\searrow & \downarrow {\scriptstyle pr_2} \\ & & \Delta \end{array} \tag{7}$$

where j is defined by the sections of $\mathcal{L}$.

We shall use the following notation: $\Delta^\circ := \Delta \setminus \{o\}$; $\mathcal{S}^\circ = \mathcal{S} \setminus p^{-1}(o)$; $p^\circ = p|_{\mathcal{S}^\circ}$; $\mathcal{L}^\circ = \mathcal{L}|_{\mathcal{S}^\circ}$.

A *relative extension* of $\mathcal{S} \subset \mathbf{P}^g \times \Delta$ consists of an $\mathcal{X} \subset \mathbf{P}^{g+1} \times \Delta$, flat over Δ, together with a relative hyperplane $\mathcal{H} \cong \mathbf{P}^g \times \Delta \subset \mathbf{P}^{g+1} \times \Delta$ such that $\mathcal{X} \cap \mathcal{H} = \mathcal{S}$ and $\mathcal{X}(t)$ is not a cone over $\mathcal{S}(t)$ for all $t \in \Delta$. Similarly, one defines relative extensions of $\mathcal{S}^\circ \subset \mathbf{P}^g \times \Delta^\circ$.

Theorem 5.2 *Let $S_0 \subset \mathbf{P}^g$ be a $K3$ surface, possibly with canonical singularities, and $V_0 \subset \mathbf{P}^{g+1}$ be an extension of S_0. Let $p : \mathcal{S} \to \Delta$ be a smoothing of S_0 in $\mathbf{P}^g$ as above, and assume that the following conditions hold:*

(a) $g \geq 11$, and for all $t \in \Delta^\circ$ we have $\mathrm{Cliff}(S_t) > 2$*;*
(b) S_0 has the N_2 property;
(c) $t \in \Delta \mapsto \alpha(S_t)$ is constant.

Then there exists a deformation of V_0 in $\mathbf{P}^{g+1}$ which is a relative extension of $\mathcal{S} \subset \mathbf{P}^g \times \Delta$.

Proof We have a base change map [20]:

$$\tau(o) : \mathcal{E}xt^1_p(\Omega^1_{\mathcal{S}/\Delta}, \mathcal{L}^{-1})_o \otimes k(o) \longrightarrow \mathrm{Ext}^1_{S_0}\big(\Omega^1_{S_0}, L_0^{-1}\big)$$

with $L_0 = \mathcal{L}(o) = \mathcal{O}_{S_0}(1)$ (note that $\Omega^1_{\mathcal{S}/\Delta}$ is Δ-flat because p is flat and has reduced fibers). By our assumption (c) and the results in Sect. 4, the function

$$t \in \Delta \longmapsto \dim\big[\mathrm{Ext}^1_{\mathcal{S}(t)}(\Omega^1_{\mathcal{S}(t)}, \mathcal{L}(t)^{-1})\big]$$

is constant, hence $\mathcal{E}xt^1_p(\Omega^1_{\mathcal{S}/\Delta}, \mathcal{L}^{-1})$ is locally free and $\tau(o)$ is an isomorphism. It follows that there exists a section $E \in \mathcal{E}xt^1_p(\Omega^1_{\mathcal{S}/\Delta}, \mathcal{L}^{-1})$ such that $\tau(o)(E) = e_{S_0/V_0}$ (see 4.6 for the definition of e_{S_0/V_0}).

For all $t \in \Delta^\circ$, the smooth $K3$ surface $\mathcal{S}(t) \subset \mathbf{P}^g$ satisfies the assumptions of Theorem 3.4 by (a), and therefore there exists a unique extension $\mathcal{S}(t) \subset \mathcal{X}(t) \subset \mathbf{P}^{g+1}$ such that $e_{\mathcal{S}(t)/\mathcal{X}(t)} = E(t)$. We then consider $\mathcal{X}^\circ := \bigcup_{t\in\Delta^\circ} \mathcal{X}(t)$: it is a relative extension of $\mathcal{S}^\circ$, and its Zariski closure $\mathcal{X} = \overline{\mathcal{X}^\circ} \subset \mathbf{P}^{g+1} \times \Delta$ is a relative extension of $\mathcal{S}$.

Let $X_0 = \mathcal{X}(o)$. One has $e_{S_0/X_0} = E(o) = e_{S_0/V_0}$, so assumption (b) and Theorem 4.8 imply that $X_0 = V_0$, which ends the proof. □

We now set up the situation in which we will apply the above Theorem 5.2. The notation is the same as in 2.3.

5.3 Consider $\mathbf{P} = \mathbf{P}(a_0, a_1, a_2, a_3)$ a weighted projective space with Gorenstein canonical singularities, and (S, L) a general anticanonical divisor of $\mathbf{P}$, so $L = -K_{\mathbf{P}}|_S$. Let i_S be the divisibility of L in $\mathrm{Pic}(S)$, and L_1 be the primitive line bundle on S such that $L = i_S L_1$. Thus $(S, L_1) \in \mathcal{K}_{g_1}^{\mathrm{prim}}$, where $g_1 = h^0(L_1) - 1$.

We may then consider a deformation $\big(p : \mathcal{S} \to (\Delta, o), \mathcal{L}_1\big)$ of (S, L_1) to general primitive polarized smooth $K3$ surfaces of genus g_1. To such a smoothing, there is associated a flat family of surfaces in $\mathbf{P}^{g_1}$ as in (7) and also an analogous family in $\mathbf{P}^g$ defined by the sections of $\mathcal{L}_1^{i_S}$:

$$
\begin{array}{ccc}
\mathcal{S} & \overset{j_{i_S}}{\hookrightarrow} & \mathbf{P}^{g_{\mathbf{P}}} \times \Delta \\
& \underset{p}{\searrow} & \downarrow {\scriptstyle pr_2} \\
& & \Delta
\end{array}
\qquad (8)
$$

where $g = h^0(S, L) - 1$ (j_{i_S} is the i_S-uple Veronese re-embedding of $\mathcal{S}$).

We shall apply Theorem 5.2 to $S_0 = S \subset \mathbf{P}^g$, and $V_0 = \mathbf{P} \subset \mathbf{P}^{g+1}$ in its anticanonical embedding. In this case, assumption (a) is always satisfied, as a direct computation shows. Assumption (b) is always satisfied as well, because $\mathbf{P}$ has the property N_2 by Proposition 6.1 below. Assumption (c), however, does not hold in all cases: we compute $\alpha(S, L)$ in Proposition 6.2 below, and compare it with $\alpha(S', L')$ for a general $(S', L') \in \mathcal{K}_g^{i_S}$, in other words with $\alpha(S', (L_1')^{i_S})$ for a general $(S', L_1') \in \mathcal{K}_{g_1}^{\mathrm{prim}}$.

When $\alpha(S, L) = \alpha(S', L')$ holds, we conclude that $\mathbf{P}$ deforms to a threefold extension of a general $K3$ surface $(S', L') \in \mathcal{K}_g^{i_S}$. This happens exactly for cases $\#i$, $i \in \{1, \dots, 7, 9\}$, see Table 3. We refer to Sect. 7 for a more precise description of the output in each of these cases.

6 Explicit Computations on WPS

The main object of this section is to analyze which Gorenstein projective spaces enjoy the required properties for Theorem 5.2 to apply, as described in 5.3 above. We carry this out by explicit computations using the software Macaulay2 [23]. As a

Table 2 First Betti numbers of Gorenstein weighted projective spaces

#	Weights	g_1	i_S	g	β_1	β_2
1	(1, 1, 1, 3)	2	6	37	595	13056
2	(1, 1, 4, 6)	2	6	37	595	13056
3	(1, 2, 2, 5)	2	5	26	276	4025
4	(1, 1, 1, 1)	3	4	33	465	8960
5	(1, 1, 2, 4)	3	4	33	465	8960
6	(1, 3, 4, 4)	3	3	19	136	1344
7	(1, 1, 2, 2)	4	3	28	325	5175
8	(1, 2, 6, 9)	4	3	28	325	5175
9	(2, 3, 3, 4)	4	2	13	55	320
10	(1, 4, 5, 10)	6	2	21	171	1920
11	(1, 2, 3, 6)	7	2	25	253	3520
12	(1, 3, 8, 12)	7	2	25	253	3520
13	(2, 3, 10, 15)	16	1	16	91	715
14	(1, 6, 14, 21)	22	1	22	190	2261

bonus, we obtain the number of times each Gorenstein weighted projective 3-space is extendable.

Proposition 6.1 *Let* $\mathbf{P} \subset \mathbf{P}^{g+1}$ *be a* 3*-dimensional Gorenstein weighted projective space, considered in its anticanonical embedding. Then* $\mathbf{P}$ *is projectively normal, and its homogeneous coordinate ring* $A = R/I_{\mathbf{P}}$ *has a minimal resolution of the form*

$$\cdots \longrightarrow R(-3)^{\beta_2} \longrightarrow R(-2)^{\beta_1} \longrightarrow R \longrightarrow A \longrightarrow 0$$

with β_1, β_2 *as indicated in Table 2. In particular* $\mathbf{P}$ *has the* N_2 *property.*

Of course $\beta_1 = \binom{g-2}{2}$, since curve linear sections of $\mathbf{P}$ are canonical curves of genus g.

Proof The projective normality follows from the fact that $\mathbf{P}$ has canonical curves as linear sections, see [8, Theorem 5.1]. For property N_2, we explicitly compute the ideal of $\mathbf{P}$ in $\mathbf{P}^{g+1}$ using Macaulay2, then compute the first syzygies of this ideal, and eventually check that they are of the asserted shape. This computation goes as follows.

Let $\mathbf{P} = \mathbf{P}(a_0, a_1, a_2, a_3)$ endowed with weighted homogeneous coordinates $\mathbf{x} = (x_0, x_1, x_2, x_3)$. First, one writes down the list $(M_0, \ldots, M_{g+1})$ of all monomials in $\mathbf{x}$ of weighted degree $s = a_0 + a_1 + a_2 + a_3$, which form a basis of $H^0(\mathbf{P}, -K_{\mathbf{P}})$. Then the ideal of the graph $\Gamma \subset \mathbf{P} \times \mathbf{P}^{g+1}$ of the embedding $\mathbf{P} \subset \mathbf{P}^{g+1}$ is

$$I_\Gamma = \big(y_i - M_i(\mathbf{x}), \quad i = 0, \ldots, g+1\big),$$

with $(y_0, \ldots, y_{g+1})$ homogeneous coordinates on $\mathbf{P}^{g+1}$. One obtains the ideal $I_{\mathbf{P}}$ of $\mathbf{P} \subset \mathbf{P}^{g+1}$ by eliminating $\mathbf{x}$ from I_Γ, which may be performed efficiently using a Gröbner basis algorithm. Eventually, there is a Macaulay2 function which computes step by step the syzygies of this ideal. We provide the explicit Macaulay2 commands implementing this procedure at the end of the arXiv version of this article. □

In principle, one may use any basis of $H^0(\mathbf{P}, -K_{\mathbf{P}})$ to compute the ideal, but the computations turn out to work faster with a monomial basis. In fact doing so one takes advantage of $\mathbf{P}$ being a toric variety. There is also a Macaulay2 function computing the whole resolution of a graded ideal, but we have not been able to run these computations successfully for $I_{\mathbf{P}}$ (apart for #13) because the complexity was too large.

In principle, it is possible to compute all Betti numbers of any lattice ideal I_Λ as the dimensions of the reduced homology groups of a simplicial complex explicitly construct from the lattice Λ, see, e.g., [25, Theorem 9.2] or [26, Chap. 5]. It seems to us however that this leaves non-trivial computations to be performed, which we haven't tried to carry out.

Proposition 6.2 *Let* $\mathbf{P} \subset \mathbf{P}^{g+1}$ *be a 3-dimensional Gorenstein weighted projective space in its anticanonical embedding, and* (S, L) *be a general hyperplane section of* $\mathbf{P}$. *We write* i_S *for the divisibility of* $L = -K_{\mathbf{P}}|_S$ *in* Pic(S). *Let* (S', L') *be a general member of* $\mathcal{K}_g^{i_S}$. *Then the values of* $\alpha(S)$ *and* $\alpha(S')$ *are as indicated in Table 3. Moreover,* $\alpha(\mathbf{P}) = \alpha(S) - 1$, *and* $\alpha(C) = \alpha(S) + 1$ *for* C *a general curve linear section of* $\mathbf{P}$.

Table 3 Dimension of the weight -1 piece of T^1

#	Weights	g_1	i_S	$\alpha(S)$	$\alpha(S')$	3-fold
1	(1, 1, 1, 3)	2	6	1	1	$\mathbf{P}(1^3, 3)$
2	(1, 1, 4, 6)	2	6	1	1	$\mathbf{P}(1^3, 3)$
3	(1, 2, 2, 5)	2	5	3	3	$H_6 \subset \mathbf{P}(1^3, 3, 5)$
4	(1, 1, 1, 1)	3	4	1	1	$\mathbf{P}^3$
5	(1, 1, 2, 4)	3	4	1	1	$\mathbf{P}^3$
6	(1, 3, 4, 4)	3	3	4	4	$H_4 \subset \mathbf{P}(1^4, 3)$
7	(1, 1, 2, 2)	4	3	1	1	$\mathbf{Q}$
8	(1, 2, 6, 9)	4	3	2	1	$\mathbf{Q}$
9	(2, 3, 3, 4)	4	2	6	6	$H_3 \subset \mathbf{P}^4$
10	(1, 4, 5, 10)	6	2	3	1	V_5
11	(1, 2, 3, 6)	7	2	1	0	Does not exist
12	(1, 3, 8, 12)	7	2	2	0	Does not exist
13	(2, 3, 10, 15)	16	1	3	0	Does not exist
14	(1, 6, 14, 21)	22	1	2	0	Does not exist

In the table, we also indicate the general 3-fold extension of S', with the following notation: $\mathbf{Q}$ denotes the smooth 3-dimensional quadric in $\mathbf{P}^4$; H_d denotes a general degree d hypersurface in the specified projective space; V_5 denotes the degree 5 Del Pezzo threefold, i.e., the section of the Grassmannian $\mathbf{G}(2,5)$ by a general $\mathbf{P}^6$ in the Plücker embedding. We refer to [5, 6] for these matters.

We will need the following lemma for the proof, which is a generalization of a well-known fact when all involved varieties are smooth.

Lemma 6.3 *Let $\mathbf{P} \subset \mathbf{P}^{g+1}$ be a Gorenstein weighted projective space, S a hyperplane section of $\mathbf{P}$, and C a hyperplane section of S. Then one has*

$$\alpha(C) \geq \alpha(S) + 1 \geq \alpha(\mathbf{P}) + 2.$$

Proof We first compare $\alpha(\mathbf{P})$ and $\alpha(S)$. Since S is a hyperplane section of $\mathbf{P}$, one has $N_{\mathbf{P}/\mathbf{P}^{g+1}}\big|_S = N_{S/\mathbf{P}^g}$. We thus have the following exact sequence, where the rightmost map is the restriction map:

$$0 \to N_{\mathbf{P}/\mathbf{P}^{g+1}}(-2) \longrightarrow N_{\mathbf{P}/\mathbf{P}^{g+1}}(-1) \longrightarrow N_{S/\mathbf{P}^g}(-1) \to 0 \tag{9}$$

with $\mathcal{O}(1)$ the line bundle induced by the embedding in $\mathbf{P}^{g+1}$. By Proposition 6.1, $\mathbf{P} \subset \mathbf{P}^{g+1}$ has the property N_2, hence $H^0(N_{\mathbf{P}/\mathbf{P}^{g+1}}(-2)) = 0$ (see [18, Lemma 1.1] and the references therein). So the long exact sequence induced by (9) shows the inequality

$$h^0\big(N_{\mathbf{P}/\mathbf{P}^{g+1}}(-1)\big) \leq h^0\big(N_{S/\mathbf{P}^g}(-1)\big).$$

By the definition of α in Theorem 3.3, this ends the proof. The inequality between $\alpha(S)$ and $\alpha(C)$ is obtained in the same way. □

Proof of Proposition 6.2 We know the ideal $I_{\mathbf{P}}$ of $\mathbf{P} \subset \mathbf{P}^{g+1}$ from the proof of Proposition 6.1. Using the Macaulay2 package "VersalDeformations" [17] one can then compute $\dim(T^1_{C\mathbf{P},-1})$, and this equals $\alpha(\mathbf{P})$ by Corollary 4.4.

Next, we choose two explicit (see below) linear functionals l_0 and l_1 defining hyperplanes H_0 and H_1 in $\mathbf{P}^{g+1}$, and consider $S_0 = \mathbf{P} \cap H_0 \subset \mathbf{P}^g$ and $C_0 = S_0 \cap H_1 \subset \mathbf{P}^{g-1}$. Using the same procedure we compute $\dim(T^1_{CS_0,-1})$ and $\dim(T^1_{CC_0,-1})$, and find out that

$$\dim(T^1_{CS_0,-1}) = \alpha(\mathbf{P}) + 1 \quad \text{and} \quad \dim(T^1_{CC_0,-1}) = \alpha(\mathbf{P}) + 2.$$

Again, the explicit Macaulay2 commands implementing this procedure are given at the end of the arXiv version of this article.

Let S be a general hyperplane section of $\mathbf{P}$. Then on the one hand one has $\alpha(S) \geq \alpha(\mathbf{P}) + 1$ by Lemma 6.3 above, and on the other hand one has $\alpha(S) \leq \alpha(S_0)$ by semicontinuity since $\alpha(S) = h^0(N_{S/\mathbf{P}^g}(-1)) - g - 1$ by definition, and $\alpha(S_0) \leq \dim(T^1_{CS_0,-1})$ by Corollary 4.3. Hence $\alpha(S) = \alpha(\mathbf{P}) + 1$. Similar reasoning yields $\alpha(C) = \alpha(\mathbf{P}) + 2$ for a general curve linear section of $\mathbf{P}$. □

In practice, if one chooses random linear functionals l_0 and l_1 then the complexity of the computation of the weight -1 piece of T^1 is too high and one cannot get an answer. We chose

$$l_0 = x_7 + x_{g+1} \quad \text{and} \quad l_1 = x_3 + x_g,$$

so that the corresponding linear sections are again toric; in particular S_0 is not a $K3$ surface and C_0 is singular. This is the reason why we have to resort to Corollary 4.3 in the proof; note that we cannot guarantee either that $\dim(T^1_{CS,-1})$ is semi-continuous as S approaches S_0. In principle, Macaulay2 can compute $h^0(N_{S_0/\mathbf{P}^g}(-1))$ directly, but in practice, it is not able to return an answer.

Corollary 6.4 *Let $\mathbf{P} \subset \mathbf{P}^{g+1}$ be a 3-dimensional Gorenstein weighted projective space in its anticanonical embedding. Then $\mathbf{P}$ is extendable exactly $\alpha(\mathbf{P})$ times.*

(Recall that $\alpha(\mathbf{P}) = \alpha(S) - 1$ with $\alpha(S)$ as in Table 3).

Proof First note that by Lvovski's Theorem 3.3, applied to a general (smooth) curve linear section of $\mathbf{P}$, $\mathbf{P} \subset \mathbf{P}^{g+1}$ is at most $\alpha(\mathbf{P})$-extendable. To prove the converse, let us consider C a general curve linear section of $\mathbf{P}$. It is a smooth canonical curve of genus $g \geq 11$ and Clifford index strictly larger than 2, hence liable to Theorem 3.4. So there exists a universal extension of C, which is an $(\alpha(C) + 1)$-dimensional variety $X \subset \mathbf{P}^{g-1+\alpha(C)}$, i.e., an $(\alpha(\mathbf{P}) + 3)$-dimensional variety $X \subset \mathbf{P}^{g+1+\alpha(\mathbf{P})}$.

The pencil of hyperplanes in $\mathbf{P}^{g+1}$ containing C cuts out on $\mathbf{P}$ a pencil of $K3$ surfaces, which are not all isomorphic by [27, Proposition 1.7] (as observed in [7], the latter statement in fact applies to all varieties different from cones). By the universality of X, this implies that $\mathbf{P}$ is a linear section of X, hence it is $\alpha(\mathbf{P})$-extendable. □

7 Examples

In this section, we describe explicitly the output of Theorem 5.2 and make additional remarks. We first list the cases to which Theorem 5.2 applies; see also Remarks 7.7 and 7.8 for another point of view on these examples. The notation is that of Table 3.

Example 7.1 (#1 and #2) The general member of $\mathcal{K}^6_{37}$ extends to $\mathbf{P}(1^3, 3)$, hence the application of Theorem 5.2 to #1 is trivial. On the other hand, the application to #2 tells us that there exists a deformation of $\mathbf{P}(1, 1, 4, 6)$ to $\mathbf{P}(1^3, 3)$. Note that these are the only Fano varieties with canonical Gorenstein singularities of genus 37, the maximal possible value, by [30]. $\mathbf{P}(1^3, 3) \subset \mathbf{P}^{38}$ is the 2-Veronese reembedding of the cone in $\mathbf{P}^{10}$ over the Veronese variety $v_2(\mathbf{P}^3)$; in particular it is rigid. Thus the deformation of $\mathbf{P}(1, 1, 4, 6)$ to $\mathbf{P}(1^3, 3)$ exhibits a jump phenomenon.

Example 7.2 (#3) Theorem 5.2 tells us in this case that $\mathbf{P}(1, 2, 2, 5) \subset \mathbf{P}^{27}$ deforms to a general 6-ic hypersurface $H_6 \subset \mathbf{P}(1^3, 3, 5)$ in its anticanonical embedding by $\mathcal{O}(5)$. Such an H_6 is singular, and its singularities may be listed following [13]; in particular as $5 \nmid 6$, H_6 passes through the point $P_4 = (0:0:0:0:1)$ and one finds it has a quotient singularity of type $\frac{1}{5}(1, 1, 3)$ there.

Corollary 6.4 tells us that $\mathbf{P}(1, 2, 2, 5) \subset \mathbf{P}^{27}$ is 2-extendable, as is $H_6 \subset \mathbf{P}(1^3, 3, 5)$. The same deformation argument as that given to prove Theorem 5.2 shows that the 2-extension of $\mathbf{P}(1, 2, 2, 5)$ deforms to that of H_6, which is a sextic hypersurface $\tilde{H}_6 \subset \mathbf{P}(1^3, 3, 5^3)$ embedded by $\mathcal{O}(5)$, that is $-\frac{1}{3}K_{\tilde{H}_6}$, see [5].

Example 7.3 (#4 and #5) #4 is of course the Veronese variety $v_4(\mathbf{P}^3)$, which is rigid and extends the general member of $\mathcal{K}^4_{33}$; Theorem 5.2 is trivial in this case. The application to #5 however tells us that $\mathbf{P}(1, 1, 2, 4) \subset \mathbf{P}^{34}$ smoothes to $v_4(\mathbf{P}^3)$, which may be seen elementarily as follows.

Spelling out a monomial basis of $H^0(\mathbf{P}(1, 1, 2, 4), \mathcal{O}(4))$, one sees that $\mathcal{O}(4)$ induces an embedding of $\mathbf{P}(1, 1, 2, 4)$ as a cone over $\mathbf{P}(1, 1, 2)$ embedded by its own $\mathcal{O}(4)$, with vertex a point, in $\mathbf{P}^9$. In turn $\mathbf{P}(1, 1, 2, 4) \subset \mathbf{P}^{34}$ is embedded by $\mathcal{O}(8)$, hence it is the 2-Veronese reembedding of the latter cone in $\mathbf{P}^9$. In the same way, the embedding of $\mathbf{P}(1, 1, 2)$ by $\mathcal{O}(4)$ is the 2-Veronese reembedding of a quadric cone (of rank 3) in $\mathbf{P}^3$.

Thus in the embedding by $\mathcal{O}(4)$, $\mathbf{P}(1, 1, 2, 4)$ is the cone over a section of the Veronese variety $v_2(\mathbf{P}^3)$ by a tangent hyperplane. This deforms to the cone over a section by a transverse hyperplane (this corresponds to smoothing the quadric in $\mathbf{P}^3$ image of $\mathbf{P}(1, 1, 2)$ by $\mathcal{O}(2)$). In turn, this deforms to the Veronese variety $v_2(\mathbf{P}^3)$ itself by "sweeping out the cone" (see 4.6). In its anticanonical embedding, $\mathbf{P}(1, 1, 2, 4)$ correspondingly deforms to the 2-Veronese re-embedding of $v_2(\mathbf{P}^3)$, which is the Veronese variety $v_4(\mathbf{P}^3)$.

Example 7.4 (#6) This case is similar to #3 and we will be brief. Theorem 5.2 provides a deformation of $\mathbf{P}(1, 3, 4, 4) \subset \mathbf{P}^{20}$ to the anticanonical embedding by $\mathcal{O}(3)$ of a general 4-ic $H_4 \subset \mathbf{P}(1^4, 3)$. The latter is singular; in particular as $3 \nmid 4$, H_4 always passes through the coordinate point P_4 and has a quotient singularity of type $\frac{1}{3}(1, 1, 1)$ there, i.e., , it is locally isomorphic to the cone over the Veronese variety $v_3(\mathbf{P}^2)$.

The argument of Theorem 5.2 shows that the 3-extension of $\mathbf{P}(1, 3, 4, 4) \subset \mathbf{P}^{20}$ deforms to that of H_4, which is a 4-ic hypersurface $\tilde{H}_4 \subset \mathbf{P}(1^4, 3^4)$ embedded by $\mathcal{O}(3)$, see [6, Sect. 3].

Example 7.5 (#7) Theorem 5.2 provides a smoothing of $\mathbf{P}(1, 1, 2, 2) \subset \mathbf{P}^{29}$ to a smooth quadric $\mathbf{Q}$, in its canonical embedding. This smoothing may be elementarily found, noting (as we did for case #5) that $\mathcal{O}(2)$ realizes $\mathbf{P}(1, 1, 2, 2)$ as a rank 3 quadric in $\mathbf{P}^4$.

Example 7.6 (#9) This case is similar to #3 and #6, and in fact easier, so we will be very brief. Theorem 5.2 proves that $\mathbf{P}(2, 3, 3, 4)$ deforms to a general cubic hypersurface H_3 in $\mathbf{P}^4$, in particular this is a smoothing. The 5-extension of $\mathbf{P}(2, 3, 3, 4)$ deforms to that of H_3, which is a complete intersection $\tilde{H}_2 \cap \tilde{H}_3$ in $\mathbf{P}(1^5, 2^6)$, see [6, Sect. 3].

Remark 7.7 The degeneration of $\mathbf{P}(1^3, 3)$ to $\mathbf{P}(1^2, 4, 6)$ may be seen explicitly as follows; this has been shown to us by the referee, inspired by [15, Sect. 11]. The weighted projective space $\mathbf{P}(1^2, 4, 6)$ is $\mathrm{Proj}(R)$ with R the graded algebra $\mathbf{C}[x, y, z, w]$ in which x, y, z, w have respective weights 1, 1, 4, 6. It is isomorphic to $\mathrm{Proj}(R^{(2)})$ where $R^{(2)}$ is the algebra determined by $\mathcal{O}(2)$ on $\mathbf{P}(1^2, 4, 6)$, i.e., the graded piece $R^{(2)}_n$ is R_{2n} for all $n \in \mathbf{Z}$ by definition.

We claim that $\mathrm{Proj}(R^{(2)})$ is naturally a quadric in $P(1^3, 2, 3)$. To see this we note that $R^{(2)}$ is generated as a $\mathbf{C}$-algebra by

$$x^2, xy, y^2, z, w,$$

which have weights 2, 2, 2, 4, 6 in R, hence 1, 1, 1, 2, 3 in $R^{(2)}$. The only relation between them is $x^2 \cdot y^2 = (xy)^2$, so

$$R^{(2)} \cong \frac{\mathbf{C}[a, b, c, u, v]}{(ac - b^2)},$$

the isomorphism being given by mapping x^2, xy, y^2, z, w to a, b, c, u, v, respectively. Therefore, $\mathrm{Proj}(R^{(2)})$ is the quadric $ac = b^2$ in $\mathrm{Proj}(\mathbf{C}[a, b, c, u, v]) = \mathbf{P}(1^3, 2, 3)$.

The degeneration is then gotten by noting that $\mathbf{P}(1^3, 3)$ is the quadric $u = 0$ in $\mathbf{P}(1^3, 2, 3)$. In the pencil of quadrics

$$ac - b^2 + \lambda u = 0,$$

the member given by $\lambda = 0$ is $\mathbf{P}(1^2, 4, 6)$ and all the others are isomorphic to $\mathbf{P}(1^3, 3)$.

Remark 7.8 In fact, following a suggestion of the referee, we have found that all our examples above may be understood as in the previous Remark 7.7. In general, we shall consider the d-Veronese embedding, i.e., the graded ring $R^{(d)}$, with $d = s/i_S$, where s is the sum of the weights so that $\omega_{\mathbf{P}}^{-1} = \mathcal{O}(s)$, and i_S is as in Table 3. Let us briefly indicate the explicit computations. We take $R = \mathbf{C}[x, y, z, w]$ the graded algebra giving the weighted projective space under consideration, as in Remark 7.7.

Example 7.2: $\mathbf{P}(1, 2, 2, 5)$ is itself a sextic hypersurface in $P(1^3, 3, 5)$. Indeed the algebra $R^{(2)}$ is generated by

$$x^2, y, z, xw, w^2$$

which have weights 2, 2, 2, 6, 10 in R, hence 1, 1, 1, 3, 5 in $R^{(2)}$. The only relation between these generators is $x^2 \cdot w^2 = (xw)^2$, which is in weight 6 in $R^{(2)}$, so that $\mathrm{Proj}(R^{(2)})$ is naturally a sextic hypersurface in $P(1^3, 3, 5)$.

Example 7.3: $\mathbf{P}(1^2, 2, 4)$ is isomorphic to a quadric in $\mathbf{P}(1^4, 2)$, hence it is a degeneration of $\mathbf{P}^3$. Indeed the algebra $R^{(2)}$ is generated by

$$x^2, xy, y^2, z, w$$

which have weights $2, 2, 2, 2, 4$ in R, hence $1, 1, 1, 1, 2$ in $R^{(2)}$. The only relation between these generators is $x^2 \cdot y^2 = (xy)^2$, which has weight 2 in $R^{(2)}$.

Example 7.4: $\mathbf{P}(1, 3, 4, 4)$ is itself a quartic in $\mathbf{P}(1^4, 3)$. Indeed the algebra $R^{(4)}$ is generated by

$$x^4, xy, y^4, v, w$$

which have weights $4, 4, 12, 4, 4$ in R, hence $1, 1, 3, 1, 1$ in $R^{(4)}$. The only relation between these generators is $x^4 \cdot y^4 = (xy)^4$, which has weight 4 in $R^{(2)}$.

Example 7.5: we have already noted that $\mathcal{O}(2)$ realizes $\mathbf{P}(1, 1, 2, 2)$ as a rank 3 quadric in $\mathbf{P}^4$.

Example 7.6: $\mathbf{P}(2, 3, 3, 4)$ is the complete intersection of a quadric and a cubic in $\mathbf{P}(1^5, 2)$, and thus it is a degeneration of a cubic in $\mathbf{P}^4$. The algebra $R^{(6)}$ is generated by

$$x^3, xw, w^3, y^2, yz, z^2$$

which have weights $6, 6, 12, 6, 6, 6$ in R, hence $1, 1, 2, 1, 1, 1$ in $R^{(6)}$. This time there are two relations, namely

$$x^3 \cdot w^3 = (xw)^3 \quad \text{and} \quad y^2 \cdot z^2 = (yz)^2,$$

which have respectively degrees 3 and 2 in $R^{(6)}$.

We conclude with some remarks on the cases in which Theorem 5.2 does not apply because $\alpha(S) > \alpha(S')$; the notation is still that of Table 3.

Proposition 7.9 *Let $\mathbf{P} = \mathbf{P}(a_0, a_1, a_2, a_3)$ be a Gorenstein weighted projective 3-space of type #i with $i \in \{8, 10, 11, 12, 13, 14\}$. Then an anticanonical divisor of $\mathbf{P}$ is a double cover of the weighted projective plane $\mathbf{P}(a_0, a_1, a_2)$, branched over a bianticanonical divisor $B \in |-2K_{\mathbf{P}(a_0,a_1,a_2)}|$. In all cases $2K_{\mathbf{P}(a_0,a_1,a_2)}$ is invertible, whereas in all cases but #11 the canonical sheaf $K_{\mathbf{P}(a_0,a_1,a_2)}$ is not invertible.*

Proof The key fact is that in all cases one has $a_3 = a_0 + a_1 + a_2$. Then in homogeneous coordinates $(x_0 : \cdots : x_3)$, a degree $s = a_0 + a_1 + a_2 + a_3$ homogeneous polynomial is of the form

$$x_3^2 + x_3 \cdot f_{a_3}(x_0, x_1, x_2) + f_s(x_0, x_1, x_2),$$

with f_d homogeneous of degree d. We may change weighted homogeneous coordinates by setting $x_3' = x_3 + \frac{1}{2} f_{a_3}$. This gives the polynomial

$$(x_3')^2 + f_s'(x_0, x_1, x_2)$$

where $f_s' = f_s - f_{a_3}^2$, which defines a double cover of $\mathbf{P}(a_0, a_1, a_2)$ as asserted. The last affirmation is readily checked using the statements of Sect. 2. $\square$

Example 7.10 (#11) The anticanonical divisors in $\mathbf{P}(1, 2, 3, 6)$ are double covers of $\mathbf{P}(1, 2, 3)$, which is in fact a Del Pezzo surface of degree 6, with one A_1 and one A_2 double points, which may be constructed by blowing up the plane $\mathbf{P}^2$ along three aligned infinitely near points.

In the embedding by $\mathcal{O}(6) = -\frac{1}{2}K_{\mathbf{P}}$, $\mathbf{P}(1, 2, 3, 6)$ is the cone over this toric Del Pezzo surface, in its anticanonical embedding in $\mathbf{P}^6$. It follows that $\mathbf{P}(1, 2, 3, 6) \subset \mathbf{P}^7$ is a limit of cones over smooth Del Pezzo surfaces of degree 6. Every such cone T is obstructed in $\mathrm{Hilb}^{\mathbf{P}^7}$, being in the closure of two components, one parametrizing embedded $\mathbf{P}^1 \times \mathbf{P}^1 \times \mathbf{P}^1$'s and the other hyperplane sections of $\mathbf{P}^2 \times \mathbf{P}^2$, as observed in [10, Example 4.5]. Therefore $[\mathbf{P}(1, 2, 3, 6)] \in \mathrm{Hilb}^{\mathbf{P}^7}$ is obstructed as well. In fact the embedded versal deformation of $\mathbf{P}(1, 2, 3, 6) \subset \mathbf{P}^7$ has been explicitly computed, see [1, 4].

On the other hand, $\mathcal{O}(12) = \mathcal{O}(K_{\mathbf{P}})$ embeds $\mathbf{P}(1, 2, 3, 6) \subset \mathbf{P}^{26}$ and $\alpha(\mathbf{P}, K_{\mathbf{P}}) = 0$, while its general anticanonical divisor S satisfies $\alpha(S, -K_{\mathbf{P}}|_S) = 1$ (see Table 3). Therefore, by Theorem 4.8, $\overline{CS}$ has a unique 1-parameter deformation to $\mathbf{P}$.

Acknowledgements ES thanks Alessio Corti and Massimiliano Mella for enlightening conversations. ThD thanks Laurent Busé for having shown him the elimination technique enabling the computation of the ideal of a weighted projective space, and Enrico Fatighenti for introducing him to the Macaulay2 package "VersalDeformations". We thank the referee for useful comments and bibliographical suggestions.

References

1. K. Altmann, One parameter families containing three dimensional toric Gorenstein singularities, in *Explicit Birational Geometry of 3-Folds*, LMS Lecture Notes 281 (Cambridge University Press, Cambridge, 2000), pp. 21–50
2. E. Arbarello, A. Bruno, E. Sernesi, On hyperplane sections of K3 surfaces. Algebr. Geom. **4**, 562–596 (2017)
3. M. Beltrametti, L. Robbiano, Introduction to the theory of weighted projective spaces. Expo. Math. **4**, 111–162 (1986)
4. G. Brown, M. Reid, J. Stevens, Tutorial on Tom and Jerry: the two smoothings of the anticanonical cone over $\mathbf{P}(1, 2, 3)$. EMS Sunv. Math. Sci. 8 (2021), no 1/2, 25–38 arXiv:1812.02594
5. C. Ciliberto, T. Dedieu, Double covers and extensions. to appear in Kyoto J. of Math. arXiv:2008.03109
6. C. Ciliberto, T. Dedieu, $K3$ curves with index $k > 1$. Bollettino UMI 15 (2022), 87–115 arXiv:2012.10642
7. C. Ciliberto, Th. Dedieu, C. Galati, A.L. Knutsen, Moduli of curves on Enriques surfaces. Adv. Math. 365 (2020)
8. C. Ciliberto, T. Dedieu, E. Sernesi, Wahl maps and extensions of canonical curves and K3 surfaces. J. Reine Angew. Math. **761**, 219–245 (2020)
9. C. Ciliberto, A.F. Lopez, R. Miranda, Classification of varieties with canonical curve section via gaussian maps on canonical curves. Amer. J. of Math. **120**, 1–21 (1998)
10. S. Coughlan, T. Sano, Smoothing cones over K3 surfaces. EPIGA 2 (2018), Nr. 15
11. I. Dolgachev, Weighted projective varieties, in *Springer Lecture Notes in Mathematics*, vol. 956 (1982), pp. 34–71
12. R. Donagi, D.R. Morrison, Linear systems on K3 sections. J. Differ. Geom. **29**, 49–64 (1989)

13. A.R. Iano-Fletcher, Working with weighted complete intersections, in *Explicit Birational Geometry of 3-Folds*, Mathematical Society Lecture Note Series, vol. 281 (Cambridge University Press, Cambridge, 2000), pp. 101–173
14. M. Green, R. Lazarsfeld, Special divisors on curves on a K3 surface. Inventiones Math. **89**, 357–370 (1987)
15. P. Hacking, Compact moduli of plane curves. Duke Math. J. **124**(2), 213–257 (2004)
16. R. Hartshorne, Stable reflexive sheaves. Math. Ann. **254**, 121–176 (1980)
17. N. Ilten, Versal deformations and local Hilbert schemes. J. Softw. Algebra Geom. **4**, 12–16 (2012)
18. A.L. Knutsen, Global sections of twisted normal bundles of K3 surfaces and their hyperplane sections. Rend. Lincei Math. Appl. **31**, 57–79 (2020)
19. J. Kollar, *Rational Curves on Algebraic Varieties*. Ergebnisse b. 32 (Springer, 1999)
20. H. Lange, Universal families of extensions. J. Algebra **83**, 101–112 (1983)
21. A.F. Lopez, On the extendability of projective varieties: a survey (with an appendix by Thomas Dedieu) in this volume. arXiv:2102.04431
22. S. Lvovski, Extensions of projective varieties and deformations. I, II. Michigan Math. J. **39**(1), 41–51, 65–70 (1992)
23. D.R. Grayson, M.E. Stillman, Macaulay2, a software system for research in algebraic geometry. http://www.math.uiuc.edu/Macaulay2/
24. M. Manetti, Normal degenerations of the complex projective plane. J. Reine Angew. Math. **419**, 89–118 (1991)
25. E. Miller, B. Sturmfels, *Combinatorial Commutative Algebra*. GTM 227 (Springer, 2005)
26. D. Maclagan, R.R. Thomas, S. Faridi, L. Gold, A.V. Jayanthan, A. Khetan, T. Puthenpurakal, Computational algebra and combinatorics of toric ideals, in *Commutative Algebra and Combinatorics* Ramanujan Mathematical Society. Lecture Notes Series 4 (2007)
27. R. Pardini, Some remarks on varieties with projectively isomorphic hyperplane sections. Geom. Dedicata. **52**, 15–32 (1994)
28. H. Pinkham, Deformations of algebraic varieties with $\mathbf{C}^*$ action. Asterisque 20 (1974)
29. Y.G. Prokhorov, A remark on Fano threefolds with canonical Gorenstein singularities, in *The Fano Conference* (University of Torino, Turin, 2004), pp. 647–657
30. Y.G. Prokhorov, The degree of Fano threefolds with canonical Gorenstein singularities. Math. Sb. **196**, 81–122 (2005)
31. M. Reid, Special linear systems on curves lying on a K3 surface. J. Lond. Math. Soc. **13**, 454–458 (1976)
32. M. Reid, Canonical 3-folds, in *Algebraic Geometry - Angers 1979* (Sijthoff & Noordhoff, 1980), pp. 273–310
33. M. Reid, Projective morphisms according to Kawamata (University of Warwick, 1983)
34. M. Schlessinger, On rigid singularities. Rice Univ. Studies **59**, 147–162 (1973)
35. J. Wahl, On cohomology of the square of an ideal sheaf. J. Algebraic Geom. **6**, 481–511 (1997)

Intersection Cohomology and Severi Varieties

Vincenzo Di Gennaro and Davide Franco

Abstract Let $X^{2n} \subseteq \mathbb{P}^N$ be a smooth projective variety. Consider the intersection cohomology complex of the local system $R^{2n-1}\pi_*\mathbb{Q}$, where π denotes the projection from the universal hyperplane family of X^{2n} to $(\mathbb{P}^N)^\vee$. We investigate the cohomology of the intersection cohomology complex $IC(R^{2n-1}\pi_*\mathbb{Q})$ over the points of a Severi variety, parametrizing nodal hypersurfaces, whose nodes impose independent conditions on the very ample linear system giving the embedding in $\mathbb{P}^N$.

Keywords Intersection cohomology · Decomposition Theorem · Normal functions · Hodge conjecture · Severi varieties

MSC2010 Primary 14B05 · Secondary 14E15 · 14F05 · 14F43 · 14F45 · 14M15 · 32S20 · 32S60 · 58K15

1 Introduction

In the last years a great deal of work has been devoted to focusing on the deep relationship among Hodge conjecture and singularities of normal functions (compare, e.g., with [25, 30] and references therein). The theory of normal functions, which dates back to Poincaré, Lefschetz and Hodge, had a renewed interest in last years after a crucial remark of Green and Griffiths [25] that the Hodge conjecture is equivalent

---

Dedicated to Ciro Ciliberto on his seventieth birthday.

V. Di Gennaro (✉)
Dipartimento di Matematica, Università di Roma "Tor Vergata", Via della Ricerca Scientifica, 00133 Roma, Italy
e-mail: digennar@axp.mat.uniroma2.it

D. Franco
Dipartimento di Matematica e Applicazioni "R. Caccioppoli", Università di Napoli "Federico II", Via Cintia, 80126 Napoli, Italy
e-mail: davide.franco@unina.it

T. Dedieu et al. (eds.), *The Art of Doing Algebraic Geometry*, Trends in Mathematics,
https://doi.org/10.1007/978-3-031-11938-5_6

to the existence of appropriately defined singularities for the normal function defined by means of a primitive Hodge cycle.

One of the starting points of this remark is a fundamental result of Kleiman concerning the smoothability of algebraic cycles of intermediate dimension ([31], [23, Example 15.3.2]). In light of this, R. Thomas showed inductively that the Hodge conjecture reduces to the following statement concerning middle dimensional Hodge cycles: *for all even dimensional smooth complex projective varieties* $(X^{2n}, \mathcal{O}(1))$ *and any class* $A \in H^{n,n}(X, \mathbb{C}) \cap H^{2n}(X, \mathbb{Q})$, *there is a nodal hypersurface* $D \subset X$ *in* $| \mathcal{O}_X(l) |$ *for some* l, *such that the Poincaré dual of* A *is in the image of the pushforward map* $H_{2n}(D, \mathbb{Q}) \to H_{2n}(X, \mathbb{Q})$ [39].

In their fundamental work [25], Green and Griffiths further clarified the question by relating it to the singularities of normal functions. Specifically, they showed that the above hypersurface D would be a singular point of the normal function associated to the Hodge class A, thus reducing the Hodge conjecture to the existence of such singularities.

In the paper [25], one can find various definitions of singularities of normal functions. Some of them are formalized by de Cataldo and Migliorini by means of the Decomposition Theorem [7]. More precisely, let $X = X^{2n}$ be a $2n$-dimensional smooth complex projective variety embedded in $\mathbb{P}^N_{\mathbb{C}} \equiv \mathbb{P}$ via a complete linear system $| H |$. Denote by $X^\vee$ the *dual variety* of X and consider the *universal hyperplane family*

$$X \xleftarrow{q} \mathcal{X} \xrightarrow{\pi} \mathbb{P}^\vee, \qquad \dim \mathcal{X} = 2n - 1 + N.$$

The hyperplane sections are the fibers of π:

$$\mathcal{X}_H := \pi^{-1}(H), \qquad \dim \mathcal{X}_H = 2n - 1.$$

We observe that the projection from the universal hyperplane family $\mathcal{X} \xrightarrow{\pi} \mathbb{P}^\vee$ is a proper map with equidimensional fibres and it is smooth outside the dual variety $X^\vee$.

As explained in [7, Sect. 2], the Decomposition Theorem for π provides a non-canonical decomposition

$$R\pi_*\mathbb{Q}_{\mathcal{X}} \cong \bigoplus_{i\in\mathbb{Z}} \bigoplus_{j\in\mathbb{N}} IC(L_{ij})[-i - (2n - 1 + N)], \quad \text{in } D^b_c(\mathbb{P}^\vee), \tag{1}$$

where L_{ij} denotes a local system on a suitable stratum of codimension j in $\mathbb{P}^\vee$. By [7], the most important summand for our purposes is the one related to $L_{00} = (R^{2n-1}\pi_*\mathbb{Q}_{\mathcal{X}})_{|U}$ $(U := \mathbb{P}^\vee \backslash X^\vee)$ [7, 2.5]. Specifically, if one fix an intermediate primitive Hodge class ξ, de Cataldo and Migliorini define the *singular locus* of the normal function of ξ as the support of the component of ξ belonging to $\mathcal{H}^{-N+1}IC(R^{2n-1}\pi_*\mathbb{Q}_{\mathcal{X}})$ in the decomposition above (which is well defined in view of the perverse filtration) [7, Definition 3.3, Remark 3.4]. The main result of [7] says such a singularity is able to detect the local triviality of ξ.

In view of the aforementioned work of Thomas, it is particularly interesting to take a closer look at the cohomology sheaf $\mathcal{H}^{-N+1}IC(R^{2n-1}\pi_*\mathbb{Q}_{\mathcal{X}})$, especially in correspondence of nodal hypersurfaces. This paper, which is a small step in this direction, is devoted to the study of $IC(R^{2n-1}\pi_*\mathbb{Q}_{\mathcal{X}})_D$ with D *nodal divisor whose nodes impose independent conditions on the linear system giving the embedding* $X \subseteq \mathbb{P}$. Under this hypothesis, what we are going to do is *to compute the cohomology of the complex* $IC(R^{2n-1}\pi_*\mathbb{Q}_{\mathcal{X}})_D$ *and, above all, to give it a geometric interpretation.*

The hypothesis that the nodes impose independent conditions to the hypersurfaces of a very ample linear system has long been investigated in relation to the study of Severi varieties, parametrizing irreducible nodal curves on a smooth algebraic surface (compare, e.g., with [4, 37, 41] and references therein). In particular, in the paper [37], Sernesi proves that such a condition implies that Severi variety is smooth of the expected dimension, by using deformation theory.

Our first result consists of a different proof of the same result, extended to smooth algebraic varieties of even dimension (compare with Theorem 3.3 and Remark 3.4). Our approach, which is independent of deformation theory and consists in a careful local study of the conormal map, allows us to prove that *the dual variety, in a neighborhood of a nodal hypersurface whose nodes impose independent conditions, is a divisor with normal crossings* (cf. Theorem 3.3).

Understanding the local structure of the dual manifold allows us in Sect. 3 to compute the cohomology of the complex $IC(R^{2n-1}\pi_*\mathbb{Q}_{\mathcal{X}})_D$, where D denotes a nodal hypersurface whose nodes pose independent conditions, and above all to give it a geometric interpretation. More precisely, we see that the cohomology is concentrated in degrees $-N$ and $-N+1$, that $\mathcal{H}^{-N}IC(R^{2n-1}\pi_*\mathbb{Q}_{\mathcal{X}})_D$ is naturally isomorphic to $H^{2n-1}(D)$, and that $\mathcal{H}^{-N+1}IC(R^{2n-1}\pi_*\mathbb{Q}_{\mathcal{X}})_D$ is related with the *defect* of the nodes (Remark 4.7). Furthermore, we prove that *the perverse filtration of* $H^{2n}(D)$ *is as simple as possible because it consists of only two pieces that vary in local systems over any component of the Severi variety* (Corollary 4.6).

As a by-product, we see that *under the hypothesis that a hyperplane cuts X in a set of nodes imposing independent conditions, the pull-back of a primitive Hodge cycle coincides with its local Green–Griffiths invariant.* In particular, we get a different proof of the fact that *the local Green–Griffiths invariant detects the local triviality of a Hodge cycle*, in our context [7, Proposition 3.8 (ii)].

Last but not least, in Sect. 4 we provide several examples of even-dimensional smooth projective varieties equipped with linear systems containing nodal hypersurfaces D, whose nodes impose independent conditions and such that $\mathcal{H}^{-N+1}IC(R^{2n-1}\pi_*\mathbb{Q}_{\mathcal{X}})_D$ is non-trivial.

2 Notations and Basic Facts

Notation 2.1 From now on, unless otherwise stated, all cohomology and intersection cohomology groups are with $\mathbb{Q}$-coefficients.

(1) For a complex algebraic variety X, we denote by $H^l(X)$ and $IH^l(X)$ its cohomology and intersection cohomology groups. Let $D^b_c(X)$ be the constructible derived category of sheaves of $\mathbb{Q}$-vector spaces on X. For a complex of sheaves $\mathcal{F} \in D^b_c(X)$, we denote by $\mathcal{H}^l(\mathcal{F})$ the lth cohomology sheaf of $\mathcal{F}$ and by $\mathbb{H}^l(\mathcal{F})$ the lth hypercohomology group of $\mathcal{F}$. Let IC_X denotes the intersection cohomology complex of X. If X is nonsingular, we have $IC_X \cong \mathbb{Q}_X[\dim_{\mathbb{C}} X]$, where $\mathbb{Q}_X$ is the constant sheaf $\mathbb{Q}$ on X.

(2) More generally, let $i : S \to X$ be a locally closed embedding of a smooth irreducible subspace of X and let L be a local system on S. Denote by $IC_{\overline{S}}(L) := i_{!*}L[\dim S] \in D^b_c(X)$ the *intersection cohomology complex* of L [18, Sect. 5.2], [6, Sect. 2.7]. It is defined as the *intermediary extension* of L, that is the unique extension of L in $D^b_c(X)$ with neither subobjects nor quotients supported on $\overline{S}\backslash S$.

In this paper, X denotes a $2n$-dimensional smooth complex projective variety embedded in $\mathbb{P}^N_{\mathbb{C}} \equiv \mathbb{P}$ via a complete linear system $\mid H \mid$. Denote by $X^\vee$ the *dual variety* of X and consider the *universal hyperplane family* $\mathcal{X} \subset X \times \mathbb{P}^\vee$. We have natural projections:

$$X \xleftarrow{q} \mathcal{X} \xrightarrow{\pi} \mathbb{P}^\vee, \qquad \dim \mathcal{X} = 2n - 1 + N.$$

The hyperplane sections are the fibers of π:

$$\mathcal{X}_H := \pi^{-1}(H) = X \cap H, \qquad \dim \mathcal{X}_H = 2n - 1.$$

Let $\mathrm{Con}(X) \subset \mathcal{X}$ be the *conormal variety of* X:

$$\mathrm{Con}(X) := \{(p, H) \in X \times \mathbb{P}^\vee : TX_p \subseteq H\}, \qquad \dim \mathrm{Con}(X) = N - 1,$$

where TX_p denotes the embedded tangent space to X at p. We denote by $\pi_1 : \mathrm{Con}(X) \to \mathbb{P}^\vee$ the restriction of $\pi : \mathcal{X} \longrightarrow \mathbb{P}^\vee$ to $\mathrm{Con}(X)$. We have $X^\vee = \pi_1 (\mathrm{Con}(X))$. If $\dim X^\vee < N - 1$, then, for a general $H \in X^\vee$, the fiber $\pi_1(H)^{-1}$ has positive dimension. This means that the general tangent hyperplane to X is tangent along a subvariety of positive dimension. Actually, the contact locus is a linear space by the bi-duality theorem, and for this reason, by substituting H with any multiple of it, one obtains a variety X whose dual variety is a hypersurface. So we can always assume that $X^\vee$ *is a hypersurface* of $\mathbb{P}^\vee$.

If $H \in U := \mathbb{P}^\vee\backslash X^\vee$, then $\mathcal{X}_H$ is smooth hence $\pi : \pi^{-1}(U) \to U$ is a smooth fibration. This implies that the sheaf $R^{2n-1}\pi_*\mathbb{Q}_{\mathcal{X}}$ restricts to a local system on U.

As explained in [7, Sect. 2], the Decomposition Theorem for π provides a noncanonical decomposition

$$R\pi_*\mathbb{Q}_{\mathcal{X}} \cong \bigoplus_{i\in\mathbb{Z}} \bigoplus_{j\in\mathbb{N}} IC(L_{ij})[-i - (2n - 1 + N)], \quad \text{in } D^b_c(\mathbb{P}^\vee), \tag{2}$$

where L_{ij} denote a local system on a suitable stratum of codimension j in $\mathbb{P}^\vee$. By [7], the most important summand for the purpose of detecting the primitive Hodge classes of X, is the one related to $L_{00} = (R^{2n-1}\pi_*\mathbb{Q}_{\mathcal{X}})_{|U}$ [7, 2.5].

The main aim of this paper is to investigate the cohomology sheaves of the complex $IC((R^{2n-1}\pi_*\mathbb{Q}_{\mathcal{X}})_{|U})[-N] \in D^b_c(\mathbb{P}^\vee)$, near the points corresponding to *nodal divisors*.

One can find different approaches to the Decomposition Theorem [2, 5, 6, 35, 42], which is a very general result but also rather implicit. On the other hand, there are many special cases for which the decomposition theorem admits a simplified and explicit approach. One of these is the case of varieties with isolated singularities [15–17, 22, 34]. For instance, in the work [15], one can find a simplified approach to the Decomposition Theorem for varieties with isolated singularities, in connection with the existence of a *natural Gysin morphism*, as defined in [13, Definition 2.3]. One of the main ingredients of these arguments is a generalization of the Leray–Hirsch theorem in a categorical framework ([40, Theorem 7.33], [14, Lemma 2.5]).

3 Local Study of the Dual Hypersurface

Let X be a $2n$-dimensional smooth complex projective variety embedded in $\mathbb{P}^N_{\mathbb{C}} \equiv \mathbb{P}$ via a complete linear system $\mid H \mid$. Assume moreover that $X^\vee$, the *dual variety* of X, is a hypersurface in $\mathbb{P}^\vee$ (compare with the previous section).

With notations as above, for any $H \in X^\vee$, the corresponding hyperplane cuts X in a singular divisor $\mathcal{X}_H := \pi^{-1}(H) = X \cap H$. A singular point of the hypersurface $\mathcal{X}_H \subset X$ is called an *ordinary double point*, or *node*, if, in local analytic coordinates $(u_1, \dots, u_{2n})$, $\mathcal{X}_H$ is given by $u_1^2 + \cdots + u_{2n}^2 = 0$ [18, p. 17].

Definition 3.1 Let $X_r^\vee \subseteq X^\vee$ be the *locally closed* Zariski subset of $X^\vee$ containing all the hyperplanes H s.t. Sing $\mathcal{X}_H$ consists in exactly r nodes. An irreducible component of $X_r^\vee$ is said to have the *expected dimension* if its dimension is equal to $N - r$ (in what follows we assume $r \leq N$). If $H \in X_r^\vee$, then we say that $X_r^\vee$ *has the expected dimension at* H, if it belongs to a component having the expected dimension and not in components having bigger dimension.

Remark 3.2 A priori, according to the previous definition, $X_r^\vee$ may have the expected dimension at a singular point H. However, in next Theorem 3.3, we will see that if Sing$(\mathcal{X}_H)$ imposes independent conditions to $\mid H \mid$, then $X_r^\vee$ is smooth at H.

The following Theorem 3.3 is probably well known via deformation theory. Here we include a slightly different proof of it in the attempt of making the present paper reasonably self-contained. Our argument is very classical and similar to the one used by Severi to study the local geometry of Severi varieties of plane nodal curves: the Zariski closure of the Severi variety of 1-nodal degree n plane curves contains the locus of δ-nodal degree n curves and, at every point of this locus, it is the intersection of δ smooth branches meeting transversally [1, 38].

Theorem 3.3 *Assume that* $\Delta := \mathrm{Sing}(\mathcal{X}_H)$ *consists of* δ *nodes. If* Δ *imposes independent conditions to* $\mid H \mid$, *that is if* $H^1(\mathcal{I}_{\Delta,X}(1)) = 0$, *then for every small ball* $B \subseteq \mathbb{P}^\vee$ *containing* H *we have:*

(1) $B \cap X^\vee$ *is a divisor of* B *with normal crossings;*
(2) for every $r \leq \delta$, $B \cap X_r^\vee$ *is non-empty smooth of pure dimension* $N - r$.

In particular, $X_\delta^\vee$ *is smooth and has the expected dimension at* H.

Proof By [29, Proposition 3.3], the projection $\pi_1 : \mathrm{Con}(X) \to \mathbb{P}^\vee$ is unramified at (x_i, H), where the x_i's are the nodes of X_H (observe that $(x_i, H) \in \mathrm{Con}(X)$, $\forall 1 \leq i \leq \delta$). Hence, π_1 provides an embedding of a suitable analytic neighborhood U_i of (x_i, H) in $\mathbb{P}^\vee$:

$$(x_i, H) \in U_i \subset \mathrm{Con}(X).$$

Hence, the image $\mathcal{U}_i := \pi_1(U_i)$ provides a branch of $X^\vee$ passing through H, $\forall i$ such that $1 \leq i \leq \delta$. Notice that $\mathcal{U}_i$ is smooth at H because π_1 has injective differential.

Since having nodes is a locally open property and $X^\vee \backslash \mathrm{Sing}(X^\vee)$ is dense in $X^\vee$ [29, (4.1) and Lemme 4.1.2, p. 235], each $\mathcal{U}_i$ intersects $X^\vee \backslash \mathrm{Sing}(X^\vee)$. Since $X^\vee \backslash \mathrm{Sing}(X^\vee)$ consists of the hyperplanes whose intersection with X has exactly one singular point, which is a node [29, Proposition 3.2, p. 227], it follows that π_1 induces a bijection between:

$$\pi_1^{-1}(X^\vee \backslash \mathrm{Sing}(X^\vee)) = \{(p, L) \in \mathrm{Con}(X) : p \text{ is a node for } X \cap L \text{ and } \mathrm{Sing}(X \cap L) = \{p\}\},$$

and $X^\vee \backslash \mathrm{Sing}(X^\vee)$. Therefore, we can find a sequence $(p_n^i, H_n^i) \in U_i \cap \pi_1^{-1}(X^\vee \backslash \mathrm{Sing}(X^\vee))$ such that

$$H_n^i \in X^\vee \backslash \mathrm{Sing}(X^\vee) \quad \text{with} \quad (p_n^i, H_n^i) \quad \text{specializing to} \quad (x_i, H).$$

By [27, p. 209], the embedded tangent space of $\mathcal{U}_i$ at H_n^i is p_n^i viewed as a hyperplane of $\mathbb{P}^\vee$. Hence, the embedded tangent space of the branch $\mathcal{U}_i$ at H (in the following denoted by $T_{H,\mathcal{U}_i}$) is x_i (viewed as a hyperplane of $\mathbb{P}^\vee$).

On the other hand, the hypothesis $H^1(\mathcal{I}_{\Delta,X}(1)) = 0$, applied to the short exact sequence

$$0 \to \mathcal{I}_{\Delta,X}(1) \to \mathcal{O}_X(1) \to \mathcal{O}_\Delta(1) \to 0,$$

implies that $h^0(\mathcal{I}_{\Delta,X}(1)) = N + 1 - \delta$ and the nodes span a linear subspace of dimension $\delta - 1$ in $\mathbb{P}$. By our previous argument, we have

$$\bigcap_{i=1}^{\delta} T_{H,\mathcal{U}_i} = < x_1, \ldots, x_\delta >^\vee = \mathbb{P}(H^0(\mathcal{I}_{\Delta,X}(1))) \quad \text{and} \quad \dim \bigcap_{i=1}^{\delta} T_{H,\mathcal{U}_i} = N - \delta.$$

In other words, the branches of $X^\vee$ at H are independent and $X^\vee$ is a divisor with normal crossings around H.

Consider now a subset $I \subseteq \{1, \dots, \delta\}$ and set

$$X_I^\vee := \bigcap_{i \in I} \mathcal{U}_i .$$

As the branches $\mathcal{U}_i$ have independent tangent hyperplanes at H, then for every sufficiently small ball $B \subseteq \mathbb{P}^\vee$ containing H, $B \cap X_I^\vee$ is a smooth complete intersection. Furthermore, we have

$$\dim B \cap X_I^\vee = N - \mid I \mid \quad \text{and} \quad \dim B \cap \mathcal{U}_j \cap X_I^\vee = N - 1 - \mid I \mid, \quad \forall j \notin I. \tag{3}$$

We point out that one could also deduce (3) from the factoriality of the ring of holomorphic functions defined in B [26, p. 10]. Indeed, by factoriality, each branch $\mathcal{U}_i$ is defined in B by an analytic function f_i. Since $X^\vee$ is a divisor with normal crossings around H, the sequence $f_1, \dots, f_\delta$ *is a regular sequence of analytic functions* in B. Thus, any subset of $f_1, \dots, f_\delta$ is regular as well and (3) follows at once.

Finally, the following locally closed subset of $X^\vee$

$$B \cap X_I^\vee \backslash \left(\bigcup_{j \notin I} B \cap \mathcal{U}_j \cap X_I^\vee \right)$$

is a non-empty analytic subspace of $X_{|I|}^\vee$, with the expected dimension. □

Remark 3.4 The final part of the proof of Theorem 3.3 shows that under the hypothesis $H^1(\mathcal{I}_{\Delta,X}(1)) = 0$, *the nodes of H can be independently smoothed.* This fact is usually proved by means of deformation theory (compare with [4, 37, 41]). We preferred to use a different approach here, because it is more suited to our purposes.

4 Intersection Cohomology Complex on Severi Varieties

Notation 4.1 (1) Assume now that the hypotheses of Theorem 3.3 are verified. In particular, there exists a tubular neighborhood T of some connected component $\mathcal{C}$ of $X_\delta^\vee$ such that $T \cap X^\vee$ is a *divisor with normal crossings* in T. Set $T^0 := T \backslash (T \cap X^\vee)$. The local system $(R^{2n-1}\pi_*\mathbb{Q})_{|T^0}$ has a canonical extension to a holomorphic vector bundle $\mathcal{V}$ on T (see [8, 36]).

(2) Fix $H \in U$. Combining Hard Lefschetz Theorem with Lefschetz Hyperplane Theorem we have

$$H^{2n}(\mathcal{X}_H) \cong H^{2n-2}(\mathcal{X}_H) \cong H^{2n-2}(X).$$

Hence, $R^{2n}\pi_*\mathbb{Q}_{\mathcal{X}}$ is a *constant system* on U. Put $h := h^{2n-2}(X)$.

Remark 4.2 By Thom's first isotopy lemma [24, Theorem 5.2], the family $\mathcal{X}\,|_{\mathcal{C}}$ is locally trivial thus both $(R^{2n-1}\pi_*\mathbb{Q}_{\mathcal{X}})_{|\mathcal{C}}$ and $(R^{2n}\pi_*\mathbb{Q}_{\mathcal{X}})_{|\mathcal{C}}$ are local systems over $\mathcal{C}$. Nevertheless, in the following theorem, we give a direct proof of this result, that we believe in independent interest.

Theorem 4.3 *Assume the hypotheses of Theorem 3.3 are verified for $H \in X^\vee_\delta$. Fix a connected component $\mathcal{C}$ of $X^\vee_\delta$ containing H. With notations as above, we have an isomorphism of local systems on $\mathcal{C}$*

$$\mathcal{H}^0(IC((R^{2n-1}\pi_*\mathbb{Q}_{\mathcal{X}})_{|U})[-N])_{|\mathcal{C}} \cong (R^{2n-1}\pi_*\mathbb{Q}_{\mathcal{X}})_{|\mathcal{C}}, \tag{4}$$

and a short exact sequence of local systems on $\mathcal{C}$

$$0 \to \mathcal{H}^1(IC((R^{2n-1}\pi_*\mathbb{Q}_{\mathcal{X}})_{|U})[-N])_{|\mathcal{C}} \to (R^{2n}\pi_*\mathbb{Q}_{\mathcal{X}})_{|\mathcal{C}} \to \mathbb{Q}^h \to 0. \tag{5}$$

Furthermore, we have

$$\mathcal{H}^i(IC((R^{2n-1}\pi_*\mathbb{Q}_{\mathcal{X}})_{|U})[-N])_{|\mathcal{C}} = 0, \quad \forall i \geq 2. \tag{6}$$

In particular, for any nodal divisor $H \in X^\vee_\delta$, we have

- $h^0(IC((R^{2n-1}\pi_*\mathbb{Q}_{\mathcal{X}})_{|U})[-N])_H = h^{2n-1}(\mathcal{X}_H)$,
- $h^1(IC((R^{2n-1}\pi_*\mathbb{Q}_{\mathcal{X}})_{|U})[-N])_H = h^{2n}(\mathcal{X}_H) - h$,
- $h^i(IC((R^{2n-1}\pi_*\mathbb{Q}_{\mathcal{X}})_{|U})[-N])_H = 0$, $\forall i \geq 2$.

Remark 4.4 Observe that, for every i, one has

$$\mathcal{H}^i(IC((R^{2n-1}\pi_*\mathbb{Q}_{\mathcal{X}})_{|U})[-N]) = \mathcal{H}^{i-N}(IC((R^{2n-1}\pi_*\mathbb{Q}_{\mathcal{X}})_{|U})).$$

Proof of Theorem 4.3 Assume that the hypotheses of Theorem 3.3 are verified. In particular, there exists a tubular neighborhood T of $\mathcal{C}$ such that $T \cap X^\vee$ is a divisor with normal crossings in T. Set $T^0 := T\backslash(T \cap X^\vee)$. The local system $(R^{2n-1}\pi_*\mathbb{Q})_{|T^0}$ has a canonical extension to a local system $\mathcal{V}$ on T (see [8, 36]).

Further, in a suitable neighborhood of $H \in \mathcal{C}$, the equation of $X^\vee$ has the form $t_1 \dots t_\delta = 0$ and the local system $(R^{2n-1}\pi_*\mathbb{Q}_{\mathcal{X}})_{|T^0}$ has monodromy operators $T_1, \dots, T_\delta$, with T_i given by moving around the hyperplane $t_i = 0$. If we denote by N_i the *logarithm of the monodromy operator* T_i, by [3] and [30, p. 322] the cohomology

$$\mathcal{H}^i(IC((R^{2n-1}\pi_*\mathbb{Q}_{\mathcal{X}})_{|U})[-N])_H$$

of the intersection cohomology complex at $H \in \mathcal{C}$ can be computed as the ith cohomology of the complex of finite-dimensional vector spaces

$$B^p := \bigoplus_{i_1 < i_2 < \cdots < i_p} N_{i_1} N_{i_2} \ldots N_{i_p} \mathcal{V}_H,$$

with differential acting on the summands by the rule

$$N_{i_1} \ldots \hat{N}_{i_l} \ldots N_{i_{p+1}} \mathcal{V}_H \stackrel{(-1)^{l-1} N_{i_l}}{\longrightarrow} N_{i_1} \ldots N_{i_l} \ldots N_{i_{p+1}} \mathcal{V}_H.$$

Fix $H \in X_\delta^\vee$. Since $\mathcal{X}_H$ is nodal, the logarithm of the monodromy operators N_i act according to the *Picard–Lefschetz formula*. Furthermore, as $\mathcal{X}_H$ has δ ordinary double points, the vanishing spheres are disjoint to each other and we have

$$N_i N_j = 0, \quad \text{for any} \quad i \neq j.$$

So the complex above is concentrated in degrees 0 and 1. We have

$$\mathcal{H}^i(IC((R^{2n-1}\pi_*\mathbb{Q}_\mathcal{X})_{|U})[-N])_H = 0, \quad \forall H \in X_\delta^\vee, \quad \forall i \geq 2.$$

Thus (6) is proved.

Furthermore, we have the following exact sequence

$$\begin{aligned} 0 \to \mathcal{H}^0(IC((R^{2n-1}\pi_*\mathbb{Q}_\mathcal{X})_{|U})[-N])_H \to \mathcal{V}_H \to \bigoplus_i N_i \mathcal{V}_H \to \\ \to \mathcal{H}^1(IC((R^{2n-1}\pi_*\mathbb{Q}_\mathcal{X})_{|U})[-N])_H \to 0. \end{aligned} \tag{7}$$

On the other hand, consider a hyperplane $H_t \in U$ very near to H, such that $\mathcal{X}_{H_t}$ is smooth, and denote by B_i a small ball around the ith node of $\mathcal{X}_H$. By excision, we have

$$H^l(\mathcal{X}_H, \cup_i(\mathcal{X}_H \cap B_i)) \cong H^l(\mathcal{X}_{H_t}, \cup_i(\mathcal{X}_{H_t} \cap B_i)).$$

From the exact sequence

$$\cdots \to H^{l-1}(\cup_i(\mathcal{X}_H \cap B_i)) \to H^l(\mathcal{X}_H, \cup_i(\mathcal{X}_H \cap B_i)) \to H^l(\mathcal{X}_H) \to H^l(\cup_i(\mathcal{X}_H \cap B_i)) \to \cdots,$$

and recalling the conic nature of isolated singularities of a divisor [33], we get

$$H^l(\mathcal{X}_H) \cong H^l(\mathcal{X}_H, \cup_i(\mathcal{X}_H \cap B_i)) \cong H^l(\mathcal{X}_{H_t}, \cup_i(\mathcal{X}_{H_t} \cap B_i)), \text{ if } l \geq 2.$$

Inserting in the relative cohomology sequence for the pair $(\mathcal{X}_{H_t}, \cup_i(\mathcal{X}_{H_t} \cap B_i))$, we find the exact sequence

$$0 \to H^{2n-1}(\mathcal{X}_H) \to H^{2n-1}(\mathcal{X}_{H_t}) \to H^{2n-1}(\cup_i(\mathcal{X}_{H_t} \cap B_i)) \cong \mathbb{Q}^\delta \to \\ \to H^{2n}(\mathcal{X}_H) \to H^{2n-2}(X) \to 0, \tag{8}$$

where we have also taken into account the isomorphism (compare with Notations 4.1)

$$H^{2n-2}(X) \cong H^{2n}(\mathcal{X}_{H_t}).$$

By the Picard–Lefschetz formula, the map $H^{2n-1}(\mathcal{X}_{H_t}) \to \mathbb{Q}^\delta$ coincides, up to some irrelevant sign, with the map $\mathcal{V}_H \to \bigoplus_i N_i \mathcal{V}_H$ of the sequence (7). Hence, comparing (7) and (8), we find

$$\mathcal{H}^0(IC((R^{2n-1}\pi_*\mathbb{Q}_\mathcal{X})_{|U})[-N])_H \cong H^{2n-1}(\mathcal{X}_H) \cong (R^{2n-1}\pi_*\mathbb{Q}_\mathcal{X})_H$$

and

$$0 \to \mathcal{H}^1(IC((R^{2n-1}\pi_*\mathbb{Q}_\mathcal{X})_{|U})[-N])_H \to (R^{2n}\pi_*\mathbb{Q}_\mathcal{X})_H \to \mathbb{Q}^h \to 0.$$

Thus, (4) and (5) are proved for any $H \in \mathcal{C}$ and we need only to prove a similar result at the level of local systems on $\mathcal{C}$. First of all, the fact that $T \cap X^\vee$ is a divisor with normal crossings implies that, if we fix a small neighborhood B of H in $\mathbb{P}^\vee$, then the local fundamental group $\pi_1(B\backslash(B \cap X^\vee), H) \cong \mathbb{Z}^\delta$ is independent of $H \in X^\vee_\delta$. So, the rank of $\mathcal{V}_{|X^\vee_\delta} \to \bigoplus_i N_i \mathcal{V}_{|X^\vee_\delta}$ does not change as long as we let H to vary in $\mathcal{C}$, and its kernel is a vector bundle on $\mathcal{C}$. Furthermore, from the description of the canonical extension given, e.g., in [36, Sect. 2], one infers that such a kernel is the $\mathbb{Z}^\delta$-invariant part of the local system $(R^{2n-1}\pi_*\mathbb{Q}_\mathcal{X})_{|T^0} \otimes \mathbb{C}$. Moreover, since the tubular neighborhood T is homeomorphic to a fiber bundle on $X^\vee_\delta$, the long exact sequence of homotopy groups of T^0

$$\cdots \to \mathbb{Z}^\delta \to \pi_1(T^0, H) \to \pi_1(\mathcal{C}, H) \to 0$$

shows that the kernel of $\mathcal{V}_{|\mathcal{C}} \to \bigoplus_i N_i \mathcal{V}_{|\mathcal{C}}$ descends to a local system on $\mathcal{C}$, consisting in the $\mathbb{Z}^\delta$-invariant part of the local system $(R^{2n-1}\pi_*\mathbb{Q}_\mathcal{X})_{|T^0} \otimes \mathbb{C}$, i.e., with $(R^{2n-1}\pi_*\mathbb{Q}_\mathcal{X})_{|X^\vee_\delta} \otimes \mathbb{C}$ by taking also into account of (8). This concludes the proof of (4).

As for the proof of (5), we argue in a similar way. First of all we observe that the vector space $\bigoplus_i N_i \mathcal{V}_H$ is contained in the $\mathbb{Z}^\delta$-invariant part of the local system $(R^{2n-1}\pi_*\mathbb{Q}_\mathcal{X})_{|T^0} \otimes \mathbb{C}$ as well. This follows just combining [32, p. 42] (recall that the dimension of $\mathcal{X}_{H_t}$ is odd), with the fact that the vanishing spheres are disjoint to each other. By the same argument as above, $\bigoplus_i N_i \mathcal{V}_H$ is the stalk at H of a local system $\mathcal{C}$, on which $\pi_1(\mathcal{C}, H)$ acts by "exchanging the branches". By (7), $\mathcal{H}^1(IC((R^{2n-1}\pi_*\mathbb{Q}_\mathcal{X})_{|U})[-N])_H$ is a local system $\mathcal{C}$ as well and (5) follows by comparison with (8). □

Remark 4.5 Theorem 4.3 implies that both $(R^{2n-1}\pi_*\mathbb{Q}_{\mathcal{X}})_{|C}$ and $(R^{2n}\pi_*\mathbb{Q}_{\mathcal{X}})_{|C}$ are local systems on any component $\mathcal{C}$ of $X^\vee_\delta$ having the expected dimension. We observed that this also follows Thom's first isotopy lemma.

Fix again a nodal hypersurface $\mathcal{X}_H$ such that $H \in \mathcal{C}$. As explained in [7, 2.11], the decomposition (2) produces a *perverse filtration* on the groups $H^l(\mathcal{X}_H) = H^l(X \cap H)$. By [7, Lemma 3.1], $H^{2n}(\mathcal{X}_H) = H^{2n}_{\leq 1}(\mathcal{X}_H)$ and the graded piece $H^{2n}_1(\mathcal{X}_H)$ coincides with the stalk at H of the constant system $(R^{2n}\pi_*\mathbb{Q}_{\mathcal{X}})_U$. Furthermore, in the proof of [7, Lemma 3.1] it is shown that $\mathcal{H}^1(IC((R^{2n-1}\pi_*\mathbb{Q}_{\mathcal{X}})_{|U})[-N])_H$ is one of the summands of the graded piece $H^{2n}_0(\mathcal{X}_H)$. By (5) of Theorem 4.3, we conclude that these are the only non-trivial summands of the perverse filtration of $H^{2n}(\mathcal{X}_H) = H^{2n}(X \cap H)$. In a nutshell, we have proved that *the perverse filtration is as simple as possible because it consists of only two pieces that vary in local systems over any component of the Severi variety.*

Corollary 4.6 *With notations as above, let $H \in \mathcal{C}$. In the perverse filtration of $H^{2n}(\mathcal{X}_H)$ we have $H^{2n}_{\leq -1}(\mathcal{X}_H) = 0$ and $H^{2n}_0(\mathcal{X}_H) \cong \mathcal{H}^1(IC((R^{2n-1}\pi_*\mathbb{Q}_{\mathcal{X}})_{|U})[-N])_H$. In other words, the sequence*

$$0 \to \mathcal{H}^1(IC((R^{2n-1}\pi_*\mathbb{Q}_{\mathcal{X}})_{|U})[-N])_{|C} \to (R^{2n}\pi_*\mathbb{Q}_{\mathcal{X}})_{|C} \to \mathbb{Q}^h \to 0,$$

represents the perverse filtration of the local system $(R^{2n}\pi_*\mathbb{Q}_{\mathcal{X}})_{|C}$.

By [7, Lemma 3.1 (ii)], any primitive Hodge class belongs to $H^{2n}_{\leq 0}(\mathcal{X}_H)$. In view of previous Corollary, we have $H^{2n}_{\leq 0}(\mathcal{X}_H) = H^{2n}_0(\mathcal{X}_H) \cong \mathcal{H}^1(IC((R^{2n-1}\pi_*\mathbb{Q}_{\mathcal{X}})_{|U})[-N])_H$. This implies that, *under the hypothesis that our hyperplane cuts X in a set of nodes imposing independent conditions, the pull-back of a primitive Hodge cycle coincide with its local Green–Griffiths invariant.* In particular, we get a different proof of the fact that *the local Green–Griffiths invariant detects the local triviality of an Hodge cycle* [7, Proposition 3.8 (ii)].

Remark 4.7 Since $h^{2n}(\mathcal{X}_H) - h = h^{2n}(\mathcal{X}_H) - h^{2n-2}(\mathcal{X}_H)$, the difference $h^{2n}(\mathcal{X}_H) - h$, in some sense, measures how much Poincaré Duality fails for $\mathcal{X}_H$. We call it the *defect* of $\mathcal{X}_H$. By Theorem 4.3 we see that $h^1(IC((R^{2n-1}\pi_*\mathbb{Q}_{\mathcal{X}})_{|U})[-N])_H$ coincides with the defect of $X \cap H$, which is constant on some connected component of $\mathcal{X}_\delta$. We note in passing that a great deal of work has been devoted to the study of the defect of a nodal hypersurface, in relation to the position and the number of the nodes. In fact, when X is a projective space, and, in certain cases, when X is a fourfold, one may interpret the defect in terms of the Hilbert function of the set of the nodes of $\mathcal{X}_H$ [18, Sect. 6.4, p. 209], [10–12]. In the paper [9] one can find a construction of factorial complete intersections of codimension 2, having (asymptotically) a maximal number of nodes (see also [20, 21] for other results concerning smooth projective varieties of codimension 2).

5 Examples

In this section we provide several examples of even-dimensional smooth projective varieties equipped with linear systems containing nodal hypersurfaces D, whose nodes impose independent conditions.

5.1 *Curves in a Projective Surface*

Let X denotes a smooth projective surface embedded in some projective space via a very ample divisor H. Let $C \subset X$ be a smooth curve in X and let K_X denotes the canonical divisor of X. If $n \gg 0$, we can assume

$$H^1(\mathcal{I}_{C,X}(n)) = 0, \text{ and } \mid nH - C \mid \text{ very ample on X.} \tag{9}$$

Under these conditions, the general curve $R \in\mid nH - C \mid$ is smooth, $\Delta := R \cap C$ is reduced and $R \cup C \sim nH$ has only ordinary double points.

Proposition 5.1 *With notations as above, assume conditions (9) verified. Assume additionally that $C \cdot K_X < 0$. Then the set of nodes $\Delta = R \cap C$ imposes independent conditions to the linear system $\mid nH \mid$, that is to say $H^1(\mathcal{I}_{\Delta,X}(n)) = 0$.*

Proof From the cohomology exact sequence deduced from the following short exact sequence

$$0 \to \mathcal{I}_{C,X}(n) \to \mathcal{I}_{\Delta,X}(n) \to \mathcal{I}_{\Delta,C}(n) \to 0,$$

and taking into account of $H^1(\mathcal{I}_{C,X}(n)) = 0$ (remember (9)), we see that it suffices to prove $H^1(\mathcal{I}_{\Delta,C}(n)) = 0$. On the other hand, by adjunction we have

$$\omega_C \cong \mathcal{O}_C(K_X + C) \cong \mathcal{O}_C(K_X + nH - R) \cong \mathcal{I}_{\Delta,C}(K_X + nH).$$

In view of the hypothesis $C \cdot K_X < 0$, we conclude at once:

$$H^1(\mathcal{I}_{\Delta,C}(n)) \cong H^1(\omega_C(-K_X)) \cong H^0(\mathcal{O}_C(K_X)) = 0.$$

□

Remark 5.2 We observe that the hypothesis of proposition above is satisfied when either K_X is negative or when C is an exceptional curve.

5.2 Defective Hypersurfaces of $\mathbb{P}^{2n}$

Let $X = \mathbb{P}^{2n}$. Let $L = \mathbb{P}^n$ be the linear subspace of dimension $n \geq 2$, given by $x_0 = x_1 = \cdots = x_{n-1} = 0$. Let $a_0, a_1, \ldots, a_{n-1} \in \mathbb{C}[x_0, x_1, \ldots, x_{2n}]_{k-1}$ be general homogeneous polynomials of degree $k - 1 \geq 1$. Let D be the hypersurface given by the equation

$$a_0x_0 + a_1x_1 + \cdots + a_{n-1}x_{n-1} = 0.$$

D is a general hypersurface of degree $k \geq 2$, containing L. The singular locus Δ of D is a set of $\delta = (k-1)^n$ nodes, complete intersection of type $(k-1, \ldots, k-1)$ in L. Let $\mathcal{I}_\Delta$ the ideal sheaf of Δ in L. Then, the Koszul complex of Δ in L is

$$0 \to \wedge^n E^* \to \wedge^{n-1} E^* \to \ldots E^* \to \mathcal{I}_\Delta \to 0,$$

with $E = \mathcal{O}_{\mathbb{P}^n}(k-1)^n$. We observe that $\wedge^n E^* = \mathcal{O}_{\mathbb{P}^n}(n(1-k))$.

The set Δ imposes independent conditions to the linear system $|\mathcal{O}_X(k)|$ if and only if $h^1(\mathbb{P}^n, \mathcal{I}_\Delta(k)) = 0$. Using kernels and images, we may split the Koszul complex above into short exact sequences. Taking the cohomology long exact sequence of these short sequences, one may reduce the vanishing $h^1(\mathbb{P}^n, \mathcal{I}_\Delta(k)) = 0$ to the vanishing of

$$h^n(\mathbb{P}^n, \mathcal{O}_{\mathbb{P}^n}(n(1-k) + k)).$$

Hence, in order to have $h^1(\mathbb{P}^n, \mathcal{I}_\Delta(k)) = 0$, it suffices that

$$k < \frac{2n+1}{n-1}.$$

This is true when $n = 2$ e $k \leq 4$, $n = 3$ and $k \leq 3$, $n \geq 4$ e $k = 2$.

Therefore, in all these cases the nodes impose independent conditions.

Moreover, we also have that $\mathcal{H}^1 IC(R^{2n-1}\pi_*\mathbb{Q})_D[-N]$ is non-trivial, because a nodal hypersurface $D \subset X = \mathbb{P}^{2n}$, containing a $\mathbb{P}^n$, has defect. In fact, suppose, on the contrary, that $H_{2n}(D) \cong \mathbb{Q}$ (compare with Remark 4.7). Then the class $[L] \in H_{2n}(D)$ of $L = \mathbb{P}^n$ is a rational multiple of the class of a linear section of D. Let $D' \subset X' = \mathbb{P}^{2n-1}$ be a general (hence smooth) hyperplane section of D. The Gysin map $H_{2n}(D) \to H_{2n-2}(D')$ [23, p. 382, Example 19.2.1.] sends $[L]$ in the class $[L']$ of a $L' = \mathbb{P}^{n-1} \subset D'$, which is a rational multiple of the class of a linear section of D', because so is $[L]$ in $H_{2n}(D)$. On the other hand, by Poincaré Duality and Lefschetz Hyperplane Theorem, the Gysin map $H_{2n}(X', \mathbb{Z}) \to H_{2n-2}(D', \mathbb{Z})$ has torsion-free cokernel. It follows that $[L']$ actually is an *integral* multiple of the class of a general linear section of D'. This is impossible, because $\deg L' = 1$.

5.3 Defective Hypersurfaces in a Complete Intersection of Quadrics

Let $L = \mathbb{P}^n \subset \mathbb{P}^{2n+h}$ be the linear subspace of dimension $n \geq 2$, given by $x_0 = x_1 = \cdots = x_{n+h-1} = 0$. Let

$$A := \begin{bmatrix} a_0^1 & a_1^1 & \dots & a_{n+h-1}^1 \\ a_0^2 & a_1^2 & \dots & a_{n+h-1}^2 \\ \dots & \dots & & \dots \\ a_0^h & a_1^h & \dots & a_{n+h-1}^h \\ \alpha_0 & \alpha_1 & \dots & \alpha_{n+h-1} \end{bmatrix}$$

be a matrix with $h+1$ rows and $n+h$ columns, whose entries are given by general linear forms in $\mathbb{C}[x_0, x_1, \dots, x_{2n+h}]_1$. Let $X = X^{2n} \subset \mathbb{P}^{2n+h}$ given by the equations

$$X = \begin{cases} a_0^1 x_0 + a_1^1 x_1 + \dots + a_{n+h-1}^1 x_{n+h-1} = 0 \\ a_0^2 x_0 + a_1^2 x_1 + \dots + a_{n+h-1}^2 x_{n+h-1} = 0 \\ \dots \\ a_0^h x_0 + a_1^h x_1 + \dots + a_{n+h-1}^h x_{n+h-1} = 0. \end{cases}$$

$X = X^{2n} \subset \mathbb{P}^{2n+h}$ is a general complete intersection of h quadrics containing $L = \mathbb{P}^n$. Let $D = X \cap Q$ be the hypersurface of X given cutting X with a further general quadric Q containing L, defined by the equation

$$\alpha_0 x_0 + \alpha_1 x_1 + \dots + \alpha_{n+h-1} x_{n+h-1} = 0.$$

Let

$$\phi : \mathcal{O}_{\mathbb{P}^n}^{h+1} \to \mathcal{O}_{\mathbb{P}^n}(1)^{h+n}$$

be the map defined by the transpose of the matrix A. The map ϕ is a general morphism, whose degeneracy locus $\Delta = D_h(\phi)$ is the singular locus of $D = X \cap Q$, consisting of $\delta = \binom{n+h}{n}$ nodes. The Eagon–Northcott complex of ϕ is

$$0 \to S^{n-1}\mathcal{O}_{\mathbb{P}^n}^{h+1} \to S^{n-2}\mathcal{O}_{\mathbb{P}^n}^{h+1} \otimes \mathcal{O}_{\mathbb{P}^n}(1)^{h+n} \to S^{n-3}\mathcal{O}_{\mathbb{P}^n}^{h+1} \otimes \wedge^2 \mathcal{O}_{\mathbb{P}^n}(1)^{h+n} \to \cdots$$
$$\cdots \to \wedge^{n-1}\mathcal{O}_{\mathbb{P}^n}(1)^{h+n} \to \mathcal{I}_\Delta(h+n) \to 0,$$

where $\mathcal{I}_\Delta$ is the ideal sheaf of Δ in L.

The set Δ imposes independent conditions to the linear system $|\mathcal{O}_X(2)|$ if and only if $h^1(\mathbb{P}^n, \mathcal{I}_\Delta(2)) = 0$. As in the previous example, splitting the Eagon–Northcott complex above in short exact sequences, and taking into account that $S^{n-1}\mathcal{O}_{\mathbb{P}^n}^{h+1}$ is a direct sum of $\mathcal{O}_{\mathbb{P}^n}$, we deduce that $h^1(\mathbb{P}^n, \mathcal{I}_\Delta(2)) = 0$ as soon as

$$h^n(\mathbb{P}^n, \mathcal{O}_{\mathbb{P}^n}(-h-n+2)) = 0,$$

namely if

$$h^0(\mathbb{P}^n, \mathcal{O}_{\mathbb{P}^n}(h-3)) = 0.$$

The last condition is satisfied if $1 \leq h \leq 2$. When $h = 1$ we have $n + 1$ nodes in $\mathbb{P}^n$, if $h = 2$ we have $\frac{1}{2}(n+2)(n+1)$ nodes in $\mathbb{P}^n$.

Also in this case, the nodes impose independent conditions, and $\mathcal{H}^1 IC(R^{2n-1}\pi_*\mathbb{Q})_D[-N]$ is non-trivial, because, with the same argument as in the previous case of hypersurfaces in $\mathbb{P}^{2n}$, one sees that D, containing a $\mathbb{P}^n$, has defect.

Acknowledgements We thank the Referee for his valuable suggestions which allowed us to substantially improve the presentation.

References

1. G. Albanese, Sui sistemi continui di curve piane algebriche, in *Collected Papers of G. Albanese*, with a biography of Albanese in English by Ciro Ciliberto and Edoardo Sernesi, ed. by Ciliberto, P. Ribenboim and Sernesi. Queen's Papers in Pure and Applied Mathematics, vol. 103 (Queen's University, Kingston, 1996), xii+408 pp
2. A. Beilinson, J. Bernstein, P. Deligne, *Faisceaux pervers*, Analysis and Topology on Singular Spaces, I (Luminy, 1981), Astérisque, vol. 100 (Société Mathématique de France, Paris, 1982), pp. 5–171
3. E. Cattani, A. Kaplan, W. Schmidt, L^2 and intersection cohomologies of polarizable variation of Hodge structures. Invent. Math. **87**, 217–252 (1987)
4. L. Chiantini, E. Sernesi, Nodal curves on surfaces of general type. Math. Ann. **307**, 41–56 (1997)
5. M.A. de Cataldo, L. Migliorini, The Hodge theory of algebraic maps. Ann. Sci. École Norm. Sup. 4, **38**(5), 693–750 (2005)
6. M.A. de Cataldo, L. Migliorini, The decomposition theorem, perverse sheaves and the topology of algebraic maps. Bull. Amer. Math. Soc. (N.S.) **46**(4), 535–633 (2009)
7. M.A. de Cataldo, L. Migliorini, On singularities of primitive cohomology classes. PAMS **137**, 3593–3600 (2009)
8. P. Deligne, Equations differentielles a points singuliers reguliers. LNM **163** (1970)
9. V. Di Gennaro, D. Franco, Factoriality and Néron-Severi groups. Commun. Contemp. Math. **10**(5), 745–764 (2008)
10. I.A. Chel'tsov, The factoriality of nodal threefolds and connectedness of the set of log canonical singularities. Math. Sb. **197**(3), 87–116 (2006). Translation in Sb. Math. **197**(3–4), 387–414 (2006)
11. S. Cynk, Defect of a nodal hypersurface. Manuscripta Math. **104**(3), 325–331 (2001)
12. S. Cynk, Defect formula for nodal complete intersection threefolds. Internat. J. Math. **30**(4), 1950020, 14 pp (2019)
13. V. Di Gennaro, D. Franco, On the existence of a Gysin morphism for the Blow-up of an ordinary singularity. Ann. Univ. Ferrara, Sez. VII, Sci. Mat. **63**(1), 75–86, (2017). https://doi.org/10.1007/s11565-016-0253-z
14. V. Di Gennaro, D. Franco, Néron-Severi group of a general hypersurface. Commun. Contemp. Math. **19**(01), Article ID 1650004, 15 p. (2017). https://doi.org/10.1142/S0219199716500048
15. V. Di Gennaro, D. Franco, On the topology of a resolution of isolated singularities. J. Singul. **16**, 195–211 (2017). https://doi.org/10.5427/jsing.2017.16j
16. V. Di Gennaro, D. Franco, On a resolution of singularities with two strata. Results Math. **74**(3), 74:115 (2019). https://doi.org/10.1007/s00025-019-1040-9

17. V. Di Gennaro, D. Franco, On the topology of a resolution of isolated singularities, II. J. Singul. **20**, 95–102 (2020). https://doi.org/10.5427/jsing.2020.20e
18. A. Dimca, *Singularity and Topology of Hypersurfaces* (Springer Universitext, New York, 1992)
19. A. Dimca, *Sheaves in Topology* (Springer Universitext, 2004)
20. Ph. Ellia, D. Franco, On codimension two subvarieties of $\mathbb{P}^5$ and $\mathbb{P}^6$. J. Algebraic Geom. **11**(3), 513–533 (2002)
21. Ph. Ellia, D. Franco, L. Gruson, Smooth divisors of projective hypersurfaces. Comment. Math. Helv. **83**(2), 371–385 (2008)
22. D. Franco, Explicit decomposition theorem for special Schubert varieties. Forum Math. **32**(2), 447–470 (2020)
23. W. Fulton, *Intersection theory*, Ergebnisse der Mathematik und ihrer Grenzgebiete; 3.Folge, Bd. 2 (Springer, 1984)
24. C.G. Gibson, K. Wirthmuller, A.A. du Plessis, E.J.N. Looijenga, Topological stability of smooth mappings. LNM **552** (1976)
25. M. Green, P. Griffiths, Algebraic cycles and singularities of normal functions, in *Algebraic Cycles and Motives*, LMS, vol. 343 (Cambridge University Press, Cambridge, 2007), pp. 206–263
26. P. Griffiths, J. Harris, *Principles of Algebraic Geometry* (Wiley, New York, 1978)
27. J. Harris, *Algebraic Geometry - A Firs Course*, vol. 133 (Springer GTM, 1992)
28. J. Harris, On the Severi problem. Invent. Math. **84**, 445–461 (1986)
29. N. Katz, *Pinceaux de Lefschetz: théorème d'existence*, Séminaire de Géométrie Algébrique du Bois Marie - 1967-69 - Groupes de monodromie en géométrie algébrique - SGA7 II (1973), pp. 212–253
30. M. Kerr, G. Pearlstein, An exponential history of functions with logarithmic growth. Topol. Stratif. Spaces MSRI Publ . **58**, 281–374 (2011)
31. S. Kleiman, Geometry on grassmannians and applications to splitting bundles and smoothing cycles. Publ. Math. Inst. Hautes Études Sci. **36**, 281–297 (1969)
32. E.J.N. Looijenga, *Isolated Singular Points on Complete Intersections*. London Mathematical Society Lecture Note Series 77 (Cambridge University Press, Cambridge, 1984)
33. J. Milnor, *Singular Points of Complex Hypersurfaces*. Annals of Mathematics Studies (1968)
34. V. Navarro Aznar, Sur la théorie de Hodge des variétés algébriques à singularités isolées. Astérisque **130**, 272–307 (1985)
35. M. Saito, Mixed Hodge modules. Publ. RIMS, Kyoto Univ. **26**, 221–333 (1990)
36. Ch. Schnell, Computing cohomology of local systems, https://www.math.stonybrook.edu/~cschnell/pdf/notes/locsys.pdf
37. E. Sernesi, On the existence of certain families of curves. Invent. Math. **75**, 25–57 (1984)
38. F. Severi, *Vorlesungen über algebraische Geometrie* (Teuner, Leipzig, 1921)
39. R.P. Thomas, Nodes and the Hodge conjecture. J. Alg. Geom. **14**, 177–185 (2005)
40. C. Voisin, *Hodge Theory and Complex Algebraic Geometry, I*. Cambridge Studies in Advanced Mathematics 76 (Cambridge University Press, Cambridge, 2002)
41. J. Wahl, Deformations of plane curves with nodes and cusps. Amer. J. Math. **96**, 529–577 (1974)
42. G. Williamson, Hodge theory of the decomposition theorem [after M.A. de Cataldo and L. Migliorini]. Séminaire BOURBAKI, 2015–2016, n. 1115, p. 31

Cremona Orbits in $\mathbb{P}^4$ and Applications

Olivia Dumitrescu and Rick Miranda

Abstract This article is motivated by the authors' interest in the geometry of the Mori dream space $\mathbb{P}^4$ blown up in 8 general points. In this article, we develop the necessary technique for determining Weyl orbits of linear cycles for the four-dimensional case, by explicit computations in the Chow ring of the resolution of the standard Cremona transformation. In particular, we close this paper with applications to the question of the dimension of the space of global sections of effective divisors having at most 8 base points.

1 Introduction

Let X^n_s be the projective space $\mathbb{P}^n$ blown up at s general points. Motivated by the study of the dimensionality problem for effective divisors on X^n_s, we analyze the standard Cremona action on X^4_8 and give several applications. We first establish the terminology we use throughout the paper. We call a *Weyl line/Cremona line (Weyl hyperplane, respectively)* to be the orbit under the Weyl group action of a line passing through two of the s points (hyperplane passing through n of the points). In dimension two, the *Weyl lines* are also known in the literature as (-1) curves; via a theorem of Nagata [15, Theorem 2a] they can be described via numerical properties as irreducible classes with self-intersection -1 and anticanonical degree 1. In [10], the authors noticed that Nagata's work can be generalized, and similar numerical properties via the Dolgachev–Mukai bilinear form are equivalent to *Weyl divisors*. In dimension three, the Weyl group action on curves was analyzed by Laface and

The first author is supported by NSF grant DMS1802082.

O. Dumitrescu
University of North Carolina at Chapel Hill, Chapel Hill, NC 27599-3250, USA

Simion Stoilow Institute of Mathematics, Romanian Academy, 010702 Bucharest, Romania

R. Miranda (✉)
Colorado State University, Fort Collins, CO 80523, USA
e-mail: Rick.Miranda@colostate.edu

T. Dedieu et al. (eds.), *The Art of Doing Algebraic Geometry*, Trends in Mathematics,
https://doi.org/10.1007/978-3-031-11938-5_7

Ugaglia in [13]. Finally, in arbitrary dimension, the Weyl group action on curves in X^n_s and their connection to the (-1)-curves introduced by Kontsevich is analyzed by the two authors in a forthcoming paper [11].

In the planar case, the Gimigliano-Harbourne–Hirschowitz conjecture, still open, predicts that the dimension of the space of global sections of an effective divisor depends on the Euler characteristic and the multiplicity of containment of Weyl lines in the base locus of the divisor. In $\mathbb{P}^3$, the conjecture of Laface–Ugaglia [13] predicts that this dimension depends on the multiplicity of containment of Weyl lines, Weyl hyperplanes, and Weyl orbit of the unique quadric in X^3_s passing through nine general points.

In general, for a small number of points, X^n_{n+2}, it was proved that this dimension depends on the Euler characteristic and the multiplicity of containment of *linear cycles* spanned by the fixed points in the base locus of the divisor D as in [3, Theorem 2.3]. Moreover, the birational geometry of the space X^n_{n+3}, studied in several publications (e.g., [1, 2, 4]), namely the effective and movable cone of divisors, their Mori chamber decompositions together with the dimension of space of global sections is determined by secant varieties to the rational normal curve of degree n passing through $n+3$ general points together with their joins. In general, the case X^n_{n+4} seems to be mysterious.

We dedicate this paper to study X^4_8, which is a Mori Dream Space, whose birational geometry is not totally explained in the literature. In this paper, together with [5] we define and classify the varieties that determine combinatorial data describing the geometry of X^4_8.

The two spaces $X_{2,8}$ and $X_{4,8}$ are related by Gale duality as described in [14]. The precise relation between $X_{2,8}$ and $X_{4,8}$ was established in the following theorem of Mukai (semistability refers to semistability in the sense of Gieseker–Maruyama): $X_{4,8}$ *is isomorphic to the moduli space of rank* 2 *torsion free sheaves* F *on* $X_{2,8}$ *for which* $c_1(F) = -K_S$ *and* $c_2(F) = 2$. Via Mukai's correspondence, Casagrande et al. describe in [7] the five types of surfaces in $X_{4,8}$ playing a special role in the Mori program. In this paper, we rediscover these surfaces as *Weyl planes*, defined below analogously to Weyl lines and hyperplanes.

The Weyl group of X^4_8 is generated by the standard Cremona transformations together with permutations of the base points. In order to define and construct Weyl planes, we introduce Y^4_8 to denote the blowup of X^4_8 along all lines joining any two points and the eight rational normal curves of degree 4 passing through 7 points. (These curves are all disjoint in X^4_8.)

Definition 1.1 A *Weyl plane* is the Weyl orbit of the proper transform of a plane through three fixed points under the blowup of the three lines joining any two points in Y^4_8.

It is important to remark that *Weyl planes* live on the space Y^4_8. We emphasize that this orbit is different (in the Chow ring) than the Weyl orbit of *planes through three points*. Moreover, in [5], the authors introduce and classify the notions of *Weyl curves* and *Weyl surfaces* in X^4_8 as the intersection of two distinct *Weyl divisors* that are orthogonal with respect to the Dolgachev–Mukai bilinear pairing. Since the classification of *Weyl surfaces* [5] in X^4_8 is the same with the classification of Proposition 7.3, we can deduce that the two definitions of *Weyl planes* (1.1) and *Weyl surfaces* [5] are equivalent in X^4_8. By definition, *Weyl lines* coincide with *Weyl curves* in the projective plane X^2_s, but the explicit relation between the two definitions, in general, will be studied in a different paper.

In this paper, Corollaries 5.3 and 7.2 enable us to determine the Weyl action on
(a) 1-cycles (i.e., curves) on the Chow ring of blowup of X^4_s;
(b) 2-cycles (i.e., surfaces) on the Chow ring of Y^4_8.

As a consequence, Proposition 7.3 determines the complete list of *Weyl planes* and *Weyl divisors* on X^4_8, and it also gives the formulas for all *Weyl lines* on X^4_s, (for arbitrary number of fixed points s). In particular, for X^4_8, *the only Weyl lines are lines through two fixed points and the rational normal curve of degree* 4 *passing through* 7 *of the* 8 *points*. In fact, in a forthcoming paper [11], we prove that this statement holds for all Mori Dream Spaces. Let Q_i denote *Weyl line* of degree 4 (the rational normal quartic) skipping only the ith point. In particular, we prove that on X^4_8, there are 5 types of Weyl planes (modulo permutation of points), matching computations in [7, Theorem 8.7] and [5]:

- The 56 planes $S_1(ijk)$ through three of the eight points (p_i, p_j, p_k); it has multiplicity one along the three lines L_{ij}, L_{ik}, and L_{jk}.
- The 56 cubic surfaces $S_3(i, j)$ triple at p_i, passing through all other points except p_j; it has multiplicity one along the lines L_{ik} for $k \neq i, j$, and along Q_j
- The 56 sextic surfaces $S_6(ijk)$ passing through p_i, p_j, and p_k and triple at the other five points; it has multiplicity one along all lines joining two of the five points, and along Q_i, Q_j, and Q_k
- The 28 surfaces $S_{10}(ij)$ of degree 10 having two points p_i and p_j of multiplicity 6 and triple at the other six points; it has multiplicity 3 along the line L_{ij}, multiplicity one along all lines L_{ik} and L_{jk} for $k \neq i, j$, and multiplicity one along the curves Q_k for $k \neq i, j$
- The 8 surfaces $S_{15}(i)$ of degree 15 having one point p_i with multiplicity 3 and having multiplicity 6 at the other seven points; it has multiplicity one along all lines L_{jk} for $j, k \neq i$, multiplicity one along each Q_j for $j \neq i$, and multiplicity 3 along Q_i.

In addition to the multiplicities at the points p_i, the reader will note that for all of these surfaces we also compute the multiplicities along the lines L_{ij} and along the rational normal quartics (through 7 of the 8 points). This is important for computations in the Chow ring: unless one takes into account that these surfaces have multiplicity along these curves, one does not fully capture the intersection behavior of these surfaces after one blows up the points (and in general, the curves and surfaces that appear as base loci of linear systems of divisors). It is also critical for computations of the dimensions of the linear systems: it is one of the principles of this article that the multiplicities along these curves must be taken into account in determining the difference between the virtual dimension and the actual dimension of linear systems. Indeed, for certain purposes, it is useful to consider not only the blowup X_8^4 of $\mathbb{P}^4$ at the 8 general points but also then the further blowup Y_8^4 of all of the proper transforms of the lines L_{ij} and the rational normal quartics Q_k; these are easily seen to be disjoint in X_8^4 and therefore Y_8^4 is smooth.

Remark 1.2 In paper [5], the authors use a different notation for the Chow ring basis. For example, $\{h, e_i, e_{ij}\}$ and $\{h^1, e_i^1\}$ of [5] represent here $\{S, S_i, G_{ij}\}$ and $\{l, l_i\}$, respectively. In [5], surfaces denoted above by $S_1(ijk)$, $S_3(i, j)$, $S_6(ijk)$, $S_{10}(ij)$, and $S_{15}(i)$ are denoted by H_{ijk}, $S^3_{i,\widehat{j}}$, S^6_{ijk}, S^{10}_{ij}, and S^{15}_i, respectively.

We predict that the birational geometry of X_8^4 is determined not only by Weyl hyperplanes but also Weyl lines and Weyl planes classified in Proposition 7.3. Finally, in Sect. 8, we present applications to the vanishing conjecture and dimensionality problem.

2 The Standard Cremona Transformation and Its Resolution

The standard Cremona transformation of $\mathbb{P}^n$ can be elegantly factored into a series of blowups at the proper transforms of the coordinate linear spaces, followed by a series of symmetric blowdowns.

Fix coordinates $[x_0 : x_1 : \cdots : x_n]$ in $\mathbb{P}^n$, and consider the standard Cremona involution

$$[x_0 : x_1 : \cdots : x_n] \longrightarrow [x_0^{-1} : x_1^{-1} : \cdots : x_n^{-1}]$$

which simply inverts all the coordinates. This is well defined on the torus where all coordinates are non-zero, and has a fundamental locus the union of the coordinate hyperplanes. The transformation is relatively straightforward to resolve in a sequence of blowups and blowdowns, as follows.

Let $p_0, p_1, \ldots, p_n$ be the coordinate points of $\mathbb{P}^n$. For an index set $I \subset \{0, 1, \ldots, n\}$, denoted by L_I, the linear span of the coordinate points indexed by I: $L_I = \text{span}\{p_i \mid i \in I\}$. We have that $\dim L_I = |I| - 1$.

We set $\mathbb{X}_0^n = \mathbb{P}^n$, and define $\pi_j : \mathbb{X}_j^n \to \mathbb{X}_{j-1}^n$ to be the blowup of the proper transforms of all L_I with $|I| = j$. Hence, π_1 is the blowup of all the coordinate points in $\mathbb{P}^n$; π_2 is the blowup of the (proper transforms of the) coordinate lines L_{ij}, etc. Note that the sequence of blowups stops with π_{n-1}, the blowup of the codimension two coordinate linear spaces, creating the space $\mathbb{X}_{n-1}^n$. We will denote by E_I the exceptional divisor created when L_I is blown up. E_I is created on $\mathbb{X}_{|I|}^n$, and we will use the notation E_I for the proper transform on subsequent blowups too. If $|I| = n$, then L_I is a coordinate hyperplane in $\mathbb{P}^n$; we will denote its proper transform in $\mathbb{X}_{n-1}^n$ by E_I as well.

We note that, at this point , on $\mathbb{X}_{n-1}^n$, the nature and configuration of the divisors E_I are completely symmetric, with respect to taking complements; in other words, we have an isomorphism of $\mathbb{X}_{n-1}^n$ that switches the roles of E_I and E_J when I and J

are complementary in $\{0, 1, \ldots, n\}$. Hence, we can reverse the sequence of blowups with the complementary divisors, and blow down to $\mathbb{P}^n$ "the other way": first blow down the E_I with $|I| = 2$, then the E_I with $|I| = 3$, etc., finishing by blowing down the proper transforms of the coordinate hyperplanes $E_{|I|}$ with $|I| = n$. This is the resolution of the birational involution.

We note that:

- On $\mathbb{X}^n_{|I|-1}$ when the L_I are blown up, they are all disjoint.
- Each linear space L_I experiences a sequence of blowups (by the earlier blowups); on $\mathbb{X}^n_{|I|-1}$, the proper transform of each L_I is isomorphic to $\mathbb{X}^{|I|-1}_{|I|-2}$.
- By induction, this proper transform has both the hyperplane divisor class H (the pullback of the hyperplane divisor class on $\mathbb{X}^{|I|-1}_0 = \mathbb{P}^{|I|-1}$) and its Cremona involution image H'.
- On $\mathbb{X}^{|I|-1}_{|I|-2}$, the normal bundle of the proper transform of L_I is isomorphic to

$$\mathcal{O}(-H')^{\oplus n-|I|+1}.$$

- Since the normal bundle of the proper transform of L_I splits as a direct product of identical line bundles, when E_I is created on $\mathbb{X}^n_{|I|}$, it is isomorphic to a product $\mathbb{X}^{|I|-1}_{|I|-2} \times \mathbb{P}^{n-|I|}$.
- E_I experiences further blowups on its way to $\mathbb{X}^n_{n-1}$, and there it is isomorphic to $\mathbb{X}^{|I|-1}_{|I|-2} \times \mathbb{X}^{n-|I|}_{n-|I|-1}$, where it has a normal bundle isomorphic to the tensor product of the anti-Cremona-hyperplane bundles coming from the two factors.

This construction generalizes the familiar construction of the quadratic Cremona transformation of $\mathbb{P}^2$, which is obtained by blowing up the three coordinate points L_0, L_1, and L_2 (obtaining $\mathbb{X}^2_1$) and then blowing down the three coordinate lines L_{01}, L_{02}, and L_{12}.

3 The Case of Three Space

For three space, the sequence of iterated blowups, in this case, involves two sets of blowups:

$$\mathbb{X}^3_2 \xrightarrow{\pi_2} \mathbb{X}^3_1 \xrightarrow{\pi_1} \mathbb{X}^3_0 = \mathbb{P}^3$$

where π_1 blows up the four coordinate points $p_i = L_i$ and π_2 blows up the six proper transforms of the coordinate lines L_{ij}. The exceptional divisors E_i start out as $\mathbb{P}^2$'s in $\mathbb{X}^3_1$, and then are further blown up to become isomorphic to $\mathbb{X}^2_1$'s in $\mathbb{X}^3_2$. The coordinate lines start in $\mathbb{P}^2$ having normal bundle of bidegree $(1, 1)$; after blowing up the two coordinate points on each, the proper transforms have normal bundles with bidegree $(-1, -1)$ in $\mathbb{X}^3_1$. They are then blown up to $E_{ij} \cong \mathbb{P}^1 \times \mathbb{P}^1$ in $\mathbb{X}^3_2$. Finally the coordinate hyperplanes L_{ijk} are each blown up three times by π_1, and then not blown up further by π_2, and so arrive at $\mathbb{X}^3_2$ as surfaces isomorphic to $\mathbb{X}^2_1$.

The blowing down proceeds by blowing down the E_{ij} via the other ruling, which blows down each L_{ijk} to a $\mathbb{P}^2$; one then blows down each of these to points, finishing the process.

If one is interested in intersection phenomena related to these coordinate subspaces, the Chow ring is the appropriate tool; it is useful primarily for recording two different kinds of phenomena. One is *containment* (with multiplicity) by a given subvariety of one of the blowup centers. In $\mathbb{P}^3$, for divisors, this is the multiplicity of the divisor at one of the coordinate points, and the multiplicity of containment along one of the coordinate lines. For curves, this is the multiplicity of the curve at one of the coordinate points. For a divisor written in the form $D = dH - \sum_i m_i E_i - \sum_{ij} n_{ij} E_{ij}$, the coefficient d is the degree; m_i is the multiplicity at the coordinate point L_i; and m_{ij} is the multiplicity along the line L_{ij}.

The other phenomenon which the Chow ring coefficients can record is the higher-dimensional *contact* that the given subvariety may have with one of the blowup centers. (Higher-dimensional contact in the sense of higher than expected dimension.) In $\mathbb{P}^3$, for surfaces, this is not relevant for the coordinate points and lines; higher-dimensional contact is containment with multiplicity. This is also true for curves with respect to the points: the only phenomenon is that of containment. However, with curves, one can have additional contact with the lines, without containment.

The Chow ring of $\mathbb{X}^3_2$ is not difficult to compute; all the relevant tools are presented in [12], Chaps. 9 and 13. The codimension zero classes are one-dimensional, generated by $[\mathbb{X}^3_2]$ itself; the codimension three classes are also one-dimensional, generated by the class $[p]$ of a point. The codimension one classes are freely generated by the pullback H of the hyperplane class, and the exceptional divisors E_i and E_{ij}.

In codimension two, the group $A^2(\mathbb{X}^3_2)$ contains the following elements. The pullback of the general line class in $\mathbb{P}^3$ will be denoted by ℓ. The general line class inside the exceptional divisor E_i will be denoted by ℓ_i. The exceptional divisor E_{ij} is isomorphic to $\mathbb{P}^1 \times \mathbb{P}^1$, and contributes a priori two curve classes: the class f_{ij} of the fiber of the blowup π_2, and the class g_{ij} which is the horizontal ruling of E_{ij}. These are not independent though in $A^2(\mathbb{X}^3_2)$; it is an exercise to check that

$$g_{ij} = f_{ij} + \ell - \ell_i - \ell_j$$

and that this is the only relation in A^2.

For a curve class C written as $C = d\ell - \sum_i m_i \ell_i - \sum_{ij} n_{ij} f_{ij}$, the coefficient d is the degree, m_i is the multiplicity of C at the coordinate point L_i, and n_{ij} is the additional contact of C with the coordinate line L_{ij} (over and above the contact implied by the multiplicities at the two coordinate points on L_{ij}).

We have the following, where we use typical δ-notation: $\delta_{I,J} = 1$ if $I \subseteq J$ and 0 otherwise.

Proposition 3.1 *(a) A basis for the Chow ring of X_2^3 is given by*

$$\begin{array}{ll} A^0: & [X_2^3] \\ A^1: & H, E_0, E_1, E_2, E_3, E_{01}, E_{02}, E_{03}, E_{12}, E_{13}, E_{23} \\ A^2: & \ell, \ell_0, \ell_1, \ell_2, \ell_3, f_{01}, f_{02}, f_{03}, f_{12}, f_{13}, f_{23} \\ A^3: & p \end{array}$$

(b) Multiplication of these basis elements is given by

$$\begin{array}{cccc} A^1 \cdot A^1 & H & E_i & E_{ij} \\ H & \ell & 0 & f_{ij} \\ E_k & 0 & -\ell_i \delta_{ik} & f_{ij}\delta_{k,ij} \\ E_{kl} & f_{kl} & f_{kl}\delta_{i,kl} & (-2f_{ij} - \ell + \ell_i + \ell_j)\delta_{ij,kl} \end{array}$$

$$\begin{array}{cccc} A^1 \cdot A^2 & H & E_i & E_{ij} \\ \ell & p & 0 & 0 \\ \ell_k & 0 & -p\delta_{i,k} & 0 \\ f_{kl} & 0 & 0 & -p\delta_{ij,kl} \end{array}$$

The Cremona involution extends to an involution ϕ on the Chow ring; we denote the image of the involution using a superscript prime:

- $[\mathbb{X}_2^3] \leftrightarrow [\mathbb{X}_2^3]$
- $H \leftrightarrow H' = 3H - 2\sum_i E_i - \sum_{ij} E_{ij}$
- $E_l \leftrightarrow E_l' = L_{ijk} = H - E_i - E_j - E_k - E_{ij} - E_{ik} - E_{jk}$ for $i, j, k \neq l$
- $E_{ij} \leftrightarrow E_{ij}' = E_{kl}$ for $k, l \neq i, j$.
- $\ell \leftrightarrow \ell' = 3\ell - \sum_i \ell_i$
- $\ell_i \leftrightarrow \ell_i' = 2\ell - \sum_{j \neq i} \ell_j$
- $f_{ij} \leftrightarrow f_{ij}' = g_{kl} = f_{kl} + \ell - \ell_k - \ell_l$ for $k, l \neq i, j$.
- $p \leftrightarrow p$.

We leave it to the reader to check that this is a ring automorphism, and is an involution.

Proposition 3.2 *(a) Let $D = dH - \sum_i m_i E_i - \sum_{ij} n_{ij} E_{ij}$ be a general class in $A^1(\mathbb{X}_2^3)$. Then the Cremona image D' of D under the involution is $D' = d'H - \sum_i m_i' E_i - \sum_{ij} n_{ij}' E_{ij}$ where*

$$d' = D' \cdot \ell = D \cdot \ell' = D \cdot \left(3\ell - \sum_i \ell_i\right) = 3d - \sum_i m_i;$$

$$m_i' = D' \cdot \ell_i = D \cdot \ell_i' = D \cdot \left(2\ell - \sum_{j \neq i} \ell_j\right) = 2d - \sum_{j \neq i} m_j;$$

$$n_{ij}' = D' \cdot f_{ij} = D \cdot f_{ij}' = D \cdot f_{kl} + \ell - \ell_k - \ell_l = d + n_{kl} - m_k - m_l$$
$$\textit{for} \quad k, l \neq i, j$$

(b) Let $C = d\ell - \sum_i m_i \ell_i - \sum_{ij} n_{ij} f_{ij}$ *be a general class in* $A^2(\mathbb{X}_2^3)$*. Then the Cremona image* C' *of* C *under the involution is* $C' = d'\ell - \sum_i m_i' \ell_i - \sum_{ij} n_{ij}' f_{ij}$ *where*

$$d' = C' \cdot H = C \cdot H' = C \cdot (3H - 2\sum_i E_i - \sum_{ij} E_{ij}) = 3d - 2\sum_i m_i - \sum_{ij} n_{ij};$$

$$m_i' = C' \cdot E_i = C \cdot E_i' = C \cdot (H - \sum_{j \neq i} E_j - \sum_{j,k \neq i} E_{jk}) = d - \sum_{j \neq i} m_j - \sum_{j,k \neq i} n_{jk};$$

$$n_{ij}' = C' \cdot E_{ij} = C \cdot E_{ij}' = C \cdot E_{kl} = n_{kl} \quad \textit{for} \quad k, l \neq i, j$$

(In the computations above, we abuse notation and give the multiplications as integers instead of integer multiples of the point class p.)

If one is in the position of not needing to consider the contact phenomena for curves, one can simplify the formulas as follows.

Corollary 3.3 *The subspace of* $A^2(\mathbb{X}_2^3)$ *spanned by* ℓ *and the* ℓ_i*, is invariant under the Cremona involution. If* $C = d\ell - \sum_i m_i \ell_i$ *is a general class in* $A^2(\mathbb{X}_2^3)$ *in this subspace, then the Cremona image* C' *of* C *under the involution is* $C' = d'\ell - \sum_i m_i' \ell_i$ *where*

$$d' = C' \cdot H = C \cdot H' = C \cdot (3H - 2\sum_i E_i - \sum_{ij} E_{ij}) = 3d - 2\sum_i m_i;$$

$$m_i' = C' \cdot E_i = C \cdot E_i' = C \cdot (H - \sum_{j \neq i} E_j - \sum_{j,k \neq i} E_{jk}) = d - \sum_{j \neq i} m_j;$$

4 The Chow Ring for the Case of $\mathbb{P}^4$

The sequence of iterated blowups in this case involves three sets of blowups:

$$\mathbb{X}_3^4 \xrightarrow{\pi_3} \mathbb{X}_2^4 \xrightarrow{\pi_2} \mathbb{X}_1^4 \xrightarrow{\pi_1} \mathbb{X}_0^4 = \mathbb{P}^4$$

where π_1 blows up the five-coordinate points $p_i = L_i$ to divisors E_i, π_2 blows up the ten proper transforms of the coordinate lines L_{ij} to E_{ij}, and π_3 blows up the ten proper transforms of the coordinate planes L_{ijk} to E_{ijk}.

We denote by H the general hyperplane class in $\mathbb{P}^4$ (and all its pullbacks); let us denote by $S = H^2$ the class of the general 2-plane, and $\ell = H^3$ the class of the general line; the point class will be p as usual.

In this section, we'll present the Chow ring $A^*(\mathbb{X}_3^4)$, proceeding through the sequence of three blowups. In the starting fourfold $\mathbb{X}_0^4 \cong \mathbb{P}^4$, the relevant subvarieties are simply the linear spaces L_I for $I \subset \{0, 1, 2, 3, 4\}$.

After blowing up the points via π_1, we have

- The divisors $E_i \cong \mathbb{P}^3$.
- The proper transforms of the lines $L_{ij} \cong \mathbb{P}^1$.
- The proper transforms of the 2-planes $L_{ijk} \cong \mathbb{X}_1^2$.
- The proper transforms of the hyperplanes $L_{ijk\ell} \cong \mathbb{X}_1^3$.

We now blow up with π_2 the proper transforms of the ten lines L_{ij}, to the exceptional divisors E_{ij}, to obtain $\mathbb{X}_2^4$; there, we have the following descriptions of the relevant subvarieties:

- The divisors $E_i \cong \mathbb{X}_1^3$.
- The exceptional divisors $E_{ij} \cong \mathbb{P}^1 \times \mathbb{P}^2$.
- The 2-planes $L_{ijk} \cong \mathbb{X}_1^2$.
- The hyperplane threefolds $L_{ijk\ell} \cong \mathbb{X}_2^3$.

Finally, we blow up the proper transforms of the ten surfaces L_{ijk}, to the exceptional divisors E_{ijk}, to obtain $\mathbb{X}_3^4$; there, the relevant subvarieties are:

- The divisors $E_i \cong \mathbb{X}_2^3$.
- The divisors $E_{ij} \cong \mathbb{P}^1 \times \mathbb{X}_1^2$.
- The exceptional divisors $E_{ijk} \cong \mathbb{X}_1^2 \times \mathbb{P}^1$.
- The hyperplane threefolds $L_{ijk\ell} \cong \mathbb{X}_2^3$.

The codimension one classes in $A^1(\mathbb{X}_3^4)$ are freely generated by the pullback H of the hyperplane class in $\mathbb{P}^4$ and the exceptional divisors E_i, E_{ij}, and E_{ijk}; there are no relations among these.

In the group $A^2(\mathbb{X}_3^4)$ of codimension two classes, we have the class $S = H^2$ of the pullback of a general 2-plane in $\mathbb{P}^4$. The other classes that will generate A^2 are supported in the exceptional divisors.

In E_i, which starts in $\mathbb{X}_1^4$ as a $\mathbb{P}^3$, we have the general 2-plane; pulled back to $\mathbb{X}_3^4$ this gives a class S_i for each i.

The divisor E_{ij} starts in $\mathbb{X}_2^4$ as isomorphic to the product $\mathbb{P}^1 \times \mathbb{P}^2$. This contributes to two surface classes: the fiber $\{\text{point}\} \times \mathbb{P}^2$ of the blowup, and the product $\mathbb{P}^1 \times \{\text{general line in } \mathbb{P}^2\}$. Denote by F_{ij} the pullback to $\mathbb{X}_3^4$ of the former, the fiber class; and by G_{ij} the pullback to $\mathbb{X}_3^4$ of the latter.

Finally, the divisor E_{ijk} is isomorphic to $\mathbb{X}_1^2 \times \mathbb{P}^1$, and contributes five surface classes. One is $M_{ijk} = \mathbb{X}_1^2 \times \{\text{point}\}$, a cross section of the blowup map. The others come from products of curve classes in $L_{ijk} \cong \mathbb{X}^2_1$ with the fiber $\mathbb{P}^1$. The curve classes in L_{ijk} are generated by the pullback (from $\mathbb{P}^2$) of the general line class ℓ_{ijk} and the three exceptional curves $e_{ijk,i}$, $e_{ijk,j}$, and $e_{ijk,k}$ which are (in $\mathbb{X}^4_2$) the intersection of L_{ijk} with the three divisors E_i, E_j, and E_k respectively. These four classes give classes $H_{ijk} = \ell_{ijk} \times \mathbb{P}^1$ and $V_{ijk,i}$, $V_{ijk,j}$, and $V_{ijk,k}$ where $V_{ijk,i}$ comes from the product of $e_{ijk,i} \times \mathbb{P}^1$ and the same for the other two.

It is useful to introduce two new classes, for notational convenience. These are:

$$P_{ij} = G_{ij} - F_{ij} \quad \text{and} \quad \Lambda_{ijk} = 2H_{ijk} - V_{ijk,i} - V_{ijk,j} - V_{ijk,k}; \tag{4.1}$$

we note that Λ_{ijk} is the pullback of the Cremona image of the line class on the 2-plane L_{ijk}. This will allow us to replace G_{ij} by P_{ij} among the generators for A^2.

There is a single relation among these codimension two classes beyond the definitional ones of (4.1). It is that

$$M_{ijk} = S - S_i - S_j - S_k - P_{ij} - P_{ik} - P_{jk} + \Lambda_{ijk}. \tag{4.2}$$

Finally, we have the classes of the curves, the codimension three classes in $A^3(\mathbb{X}^4_3)$. We again have the pullback ℓ of the general line class in $\mathbb{P}^4$, and the classes ℓ_i of the general lines in the E_i.

The curve classes supported on E_{ij} (which when it is created on $\mathbb{X}^4_2$ is isomorphic to $\mathbb{P}^1 \times \mathbb{P}^2$) are generated by the class $\ell_{ij} = \{\text{point}\} \times \{\text{general line in}\mathbb{P}^2\}$ and $h_{ij} = \mathbb{P}^1 \times \{\text{point}\}$.

The curve classes coming from E_{ijk} are the 'horizontal' ones living in L_{ijk}, crossed with a point; these we can denote again by ℓ_{ijk} and $e_{ijk,i}$, $e_{ijk,j}$, and $e_{ijk,k}$ as before. The final one is a general fiber of the blowup f_{ijk}.

There are relations among these curve classes also; these are:

$$h_{ij} = \ell_{ij} + \ell - \ell_i - \ell_j; \quad \ell_{ijk} = 2f_{ijk} + \ell - \ell_{ij} - \ell_{ik} - \ell_{jk}; \tag{4.3}$$

$$e_{ijk,i} = f_{ijk} + \ell_i - \ell_{ij} - \ell_{ik}; \quad e_{ijk,j} = f_{ijk} + \ell_j - \ell_{ij} - \ell_{jk}; \quad e_{ijk,k} = f_{ijk} + \ell_k - \ell_{ik} - \ell_{jk}. \tag{4.4}$$

(Hence, we can dispense with these to generate $A^3(\mathbb{X}^4_3)$.)

It is the case that, for a surface class T, one measures multiplicity along the line L_{ij} by the intersection with F_{ij}, and one measures higher-dimensional contact with L_{ij} by the intersection with G_{ij}. Hence, if the coefficients of T include the terms $-mP_{ij} - nF_{ij}$, then m is the multiplicity of T along the line and n is the additional contact of T with the line, so that one can read off these geometric phenomena from the coefficients directly. (P and F are the dual basis to F and G in A^2.)

We can similarly observe that a general surface class T should meet the 2-plane L_{ijk} in a finite number of points. The coefficients of H_{ijk} and $V_{ijk,i}$, $V_{ijk,j}$, and $V_{ijk,k}$ (which generate the Picard group of the blown-up L_{ijk}) record the higher-dimensional contact of a surface with L_{ijk}, namely, contact in a curve class rather than in a finite number of points. Hence, if the coefficients of T include the terms $-\alpha H_{ijk} + \beta_{ijk,i}V_{ijk,i} + \beta_{ijk,j}V_{ijk,j} + \beta_{ijk,k}V_{ijk,k}$ then the higher-dimensional contact of T with

L_{ijk} (away from the coordinate lines) is a curve in the class $\alpha\ell_{ijk} - \beta_{ijk,i}e_{ijk,i} - \beta_{ijk,j}e_{ijk,j} - \beta_{ijk,k}e_{ijk,k}$.

Having described the generators for the Chow ring $A^*(\mathbb{X}_3^4)$, we can now present the ring structure. The computations are relatively straightforward, using, for example, the formulas for the Chow rings of blowups presented in [12], Chap. 13. (The computation is iterative, first computing $A^*(\mathbb{X}_1^4)$, then using that to compute $A^*(\mathbb{X}_2^4)$, and finally $A^*(\mathbb{X}_3^4)$,)

Proposition 4.5 *The Chow ring of $\mathbb{X}_3^4$ can be described as follows.*

(a) A basis for the Chow ring $A(\mathbb{X}_3^4)$ is given by the classes:

$$\begin{array}{rc} A^0: & [\mathbb{X}_3^4] = 1 \\ A^1: & H,\, E_i,\, E_{ij},\, E_{ijk} \\ A^2: & S,\, S_i,\, P_{ij},\, F_{ij},\, H_{ijk},\, V_{ijk,i} \\ A^3: & \ell,\, \ell_i,\, \ell_{ij},\, f_{ijk} \\ A^4: & p \end{array}$$

(b) Multiplication of basis elements is given in the following tables.

$A^1 \cdot A^1$	H	E_i	E_{ij}	E_{ijk}
H	S	0	F_{ij}	H_{ijk}
E_m	0	$-S_i\delta_{i,m}$	$F_{ij}\delta_{m,ij}$	$V_{ijk,m}\delta_{m,ijk}$
E_{mn}	F_{mn}	$F_{mn}\delta_{i,mn}$	$-(P_{ij} + 2F_{ij})\delta_{ij,mn}$	$(H_{ijk} - V_{ijk,m} - V_{ijk,n})\delta_{mn,ijk}$
E_{mnr}	H_{mnr}	$V_{mnr,i}\delta_{i,mnr}$	$(H_{mnr} - V_{mnr,i} - V_{mnr,j})\delta_{ij,mnr}$	$-(M_{ijk} + \Lambda_{ijk})\delta_{ijk,mnr}$

$A^1 \cdot A^2$	H	E_i	E_{ij}	E_{ijk}
S	ℓ	0	0	f_{ijk}
S_m	0	$-\ell_i\delta_{i,m}$	0	$f_{ijk}\delta_{m,ijk}$
P_{mn}	ℓ_{mn}	$\ell_{mn}\delta_{i,mn}$	$(-\ell_{ij} - \ell + \ell_i + \ell_j)\delta_{ij,mn}$	$-f_{ijk}\delta_{mn,ijk}$
F_{mn}	0	0	$-\ell_{ij}\delta_{ij,mn}$	$f_{ijk}\delta_{mn,ijk}$
G_{mn}	ℓ_{mn}	$\ell_{mn}\delta_{i,mn}$	$(-2\ell_{ij} - \ell + \ell_i + \ell_j)\delta_{ij,mn}$	0
H_{mnr}	f_{ijk}	0	$f_{mnr}\delta_{ij,mnr}$	$(-4f_{ijk} - \ell + \ell_{ij} + \ell_{ik} + \ell_{jk})\delta_{ijk,mnr}$
$V_{mnr,m}$	0	$-f_{mnr}\delta_{i,m}$	$f_{mnr}\delta_{m,ij}$	$(-2f_{mnr} - \ell_m + \ell_{mn} + \ell_{mr})\delta_{ijk,mnr}$

$A^1 \cdot A^3$	H	E_i	E_{ij}	E_{ijk}
ℓ	p	0	0	0
ℓ_m	0	$-p\delta_{i,m}$	0	0
ℓ_{mn}	0	0	$-p\delta_{ij,mn}$	0
f_{mnr}	0	0	0	$-p\delta_{ijk,mnr}$

$A^2 \cdot A^2$	S	S_i	P_{ij}	F_{ij}	G_{ij}	H_{ijk}	$V_{ijk,i}$
S	p	0	0	0	0	0	0
S_m	0	$-p\delta_{i,m}$	0	0	0	0	0
P_{mn}	0	0	$p\delta_{ij,mn}$	$-p\delta_{ij,mn}$	0	0	0
F_{mn}	0	0	$-p\delta_{ij,mn}$	0	$-p\delta_{ij,mn}$	0	0
G_{mn}	0	0	0	$-p\delta_{ij,mn}$	$-p\delta_{ij,mn}$	0	0
H_{mnr}	0	0	0	0	0	$-p\delta_{ijk,mnr}$	0
$V_{mnr,m}$	0	0	0	0	0	0	$p\delta_{ijk,mnr}\delta_{i,m}$

5 The Cremona Involution on $\mathbb{P}^4$

Consider now the Cremona involution

$$[x_0 : x_1 : x_2 : x_3 : x_4] \longrightarrow [\frac{1}{x_0} : \frac{1}{x_1} : \frac{1}{x_2} : \frac{1}{x_3} : \frac{1}{x_4}]$$
$$= [x_1x_2x_3x_4 : x_0x_2x_3x_4 : x_0x_1x_3x_4 : x_0x_1x_2x_4 : x_0x_1x_2x_3]$$

which lifts to a biregular automorphism of $\mathbb{X}^4_3$. The induced action ϕ on the Chow ring $A(\mathbb{X}^4_3)$ is given as follows.

Proposition 5.1

$$\phi(H) = 4H - 3\sum_i E_i - 2\sum_{ij} E_{ij} - \sum_{ijk} E_{ijk}$$
$$\phi(E_i) = [L_{jkmn\neq i}] = H - \sum_{m\neq i} E_m - \sum_{mn\neq i} E_{mn} - \sum_{mnr\neq i} E_{mnr}$$
$$\phi(E_{ij}) = E_{mnr\neq i,j}$$
$$\phi(E_{ijk}) = E_{mn\neq i,j,k}$$
$$\phi(S) = 6S - 3\sum_i S_i - \sum_{ij} P_{ij}$$
$$\phi(S_m) = 3S - 2\sum_{i\neq m} S_i - \sum_{ij\neq m} P_{ij}$$
$$\phi(F_{mn}) = M_{ijk\neq mn} = S - S_i - S_j - S_k + F_{ij} + F_{ik} + F_{jk} - G_{ij} - G_{ik} - G_{jk} + \Lambda_{ijk}$$
$$= S - S_i - S_j - S_k - P_{ij} - P_{ik} - P_{jk} + 2H_{ijk} - V_{ijk,i} - V_{ijk,j} - V_{ijk,k}$$
$$\phi(G_{mn}) = \Lambda_{ijk\neq mn} = 2H_{ijk} - V_{ijk,i} - V_{ijk,j} - V_{ijk,k}$$
$$\phi(P_{mn}) = -S + S_i + S_j + S_k + P_{ij} + P_{ik} + P_{jk} (ijk \neq mn)$$
$$\phi(H_{mnr}) = 2G_{ij} - (H_{ijm} - V_{ijm,i} - V_{ijm,j}) - (H_{ijn} - V_{ijn,i} - V_{ijn,j}) - (H_{ijr} - V_{ijr,i} - V_{ijr,j})$$
$$= 2P_{ij} + 2F_{ij} - (H_{ijm} - V_{ijm,i} - V_{ijm,j}) - (H_{ijn} - V_{ijn,i} - V_{ijn,j}) - (H_{ijr} - V_{ijr,i} - V_{ijr,j})$$
$$\text{for } i, j \neq m, n, r$$
$$\phi(V_{mnr,m}) = G_{ij} - (H_{ijn} - V_{ijn,i} - V_{ijn,j}) - (H_{ijr} - V_{ijr,i} - V_{ijr,j})$$
$$= P_{ij} + F_{ij} - (H_{ijn} - V_{ijn,i} - V_{ijn,j}) - (H_{ijr} - V_{ijr,i} - V_{ijr,j})$$

$$\begin{aligned}
&\textit{for} \quad i, j \neq m, n, r\\
\phi(\ell) &= 4\ell - \sum_i \ell_i\\
\phi(\ell_m) &= 3\ell - \sum_{i \neq m} \ell_i\\
\phi(\ell_{mn}) &= 2\ell - \ell_i - \ell_j - \ell_k + f_{ijk} \textit{ for } i, j, k \neq m, n\\
\phi(f_{mnr}) &= h_{ij} = \ell - \ell_i - \ell_j + \ell_{ij} \textit{ for } i, j \neq m, n, r
\end{aligned}$$

Proposition 5.2 *(a) Let $D = dH - \sum_i m_i E_i - \sum_{ij} m_{ij} E_{ij} - \sum_{ijk} m_{ijk} E_{ijk}$ be a general class in $A^1(\mathbb{X}_3^4)$. Then the Cremona image $\phi(D)$ of D under the involution is*

$$\phi(D) = d'H - \sum_i m'_i E_i - \sum_{ij} m'_{ij} E_{ij} - \sum_{ijk} m'_{ijk} E_{ijk}$$

where

$$\begin{aligned}
d' &= \phi(D) \cdot \ell = D \cdot \phi(\ell) = D \cdot (4\ell - \sum_r \ell_r) = 4d - \sum_r m_r;\\
m'_i &= \phi(D) \cdot \ell_i = D \cdot \phi(\ell_i) = D \cdot (3\ell - \sum_{r \neq i} \ell_r) = 3d - \sum_{r \neq i} m_r\\
m'_{ij} &= \phi(D) \cdot \ell_{ij} = D \cdot \phi(\ell_{ij}) = D \cdot (2\ell - \sum_{r \neq ij} \ell_r + f_{rst \neq ij}) = 2d - \sum_{r \neq ij} m_r + m_{rst \neq ij}\\
m'_{ijk} &= \phi(D) \cdot f_{ijk} = D \cdot \phi(f_{ijk}) = D \cdot (\ell - \sum_{r \neq ijk} \ell_r + \ell_{rs \neq ijk}) = d - \sum_{r \neq ijk} m_r + m_{rs \neq ijk}
\end{aligned}$$

(b) Let $T = dS - \sum_i m_i S_i - \sum_{ij} m_{ij} P_{ij} - \sum_{ij} n_{ij} F_{ij} - \sum_{ijk} m_{ijk} H_{ijk} + \sum_{ijk} (n_{ijk,i} V_{ijk,i} + n_{ijk,j} V_{ijk,j} + n_{ijk,k} V_{ijk,k})$ be a general class in $A^2(\mathbb{X}_3^4)$. Then the Cremona image $\phi(T)$ of T under the involution is

$$\begin{aligned}
\phi(T) = d'S &- \sum_i m'_i S_i - \sum_{ij} m'_{ij} P_{ij} - \sum_{ij} n'_{ij} F_{ij}\\
&- \sum_{ijk} m'_{ijk} H_{ijk} + \sum_{ijk} (n'_{ijk,i} V_{ijk,i} + n'_{ijk,j} V_{ijk,j} + n'_{ijk,k} V_{ijk,k})
\end{aligned}$$

where

$$\begin{aligned}
d' &= \phi(T) \cdot S = T \cdot \phi(S) = T \cdot (6S - 3\sum_i S_i - \sum_{ij} P_{ij})\\
&= 6d - 3\sum_i m_i + \sum_{ij} (m_{ij} - n_{ij})\\
m'_i &= \phi(T) \cdot S_i = T \cdot \phi(S_i) = T \cdot (3S - 2\sum_{r \neq i} S_r - \sum_{rs \neq i} P_{rs})
\end{aligned}$$

$$= 3d - 2\sum_{r\neq i} m_r + \sum_{rs\neq i}(m_{rs} - n_{rs})$$

$$m'_{ij} = \phi(T)\cdot F_{ij} = T\cdot\phi(F_{ij})$$
$$= T\cdot(S - S_r - S_s - S_t - P_{rs} - P_{rt} - P_{st} + 2H_{rst} - V_{rst,r} - V_{rst,s} - V_{rst,t})$$
$$= d - m_r - m_s - m_t + m_{rs} + m_{rt} + m_{st} - n_{rs} - n_{rt} - n_{st} + 2m_{rst} - n_{rst,r} - n_{rst,s} - n_{rst,t}$$

$$n'_{ij} = \phi(T)\cdot G_{ij} = T\cdot\phi(G_{ij}) = T\cdot(\Lambda_{rst\neq ij}) = T\cdot(2H_{rst} - V_{rst,r} - V_{rst,s} - V_{rst,t})$$
$$= 2m_{rst} - n_{rst,r} - n_{rst,s} - n_{rst,t}$$

$$m'_{ijk} = \phi(T)\cdot H_{ijk} = T\cdot\phi(H_{ijk})$$
$$= T\cdot(2G_{rs} - (H_{rsi} - V_{rsi,r} - V_{rsi,s}) - (H_{rsj} - V_{rsj,r} - V_{rsj,s}) - (H_{rsk} - V_{rsk,r} - V_{rsk,s})$$
$$\text{for } rs\neq ijk$$
$$= 2n_{rs} - (m_{rsi} - n_{rsi,r} - n_{rsi,s}) - (m_{rsj} - n_{rsj,r} - n_{rsj,s}) - (m_{rsk} - n_{rsk,r} - n_{rsk,s}$$

$$n'_{ijk,i} = \phi(T)\cdot V_{ijk,i} = T\cdot\phi(V_{ijk,i})$$
$$= T\cdot(G_{rs} - (H_{rsj} - V_{rsj,r} - V_{rsj,s}) - (H_{rsk} - V_{rsk,r} - V_{rsk,s}))$$
$$= n_{rs} - (m_{rsj} - n_{rsj,r} - n_{rsj,s}) - (m_{rsk} - n_{rsk,r} - n_{rsk,s})$$

(c) Let $C = d\ell - \sum_i m_i\ell_i - \sum_{ij} m_{ij}\ell_{ij} - \sum_{ijk} m_{ijk}f_{ijk}$ be a general class in $A^3(\mathbb{X}^4_3)$. Then the Cremona image $\phi(C)$ of C under the involution is

$$\phi(C) = d'\ell - \sum_i m'_i\ell_i - \sum_{ij} m'_{ij}\ell_{ij} - \sum_{ijk} m'_{ijk}f_{ijk}$$

where

$$d' = \phi(C)\cdot H = C\cdot\phi(H) = C\cdot(4H - 3\sum_i E_i - 2\sum_{ij} E_{ij} - \sum_{ijk} E_{ijk})$$
$$= 4d - 3\sum_i m_i - 2\sum_{ij} m_{ij} - \sum_{ijk} m_{ijk};$$

$$m'_i = \phi(C)\cdot E_i = C\cdot\phi(E_i) = C\cdot(H - \sum_{r\neq i} E_r - \sum_{rs\neq i} E_{rs} - \sum_{rst\neq i} E_{rst})$$
$$= d - \sum_{r\neq i} m_r - \sum_{rs\neq i} m_{rs} - \sum_{rst\neq i} m_{rst};$$

$$m'_{ij} = \phi(C)\cdot E_{ij} = C\cdot\phi(E_{ij}) = C\cdot E_{rst\neq ij} = m_{rst\neq ij}$$

$$m'_{ijk} = \phi(C)\cdot E_{ijk} = C\cdot\phi(E_{ijk}) = C\cdot E_{rs\neq ijk} = m_{rs\neq ij}$$

We note that, for surface classes in $A^2(\mathbb{X}^4_3)$, higher-dimensional contact is observed by having nonzero coefficients in the F, H, and V basis elements. For curve classes in A^3, this higher-dimensional contact corresponds to nonzero coefficients in the ℓ_{ij} and the f_{ijk} basis elements (corresponding to a curve meeting a coordinate line or a coordinate plane). The formulas above show that a similar phenomenon happens as in the $\mathbb{P}^3$ case: if these are all zero, that is preserved under the involution.

Corollary 5.3 *(a) The subspace of $A^2(\mathbb{X}_3^4)$ spanned by S, the S_i, and the P_{ij} is invariant under the Cremona involution ϕ. If $T = dS - \sum_i m_i S_i - \sum_{ij} m_{ij} P_{ij}$ is an element in this subspace, then $\phi(T) = d'S - \sum_i m_i' S_i - \sum_{ij} m_{ij}' P_{ij}$ where*

$$d' = 6d - 3\sum_i m_i + \sum_{ij} m_{ij}$$

$$m_i' = 3d - 2\sum_{r \neq i} m_r + \sum_{rs \neq i} m_{rs}$$

$$m_{ij}' = d - m_r - m_s - m_t + m_{rs} + m_{rt} + m_{st} \quad \textit{for} \quad r, s, t \neq i, j$$

(b) The subspace of $A^3(\mathbb{X}_3^4)$ spanned by ℓ, ℓ_i is invariant under the Cremona involution ϕ. If $C = d\ell - \sum_i m_i \ell_i$ is an element in this subspace, then $\phi(C) = d'\ell - \sum_i m_i' \ell_i$ where

$$d' = 4d - 3\sum_i m_i$$

$$m_i' = d - \sum_{r \neq i} m_r$$

For divisors, the natural subspace invariant under the involution is the one generated by the E_{ij}'s and E_{ijk}'s. If we are only interested in the multiplicity conditions at the points, we can therefore mod out by this subspace of A^1, and obtain the following.

Corollary 5.4 *The subspace of $A^1(\mathbb{X}_3^4)$ spanned by the E_{ij}'s and E_{ijk}'s is invariant under the Cremona involution ϕ. Denote by $\bar{A}^1$ the quotient of A^1 by this subspace; the involution ϕ descends to an involution of $\bar{A}^1$. If $\bar{D} = dH - \sum_i m_i E_i$ represents a coset in this subspace, then $\phi(\bar{D}) = d'H - \sum_i m_i' E_i$ where*

$$d' = 4d - \sum_i m_i \quad \textit{and} \quad m_i' = 3d - \sum_{r \neq i} m_r.$$

6 Six and Seven Points in $\mathbb{P}^4$

The formulas for how degrees and multiplicities change for curves, surfaces, and divisors in $\mathbb{P}^4$ under the standard Cremona transformation can be used to analyze compositions of such Cremona transformations based at more than five points. We will present the orbits of the linear subspaces spanned by subsets of the points in this section.

If we first consider six general points in $\mathbb{P}^4$, it is easy to see using the formulas above that any line through 2 of the six points, 2-plane through 3 of them, or a hyperplane through 4, is either contracted by the Cremona transformation or is sent to itself.

The case of seven general points in $\mathbb{P}^4$ is one step more interesting. In this case, for a line through two of the seven points, it is either contracted by the Cremona transformation based at five of the points (if the two points are a subset of the five), is sent to itself (if one of the two is a subset of the five) or is sent to the rational normal quartic (RNQ) through all seven points (if neither of the two is among the five).

The iteration of Cremona now leads us to consider the transformation of the RNQ; applying Cremona at any five yields back the line joining the other two (since the Cremona is an involution).

Hence the Cremona orbit of the line through two points is the collection of all of the 21 lines, plus the rational normal quartic through all seven points.

Now consider the 2-plane spanned by three of the 7 points. Performing a Cremona transformation at 5 of the 7 points, we see that if all three points are among the 5, the plane is contracted as part of the fundamental locus. If two of the three points are among the five, the plane is sent to itself. If only one of the three points is among the five, then the Cremona image is a surface of degree three, with a point of multiplicity 3 at that one point, and multiplicity 1 at the other six points. It contains the line joining that one point to the other six, with a multiplicity of one each, and no other lines joining the points. It also contains the RNQ with multiplicity one. This cubic surface is a cone over a twisted cubic in $\mathbb{P}^3$.

Iterating the Cremona by applying it to this cone, we see that if the five points contain the vertex, it will be transformed back into the 2-plane. If it does not, it is preserved.

Hence, the Cremona orbit of the 2-plane through 3 points in $\mathbb{P}^4$ consists of the 35 planes and the 7 cubic cones.

For the hyperplanes through 4 of the seven points, there are four cases to consider. We choose five of the seven to perform the Cremona transformation at. If all 4 of the hyperplane points are among the five, then the hyperplane is contracted to a point. If 3 of the hyperplane points are among the five, then the hyperplane is transformed to another hyperplane. If 2 of the hyperplane points are among the five, it is transformed into a quadric double cone: a cone over a smooth conic with vertex a line (the line corresponding to the two points). To be explicit, take the line joining the two points, and a complementary plane; projection from the line to the plane sends the other five points to five general points in the plane, and there is a unique conic in that plane through those five points. The threefold is obtained as the cone over the conic with vertex the line. The surfaces contain all the lines joining the two points with the other five, as well as containing the RNQ too.

If we apply a second Cremona transformation to this quadric, we either return to the hyperplane, preserve the quadric, or (if we use as the base points the five points not on the vertex line) we obtain a cubic surface double at all seven points. It is also double all along the RNQ; this cubic surface is the secant variety to the RNQ, in fact.

Further applications of Cremona to this cubic surface lower the degree and return us to the quadric double cone; we see then that the orbit of the hyperplane consists of the set of 35 hyperplanes, the 21 quadric double cones, and the cubic secant variety to the RNQ.

It is interesting that the two special linear systems with irreducible members in $\mathbb{P}^4$ imposing only double points appear here: the quadrics double at two points and the cubics double at 7.

7 Eight Points in $\mathbb{P}^4$

We now consider the case of Cremona transformations based at 8 general points $p_1, \ldots, p_8$ in $\mathbb{P}^4$. Denote by L_{ij} the line joining p_i and p_j as usual. Denote by Q_i the rational normal quartic curve passing through all eight points except p_i (i.e., passing through the other 7).

It is easy to see, with a parallel computation as that done above for seven points, that the orbit of a line through two points, say L_{12}, consists of all 28 such lines L_{ij}, and all 8 of the RNQ's Q_k.

We can now take up the case of surfaces, which is more involved. We will record the data for a surface of degree d, having multiplicity m_i at p_i, multiplicity n_i along Q_i, and multiplicity m_{ij} along L_{ij}, by the triangular array of numbers:

$$\begin{array}{ccccccccc}
d & m_1 & m_2 & m_3 & m_4 & m_5 & m_6 & m_7 & m_8 \\
 & n_1 & n_2 & n_3 & n_4 & n_5 & n_6 & n_7 & n_8 \\
 & & m_{12} & m_{13} & m_{14} & m_{15} & m_{16} & m_{17} & m_{18} \\
 & & & m_{23} & m_{24} & m_{25} & m_{26} & m_{27} & m_{28} \\
 & & & & m_{34} & m_{35} & m_{36} & m_{37} & m_{38} \\
 & & & & & m_{45} & m_{46} & m_{47} & m_{48} \\
 & & & & & & m_{56} & m_{57} & m_{58} \\
 & & & & & & & m_{67} & m_{68} \\
 & & & & & & & & m_{78}
\end{array} \tag{7.1}$$

Suppose we perform the five-point Cremona on the first five points $1, 2, 3, 4, 5$. Then the degree d, the multiplicities m_i for $i \leq 5$, and the m_{ij} for $i, j \leq 5$, are transformed as indicated in Corollary 5.3(a).

For multiplicity m'_{ij} with $i \leq 5$ and $j \geq 6$, we note that this line L_{ij} is left invariant under the Cremona, so that $m'_{ij} = m_{ij}$ for these indices.

For multiplicities m_{ij} with both $i, j \geq 6$, we note that this L_{ij} is the image of Q_k where $\{i, j, k\} = \{6, 7, 8\}$; k is the third index. Hence $m'_{ij} = n_k$ for $k = \{6, 7, 8\} - \{i, j\}$.

For the n'_k with $k \geq 6$, conversely we have $n'_k = m_{ij}$ where $i, j = \{6, 7, 8\} - \{k\}$. For n'_k with $k \leq 5$, since such a Q_k is fixed, we have $n'_k = n_k$. This gives the following:

Corollary 7.2 *The surface with degree and multiplicities indicated by (7.1) is transformed, under the Cremona involution based at the first five points p_1, p_2, p_3, p_4, p_5, into the surface with degree and multiplicities recorded by:*

$$\begin{array}{ccccccccc}
d' & m'_1 & m'_2 & m'_3 & m'_4 & m'_5 & m_6 & m_7 & m_8 \\
 & n_1 & n_2 & n_3 & n_4 & n_5 & m_{78} & m_{68} & m_{67} \\
 & & m'_{12} & m'_{13} & m'_{14} & m'_{15} & m_{16} & m_{17} & m_{18} \\
 & & & m'_{23} & m'_{24} & m'_{25} & m_{26} & m_{27} & m_{28} \\
 & & & & m'_{34} & m'_{35} & m_{36} & m_{37} & m_{38} \\
 & & & & & m'_{45} & m_{46} & m_{47} & m_{48} \\
 & & & & & & m_{56} & m_{57} & m_{58} \\
 & & & & & & & n_8 & n_7 \\
 & & & & & & & & n_6
\end{array}$$

where

$$d' = 6d - 3\sum_{i=1}^{5} m_i + \sum_{1\le i<j\le 5} m_{ij}$$

$$m'_i = 3d - 2\sum_{r\le 5; r\ne i} m_r + \sum_{r,s\le 5; r,s\ne i} m_{rs} \quad \text{for } i \le 5$$

$$m'_{ij} = d - m_r - m_s - m_t + m_{rs} + m_{rt} + m_{st} \quad \text{for } i, j \le 5 \text{ and } r, s, t = \{1, 2, 3, 4, 5\} - \{i, j\}$$

The Proposition below presents the orbit of L_{123}, a 2-plane through three of the points, in (b). For notational consistency with the other surfaces in this orbit, we will also denote L_{ijk} by $S_1(ijk)$. We have included in (a) the remarks above about the orbit of the line L_{12}. In (c), we present the orbit of a hyperplane; the reader can verify the computations as an exercise.

Proposition 7.3 *Fix* 8 *general points in* $\mathbb{P}^4$, *and consider Cremona transformations based at* 5 *of the* 8, *in series.*

(a) The orbit of a line through two of the 8 *points consists of the* 28 *lines* L_{ij} $(1 \le i < j \le 8)$ *through two* (p_i *and* p_j) *of the* 8 *points, and the* 8 *rational normal quartics* Q_k $(1 \le k \le 8$ *through* 7 *of the* 8 *points (through all seven except* p_k).

(b) The orbit of a plane through three of the 8 *points consists of:*

(b1) the 56 *planes* $L_{ijk} = S_1(1jk)$ *through three of the* 8 *points (namely* p_i, p_j, *and* p_k*); the plane* $L_{123} = S_1(123)$ *is recorded as*

$$\begin{array}{r} 1\ 1\ 1\ 1\ 0\ 0\ 0\ 0\ 0 \\ 0\ 0\ 0\ 0\ 0\ 0\ 0\ 0 \\ 1\ 1\ 0\ 0\ 0\ 0\ 0 \\ 1\ 0\ 0\ 0\ 0\ 0 \\ 0\ 0\ 0\ 0\ 0 \\ 0\ 0\ 0\ 0 \\ 0\ 0\ 0 \\ 0\ 0 \\ 0 \end{array}$$

(b2) *the* 56 *surfaces* $S_3(i, j)$ *of degree* 3 *with one point* p_i *of multiplicity* 3, 6 *points of multiplicity one, and one point* p_j *of multiplicity* 0. *It contains the lines joining the triple point* p_i *to all other multiplicity one points* p_k $(k \neq j)$ *and no other lines; it contains the rational normal quartic* Q_j *through the triple point and the six multiplicity one points. For example,* $S_3(8, 1)$ *is recorded as:*

$$\begin{array}{r} 3\ 0\ 1\ 1\ 1\ 1\ 1\ 1\ 3 \\ 1\ 0\ 0\ 0\ 0\ 0\ 0\ 0 \\ 0\ 0\ 0\ 0\ 0\ 0\ 0 \\ 0\ 0\ 0\ 0\ 0\ 1 \\ 0\ 0\ 0\ 0\ 1 \\ 0\ 0\ 0\ 1 \\ 0\ 0\ 1 \\ 0\ 1 \\ 1 \end{array}$$

(b3) *the* 56 *sextic surfaces* $S_6(ijk)$ *of degree* 6 *with three points* (p_i, p_j, p_k) *of multiplicity one, and the other* 5 *points of multiplicity* 3. *It contains the lines joining any two of the multiplicity* 3 *points and no other lines; It contains the rational normal quartics through the five multiplicity* 3 *points and any two of the three multiplicity one points. For example,* $S_6(678)$ *is recorded as:*

$$\begin{array}{r} 6\ 3\ 3\ 3\ 3\ 3\ 1\ 1\ 1 \\ 0\ 0\ 0\ 0\ 0\ 1\ 1\ 1 \\ 1\ 1\ 1\ 1\ 0\ 0\ 0 \\ 1\ 1\ 1\ 0\ 0\ 0 \\ 1\ 1\ 0\ 0\ 0 \\ 1\ 0\ 0\ 0 \\ 0\ 0\ 0 \\ 0\ 0 \\ 0 \end{array}$$

(b4) *the* 28 *surfaces* $S_{10}(ij)$ *of degree* 10 *with two points* (p_i *and* p_j) *of multiplicity* 6 *and the other* 6 *points of multiplicity* 3. *It contains the lines joining the multiplicity one points to the multiplicity six points (each with multiplicity one) and the line joining the two multiplicity* 6 *points with multiplicity* 3*. It contains the* 6 *rational normal quartics that pass through the two multiplicity* 6 *points and five of the six multiplicity one points. For example,* $S_{10}(78)$ *is recorded as:*

10 3 3 3 3 3 3 6 6
1 1 1 1 1 1 0 0
0 0 0 0 0 1 1
0 0 0 0 1 1
0 0 0 1 1
0 0 1 1
0 1 1
1 1
3

(b5) the 8 *surfaces* $S_{15}(i)$ *of degree* 15 *with one point* (p_i) *of multiplicity* 3 *and the other seven points of multiplicity* 6*. It contains the joining any two points of multiplicity* 6*, and no other lines. It contains all* 8 *of the rational normal quartics; the one through the seven multiplicity* 6 *points with multiplicity three, and all others with multiplicity one. For example,* $S_{15}(1)$ *is recorded as:*

15 3 6 6 6 6 6 6 6
3 1 1 1 1 1 1 1
0 0 0 0 0 0 0
1 1 1 1 1 1
1 1 1 1 1
1 1 1 1
1 1 1
1 1
1

(c) We use the notation that $(d; m_1 m_2 \cdots m_8)$ *represents a hyperplane of degree* d *having multiplicity* m_i *at* p_i*. The orbit of the hyperplane through the first four points (represented by* (1; 11110000) *consists of the following divisors, and all related divisors obtained by permutations of the eight points:*

(1; 11110000) (2; 22111110) (3; 22222220) (3; 32222111)
(4; 33332221) (4; 43222222) (5; 44333322) (6; 44444432)
(6; 54443333) (7; 55544443) (7; 64444444) (8; 65555544)
(9; 66665555) (10; 76666666)

8 Applications

Proposition 8.1 *Let R and T be two Weyl planes on $X_{4,8}$. Then $R \cdot T \in \{0, 1, 3\}$.*

Proof If we choose an element w of the Weyl group that sends Weyl plane R to the actual plane $S_1(123)$, then since the intersection form is preserved we have $R \cdot T = S_1(123) \cdot w(T)$. Hence it suffices to show that the intersection of $S_1(123)$ with any Weyl plane is in $\{0, 1, 3\}$. This one can check by hand for all of the cases.

Even easier would be to notice that, if ϕ is the Cremona transformation centered at the first five points, then by Corollary 7.2 we have $\phi(S_1(123)) = -P_{45}$ in the Chow ring. Hence it also suffices to show that $-P_{45} \cdot T \in \{0, 1, 3\}$ for all Weyl planes T. By Proposition 4.5, intersecting with $-P_{45}$ picks out exactly the multiplicity m_{45} for the Weyl plane. Hence it suffices, after taking account of permutations, to observe that for all Weyl planes, all m_{ij} are in $\{0, 1, 3\}$. □

Proposition 8.2 *Let R and T be any Weyl planes on X^4_8. If $R \cdot T \neq 3$, then there exists w in the Weyl group of X^4_8 and $i \in \{1, 4\}$ such that $w(R) = H_{123}$ and $w(T) = H_{i56}$.*

Proof It is enough to prove the statement for $R \neq T$. One can use the same technique as in Proposition 8.1 and reduce one Weyl surface to $-P_{45}$ and select Weyl surfaces from the list of Proposition 7.3(b) that have $m_{45} \in \{0, 1\}$. Then applying the Cremona transformation ϕ centered at the first five points, we have the first Weyl surface being $S_1(123)$ and the other on the following lists (up to permutations that fix $\{1, 2, 3\}$):

1. Case $S_1(123) \cdot T = 1$:

   (a) $S_1(123) \cdot S_1(456) = 1$
   (b) $S_1(123) \cdot S_3(4, 1) = 1$
   (c) $S_1(123) \cdot S_6(126) = 1$
   (d) $S_1(123) \cdot S_{10}(45) = 1$
   (e) $S_1(123) \cdot S_{15}(1) = 1$

   We are done in the first case of course. In the other cases it suffices to find five indices, two of them among $\{1, 2, 3\}$, so that the corresponding Cremona transformation reduces the degree of the second surface; such a Cremona will fix $S_1(123)$ and we proceed then by induction on the degree.
   To reduce the cubic surface, $\{2, 3, 4, 7, 8\}$ will work; for the sextic, $\{1, 3, 5, 7, 8\}$ works. For the surface of degree 10, $\{1, 2, 4, 5, 6\}$ suffices; finally for the last surface of degree 15, $\{2, 3, 6, 7, 8\}$ works.
2. Case $S_1(123) \cdot T = 0$: In this case a similar approach yields the following lists to analyze:

   (a) $S_1(123) \cdot S_1(145) = 0$ or $S_1(123) \cdot S_1(124) = 0$
   (b) $S_1(123) \cdot S_3(1, 2) = 0$ or $S_1(123) \cdot S_3(1, 4) = 0$ or $S_1(123) \cdot S_3(4, 5) = 0$
   (c) $S_1(123) \cdot S_6(145) = 0$ or $S_1(123) \cdot S_6(456) = 0$
   (d) $S_1(123) \cdot S_{10}(12) = 0$ or $S_1(123) \cdot S_{10}(15) = 0$

(e) $S_1(123) \cdot S_{15}(4) = 0$

The same proof as in the prior case works; in each situation one finds five indices, two among $\{1, 2, 3\}$, that reduce the degree of the second surface. For example, $\{2, 3, 6, 7, 8\}$ works for the degree 15 surface. We leave the details of the other cases to the reader.

□

We remark that Weyl planes that intersect in three points (modulo permutations of points) are

$$S_1(123) \cdot S_6(123) = S_3(1, 8) \cdot S_3(8, 1) = 3.$$

Corollary 8.3 *Assume R and T are Weyl planes in the base locus of the linear system $|D|$ for an effective divisor $D = dH - \sum_{i=1}^{8} m_i E_i$ on X_8^4. Then $R \cdot T = 0$.*

Proof We argue by contradiction. Assume first that $R \cdot T = 1$. By Proposition 8.2, we can apply a series of Cremona transformations, which do not change the hypothesis on the base locus, and assume that $R = S_1(123)$ and $T = S_1(456)$. It follows from the results of [6], Sect. 4, and [8], Proposition 4.2, that we therefore have

$$m_1 + m_2 + m_3 - 2d > 0 \text{ and } m_4 + m_5 + m_6 - 2d > 0.$$

Hence, the system of rational normal curves of degree 4 passing through the first 6 points must be in the base locus of $|D|$; since this family of curves covers $\mathbb{P}^4$, we conclude $|D|$ is empty, a contradiction.

If the two Weyl planes intersect in three points, then they are either $S_3(1, 8)$ and $S_3(8, 1)$ or $S_1(123)$ and $S_6(123)$ (up to permutations). We will analyze the first case; the other is handled by a similar argument. Assume by contradiction that both such Weyl planes are in the base locus of the linear system $|D|$ of an effective divisor D. By Proposition 3 of [5], the multiplicity of containment of the surface $S_3(1, 8)$ in the base locus of a divisor D is $2m_1 + m_2 + \ldots + m_7 - 5d < 0$; therefore since both $S_3(1, 8)$ and $S_3(8, 1)$ are in the base locus we obtain $2(m_1 + \ldots + m_8) - 10d < 0$. This contradicts the effectivity of the divisor D because $2(m_1 + \ldots + m_6) + m_7 + 3m_8 - 10d \leq 2(m_1 + \ldots + m_8) - 10d < 0$; therefore a family of curves of degree 10 with six double points, one simple point, and one triple point meets D negatively, and so is part of the base locus also. Corollary 5.3 implies that these curves are in the Weyl orbit of a line through a point, and therefore again cover the projective space, a contradiction. The remaining case can be handled by the same argument. □

Remark 8.4 In fact, the linear equations of pencils of curves in the base locus of the linear system of an effective divisor D, that in this case are equivalent to two Weyl planes that meet in the base locus of $|D|$, give the *faces of the cone of effective divisors*. We will prove this theorem in the case of a Mori Dream Space in arbitrary dimension in [11].

Remark 8.5 In [11], we prove that a Weyl curve and a Weyl divisor that meet can not be simultaneously in the base locus of the linear system of an effective divisor D.

For any effective divisor $D \in Pic(X^4_8)$, define $\widetilde{D} \in Pic(\widehat{X^4_8})$ to be the proper transform of D after blowing up all the Weyl lines and Weyl planes in the base locus of $|D|$ to obtain $\widehat{X^4_8}$. Corollary 8.3 proves that the space $\widehat{X^4_8}$ is smooth.

We remark first that the Weyl line C has normal bundle $\oplus\mathcal{O}(-1)^3$. If $D \cdot C < 0$ then the Weyl line C is in the base locus of the linear system $|D|$. Let $D_{(1)}$ denote the proper transform of D under the blowup Y of all fixed Weyl lines in X^4_s. For each Weyl line C, define $k_C = -D \cdot C$.

Proposition 8.6 *If D be an effective divisor on X^4_8, then*

$$h^1(X^4_s, \mathcal{O}_{X^4_s}(D)) = \sum_C \binom{2+k_C}{4} + h^1(Y, \mathcal{O}_Y(D_{(1)}) - h^2(Y, \mathcal{O}_Y(D_{(1)}).$$

A general form of Proposition 8.6 for (-1)-curves in arbitrary dimension will be given in [11]. We conclude that if $k_C \geq 2$ then $h^1(X^4_s, \mathcal{O}_{X^4_s}(D)) \geq 1 + h^1(Y, \mathcal{O}_Y(D_{(1)}) - h^2(Y, \mathcal{O}_Y(D_{(1)})$.

Conjecture 8.7 Let D be an effective divisor on X^4_s, with $H^1(X^4_s, \mathcal{O}_{X^4_s}(D)) = 0$. Then $D \cdot C \geq -1$ for any Weyl line C.

Remark 8.8 For arbitrary number of points s, the converse of Conjecture (8.7) is not true. Indeed, take $D := 4H - 2\sum_{i=1}^{14} E_i \in Pic(X^4_{14})$. We can see that $D \cdot C \geq 0$ for any Weyl line C; however, the Alexander Hirschowitz Theorem implies that

$$h^1(X^4_s, \mathcal{O}_{X^4_s}(D)) = 1.$$

For every r-subset $I(r)$ of the indices $\{1, \ldots, 8\}$, let $L_{I(r)}$ be the linear span of the corresponding points. Let $k_{w(L_{I(r)})}$ be the multiplicity of containment of the Weyl cycle $w(L_{I(r)})$ in the base locus of D, for a Weyl group element w. In [5], the *Weyl expected dimension* for an effective divisor D was introduced as

$$wdim(D) := \chi(D) + \sum_{r=1}^{3} \sum_{I(r) \in \{1,\ldots,8\}} \sum_{w \in W} (-1)^{r+1} \binom{4 + k_{w(L_{I(r)})} - r - 1}{4}.$$

Moreover, in [5] it was conjectured that for every effective divisor D on $\widehat{X^4_8}$, the dimension of space of global sections of D equals the Weyl expected dimension.

Conjecture 8.9 Let D be an effective divisor on $\widehat{X^4_8}$.

1. If $D \cdot C \geq -1$ for all Weyl curves C then

$$H^1(X^4_s, \mathcal{O}_{X^4_s}(D)) = 0.$$

2. $h^0(D) = wdim(D) + \sum_{r=1}^{3}(-1)^{r+1}h^r(\widetilde{D})$.
3. For every $r \geq 1$, $h^r(\widetilde{D}) = 0$.
4. Moreover, $\widetilde{D}$ is globally generated on $\widehat{X^4_8}$.

We remark that Conjecture 8.9 part (2) implies $wdim(D) = \chi(\widetilde{D})$, while part (3) implies that conjecture of [5] regarding dimension $h^0(D)$ is true.

Remark 8.10 We remark that Conjecture 8.9 holds for effective divisors on X^n_{n+2} [3, 8, 9]; therefore it holds for X^4_6. Notice that $\widehat{X^4_9}$ is not a Mori Dream Space and in fact, there are infinitely many Weyl lines. The authors believe that Conjecture 8.9 also holds for $\widehat{X^4_9}$ with a similar construction for the Weyl planes as the one presented here.

Remark 8.11 Conjecture 8.9 fails in $\widehat{X^4_{10}}$, because for arbitrary number of points, in non Mori-dream spaces Weyl cycles are not the only obstructions. Indeed, consider the divisor

$$D := 4H - 4E_1 - 2\sum_{i=2}^{10} E_i.$$

We remark that D contains in the base locus of its linear system just double lines $k_{L_{1i}} = 2$; therefore its proper transform under the blowup of all its Weyl base locus (i.e. only lines) is

$$\widehat{D} := 4H - 4E_1 - 2\sum_{i=2}^{10} E_i - 2\sum_{i=2}^{10} E_{1i}$$

Moreover, since $k_{L_{1i}} = 2$ we have

$$\chi(D) = \binom{4+4}{4} - \binom{4+4-1}{4} - 9\binom{4+2-1}{4} = 70 - 35 - 45 = -10$$

$$wdim(D) = \chi(\widehat{D}) = \chi(D) + 9\binom{2+2}{4} = -1$$

However, this divisor is effective, and, in fact, the Alexander–Hirschowitz theorem implies that it is unique in its linear system. We conclude that $h^0(D) = 1 \neq 0 = wdim(D)$, therefore $h^1(\widehat{D}) = 1$.

Acknowledgements The collaboration was partially supported by NSF grant DMS1802082.

References

1. C. Araujo, C. Casagrande, On the Fano variety of linear spaces contained in two odd-dimensional quadrics. Geom. Topol. **21**(5), 3009–3045 (2017)
2. C. Araujo, A. Massarenti, Explicit log Fano structures on blow-ups of projective spaces. Proc. Lond. Math. Soc. (3)113, no. 4, 445–473 (2016)
3. M.C. Brambilla, O. Dumitrescu, E. Postinghel, On a notion of speciality of linear systems in $\mathbb{P}^n$. Trans. Am. Math. Soc. **367**(8), 5447–5473 (2015)
4. M.C. Brambilla, O. Dumitrescu, E. Postinghel, On the effective cone of $\mathbb{P}^n$ blown-up at $n + 3$ points. Exp. Math. **25**(4), 452–465 (2016)
5. M. C. Brambilla, O. Dumitrescu, E. Postinghel, Weyl cycles on the blow-up of $\mathbb{P}^4$ at eight points, to appear in *The Art of Doing Algebraic Geometry* (Springer)
6. S. Cacciola, M. Donten-Bury, O. Dumitrescu, A. Lo Giudice, J. Park, Cones of divisors of blow-ups of projective spaces. Le Matematiche LXVI—Fasc. II, 153–187 (2011)
7. C. Casagrande, G. Codogni, A. Fanelli, The blow-up of $\mathbb{P}^4$ at 8 points and its Fano model, via vector bundles on a del Pezzo surface. Revista Matemática Complutense **32**, 475–529 (2019)
8. O. Dumitrescu, E. Postinghel, Vanishing theorems for linearly obstructed divisors. J. Algebra **477**, 312–359 (2017)
9. O. Dumitrescu, E. Postinghel, Positivity of divisors on blown-up projective spaces II. J. Algebra **529**, 226–267 (2019)
10. O. Dumitrescu, N. Priddis, *On* (-1) *classes*. https://arxiv.org/pdf/1905.00074.pdf
11. O. Dumitrescu, R. Miranda *On* (-1) *Curves in* $\mathbb{P}^r$, in preparation
12. D. Eisenbud, J. Harris, *3264 and all that: a Second Course in Algebraic Geometry* (Cambridge University Press, 2016)
13. A. Laface, L. Ugaglia, On a class of special linear systems on $\mathbb{P}^3$. Trans. Am. Math. Soc. **358**(12), 5485–5500 (2006) (electronic)
14. S. Mukai, *Finite Generation of the Nagata Invariant Rings in A-D-E cases*. RIMS Preprint n. 1502, Kyoto (2005)
15. M. Nagata, On rational surfaces, II. Mem. Coll. Sci. Univ. Kyoto Ser. A Math. **33**, 271–293 (1960)

On Some Components of Hilbert Schemes of Curves

Flaminio Flamini and Paola Supino

Serietà, metodo, rigore, passione, generosità, umanità.
Grazie per tutto, Ciro!

Abstract Let $\mathcal{I}_{d,g,R}$ be the union of irreducible components of the Hilbert scheme whose general points parametrize smooth, irreducible, curves of degree d, genus g, which are non-degenerate in the projective space $\mathbb{P}^R$. Under some numerical assumptions on d, g and R, we construct irreducible components of $\mathcal{I}_{d,g,R}$ other than the so-called *principal* (or *distinguished*, as in [12, 13]) *component*, dominating the moduli space $\mathcal{M}_g$ of smooth genus-g curves, which are generically smooth and turn out to be of dimension higher than the expected one. The general point of any such a component corresponds to a curve $X \subset \mathbb{P}^R$ which is a suitable ramified m-cover of an irrational curve $Y \subset \mathbb{P}^{R-1}$, $m \geqslant 2$, lying in a surface cone over Y. The paper extends some of the results in [12, 13].

Keywords Hilbert scheme of curves · Brill–Noether theory · Ruled surfaces · Cones · Coverings · Gaussian–Wahl maps

2010 Mathematics Subject Classification Primary 14C05 · Secondary 14E20, 14F05, 14J10, 14J26, 14H10

F. Flamini (✉)
Dipartimento di Matematica, Università degli Studi di Roma "Tor Vergata", Viale della Ricerca Scientifica 1, 00133 Roma, Italy
e-mail: flamini@mat.uniroma2.it

P. Supino
Dipartimento di Matematica e Fisica, Università degli Studi "Roma Tre", Largo S. L. Murialdo 1, 00146 Roma, Italy
e-mail: supino@mat.uniroma3.it

T. Dedieu et al. (eds.), *The Art of Doing Algebraic Geometry*, Trends in Mathematics,
https://doi.org/10.1007/978-3-031-11938-5_8

Introduction

Projective varieties are distributed in *families*, obtained by suitably varying the coefficients of their defining equations. The study of these families and, in particular, of the properties of their parameter spaces is a central theme in algebraic geometry and sets on technical tools, like *flatness*, *base change*, etc., as well as on the existence (due to Grothendieck, with refinements by Mumford) of the so-called *Hilbert scheme*, a closed, projective scheme parametrizing closed projective subschemes with fixed numerical/projective invariants (i.e. the *Hilbert polynomial*), and having fundamental *universal* properties.

Hilbert schemes have interested several authors over the decades, owing also to deep connections with several other subjects in algebraic geometry (cf. e.g. bibliography in [39] for an overview). Indeed, results and techniques in the "projective domain" of the Hilbert schemes have frequently built bridges towards other topics in algebraic geometry, as by improving already known results, as by providing new ones. The interplay between Hilbert schemes of curves in projective spaces and the Brill–Noether theory of line bundles on curves is one of the milestones in algebraic geometry (cf. e.g. [1, 17, 30]). The construction of the moduli space $\mathcal{M}_g$ of smooth, genus-g curves (and its generalizations $\mathcal{M}_{g,n}$ of moduli spaces of smooth, n-pointed, genus-g curves), the proof of its irreducibility and the construction of a natural compactification of it deeply rely on the use of Hilbert schemes of curves (c.f. e.g. [2, 19, 31]). Similarly, together with the *Deligne–Mumford compactification* of $\mathcal{M}_g$ in [19], the use of Hilbert schemes of curves has been also fundamental in the construction of suitable compactifications of the *universal Picard variety* (cf. e.g. [10, Theorem, p. 592]).

Besides these examples, the use of Hilbert schemes has been fundamental for several other issues in algebraic geometry: unirationality and/or Torelli's type of theorems for cubic hypersurfaces and for prime Fano threefolds of given genus have been proved via the use of Hilbert schemes of lines and planes contained in such varieties (cf. e.g. [7, 18, 23, 26, 27, 34, 41]). Important connections between Hilbert schemes parametrizing k-linear spaces contained in complete intersections of hyperquadrics and intermediate Jacobians (cf. [22]) are worth to be mentioned, whereas in [8, 9] the Hilbert schemes of projective scroll surfaces have been related with families of rank-2 vector bundles as well as with moduli spaces of (semi)stable ones. Surjectivity of Gaussian–Wahl maps on curves with general moduli [15, 16] has deep reflections both on suitable Hilbert schemes of associated cones and on the extendability of such curves (especially in the $K3$-case). At last, Hilbert schemes parametrizing lines in suitable complete intersections are used either in [4], to deduce upper bounds of minimal gonality of a family of curves covering a very general projective hypersurface of high degree, or in [5, 6] to deduce new results concerning either enumerative properties or a certain "algebraic hyperbolicity" behaviour.

In the present paper, we focus on Hilbert schemes of smooth, irreducible projective curves of given degree and genus, the study of which is classical and goes back to Castelnuovo, Halphen (*Casteluovo bounds* and the *gap problem*) and Severi.

Given non-negative integers d, g and $R \geqslant 3$, we denote by $\mathcal{I}_{d,g,R}$ the union of all irreducible components of the Hilbert scheme whose general points parametrize smooth, irreducible, non-degenerate curves of degree d and genus g in the projective space $\mathbb{P}^R$. A component of $\mathcal{I}_{d,g,R}$ is said to be *regular* if it is generically smooth and of the *expected dimension*, otherwise it is said to be *superabundant* (cf. Sect. 1.1 for more details).

When the so-called *Brill–Noether number* is non-negative, it is well known that $\mathcal{I}_{d,g,R}$ has a unique irreducible component which dominates the moduli space $\mathcal{M}_g$ parametrizing (isomorphism classes of) smooth, irreducible genus-g curves (cf. [30] and Sect. 1.1). This is called the *principal component* (also the *component with general moduli* or even the *distinguished component*, as in [12, 13]) of the Hilbert scheme.

In [40], Severi claimed the irreducibility of $\mathcal{I}_{d,g,R}$ when $d \geqslant g + R$, and this was actually proved by Ein for $R = 3, 4$ in (cf. [24, 25]); further sufficient conditions on d and g ensuring the irreducibility of some $\mathcal{I}_{d,g,R}$ for $R \geqslant 5$ have been found in [3]. On the other hand, in several cases, there have been also given examples of additional *non-principal* components of $\mathcal{I}_{d,g,R}$. Some of these extra components have been constructed by using either m-sheeted covers of $\mathbb{P}^1$ (cf. e.g. [35, 37], etc.), or by using double covers of irrational curves (cf. e.g. [12, 13], etc.) or even by using non-linearly normal curves in projective space (Harris, 1984 unpublished, see e.g. [17, Chap. IV]).

In this paper, we prove the following.

Main Theorem *Let $\gamma \geqslant 10$, $e \geqslant 2\gamma - 1$, $R = e - \gamma + 1$ and $m \geqslant 2$ be integers. Set*

$$d := me \quad \text{and} \quad g := m(\gamma - 1) + \frac{m(m-1)}{2}e + 1.$$

Then $\mathcal{I}_{d,g,R}$ contains an irreducible component which is generically smooth and superabundant, having dimension

$$\lambda_{d,g,R} + \sigma_{d,g,R},$$

where

$$\lambda_{d,g,R} := (R+1)me - (R-3)\left(m(\gamma - 1) + \frac{m(m-1)}{2}e\right)$$

is the expected dimension of $\mathcal{I}_{d,g,R}$, whereas the positive integer

$$\sigma_{d,g,r} := (R-4)\left[(\gamma - 1)(m-1) + 1 + e + \frac{m(m-3)}{2}e\right] + 4(e+1) + em(m-5)$$

is the superabundance summand for the dimension of such a component.

As additional result, we explicitly describe a general point of the aforementioned superabundant component (cf. Proposition 6 and Sect. 3). We want to stress that main theorem extends some of the results in [12, 13] which deal with the case $m = 2$.

The paper consists of three sections. In Sect. 1, we remind some generalities concerning Hilbert schemes of curves and associated Brill–Noether theory (cf. Sect. 1.1), Gaussian–Wahl maps and Hilbert schemes of cones (cf. Sect. 1.2) and ramified coverings of curves (cf. Sect. 1.3), which will be used for our analysis. Section 2 deals with the construction of curves X which fill-up an open dense subscheme of the superabundant component of $\mathcal{I}_{d,g,R}$ mentioned in main theorem above. Precisely in Sect. 2.1 we more generally consider, for any $\gamma \geqslant 1$ and $e \geqslant 2\gamma - 1$, curves Y of genus γ, degree e with *general moduli*, which are non-special and projectively normal in $\mathbb{P}^{R-1}$ and which fill-up the principal component of the related Hilbert scheme $\mathcal{I}_{e,\gamma,R-1}$. Then in Sect. 2.2 we consider cones $F = F_Y$ extending in $\mathbb{P}^R$ curves Y as above, we describe abstract resolutions of cones F, together with further cohomological properties (see Proposition 5), as well as an explicit parametric description of the parameter space of such cones as Y varies in the principal component of $\mathcal{I}_{e,\gamma,R-1}$. In Sect. 2.3, we construct the desired curves X as curves sitting in cones F as m-sheeted ramified covers $\varphi : X \to Y$, where the map φ is given by the projection from the vertex of the cone. We prove that such curves X are non-degenerate and linearly normal in $\mathbb{P}^R$, we moreover compute their genus g and some other useful cohomological properties (cf. Proposition 6). We also prove Lemma 2, a technical result which deals with a more general situation involving projections and ramified covers of possibly reducible, connected, nodal curves and which is needed for a certain inductive procedure used in proving main theorem (see the proof of Lemma 3, Claim 2). Finally, Sect. 3 focuses on the proof of Main Theorem, which also involves surjectivity of suitable Gaussian–Wahl maps (cf. proof of Lemma 3, Claim 2). This explains why in this last section, as well as in Main Theorem, the hypothesis $\gamma \geqslant 10$ is required (cf. Proposition 7).

Notation and Terminology

We work throughout over the field $\mathbb{C}$ of complex numbers. All schemes will be endowed with the Zariski topology. By *variety* we mean an integral algebraic scheme and by *curve* we intend a variety of dimension 1. We say that a property holds for a *general* point x of a variety X if it holds for any point in a Zariski open non-empty subset of X. We will interchangeably use the terms rank-r vector bundle on a variety X and rank-r locally free sheaf. To ease notation and when no confusion arises, we sometimes identify line bundles with Cartier divisors, interchangeably using additive notation instead of multiplicative notation and tensor products; we moreover denote by $\sim$ the linear equivalence of divisors and by $\equiv$ their numerical equivalence. If $\mathcal{P}$ is either a parameter space of a flat family of closed subschemes of a variety X, as $\mathcal{P}$ a Hilbert scheme, or a moduli space parametrizing geometric objects modulo a given equivalence relation, as the moduli space of smooth genus-g curves, we will denote by $[Y]$ the parameter point (resp., the moduli point) corresponding to the subscheme $Y \subset X$ (resp., associated to the equivalence class of Y). For non-reminded terminology, we refer the reader to [33].

1 Generalities

We briefly recall some generalities and results which will be used in the next sections.

1.1 Hilbert Schemes and Brill–Noether Theory of Curves

Let C be a smooth, irreducible, projective curve of genus $g > 0$. Given positive integers d and r, the *Brill–Noether locus*, $W^r_d(C) \subseteq \mathrm{Pic}^d(C)$, when not empty, parametrizes degree-d line bundles L on C such that $h^0(C, L) \geqslant r + 1$. Its *expected dimension* is given by the so-called *Brill–Noether number*

$$\rho(g, r, d) := g - (r + 1)(g + r - d). \tag{1}$$

It is well known that if C *has general moduli* (i.e. when C corresponds to a general point of the moduli space $\mathcal{M}_g$ parametrizing isomorphism classes of smooth, genus-g curves) it is well known that $W^r_d(C)$ is empty if $\rho(g, r, d) < 0$, whereas it is generically smooth, of the expected dimension $\rho(g, r, d)$, otherwise. Moreover, when $\rho(g, r, d) > 0$, $W^r_d(C)$ is also irreducible and for a general L parametrized by $W^r_d(C)$ it is $h^0(C, L) = r + 1$ (cf. [1, Chaps. IV, V, VI]).

Brill–Noether theory of line-bundles on abstract projective curves C is intimately related to the study of Hilbert schemes parametrizing projective embeddings of such curves. Indeed, assume for simplicity $L \in W^r_d(C)$ very ample and such that $h^0(C, L) = r + 1$; hence, one has an embedding $C \overset{\phi_{|L|}}{\hookrightarrow} \mathbb{P}^r$ induced by the complete linear system $|L|$ determined by L, whose image $Y := \phi_{|L|}(C)$ is a smooth, irreducible curve of degree d, genus g which is non-degenerate in $\mathbb{P}^r$. If we denote by $Hilb_{d,g,r}$ the Hilbert scheme parametrizing closed subschemes of $\mathbb{P}^r$ with Hilbert polynomial $P(t) = dt + (1 - g)$, then Y corresponds to a point of $Hilb_{d,g,r}$. If we denote by $\mathcal{I}_{d,g,r}$ the union of all irreducible components of $Hilb_{d,g,r}$ whose general points parametrize smooth, irreducible, non-degenerate curves in $\mathbb{P}^r$, then Y represents a point $[Y] \in \mathcal{I}_{d,g,r}$. When $[Y]$ is a smooth point of $\mathcal{I}_{d,g,r}$, then Y is said to be *unobstructed* in $\mathbb{P}^r$.

If $N_{Y/\mathbb{P}^r}$ denotes the *normal bundle* of Y in $\mathbb{P}^r$, one has

$$T_{[Y]}(\mathcal{I}_{d,g,r}) \cong H^0(Y, N_{Y/\mathbb{P}^r}) \quad \text{and} \quad \chi(Y, N_{Y/\mathbb{P}^r}) \leqslant \dim_{[Y]} \mathcal{I}_{d,g,r} \leqslant h^0(Y, N_{Y/\mathbb{P}^r}), \tag{2}$$

where the integer $\chi(Y, N_{Y/\mathbb{P}^r}) = h^0(Y, N_{Y/\mathbb{P}^r}) - h^1(Y, N_{Y/\mathbb{P}^r})$ in (2) is the so-called *expected dimension* of $\mathcal{I}_{d,g,r}$ at $[Y]$ and the equality on the right-most side in (2) holds iff Y is unobstructed in $\mathbb{P}^r$ (for full details, cf. e.g. [39, Cor. 3.2.7, Thm. 4.3.4, 4.3.5]). The expected dimension of $\mathcal{I}_{d,g,r}$, given by $\chi(Y, N_{Y/\mathbb{P}^r})$, can be easily computed with the use of normal and Euler sequences for $Y \subset \mathbb{P}^r$, and it turns out to be

$$\lambda_{d,g,r} := \chi(Y, N_{Y/\mathbb{P}^r}) = (r + 1)d - (r - 3)(g - 1). \tag{3}$$

A component of $\mathcal{I}_{d,g,r}$ is said to be *regular* if it is both reduced (i.e. generically smooth) and of the expected dimension $\lambda_{d,g,r}$; otherwise, it is said to be *superabundant*.

By above, any component $\mathcal{I}$ of $\mathcal{I}_{d,g,r}$ has a natural rational map

$$\mu_g : \mathcal{I} \dashrightarrow \mathcal{M}_g,$$

which simply sends $[Y] \in \mathcal{I}$ general to the moduli point $[C] \in \mathcal{M}_g$ as above. The map μ_g is called the *modular morphism* of $\mathcal{I}$; with same terminology as in [38, Introduction], the dimension of $\mathrm{Im}(\mu_g)$ is called the *number of moduli of* $\mathcal{I}$. The expected dimension of $\mathrm{Im}(\mu_g)$ is $\min\{3g-3,\ 3g-3+\rho(g,r,d)\}$, where $\rho(g,r,d)$ as in (1), and it is called the *expected number of moduli* of $\mathcal{I}$. The expression of the expected number of moduli of $\mathcal{I}$ is the obvious postulation which comes from the well-known interpretation, in terms of vector bundle maps on the Picard scheme, of the existence of special line bundles on C (cf. [1, Chaps. IV, V, VI]). In this set-up, we remind the following result due to Sernesi.

Theorem 1 (cf. [38, Theorem, p. 26]) *For any integers $r \geqslant 2$, d and g such that*

$$d \geqslant r+1 \text{ and } d-r \leqslant g \leqslant \frac{r(d-r)-1}{r-1}$$

there exists a component $\mathcal{I}$ of $\mathcal{I}_{d,g,r}$ which has the expected number of moduli. *Moreover, $[Y] \in \mathcal{I}$ general corresponds to an unobstructed curve $Y \subset \mathbb{P}^r$ such that $h^1(Y, N_{Y/\mathbb{P}^r}) = 0$ and whose embedding in $\mathbb{P}^r$ is given by a complete linear system.*

Remark 1 We want to stress the "geometric counter-part" of the numerical hypotheses appearing in Theorem 1. For Y as in Theorem 1, let (C, L) be the pair consisting of a smooth, irreducible, abstract projective curve C of genus g and of $L \in \mathrm{Pic}^d(C)$ such that $Y = \phi_{|L|}(C)$. Then, condition $d \geqslant r+1$ simply means that the curve Y is of positive genus and non-degenerate in $\mathbb{P}^r$ whereas $d-r \leqslant g$, i.e. $g+r-d \geqslant 0$, simply decodes by Riemann–Roch the condition that the *index of speciality* $i(L) := g+r-d$ of L is non-negative. At last, the condition $g \leqslant \frac{r(d-r)-1}{r-1}$ reads $g - rg + rd - r^2 - 1 \geqslant 0$ which is nothing but $\rho(g,r,d) + (r+g-d) = \rho(g,r,d) + i(L) \geqslant 1$, i.e. it is a *"Brill–Noether type"* condition on the pair (C, L).

It is well known (cf. e.g. [30, p. 70]) that, when $\rho(g,r,d) \geqslant 0$, $\mathcal{I}_{d,g,r}$ has a *unique component* with a dominant modular morphism μ_g, i.e. dominating $\mathcal{M}_g$; thus such a component has maximal number of moduli $3g-3$. It is called the *principal component* of $\mathcal{I}_{d,g,r}$ (also the *component with general moduli* or even the *distinguished component*, as named in [12, 13]). In the sequel, we will denote such component by $\widehat{\mathcal{I}_{d,g,r}}$ or simply by $\widehat{\mathcal{I}}$, if no confusion arises. As a direct consequence of the uniqueness of $\widehat{\mathcal{I}}$ and of Theorem 1, one has:

Corollary 1 *For any integers $r \geqslant 2$, d and g such that*

$$d \geqslant r+1 \text{ and } d-r \leqslant g \leqslant \frac{(r+1)(d-r)-1}{r}$$

the principal component $\widehat{\mathcal{I}}$ of $\mathcal{I}_{d,g,r}$ is not empty. Its general point $[Y]$ corresponds to an unobstructed curve Y in $\mathbb{P}^r$ with $h^1(Y, N_{Y/\mathbb{P}^r}) = 0$ and whose embedding in $\mathbb{P}^r$ is given by a complete linear system. Furthermore $\widehat{\mathcal{I}}$ is regular, i.e. generically smooth and of the expected dimension $\lambda_{d,g,r}$.

Proof The condition $g \leqslant \frac{(r+1)(d-r)-1}{r}$ is equivalent to $\rho(g, r, d) \geqslant 1$. Thus, we conclude by applying Theorem 1, taking into account what discussed in Remark 1, and by applying [30, p. 70] and (2), as the condition $h^1(Y, N_{Y/\mathbb{P}^r}) = 0$ implies both the non-obstructedness of Y in $\mathbb{P}^r$ and the regularity of $\widehat{\mathcal{I}}$. □

In [40], Severi claimed the irreducibility of $\mathcal{I}_{d,g,r}$ when $d \geqslant g + r$. Severi's claim was proved by Ein for $r = 3, 4$ in (cf. [24, 25]); further sufficient conditions on d and g ensuring the irreducibility of some $\mathcal{I}_{d,g,r}$ for $r \geqslant 5$ have been found in [3]. On the other hand, in several cases, there have been also given examples of additional *non-principal* components of $\mathcal{I}_{d,g,r}$, even in the range $\rho(g, r, d) \geqslant 0$. Some of these extra components have been constructed by using either m-sheeted covers of $\mathbb{P}^1$ (cf. e.g. [35, 37], etc.), or by using double covers of irrational curves (cf. e.g. [12, 13], etc.) or even by using non-linearly normal curves in projective space (the latter approach is contained in a series of examples due to Harris, 1984 unpublished, fully described in [17, Chap. IV]). In some cases, these extra components have been also proved to be regular (cf. e.g. [17, Chap. IV], [13]).

1.2 Gaussian–Wahl Maps and Cones

Let C be a smooth, irreducible projective curve of positive genus g and L be a very ample line bundle of degree d on C. Set $Y \subset \mathbb{P}^r$ to be the embedding of C via the complete linear system $|L|$. Let F_Y (equiv., $F_{C,L}$) denote the cone in $\mathbb{P}^{r+1}$ over Y with vertex at a point $v \in \mathbb{P}^{r+1} \setminus \mathbb{P}^r$ (if no confusion arises, in the sequel we simply set F). Fundamental properties of such cones are related to the so-called *Gaussian–Wahl maps* (cf. e.g. [15, 16, 42]), as we will briefly remind.

If ω_C denotes the *canonical bundle* of C, one sets

$$R(\omega_C, L) := \operatorname{Ker}\left[H^0(C, \omega_C) \otimes H^0(C, L) \longrightarrow H^0(C, \omega_C \otimes L)\right],$$

where the previous map is a natural multiplication map among global sections. One can consider the map

$$\Phi_{\omega_C, L} : R(\omega_C, L) \to H^0(\omega_C^{\otimes 2} \otimes L), \tag{4}$$

defined locally by $\Phi_{\omega_C,L}(s \otimes t) := s\,dt - t\,ds$, which is called the *Gaussian–Wahl map*. As customary, one sets

$$\gamma_{C,L} := \mathrm{cork}(\Phi_{\omega_C,L}) = \dim \mathrm{Coker}(\Phi_{\omega_C,L}). \tag{5}$$

For reader's convenience we will remind here statement of [16, Prop. 2.1], limiting ourselves to its (2.8)–part, which will be used in Sect. 3; indeed, the full statement of [16, Prop. 2.1] is quite long, with many exceptions and dwells also on curves with low genus whereas Sect. 3 will focus on curves of genus at least 10.

Proposition 1 (cf. [16, Proposition 2.1–(2.8)]) *Let $g \geqslant 6$ be an integer. Assume that C is a smooth, projective curve of genus g with general moduli and that $L \in \mathrm{Pic}^d(C)$ is general. Then, $\gamma_{C,L} = 0$ (i.e. $\Phi_{\omega_C,L}$ is surjective) if*

$$d \geqslant \begin{cases} g+12 & \text{for } 6 \leqslant g \leqslant 8 \\ g+9 & \text{for } g \geqslant 9 \end{cases}.$$

Gaussian–Wahl maps can be used to compute the dimension of the tangent space to the Hilbert scheme of surfaces in $\mathbb{P}^{r+1}$ at points representing cones F as above (cf. e.g. [16]). Indeed, let $\mathcal{W}$ be any irreducible component of the Hilbert scheme of curves $\mathcal{I}_{d,g,r}$ and let $\mathcal{H}(\mathcal{W})$ be the variety which parametrizes the family of cones $F \subset \mathbb{P}^{r+1}$ over curves $Y \subset \mathbb{P}^r$ representing points in $\mathcal{W}$. Then, one has:

Proposition 2 (cf. [16, Cor. 2.20–(c), Prop. 2.12–(2.13) and (2.15)]) *Set notation and conditions as in Proposition 1. Let $r = d - g$, $Y \subset \mathbb{P}^r$ and $\mathcal{W}$ be any generically smooth component of $\mathcal{I}_{d,g,r}$ s.t. $[Y] \in \mathcal{W}$ is general. Then*
(i) The Gaussian–Wahl map $\Phi_{\omega_Y,\mathcal{O}_Y(1)}$ is surjective, i.e. $\gamma_{Y,\mathcal{O}_Y(1)} = 0$.
(ii) $h^0(Y, N_{Y/\mathbb{P}^r} \otimes \mathcal{O}_Y(-1)) = r+1$ whereas $h^0(Y, N_{Y/\mathbb{P}^r} \otimes \mathcal{O}_Y(-j)) = 0$, for any integer $j \geqslant 2$.
(iii) $\mathcal{H}(\mathcal{W})$ is a generically smooth component of the Hilbert scheme parametrizing surfaces of degree d and sectional genus g in $\mathbb{P}^{r+1}$. Moreover,

$$\dim \mathcal{H}(\mathcal{W}) = (r+1)(d+1) - (r-3)(g-1) = \lambda_{d,g,r} + (r+1) \tag{6}$$

and, for $[F] \in \mathcal{H}(\mathcal{W})$ general, the associated cone $F = F_Y$ is unobstructed in $\mathbb{P}^{r+1}$.

1.3 Ramified Coverings of Curves

Let Y be a scheme. A morphism $\varphi : X \to Y$ is called a *covering map of degree m* (or simply a m-cover) if $\varphi_* \mathcal{O}_X$ is a locally free $\mathcal{O}_Y$-sheaf of rank m. A map φ is a covering map (or simply *a cover*) if and only if it is surjective, finite and flat. In particular, if Y is smooth and irreducible and X is Cohen–Macaulay, then every finite, surjective morphism $\varphi : X \to Y$ is a covering map (cf. e.g. [11, p. 1361]).

When $\varphi : X \to Y$ is a covering map of degree m, one has a natural exact sequence

$$0 \to \mathcal{O}_Y \stackrel{\varphi^\sharp}{\to} \varphi_*\mathcal{O}_X \to \mathcal{T}_\varphi^\vee \to 0,$$

where $\mathcal{T}_\varphi^\vee := Coker(\varphi^\sharp)$ is the so-called *Tschirnhausen bundle* associated to the covering map φ, which is of rank $m-1$ on Y. Since Char $(\mathbb{C}) = 0$, the trace map tr: $\varphi_*\mathcal{O}_X \to \mathcal{O}_Y$ gives rise to a splitting of the previous exact sequence, so that one has $\varphi_*\mathcal{O}_X = \mathcal{O}_Y \oplus \mathcal{T}_\varphi^\vee$ (cf. e.g. [11, 13, 20, 21]).

If X and Y are, in particular, smooth, irreducible curves and $\varphi : X \to Y$ is a covering map of degree m, according to [33, Ex. IV.2.6–(d), p. 306], the *branch divisor* B_φ of φ is such that

$$\left(\bigwedge^m (\varphi_*\mathcal{O}_X)\right)^{\otimes 2} \cong \mathcal{O}_Y(-B_\varphi). \tag{7}$$

If moreover X (resp., Y) has genus g (resp., γ) then deg $B_\varphi = 2(g-1) - 2m(\gamma - 1)$. As for the *ramification divisor* R_φ, the Riemann–Hurwitz formula gives

$$\omega_X = \varphi^*(\omega_Y) \otimes \mathcal{O}_X(R_\varphi). \tag{8}$$

In this set-up, we recall the *pinching construction* described in [21, Sect. 3.1]. Let $\varphi : X \to Y$ be a covering map of degree m of smooth irreducible curves X and Y. Let Z be the reduced, reducible, connected nodal curves

$$Z := X \cup Y,$$

where X and Y are *attached nodally* at δ distinct points as follows: let $y_i \in Y$ and $x_i \in X$ be points such that $\varphi(x_i) = y_i$, $1 \leqslant i \leqslant \delta$. Set $D := \sum_{i=1}^{\delta} y_i$, $\mathcal{O}_D$ the structural sheaf of D and $\mathcal{J}$ the kernel of the map

$$\varphi_*\mathcal{O}_X \oplus \mathcal{O}_Y \to \mathcal{O}_D,$$

defined around any y_i's as

$$(f, g) \mapsto f(x_i) - g(y_i), \ \forall\ 1 \leqslant i \leqslant \delta.$$

Then $\mathcal{J} \subset \varphi_*\mathcal{O}_X \oplus \mathcal{O}_Y$ is an $\mathcal{O}_Y$-subalgebra of $\varphi_*\mathcal{O}_X \oplus \mathcal{O}_Y$ and $\mathrm{Spec}_Y(\mathcal{J}) = Z = X \cup Y$. D is called the *set of nodes* of Z. Let $\psi : Z \to Y$ be the natural induced finite and surjective map. Since Y is smooth, irreducible and Z is l.c.i. (so, in particular, Cohen–Macaulay), from what reminded above the map ψ is a covering map of degree $m+1$. In this set-up, one has the following.

Proposition 3 (cf. [21, Lemma 3.2]) *Let $\varphi : X \to Y$ be a covering map of degree-m between smooth irreducible curves X and Y. Let $\psi : Z \to Y$ be the covering map of* *degree $m+1$ induced by the pinching construction, whose set of nodes is D. Then, the following exact sequence of vector bundles on Y*

$$0 \to \mathcal{T}_{\varphi} \to \mathcal{T}_{\psi} \to \mathcal{O}_Y(D) \to 0$$

holds, where $\mathcal{T}_{\varphi}^{\vee}$ and $\mathcal{T}_{\psi}^{\vee}$ are the Tschirnhausen bundles *associated to the covering maps φ and ψ, respectively.*

2 Curves and Cones

In this section, we first construct families of non-special curves Y of any positive genus γ and of degree $e \geqslant 2\gamma - 1$ in a projective space, which turn out to fill-up the principal component $\widehat{\mathcal{I}}$ of the related Hilbert scheme (cf. Sect. 2.1). After that, we deal with the family $\mathcal{H}(\widehat{\mathcal{I}})$, as in Sect. 1.2, which parametrizes cones extending curves in $\widehat{\mathcal{I}}$, i.e. cones having curves in $\widehat{\mathcal{I}}$ as hyperplane sections. We describe an abstract resolution of a general point of $\mathcal{H}(\widehat{\mathcal{I}})$ and compute $\dim \mathcal{H}(\widehat{\mathcal{I}})$ via an explicit parametric description (cf. Sect. 2.2). To conclude the section, for cones F parametrized by $\mathcal{H}(\widehat{\mathcal{I}})$ we construct smooth, irreducible curves $X \subset F$, of suitable degree d and genus g, which turn out to be m-sheeted ramified covers of the curves Y which vary in the principal component $\widehat{\mathcal{I}}$ (cf. Sect. 2.3).

2.1 Curves in Principal Components

Let $\gamma > 0$ and $e \geqslant 2\gamma - 1$ be integers. Let C be a smooth, irreducible, projective curve of genus γ and let $\mathcal{O}_C(E) \in \mathrm{Pic}^e(C)$ be a general line bundle. Thus, $\mathcal{O}_C(E)$ is very ample and non-special (i.e. $h^1(C, \mathcal{O}_C(E)) = 0$). By Riemann–Roch, we set

$$R := h^0(C, \mathcal{O}_C(E)) = e - \gamma + 1, \tag{9}$$

so that $|\mathcal{O}_C(E)|$ defines an embedding $C \overset{\phi_{|E|}}{\hookrightarrow} \mathbb{P}^{R-1}$, whose image we denote from now on by $Y := \phi_{|E|}(C)$.

Taking into account [29, Thm. 1], Y is a smooth, projective curve of genus $\gamma > 0$, degree $e \geqslant 2\gamma - 1$, which is projectively normal in $\mathbb{P}^{R-1}$. As in Sect. 1.1, one has, in particular, that $[Y] \in \mathcal{I}_{e,\gamma,R-1}$.

If we let vary $[C] \in \mathcal{M}_{\gamma}$ and, for any such C, we let $\mathcal{O}_C(E)$ vary in $\mathrm{Pic}^e(C)$, the next proposition ensures that the corresponding curves $Y \subset \mathbb{P}^{R-1}$ fill-up the principal component $\widehat{\mathcal{I}} := \widehat{\mathcal{I}_{e,\gamma,R-1}}$ which also turns out to be regular.

Proposition 4 *Let $\gamma > 0$ and $e \geqslant 2\gamma - 1$ be integers. Let C be a smooth, projective curve of genus γ with general moduli, and let $\mathcal{O}_C(E) \in \mathrm{Pic}^e(C)$ be a general line* *bundle. Let* $Y := \phi_{|E|}(C) \subset \mathbb{P}^{R-1}$, *where* $R = e - \gamma + 1$. *Then,* Y *is a smooth, irreducible curve of degree* e *and genus* γ *which is projectively normal in* $\mathbb{P}^{R-1}$, *as an embedding of* C *via the complete linear system* $|E|$, *and such that* $h^1(Y, N_{Y/\mathbb{P}^{R-1}}) = 0$. *The curve* Y *corresponds to a general point of the* principal component $\widehat{\mathcal{I}} := \widehat{\mathcal{I}_{e,\gamma,R-1}}$ *of the Hilbert scheme* $\mathcal{I}_{e,\gamma,R-1}$, *which is regular of dimension*

$$\dim \widehat{\mathcal{I}} = \lambda_{e,\gamma,R-1} = Re - (R-4)(\gamma - 1). \tag{10}$$

Proof Numerical assumptions and [29, Thm. 1] imply that E is very ample, non-special and that $Y \subset \mathbb{P}^{R-1}$ is projectively normal, the equality $R = e - \gamma + 1$ simply following by the non-speciality of E and by Riemann–Roch.

Under our assumptions, numerical hypotheses in Corollary 1 hold true. Indeed, as explained in Remark 1, we have the following: since $Y \subset \mathbb{P}^{R-1}$ is non-degenerate and of positive genus γ, then condition $e \geqslant R$ is certainly satisfied. Concerning condition $\gamma \geqslant e - (R-1)$, i.e. $i(E) \geqslant 0$, it certainly holds from the non-speciality of E. At last, non-speciality of E gives $\rho(\gamma, R-1, e) + i(E) = \rho(\gamma, R-1, e) = \gamma \geqslant 1$, therefore $\gamma \leqslant \frac{R(e-(R-1))-1}{R-1}$ as in Corollary 1 certainly holds (cf. Remark 1).

Thus, by Corollary 1, $[Y]$ is a point in the principal component $\widehat{\mathcal{I}}$ of $\mathcal{I}_{e,\gamma,R-1}$ for which $h^1(Y, N_{Y/\mathbb{P}^{R-1}}) = 0$, i.e. $\widehat{\mathcal{I}}$ is generically smooth and of the expected dimension $\lambda_{e,\gamma,R-1}$ which equals $Re - (R-4)(\gamma-1)$, as it follows from (3). □

2.2 Cones Extending Curves in $\widehat{\mathcal{I}}$

With notation as in Sect. 1.2, we will deal here with the family of cones $\mathcal{H}(\widehat{\mathcal{I}})$, where $\widehat{\mathcal{I}} = \widehat{\mathcal{I}_{e,\gamma,R-1}}$ is the principal component in Proposition 4. For $[Y] \in \widehat{\mathcal{I}}$ general, we will denote by $F := F_Y \subset \mathbb{P}^R$ a cone over Y with general vertex $v \in \mathbb{P}^R \setminus \mathbb{P}^{R-1}$. In order to describe suitable smooth, abstract resolutions of such cones, we recall the following general facts.

Let C be a smooth, irreducible projective curve of genus $\gamma > 0$ and let $\mathcal{O}_C(E) \in \mathrm{Pic}^e(C)$ be a general line bundle of degree $e \geqslant 2\gamma - 1$. Consider the rank-two, normalized vector bundle $\mathfrak{F} := \mathcal{O}_C \oplus \mathcal{O}_C(-E)$ on C and let $S := \mathbb{P}(\mathfrak{F}) = \mathrm{Proj}_C(\mathrm{Sym}(\mathfrak{F}))$ the associated geometrically ruled surface over C. One has the structural morphism $\rho : S \to C$ such that $\rho^{-1}(p) = f_p$, for any $p \in C$, where $f_p \cong \mathbb{P}^1$ denotes the fibre of the ruling of S over the point $p \in C$. A general fibre of the ruling of S will be simply denoted by f.

S is endowed with two natural sections, C_0 and C_1, both isomorphic to C, and such that $C_0 \cdot C_1 = 0$, $C_0^2 = -C_1^2 = -e$. The section C_0 (resp., C_1) corresponds to the exact sequence

$$0 \to \mathcal{O}_C \to \mathfrak{F} \to \mathcal{O}_C(-E) \to 0 \;\; \text{(resp., } 0 \to \mathcal{O}_C(-E) \to \mathfrak{F} \to \mathcal{O}_C \to 0).$$

Moreover, one has $\mathrm{Pic}(S) \cong \mathbb{Z}[\mathcal{O}_S(C_0)] \oplus \rho^*(\mathrm{Pic}(C))$ and $\mathrm{Num}(S) \cong \mathbb{Z} \oplus \mathbb{Z}$ (cf. e.g. [33, V.2]). To ease notation, for any $D \in \mathrm{Div}(C)$, we will simply set $\rho^*(D) := D\,f$.

If K_S (resp., K_C) denotes a canonical divisor of S (resp., of C), one has (cf. e.g. [33, V.2])

$$C_1 \sim C_0 + E\,f \quad \text{and} \quad K_S \sim -2C_0 + (K_C - E)\,f. \tag{11}$$

Proposition 5 *Let C be a smooth, irreducible projective curve of genus $\gamma > 0$ and $\mathcal{O}_C(E) \in \mathrm{Pic}^e(C)$ be a general line bundle of degree $e \geqslant 2\gamma - 1$. Consider the normalized, rank-two vector bundle $\mathfrak{F} := \mathcal{O}_C \oplus \mathcal{O}_C(-E)$ on C and let $S := \mathbb{P}(\mathfrak{F})$, together with the natural sections C_0 and C_1, where $C_1 \sim C_0 + E\,f$, $C_0 \cdot C_1 = 0$, $C_0^2 = -C_1^2 = -e$. Then*
(i) The linear system $|\mathcal{O}_S(C_1)|$ is base-point-free and not composed with a pencil. It induces a morphism

$$\Psi := \Psi_{|\mathcal{O}_S(C_1)|} : S \to \mathbb{P}^R,$$

where $R = e - \gamma + 1$.
(ii) Ψ is an isomorphism, outside the section $C_0 \subset S$, onto its image $F := \Psi(S) \subset \mathbb{P}^R$, whereas it contracts C_0 at a point $v \in \mathbb{P}^R$.
(iii) F is a cone with vertex at v over $Y := \Psi(C_1) \cong C$, where $Y \subset \mathbb{P}^{R-1}$ is a hyperplane section of F not passing through v; Y is smooth, irreducible, non-degenerate, of degree e, genus γ and it is also projectively normal in $\mathbb{P}^{R-1}$.
(iv) The cone $F \subset \mathbb{P}^R$ is projectively normal, of degree $\deg F = e$, of sectional genus and speciality γ. In particular, $h^0(F, \mathcal{O}_F(1)) = R + 1 = e - \gamma + 2$ and $h^1(F, \mathcal{O}_F(1)) = \gamma$.
(v) For any $m \geqslant 2$, one has $h^0(F, \mathcal{O}_F(m)) = \frac{m(m+1)}{2}e - m(\gamma - 1) + 1$.

Proof (i) From [33, Thm. 2.17(b), p. 379], one deduces that $|\mathcal{O}_S(C_1)|$ is base-point-free and not composed with a pencil. Therefore, Ψ is a morphism and its image is a surface. Now, from (11), we have

$$h^0(S, \mathcal{O}_S(C_1)) = h^0(S, \mathcal{O}_S(C_0 + Ef)) = h^0(C, \mathcal{O}_C(E)) + h^0(C, \mathcal{O}_C) = R + 1,$$

where the second equality follows from Leray's isomorphism, projection formula, and the fact that

$$\rho_*(\mathcal{O}_S(C_0 + Ef)) = \rho_*(\mathcal{O}_S(C_0) \otimes \mathcal{O}_S(Ef)) = \mathfrak{F} \otimes \mathcal{O}_C(E) =$$
$$= (\mathcal{O}_C \oplus \mathcal{O}_C(-E)) \otimes \mathcal{O}_C(E) = \mathcal{O}_C(E) \oplus \mathcal{O}_C,$$

whereas the third equality follows from the fact that E is non-special of degree e on C of genus γ and from the definition of R in (9).
(ii) Since E is very ample on C, from [28, Propositions 2.3 and 2.14], it follows that the morphism Ψ is an isomorphism onto its image F outside the section C_0 of S. On the other hand, $C_1 \cdot C_0 = (C_0 + Ef) \cdot C_0 = -e + e = 0$, i.e. C_1 contracts the

section C_0 at a point $v \in \mathbb{P}^R$ which is off $Y = \Psi(C_1)$, the isomorphic image of the section $C_1 \cong C$.
(iii) All the fibres of the ruling of S are embedded as lines, as $C_1 \cdot f = 1$. Since C_0 is contracted to a point v and since $C_0 \cdot f = 1$, for any fibre f of the ruling of S, it follows that any line $\ell := \Psi(f)$ passes through v. Thus $F = \Psi(S)$ is a cone over Y, with vertex the point $v \in \mathbb{P}^R$. From the isomorphism $C_1 \cong C$, one also deduces that $\Psi_{|C_1} \cong \phi_{|\mathcal{O}_C(E)|}$, as the following diagram summarizes:

$$\begin{array}{ccc} S & \xrightarrow{\Psi_{|\mathcal{O}_S(C_1)|}} & F \subset \mathbb{P}^R \\ \downarrow \rho & & \downarrow \pi_v \\ C & \xrightarrow{\phi_{|\mathcal{O}_C(E)|}} & Y \subset \mathbb{P}^{R-1}, \end{array} \tag{12}$$

where π_v denotes the projection from the vertex point v. Since $\mathcal{O}_F(1)$ is induced by $\mathcal{O}_S(C_1)$, it is clear that Y is a hyperplane section of F so $Y \subset \mathbb{P}^{R-1}$ is of degree e, genus γ and it is projectively normal in $\mathbb{P}^{R-1}$, as it follows from [29, Thm. 1].
(iv) One has deg $F = Y^2 = C_1^2 = e$; moreover, since $Y \cong C$ is a hyperplane section, then F has sectional genus γ. Now, $h^0(F, \mathcal{O}_F(1)) = h^0(S, \mathcal{O}_S(C_1)) = R + 1 = e - \gamma + 2$, as computed in (i); whereas $h^1(F, \mathcal{O}_F(1)) = h^1(S, \mathcal{O}_S(C_1))$ so, from the exact sequence,

$$0 \to \mathcal{O}_S \to \mathcal{O}_S(C_1) \xrightarrow{r_{C_1}} \mathcal{O}_{C_1}(C_1) \cong \mathcal{O}_C(E) \to 0$$

one gets $h^0(S, \mathcal{O}_S(C_1)) = R + 1$, $h^0(C_1, \mathcal{O}_{C_1}(C_1)) = h^0(C, \mathcal{O}_C(E)) = R$ and $h^1(C_1, \mathcal{O}_{C_1}(C_1)) = h^1(C, \mathcal{O}_C(E)) = 0$, as E is non-special. By Leray's isomorphism and projection formula one also gets $h^1(S, \mathcal{O}_S) = h^1(C, \mathcal{O}_C) = \gamma$. Thus, the map $H^0(r_{C_1})$, induced in cohomology by the map r_{C_1}, is surjective hence, from the above exact sequence, one gets $h^1(S, \mathcal{O}_S(C_1)) = \gamma$.

At last, since F has general hyperplane section which is a projectively normal curve in $\mathbb{P}^{R-1}$, it follows that F is projectively normal in $\mathbb{P}^R$ (cf. e.g. [8, Proof of Lemma 5.7, Rem. 5.8].
(v) By the very definition of F, one has $h^0(F, \mathcal{O}_F(m)) = h^0(S, \mathcal{O}_S(mC_1))$. From [33, Ex. III.8.3, p. 253], it follows that

$$h^0(S, \mathcal{O}_S(mC_1)) = h^0(S, \mathcal{O}_S(mC_0 + mEf)) =$$

$$= \sum_{k=0}^{m} h^0(C, \mathcal{O}_C((m-k)E)) = \sum_{j=0}^{m} h^0(C, \mathcal{O}_C(jE)).$$

Since E is non-special on C, so it is any divisor jE, for any integer $1 \leqslant j \leqslant m$. Therefore, by Riemann–Roch on C, one has

$$h^0(F, \mathcal{O}_F(m)) = \sum_{j=0}^{m} h^0(C, \mathcal{O}_C(jE)) =$$

$$= 1 + [1 + 2 + 3 + \ldots + (m-1) + m]\, e - m\gamma + m = \frac{m(m+1)}{2} e - m(\gamma - 1) + 1,$$

as stated. $\square$

Proposition 5 allows to give an explicit *parametric description* of the family of cones $\mathcal{H}(\widehat{\mathcal{I}})$, where $\widehat{\mathcal{I}} := \widehat{\mathcal{I}_{e,\gamma,R-1}}$ is the principal component of the Hilbert scheme $\mathcal{I}_{e,\gamma,R-1}$ as in Proposition 4. For reader's convenience, we first report here a special case of [9, Lemma 6.3], which is needed for the parametric description of $\mathcal{H}(\widehat{\mathcal{I}})$.

Lemma 1 *With notation and assumptions as in Proposition 5, assume further that* $\mathrm{Aut}(C) = \{Id\}$ *(this, in particular, happens when C has general moduli). Let* $G_F \subset \mathrm{PGL}(R+1, \mathbb{C})$ *denote the* projective stabilizer *of F, i.e. the sub-group of projectivities of* $\mathbb{P}^R$ *which fix F as a cone.*

Then $G_F \cong \mathrm{Aut}(S)$ *and* $\dim G_F = h^0(C, \mathcal{O}_C(E)) + 1 = R + 1 = e - \gamma + 2$.

Proof There is an obvious inclusion $G_F \hookrightarrow \mathrm{Aut}(S)$; we want to show that this is actually a group isomorphism. Let $\sigma \in \mathrm{Aut}(S)$ be any automorphism of S. Since C_0 is the unique section of S with negative self-intersection, then $\sigma(C_0) = C_0$, i.e. σ induces an automorphism of $C_0 \cong C$. Assumption $\mathrm{Aut}(C) = \{Id\}$ implies that σ fixes C_0 pointwise. Now, from the fact that $C_1 \sim C_0 + E f$, it follows that $\sigma^*(C_1) \sim \sigma^*(C_0) + \sigma^*(E f) = C_0 + E f \sim C_1$. Therefore, since $|C_1|$ corresponds to the hyperplane linear system of $F = \Psi(S)$, one deduces that any automorphism $\sigma \in \mathrm{Aut}(S)$ is induced by a projective transformation of F. The rest of the proof directly follows from cases [36, Theorem 2–(2) and (3)] and from [36, Lemma 6]: indeed condition $\mathrm{Aut}(C) = \{Id\}$ implies that $\mathrm{Aut}(S) \cong \mathrm{Aut}_C(S)$; furthermore, since C_0 is the unique section of negative self-intersection on S, $\dim G_F = h^0(C, \mathcal{O}_C(E)) + 1$ follows by using the description of $\mathrm{Aut}_C(S)$ in [36, Theorem 2]. $\square$

Proposition 5 and Lemma 1 allow to exhibit a parametric description of the family of cones $\mathcal{H}(\widehat{\mathcal{I}})$, where $\widehat{\mathcal{I}}$ is the principal component of Proposition 4, as well as to compute $\dim \mathcal{H}(\widehat{\mathcal{I}})$ independently from (6).

Parametric description of $\mathcal{H}(\widehat{\mathcal{I}})$: letting $[C]$ vary in $\mathcal{M}_\gamma$ and, for any such C, letting $\mathcal{O}_C(E)$ vary in $\mathrm{Pic}^e(C)$, cones F arising as in Proposition 5 fill-up the component $\mathcal{H}(\widehat{\mathcal{I}})$, which therefore depends on the following parameters:

- $3\gamma - 3$, since $[C]$ varies in $\mathcal{M}_\gamma$, plus
- γ, which are the parameters on which $\mathcal{O}_C(E) \in \mathrm{Pic}^e(C)$ depends, plus
- $(R+1)^2 - 1 = \dim \mathrm{PGL}(R+1, \mathbb{C})$, minus
- $\dim G_F$, which is the dimension of the projectivities of $\mathbb{P}^R$ fixing a general cone F arising from the construction.

From Lemma 1, it follows that dim $G_F = R + 1$, so

$$\dim \mathcal{H}(\widehat{\mathcal{I}}) = 4\gamma - 3 + (R+1)^2 - (R+2). \tag{13}$$

From Proposition 4, we know that $[Y] \in \widehat{\mathcal{I}}$ general is such that $h^1(Y, N_{Y/\mathbb{P}^{R-1}}) = 0$; moreover, since $R = e - \gamma + 1$, it is a straightforward computation to notice that (13) equals the expression in (6), with the choice $\mathcal{W} = \widehat{\mathcal{I}} = \widehat{\mathcal{I}_{e,\gamma,R-1}}$, $r = R - 1$, $d = e$ and $g = \gamma$, namely,

$$\dim \mathcal{H}(\widehat{\mathcal{I}}) = R(e+1) - (R-4)(\gamma-1) = \lambda_{e,\gamma,R-1} + R. \tag{14}$$

Remark 2 The previous parametric description of $\mathcal{H}(\widehat{\mathcal{I}})$ can be formalized by taking into account the schematic construction of $\mathcal{H}(\widehat{\mathcal{I}})$ which deals with universal Picard varieties over $\mathcal{M}_\gamma$. To do so, we follow procedure as in [14, Sect. 2]. Let $\mathcal{M}_\gamma^0$ be the Zariski open subset of the moduli space $\mathcal{M}_\gamma$, whose points correspond to isomorphism classes of curves of genus g without non-trivial automorphisms. By definition, $\mathcal{M}_\gamma^0$ is a fine moduli space, i.e. it has a universal family $p : C \to \mathcal{M}_\gamma^0$, where C and $\mathcal{M}_\gamma^0$ are smooth schemes and p is a smooth morphism. C can be identified with the Zariski open subset $\mathcal{M}_{\gamma,1}^0$ of the moduli space $\mathcal{M}_{\gamma,1}$ of smooth, 1-pointed, genus-γ curves, whose points correspond to isomorphism classes of pairs $[(C, x)]$, with $x \in C$ a point and C a smooth curve of genus γ without non-trivial automorphisms. On $\mathcal{M}_{\gamma,1}^0$ there is again a universal family $p_1 : C_1 \to \mathcal{M}_{\gamma,1}^0$, where $C_1 = C \times_{\mathcal{M}_\gamma^0} C$. The family p_1 has a natural regular global section δ whose image is the diagonal. By means of δ, for any integer k, we have the *universal family of Picard varieties of order k over $\mathcal{M}_{\gamma,1}^0$*, i.e.

$$p_1^{(k)} : \mathcal{P}ic^{(k)} \to \mathcal{M}_{\gamma,1}^0$$

(cf. [14, Sect. 2]); for any closed point $[(C, x)] \in \mathcal{M}_{\gamma,1}^0$, its fibre via $p_1^{(k)}$ is isomorphic to $\mathrm{Pic}^{(k)}(C)$. Setting $\mathcal{Z}_k := C_1 \times_{\mathcal{M}_{\gamma,1}^0} \mathcal{P}ic^{(k)}$, we have a Poincarè line bundle $\mathcal{L}_k$ on $\mathcal{Z}_k$ (cf. a relative version of [1, pp. 166–167]).

Take $k = e \geqslant 2\gamma - 1$ and let $\pi_2 : \mathcal{Z}_e \to \mathcal{P}ic^{(e)}$ be the projection onto the second factor. For a general point $u := [(C, x), \mathcal{O}_C(E)] \in \mathcal{P}ic^{(e)}$, the restriction of $\mathcal{L}_e$ to $\pi_2^{-1}(u)$ is isomorphic to $\mathcal{O}_C(E) \in \mathrm{Pic}^e(C)$ for $[(C, x)] \in \mathcal{M}_{g,1}^0$ general; one has $\mathcal{E}_e := \mathcal{O}_{\mathcal{Z}} \oplus \mathcal{L}_e$ as a rank-two vector bundle on $\mathcal{Z}_e$.

The fibre of $\mathcal{E}_e$ over $u = [(C, x), \mathcal{O}_C(E)] \in \mathcal{P}ic^{(e)}$ is the rank-two vector bundle $\mathfrak{E}_u = \mathfrak{F}_u(E) := \mathcal{O}_C \oplus \mathcal{O}_C(E)$ on C, where $\mathfrak{F}_u = \mathcal{O}_C(-E) \oplus \mathcal{O}_C$ as in Sect. 2.2 and where $[(C, x)] \in \mathcal{M}_{\gamma,1}^0$ is general. Moreover, the sheaf $(\pi_2)_*(\mathcal{E}_e)$ is free of rank $R + 1 = e - \gamma + 2$ on a suitable dense, open subset $\mathcal{U}$ of $\mathcal{P}ic^{(e)}$; therefore, on $\mathcal{U}$, we have functions $s_0, \ldots, s_R$ such that, for each point $u \in \mathcal{U}$, $s_0, \ldots, s_R$ computed at $u = [(C, x), \mathcal{O}_C(E)]$ span the space of sections of the corresponding vector bundle $\mathfrak{E}_u = \mathfrak{F}_u(E)$. In this set-up there is a natural morphism

$$\Psi_e : \mathcal{P}ic^{(e)} \times \mathrm{PGL}(R+1, \mathbb{C}) \to \mathrm{Hilb}(e, \gamma, R),$$

where $\mathrm{Hilb}(e, \gamma, R)$ denotes the Hilbert scheme of surfaces in $\mathbb{P}^R$ of degree e and sectional genus γ: given a pair $(u, \omega) \in \mathcal{P}ic^{(e)} \times \mathrm{PGL}(R+1, \mathbb{C})$, embed $S_u := \mathbb{P}(\mathfrak{E}_u)$ to $\mathbb{P}^R$ via the sections $s_0, \ldots, s_R$ computed at u, compose with the projectivity ω and take the image. Since $\mathcal{P}ic^{(e)} \times \mathrm{PGL}(R+1, \mathbb{C})$ is irreducible, by Proposition 5, $\mathcal{H}(\widehat{\mathcal{I}})$ is the closure of the image of the above map to the Hilbert scheme. By construction, $\mathcal{H}(\widehat{\mathcal{I}})$ dominates $\mathcal{M}_\gamma$ and its general point represents a cone $F \subset \mathbb{P}^R$ as in Proposition 5. From the previous construction, for $[F] \in \mathcal{H}(\widehat{\mathcal{I}})$ general, one has $\dim \Psi_e^{-1}([F]) = \dim G_F + 1$. Since, from Lemma 1, $\dim G_F = R + 1$ then $\dim \mathcal{H}(\widehat{\mathcal{I}}) = 4\gamma - 3 + (R+1)^2 - (R+2)$ as in (13).

2.3 *Curves on Cones and Ramified Coverings*

In this section, we construct suitable ramified m-covers of $Y \subset \mathbb{P}^{R-1}$, for $[Y]$ a general point in the principal component $\widehat{\mathcal{I}} = \widehat{\mathcal{I}_{e,\gamma,R-1}}$, with the use of cones F parametrized by $\mathcal{H}(\widehat{\mathcal{I}})$. Our approach extends the strategy used in [13], which deals with double covers.

Using notation and assumptions as in Proposition 5, for any integer $m \geqslant 1$, let $C_m \in |\mathcal{O}_S(mC_1)|$ be a general member of the linear system on S and let $X_m := \Psi(C_m) \subset F$ denote its image.

Proposition 6 *For any integer $m \geqslant 1$, one has*
(i) X_m is a smooth, irreducible curve of degree $\deg X_m = me$, *which is non-degenerate and linearly normal in $\mathbb{P}^R$.*
(ii) X_m is obtained by the intersection of the cone F with a hypersurface of degree m in $\mathbb{P}^R$.
(iii) The projection π_v from the vertex $v \in F$ gives rise to a morphism $\varphi_m : X_m \to Y$, which is a covering map of degree m induced on X_m by the ruling of the cone F.
(iv) The geometric genus of X_m is

$$g_m := g(X_m) = m(\gamma - 1) + \frac{m(m-1)}{2}e + 1. \tag{15}$$

(v) For any $j \geqslant m$, the line bundle $\mathcal{O}_{X_m}(j)$ is non-special and such that

$$h^0(X_m, \mathcal{O}_{X_m}(j)) = jme - m(\gamma - 1) - \frac{m(m-1)}{2}e, \ \forall\, j \geqslant m. \tag{16}$$

Proof For $m = 1$, $X_1 \cong Y$ is a smooth, irreducible hyperplane section of F as in Proposition 5 and there is nothing else to prove. Therefore, from now on we will focus on $m \geqslant 2$.
(i) Since C_m is a smooth, irreducible curve on S and since $C_m \cdot C_0 = mC_1 \cdot C_0 = 0$, then C_m is isomorphically embedded via Ψ onto its image $X_m \subset F \subset \mathbb{P}^R$, which does not pass through the vertex $v \in F$. Moreover, $\deg X_m = C_m \cdot C_1 = mC_1 \cdot C_1 = mC_1^2 = me$.

Tensoring the exact sequence defining C_m on S by $\mathcal{O}_S(C_1)$, we get

$$0 \to \mathcal{O}_S((1-m)C_1) \to \mathcal{O}_S(C_1) \xrightarrow{r_{C_1}} \mathcal{O}_{C_m}(C_1) \cong \mathcal{O}_{X_m}(1) \to 0.$$

Since $m \geqslant 2$, then $h^0(\mathcal{O}_S((1-m)C_1)) = 0$. Moreover, by Serre duality,

$$h^1(S, \mathcal{O}_S((1-m)C_1)) = h^1(S, \omega_S \otimes \mathcal{O}_S((m-1)C_1)).$$

From the facts that Ψ is birational, $C_1^2 = e > 0$ and $(m-1) > 0$, it follows that $\mathcal{O}_S((m-1)C_1)$ is big and nef, so $h^1(S, \omega_S \otimes \mathcal{O}_S((m-1)C_1)) = 0$, by Kawamata–Viehweg vanishing theorem. Thus,

$$H^0(X_m, \mathcal{O}_{X_m}(1)) \cong H^0(C_m, \mathcal{O}_{C_m}(C_1)) \cong H^0(S, \mathcal{O}_S(C_1))$$

which implies that X_m is non-degenerate and linearly normal, as it follows from Proposition 5–(iv).
(ii) Since $C_m \sim mC_1$ on S and since C_1 induces the hyperplane section of F, it follows that $X_m \in |\mathcal{O}_F(m)|$.
(iii) Taking into account diagram (12), the projection from the vertex v induces the morphism $\varphi_m : X_m \to Y$. Since $C_m \cdot f = mC_1 \cdot f = m$ and since any fibre f is embedded by Ψ as a line of F, φ_m is induced by the ruling of the cone. As Y is smooth, irreducible and all the fibres of φ_m have constant length m, then φ_m is a surjective, finite, flat morphism from X_m to Y (cf. e.g. [39]). Therefore, φ_m is a covering map of degree m as in Sect. 1.3.
(iv) The genus of X_m equals the genus of C_m. Therefore, to compute g_m we can apply adjunction formula on S and the Riemann–Hurwitz formula as in (8) to the map $\varphi_m : C_m \to C_1$ induced by the fibres of the ruling of S (to ease notation, we use the same symbol as for the map $\varphi_m : X_m \to Y$ induced by the projection from the vertex v of the cone F).

If R_{φ_m} denotes the ramification divisor of φ_m on C_m, by (8) one has $\mathcal{O}_{C_m}(R_{\varphi_m}) \cong \mathcal{O}_{C_m}(K_{C_m} - \varphi_m^*(K_{C_1}))$. By adjunction formula, for $j = 1,\ m$, the canonical divisor K_{C_j} is induced on C_j by the divisor $K_S + C_j$ on S, which is

$$K_S + C_j \sim (j-2)C_0 + (j-1)Ef + K_C f, \quad j = 1, m.$$

Therefore, one has

$$\mathcal{O}_{C_m}(R_{\varphi_m}) \cong \mathcal{O}_{C_m}((m-1)C_1) \text{ and } \deg R_{\varphi_m} = m(m-1)e. \tag{17}$$

Using Riemann–Hurwitz formula (8), one gets therefore $2g_m - 2 = m(2\gamma - 2) + m(m-1)e$ which gives (15).

(v) Since deg $X_m = me$, then deg $\mathcal{O}_{X_m}(j) = jme$ whereas, from above, deg $\omega_{X_m} = 2g_m - 2 = 2m(\gamma - 1) + m(m-1)e$. Since $e \geqslant 2\gamma - 1$ it is a straightforward computation to notice that, for $j \geqslant m$, one has deg $\mathcal{O}_{X_m}(j) >$ deg ω_{X_m}, which implies the non-speciality of $\mathcal{O}_{X_m}(j)$ for any $j \geqslant m$. The computation of $h^0(X_m, \mathcal{O}_{X_m}(j))$ then reduces to simply apply Riemann–Roch on the curve X_m. □

We conclude this section with a general result, involving covering maps and projections, which, in particular, applies to smooth, irreducible curves X_m on F as above and which extends [13, Lemma 4, Cor. 5] to the reducible, connected case. This will be used in the proof of Main Theorem (cf. proof of Lemma 3, Claim 2).

Lemma 2 *Let $F \subset \mathbb{P}^R$ be a cone as in Proposition 5. Let $Z \subset F$ be a non-degenerate, connected, projective curve, which is possibly reducible, with at most nodes as possible singularities, where $D :=$ Sing(Z) denotes its scheme of nodes whose cardinality we denote by δ, where Z can be of the following types:*
(a) either $Z = X_{m-1} \cup Y'$, where $X_{m-1} \in |\mathcal{O}_F(m-1)|$ is a general member of the linear system on F, for some integer $m \geqslant 2$, whereas $Y' \in |\mathcal{O}_F(1)|$ is a general hyperplane section of F, in which case $D = X_{m-1} \cap Y' \neq \emptyset$ consists of $\delta = (m-1)e$ points;
(b) or $Z = X_m \in |\mathcal{O}_F(m)|$ is a general member of the linear system on F, for some integer $m \geqslant 2$, in which case $D = \emptyset$ and $\delta = 0$.
Let $\pi_v : Z \to Y \subset \mathbb{P}^{R-1}$ be the projection from the vertex v of the cone F and let R_{π_v} be the ramification divisor *of π_v.*

Then R_{π_v} is a Cartier divisor on Z. Moreover, in case (a) the following exact sequence holds

$$0 \to \mathcal{L}_Z \to N_{Z/\mathbb{P}^R} \to \pi_v^*(N_{Y/\mathbb{P}^{R-1}}) \to 0, \tag{18}$$

where $N_{Z/\mathbb{P}^R}$ denotes the normal sheaf of Z in $\mathbb{P}^R$, which is locally free on Z, and where $\mathcal{L}_Z$ is a line bundle on Z such that

$$\deg \mathcal{L}_Z = \deg Z + \deg R_{\pi_v} + (m-1)e.$$

Otherwise, in case (b), $\mathcal{L}_{X_m} \cong \mathcal{O}_{X_m}(R_{\pi_v}) \otimes \mathcal{O}_{X_m}(1)$ and (18) reads

$$0 \to \mathcal{O}_{X_m}(R_{\pi_v}) \otimes \mathcal{O}_{X_m}(1) \to N_{X_m/\mathbb{P}^R} \to \pi_v^*(N_{Y/\mathbb{P}^{R-1}}) \to 0. \tag{19}$$

Proof We first take into account case (*a*), with $Z = X_{m-1} \cup Y'$, for some $m \geqslant 2$. If $\mathcal{I}_{Z/\mathbb{P}^R}$ denotes the ideal sheaf of Z in $\mathbb{P}^R$ then, since Z is nodal, then

$$N_{Z/\mathbb{P}^R} := \mathcal{H}om(\mathcal{I}_{Z/\mathbb{P}^R}, \mathcal{O}_Z) \text{ and } T_{\mathbb{P}^R}|_Z := \mathcal{H}om(\Omega^1_{\mathbb{P}^R}, \mathcal{O}_Z)$$

are both locally free of rank $R-1$ and R, respectively (cf. [38, page 30]).

If we take into account the projection $\pi_v : Z \to Y$, one has $\pi_v^*(\mathcal{O}_Y) \cong \mathcal{O}_Z$ and $\pi_v^*(\mathcal{O}_Y(1)) \cong \mathcal{O}_Z(1)$; thus, considering the Euler sequences of Z and Y

$$0 \to \mathcal{O}_Z \to \mathcal{O}_Z(1)^{\oplus(R+1)} \to T_{\mathbb{P}^R}|_Z \to 0 \text{ and } 0 \to \mathcal{O}_Y \to \mathcal{O}_Y(1)^{\oplus R} \to T_{\mathbb{P}^{R-1}}|_Y \to 0$$

and pulling back to Z via π_v the second Euler sequence, one deduces the following exact diagram

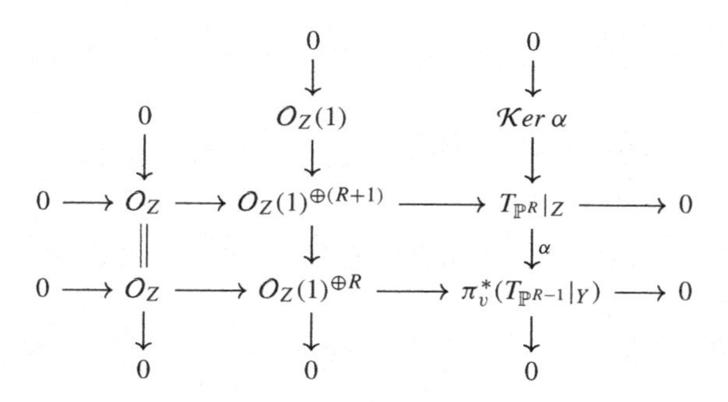

where the map α is surjective by the Snake Lemma. The exactness of the diagram implies that $\mathcal{K}er\ \alpha \cong \mathcal{O}_Z(1)$. Hence, from the right-most exact column of the diagram, we get

$$0 \to \mathcal{O}_Z(1) \to T_{\mathbb{P}^R}|_Z \to \pi_v^*(T_{\mathbb{P}^{R-1}}|_Y) \to 0. \tag{20}$$

If we set Ω^1_Z the *cotangent sheaf* (or the *sheaf of Kähler differentials*) on Z, then its dual $\Theta_Z := \mathcal{H}om(\Omega^1_Z, \mathcal{O}_Z)$ is not locally free, but it is torsion free (cf. [38]) and it is called the *sheaf of derivations* of $\mathcal{O}_Z$ (when Z is smooth and irreducible, Ω^1_Z coincides with the *canonical bundle* whereas Θ_Z with the *tangent bundle*). At last, $T^1_Z := \mathcal{E}xt^1(\Omega^1_Z, \mathcal{O}_Z)$ is called the *first cotangent sheaf* of Z, which is a torsion sheaf supported on $Sing(Z)$. By assumption on Z one has $T^1_Z \cong \mathcal{O}_D$, where D is the set of nodes of Z, and one has the exact sequence

$$0 \to \Theta_Z \to T_{\mathbb{P}^R}|_Z \to N_{Z/\mathbb{P}^R} \to T^1_Z \cong \mathcal{O}_D \to 0. \tag{21}$$

(cf. [38, page 30, (1.2)]) Putting together (20) and (21) and taking into account $\pi_v : Z \to Y$, one gets the following exact diagram:
where β is defined by the diagram. From [38], the sequence (21) splits in two exact sequences

$$0 \to \Theta_Z \to T_{\mathbb{P}^R}|_Z \to N'_Z \to 0 \text{ and } 0 \to N'_Z \to N_{Z/\mathbb{P}^R} \to T^1_Z \cong \mathcal{O}_D \to 0,$$

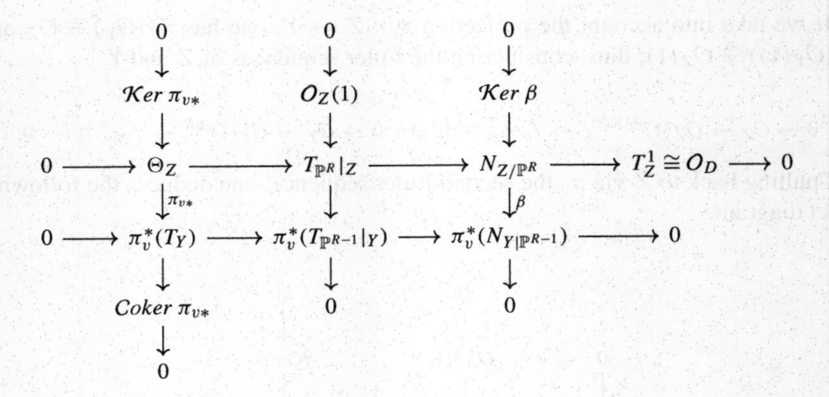

where N'_Z is the *equi-singular sheaf*. Hence, the previous exact diagram gives rise to the following:

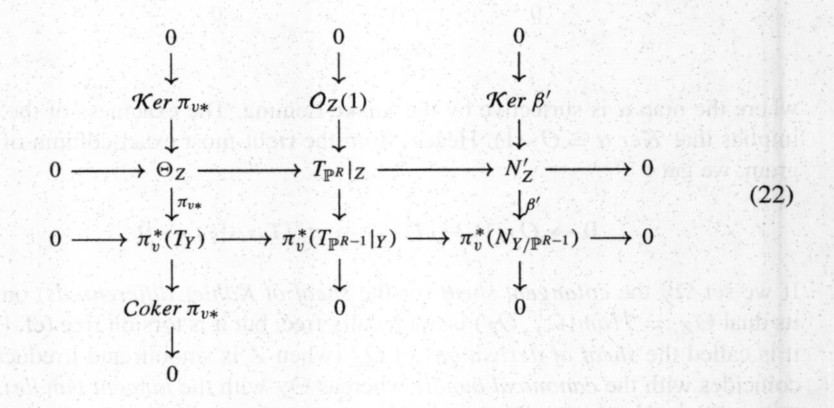

where β' is induced by β from the previous diagram. By the Snake Lemma, one has therefore

$$0 \to \mathcal{K}er\, \pi_{v*} \to \mathcal{O}_Z(1) \to \mathcal{K}er\, \beta' \to Coker\, \pi_{v*} \to 0. \tag{23}$$

Since $\pi_v^*(T_Y)$ is a line bundle on Z and Θ_Z has generically rank 1 on Z, if it were $\mathcal{K}er\, \pi_{v*} \neq 0$ then it would be a torsion sheaf, which is a contradiction by (23) and the fact that $\mathcal{O}_Z(1)$ is a line bundle on Z. Therefore (23) gives

$$0 \to \mathcal{O}_Z(1) \to \mathcal{K}er\, \beta' \to Coker\, \pi_{v*} \to 0. \tag{24}$$

Since by the right-most column of diagram (22) the sheaf $\mathcal{K}er\ \beta'$ has generically rank 1 and by the left-most column of diagram (22), $Coker\ \pi_{v*}$ is a torsion sheaf, from (24) it follows that $\mathcal{K}er\ \beta'$ is a line bundle whereas $Coker\ \pi_{v*} \cong \mathcal{O}_{R_{\pi_v}}$, and R_{π_v} is the (effective, Cartier) ramification divisor of the projection π_v. Thus, $\mathcal{K}er\ \beta'$ is a line bundle on Z such that, from (24), is isomorphic to $\mathcal{O}_Z(R_{\pi_v}) \otimes \mathcal{O}_Z(1)$. Therefore, the sequence (24) reads as

$$0 \to \mathcal{O}_Z(1) \to \mathcal{O}_Z(R_{\pi_v}) \otimes \mathcal{O}_Z(1) \to \mathcal{O}_{R_{\pi_v}} \to 0. \tag{25}$$

From diagram (22) we deduce

$$\begin{array}{ccccccccc}
 & & 0 & & 0 & & 0 & & \\
 & & \downarrow & & \downarrow & & \downarrow & & \\
0 & \longrightarrow & \mathcal{O}_Z(R_{\pi_v}) \otimes \mathcal{O}_Z(1) & \longrightarrow & \mathcal{K}er\ \beta & \longrightarrow & T^1_Z \cong \mathcal{O}_D & \longrightarrow & 0 \\
 & & \downarrow & & \downarrow & & \| & & \\
0 & \longrightarrow & N'_Z & \longrightarrow & N_{Z/\mathbb{P}^R} & \longrightarrow & T^1_Z \cong \mathcal{O}_D & \longrightarrow & 0 \\
 & & \downarrow{\scriptstyle \beta'} & & \downarrow{\scriptstyle \beta} & & \downarrow & & \\
0 & \longrightarrow & \pi_v^*(N_{Y/\mathbb{P}^{R-1}}) & = & \pi_v^*(N_{Y/\mathbb{P}^{R-1}}) & \longrightarrow & 0 & \longrightarrow & 0 \\
 & & \downarrow & & \downarrow & & \downarrow & & \\
 & & 0 & & 0 & & 0 & &
\end{array}$$

By the Snake Lemma again, $\mathcal{K}er\ \beta =: \mathcal{L}_Z$ is a line bundle too, for which

$$0 \to \mathcal{O}_Z(R_{\pi_v}) \otimes \mathcal{O}_Z(1) \to \mathcal{L}_Z \to T^1_Z \cong \mathcal{O}_D \to 0. \tag{26}$$

holds. In particular, by (26) one has

$$\deg\ \mathcal{L}_Z = \deg\ \left(\mathcal{O}_Z(R_{\pi_v}) \otimes \mathcal{O}_Z(1)\right) + \delta = \deg\ Z + \deg\ R_{\pi_v} + (m-1)e$$

as stated. Moreover, from the middle column of the above diagram, one also has

$$0 \to \mathcal{L}_Z \to N_{Z/\mathbb{P}^R} \to \pi_v^*(N_{Y/\mathbb{P}^{R-1}}) \to 0,$$

and this concludes the first part of the statement.

If otherwise $Z = X_m \subset F$ as in case (b), from the first part of the proof we have $D = \emptyset$, $\delta = 0$, $T^1_Z = 0$, $N'_Z = N_{Z/\mathbb{P}^R} = N_{X_m/\mathbb{P}^R}$ and $\mathcal{L}_{X_m} \cong \mathcal{O}_{X_m}(R_{\pi_v}) \otimes \mathcal{O}_{X_m}(1)$ so (18) becomes (19), as stated. □

3 Superabundant Components of Hilbert Schemes

This section is entirely devoted to the construction of *superabundant* components of Hilbert schemes and to the proof of our Main Theorem. To do so, we will need to deal with the surjectivity of the Gaussian–Wahl map $\Phi_{\omega_Y, \mathcal{O}_Y(1)}$ for $Y \subset \mathbb{P}^{R-1}$ as in Sect. 2.1 (cf. Lemma 3, Claim 2).

Remark 3 Recall that $\mathcal{O}_Y(1) \cong \mathcal{O}_C(E)$ is of degree $e \geqslant 2\gamma - 1$. Taking into account numerical assumptions as in Proposition 1, for $6 \leqslant \gamma \leqslant 8$, the condition $2\gamma - 1 \geqslant \gamma + 12$ cannot hold since it would give $\gamma \geqslant 13$, contradicting that $6 \leqslant \gamma \leqslant 8$; similarly, condition $2\gamma - 1 \geqslant \gamma + 9$ does not hold for $\gamma = 9$. On the contrary, $\gamma \geqslant 10$ ensures that $\deg \mathcal{O}_C(E) = e \geqslant 2\gamma - 1 \geqslant \gamma + 9$ holds true so we can apply Proposition 1 to the pair $(C, \mathcal{O}_C(E))$ giving rise to Y to prove the next result.

Proposition 7 *Let $\gamma \geqslant 10$, $e \geqslant 2\gamma - 1$ and $R = e - \gamma + 1$ be integers. Let $\widehat{\mathcal{I}_{e,\gamma,R-1}}$ be the principal component of $\mathcal{I}_{e,\gamma,R-1}$ and let $[Y] \in \widehat{\mathcal{I}_{e,\gamma,R-1}}$ be general. Then*
(i) the Gaussian-Wahl map $\Phi_{\omega_Y, \mathcal{O}_Y(1)}$ is surjective;
(ii) $h^0(Y, N_{Y/\mathbb{P}^{R-1}} \otimes \mathcal{O}_Y(-1)) = R$ whereas $h^0(Y, N_{Y/\mathbb{P}^{R-1}} \otimes \mathcal{O}_Y(-j)) = 0$, for any integer $j \geqslant 2$;
(iii) $\mathcal{H}(\widehat{\mathcal{I}_{e,\gamma,R-1}})$ is generically smooth of dimension

$$\dim \mathcal{H}(\widehat{\mathcal{I}_{e,\gamma,R-1}}) = \lambda_{e,\gamma,R-1} + R = R(e+1) - (R-4)(\gamma - 1), \tag{27}$$

and the cone F, corresponding to $[F] \in \mathcal{H}(\widehat{\mathcal{I}_{e,\gamma,R-1}})$ general, is unobstructed in $\mathbb{P}^R$.

Proof As observed in Remark 3, $\gamma \geqslant 10$ and $e \geqslant 2\gamma - 1$ imply that numerical assumptions of Proposition 1 certainly hold. Since the pair $(C, \mathcal{O}_C(E))$, giving rise to Y, is such that C is with general moduli and $\mathcal{O}_C(E) \in \mathrm{Pic}^e(C)$ is general then we are in position to apply Propositions 1 and 2, from which (i), (ii) and (iii) directly follow. □

Notice that (27) coincides with the expression (14), which has been independently computed via the parametric description of $\mathcal{H}(\widehat{\mathcal{I}_{e,\gamma,R-1}})$ in Sect. 2.2.

From Remark 3 and Proposition 7, we therefore fix from now on the following numerical assumptions:

$$\gamma \geqslant 10, \ \ e \geqslant 2\gamma - 1, \ \ R = e - \gamma + 1, \ \ m \geqslant 2. \tag{28}$$

Furthermore, to ease notation, we simply pose:

$$d := me, \ \ X := X_m, \ \ g := g_m, \tag{29}$$

where X_m and g_m are as in Proposition 6. In this set-up, therefore we have that $[X] \in \mathcal{I}_{d,g,R}$.

We now show that, as $[F]$ varies in $\mathcal{H}(\widehat{\mathcal{I}_{e,\gamma,R-1}})$, curves X fill-up an irreducible locus in $\mathcal{I}_{d,g,R}$ as follows; with notation as in Sect. 2.2, set first

$$\mathcal{U}_{e,\gamma,R-1} := \Big\{ u := ([C], \mathcal{O}_C(E), S, C_1) \mid [C] \in \mathcal{M}_\gamma \text{ and } \mathcal{O}_C(E) \in \mathrm{Pic}^e(C) \text{ general}, \ \mathfrak{F} = \mathcal{O}_C \oplus \mathcal{O}_C(-E), \ S = \mathbb{P}(\mathfrak{F}), \ C_1 \in |\mathcal{O}_S(C_0 + Ef)| \text{ general} \Big\}.$$

By construction $\mathcal{U}_{e,\gamma,R-1}$ is irreducible. Then, for any $m \geqslant 2$, consider

$$\mathcal{W}_{d,g,R} := \Big\{ (u, C_m) | \, u \in \mathcal{U}_{e,\gamma,R-1}, \ C_m \in |\mathcal{O}_S(m(C_0 + Ef))| \text{ general} \Big\} \stackrel{\pi}{\longrightarrow} \mathcal{U}_{e,\gamma,R-1},$$

where the natural projection π onto the first factor endows $\mathcal{W}_{d,g,R}$ with a structure of a non-empty, open dense subset of a projective-bundle over $\mathcal{U}_{e,\gamma,R-1}$, hence $\mathcal{W}_{d,g,R}$ is irreducible too (recall that $d = me$ and $g = g_m$ depend on m).

By the very definition of $\mathcal{W}_{d,g,R}$, one has a natural *Hilbert morphism*

$$\begin{aligned} h : \mathcal{W}_{d,g,R} &\longrightarrow \mathcal{I}_{d,g,R} \\ (u, C_m) &\mapsto [X_m] := [\Psi(C_m)], \end{aligned}$$

where Ψ is the morphism mentioned in Proposition 5, and one defines

$$\mathcal{S}_{d,g,R} := h(\mathcal{W}_{d,g,R}) \subset \mathcal{I}_{d,g,R}. \tag{30}$$

Lemma 3 *$\mathcal{S}_{d,g,R}$ is irreducible, of dimension* $\dim \mathcal{S}_{d,g,R} = \lambda_{d,g,R} + \sigma_{d,g,R}$, *where*

$$\lambda_{d,g,R} = (R+1)me - (R-3)\left(m(\gamma-1) + \frac{m(m-1)}{2}e\right)$$

is the expected dimension of $\mathcal{I}_{d,g,R}$ as in (3), *whereas the positive integer*

$$\sigma_{d,g,R} := (R-4)\left[(\gamma-1)(m-1) + 1 + e + \frac{m(m-3)}{2}e\right] + 4(e+1) + em(m-5)$$

is called the superabundance summand *of the dimension of $\mathcal{S}_{d,g,R}$. Furthermore, $\mathcal{S}_{d,g,R}$ is generically smooth.*

Proof By construction, $\mathcal{S}_{d,g,R}$ is irreducible and

$$\dim \mathcal{S}_{d,g,R} = \dim \mathcal{H}(\widehat{\mathcal{I}_{e,\gamma,R-1}}) + \dim |\mathcal{O}_F(m)|,$$

where $[F] \in \mathcal{H}(\widehat{\mathcal{I}_{e,\gamma,R-1}})$ is general. Thus, from (27) (equivalently, from (14)) and from Proposition 5 (v), the latter reads

$$\begin{aligned} \dim \mathcal{S}_{d,g,R} &= R(e+1) - (R-4)(\gamma-1) + \tfrac{m(m+1)}{2}e - m(\gamma-1) = \\ &= \lambda_{e,\gamma,R-1} + R + \tfrac{m(m+1)}{2}e - m(\gamma-1). \end{aligned} \tag{31}$$

Taking into account (3) which, in our notation, reads

$$\lambda_{d,g,R} = (R+1)me - (R-3)\left(m(\gamma-1) + \frac{m(m-1)}{2}e\right),$$

to prove the first part of the statement it suffices to showing that dim $\mathcal{S}_{d,g,R} - \lambda_{d,g,R} = \sigma_{d,g,r}$ and that the latter integer is positive.

To do so, observe that

$$\dim \mathcal{S}_{d,g,R} - \lambda_{d,g,R} = R\left(\gamma(m-1) - m + 2 + e\right) + Re\frac{m(m-3)}{2} - 4(\gamma-1)(m-1)$$

$$-em(m-1) = (R-4)\left[(\gamma-1)(m-1) + 1 + e + \frac{m(m-3)}{2}e\right] + 4(e+1) + em(m-5).$$

Notice that, since $R = e - \gamma + 1$, $e \geqslant 2\gamma - 1$ and $\gamma \geqslant 10$, then $R - 4 \geqslant 6$; moreover, since $m \geqslant 2$, the summands in square-parentheses add-up to a positive integer: the statement is clear for $m \geqslant 3$, whereas for $m = 2$ one has $\left[(\gamma-1)(m-1) + 1 + e + \frac{m(m-3)}{2}e\right] = \gamma \geqslant 10$. Concerning the summand $4(e+1)$, in our assumptions it is $4(e+1) \geqslant 8\gamma \geqslant 80$. The last summand $em(m-5)$ is non-negative for $m \geqslant 5$, whereas for $m = 2, 3, 4$ it is, respectively, $-6e, -6e, -4e$; in all the latter three sporadic cases, the negativity of the summand $em(m-5)$ does not affect the positivity of the total expression.

The previous computations show that dim $\mathcal{S}_{d,g,R} - \lambda_{d,g,R} = \sigma_{d,g,R} > 0$ so the first part of the statement is proved. Concerning the generic smoothness of $\mathcal{S}_{d,g,R}$, we have first the following:

Claim 1 *For* $[X] \in \mathcal{S}_{d,g,R}$ *general, one has*

$$h^0(X, N_{X/\mathbb{P}^R}) = \lambda_{e,\gamma,R-1} + h^0(Y, N_{Y/\mathbb{P}^{R-1}} \otimes \mathcal{T}_\varphi^\vee) + \frac{m(m+1)}{2}e - m(\gamma-1),$$

where $\mathcal{T}_\varphi^\vee$ *is the Tschirnhausen bundle associated to the covering map of degree* m $\varphi : X \to Y$ *induced by the projection* π_v *from the vertex* v *of the cone* F.

Proof of Claim 1 Consider the exact sequence (19) in Lemma 2 which, in the present notation, reads $0 \to \mathcal{O}_X(R_{\pi_v}) \otimes \mathcal{O}_X(1) \to N_{X/\mathbb{P}^R} \to \pi_v^*(N_{Y/\mathbb{P}^{R-1}}) \to 0$. From Proposition 6 (iii), the covering map of degree m, $\varphi : X \to Y$, is induced by the projection π_v from the vertex v of the cone F and, by (17), we have $\mathcal{O}_X(R_{\pi_v}) = \mathcal{O}_X(R_\varphi) \cong \mathcal{O}_X(m-1)$. Therefore, the previous exact sequence gives

$$0 \to \mathcal{O}_X(m) \to N_{X/\mathbb{P}^R} \to \varphi^*(N_{Y/\mathbb{P}^{R-1}}) \to 0.$$

From Proposition 6 (v), $\mathcal{O}_X(m)$ is non-special on X, so

$$h^0(X, N_{X/\mathbb{P}^R}) = h^0(X, \varphi^*(N_{Y/\mathbb{P}^{R-1}})) + h^0(X, \mathcal{O}_X(m)).$$

Since φ is a finite morphism, using Leray's isomorphism and projection formula, we get $h^0(X, \varphi^*(N_{Y/\mathbb{P}^{R-1}})) = h^0(Y, N_{Y/\mathbb{P}^{R-1}} \otimes \varphi_* \mathcal{O}_X)$. Moreover, from Sect. 1.3, one has $\varphi_*\mathcal{O}_X = \mathcal{O}_Y \oplus \mathcal{T}_\varphi^\vee$, where $\mathcal{T}_\varphi^\vee$ the Tschirnhausen bundle associated to φ. Thus,

$$h^0(Y, N_{Y/\mathbb{P}^{R-1}} \otimes \varphi_* \mathcal{O}_X) = h^0(Y, N_{Y/\mathbb{P}^{R-1}}) + h^0(Y, N_{Y/\mathbb{P}^{R-1}} \otimes \mathcal{T}_\varphi^\vee).$$

To sum up, one has

$$h^0(X, N_{X/\mathbb{P}^R}) = h^0(Y, N_{Y/\mathbb{P}^{R-1}}) + h^0(Y, N_{Y/\mathbb{P}^{R-1}} \otimes \mathcal{T}_\varphi^\vee) + h^0(X, \mathcal{O}_X(m)). \quad (32)$$

By (16) with $j = m$, one has

$$h^0(X, \mathcal{O}_X(m)) = \frac{m(m+1)}{2} e - m(\gamma - 1).$$

From (3) and Corollary 1, it follows that

$$h^0(Y, N_{Y/\mathbb{P}^{R-1}}) = \lambda_{e,\gamma,R-1},$$

since Y corresponds to a general point in the principal component $\widehat{\mathcal{I}_{e,\gamma,R-1}}$, which concludes the proof of Claim 1. □

To conclude that $\mathcal{S}_{d,g,R}$ is generically smooth, we need to prove the following:

Claim 2 For any $m \geqslant 2$, one has

$$h^0(Y, N_{Y/\mathbb{P}^{R-1}} \otimes \mathcal{T}_\varphi^\vee) = R. \quad (33)$$

Proof of Claim 2 To prove the statement, we will use an inductive procedure.

Assume first $m = 2$, so $X = X_2$ and $\varphi := \varphi_2 : X \to Y$ is the double cover of the curve Y, as in Proposition 6 (iii). In this case, the Tschirnhausen bundle $\mathcal{T}_\varphi^\vee$ is a line bundle on Y which, from (7) and (17), equals $\mathcal{O}_Y(-E) \cong \mathcal{O}_Y(-1)$. Since $\gamma \geqslant 10$ and $e \geqslant 2\gamma - 1$, assumptions of Proposition 1 are satisfied. Therefore, from Proposition 7 (b), we have $h^0(Y, N_{Y/\mathbb{P}^{R-1}} \otimes \mathcal{O}_Y(-1)) = R$ and (33) holds true in this case.

Take now $m \geqslant 3$ and assume that (33) holds for the covering map $\varphi := \varphi_{m-1} : X := X_{m-1} \to Y$ of degree $(m-1)$, where $X_{m-1} \in |\mathcal{O}_F(m-1)|$ general as in Proposition 6. To ease notation, the associated Tschirnhausen bundle $\mathcal{T}_\varphi^\vee$ will be simply denoted by $\mathcal{T}_{m-1}^\vee$. Let $Y' \in |\mathcal{O}_F(1)|$ be general and consider the projective, connected, non-degenerate reducible curve $Z := X \cup Y' \subset F$ which, as a Cartier divisor on F, is such that $Z \in |\mathcal{O}_F(m)|$. The singular locus of Z is $D := X \cap Y'$ and consists of δ nodes, where $\delta := (m-1)H^2 = (m-1)e$, H denoting the hyperplane section of F. As in Sect. 1.3, the curve Z is endowed with a natural covering map $\psi : Z \to Y$ of degree m, whose Tschirnhausen bundle $\mathcal{T}_{\psi}^{\vee}$ on Y will be simply denoted by $\mathcal{T}_m^{\vee}$. From Proposition 3, passing to duals, we get the exact sequence

$$0 \to \mathcal{O}_Y(-D) \to \mathcal{T}_m^{\vee} \to \mathcal{T}_{m-1}^{\vee} \to 0$$

of vector bundles on Y. Tensoring this exact sequence by $N_{Y/\mathbb{P}^{R-1}}$ gives

$$0 \to N_{Y/\mathbb{P}^{R-1}} \otimes \mathcal{O}_Y(-D) \to N_{Y/\mathbb{P}^{R-1}} \otimes \mathcal{T}_m^{\vee} \to N_{Y/\mathbb{P}^{R-1}} \otimes \mathcal{T}_{m-1}^{\vee} \to 0. \tag{34}$$

By induction, since $m-1 \geqslant 2$, one has $h^0(Y, N_{Y/\mathbb{P}^{R-1}} \otimes \mathcal{T}_{m-1}^{\vee}) = R$. Moreover, since D is cut-out on the irreducible component Y' by a hypersurface of degree $m-1$ in $\mathbb{P}^R$ and since $Y' \cong Y$, then $\mathcal{O}_Y(D) \cong \mathcal{O}_Y(m-1)$ and one has

$$h^0(Y, N_{Y/\mathbb{P}^{R-1}} \otimes \mathcal{O}_Y(-D)) = h^0(Y, N_{Y/\mathbb{P}^{R-1}} \otimes \mathcal{O}_Y(-(m-1))) = 0,$$

as it follows from Proposition 2 (iii) and from the fact that $m-1 \geqslant 2$. By (34), we deduce that $H^0(Y, N_{Y/\mathbb{P}^{R-1}} \otimes \mathcal{T}_m^{\vee})$ injects into $H^0(Y, N_{Y/\mathbb{P}^{R-1}} \otimes \mathcal{T}_{m-1}^{\vee})$, so in particular

$$h^0(Y, N_{Y/\mathbb{P}^{R-1}} \otimes \mathcal{T}_m^{\vee}) \leqslant R. \tag{35}$$

On the other hand, since $[Z] \in \overline{\mathcal{S}}_{d,g,R}$, where $\overline{\mathcal{S}}_{d,g,R}$ denotes the closure in $\mathcal{I}_{d,g,R}$ of $\mathcal{S}_{d,g,R}$, from (2) one must have

$$h^0(Z, N_{Z/\mathbb{P}^R}) = \dim\, T_{[Z]}(\mathcal{I}_{d,g,R}) \geqslant \dim\, \mathcal{S}_{d,g,R} = \lambda_{e,\gamma,R-1} + R + \frac{m(m+1)}{2}e - m(\gamma-1), \tag{36}$$

as it follows from (31). Since Z satisfies assumptions as in Lemma 2, we can consider the exact sequence (18). The line bundle $\mathcal{L}_Z$ therein has degree

$$\deg\, \mathcal{L}_Z = \deg\, Z + \deg\, R_{\psi} + \delta = me + \deg\, R_{\psi} + (m-1)e.$$

By definition of $\psi : Z \to Y$, the ramification of this map is supported on the irreducible component $X = X_{m-1}$ of Z, namely, $R_{\psi} = R_{\varphi_{m-1}}$ where $\varphi_{m-1} : X \to Y$. From (17), we therefore have $\deg\, R_{\varphi_{m-1}} = (m-1)^2 e$, so

$$\deg\, \mathcal{L}_Z = me + (m-1)^2 e + (m-1)e = m^2 e.$$

Hence, $\mathcal{L}_Z$ is a non-special line bundle on Z, Z being a reduced, connected and nodal curve of arithmetic genus $p_a(Z) = g = g_m$ as in (15) (the non-speciality of $\mathcal{L}_Z$ can be proved by applying the same numerical computation as in the proof of Proposition 6 (v), replacing the canonical bundle with the dualizing sheaf ω_Z). Thus, from (18), one gets $h^0(Z, N_{Z/\mathbb{P}^R}) = h^0(Z, \psi^*(N_{Y/\mathbb{P}^{R-1}})) + h^0(Z, \mathcal{L}_Z)$ where $h^0(Z, \mathcal{L}_Z) = \chi(Z, \mathcal{L}_Z) = \frac{m(m+1)}{2}e - m(\gamma-1)$, both equality following from the non-speciality of $\mathcal{L}_Z$. As for the summand $h^0(Z, \psi^*(N_{Y/\mathbb{P}^{R-1}}))$, we can apply projection formula and Leray's isomorphism, which gives

$$h^0(Z, \psi^*(N_{Y/\mathbb{P}^{R-1}})) = h^0(Y, N_{Y/\mathbb{P}^{R-1}}) + h^0(Y, N_{Y/\mathbb{P}^{R-1}} \otimes \mathcal{T}_m^\vee).$$

Since $h^0(Y, N_{Y/\mathbb{P}^{R-1}}) = \lambda_{e,\gamma,R-1}$ as $[Y] \in \widehat{\mathcal{I}_{e,\gamma,R-1}}$ is general (cf. Corollary 1), then comparing with (36) we deduce that $h^0(Y, N_{Y/\mathbb{P}^{R-1}} \otimes \mathcal{T}_m^\vee) \geqslant R$. Thus, using the previous inequality (35), we get $h^0(Y, N_{Y/\mathbb{P}^{R-1}} \otimes \mathcal{T}_m^\vee) = R$.

By semi-continuity on the general element $[X_m] \in \mathcal{S}_{d,g,R}$, with its natural covering map $\varphi_m : X_m \to Y$ of degree m and its associated Tschirnhausen bundle $\mathcal{T}_{\varphi_m}^\vee$, we deduce that $h^0(Y, N_{Y/\mathbb{P}^{R-1}} \otimes \mathcal{T}_{\varphi_m}^\vee) \leqslant R$. On the other hand, replacing Z with X_m in the previous computations, since

$$h^0(X_m, N_{X_m/\mathbb{P}^R}) = \dim T_{[X_m]}(\mathcal{I}_{d,g,R}) \geqslant \dim \mathcal{S}_{d,g,R} = \lambda_{e,\gamma,R-1} + R + \frac{m(m+1)}{2}e - m(\gamma - 1),$$

one can conclude by applying (19), with $\mathcal{O}_{X_m}(R_{\varphi_m}) \cong \mathcal{O}_{X_m}(m-1)$ as in (17), and reasoning as we did for Z above. This completes the proof of Claim 2. □

The previous computations show that, for $[X] \in \mathcal{S}_{d,g,R}$ general, one has

$$\dim \mathcal{S}_{d,g,R} = \lambda_{d,g,R} + \sigma_{d,g,R} = \dim T_{[X]}(\mathcal{S}_{d,g,R}) = T_{[X]}(\mathcal{I}_{d,g,R}), \qquad (37)$$

which therefore implies that $\mathcal{S}_{d,g,R}$ is generically smooth, completing the proof of the Lemma. □

We are finally in position to prove our Main Theorem.

Proof of Main Theorem The first part of Lemma 3 ensures that any irreducible component of $\mathcal{I}_{d,g,R}$ containing $\mathcal{S}_{d,g,R}$ has to be *superabundant*, having dimension at least $\dim \mathcal{S}_{d,g,R} = \lambda_{d,g,R} + \sigma_{d,g,R}$. On the other hand, the proofs of Claims 1 and 2 in Lemma 3 show that $\mathcal{S}_{d,g,R}$ is contained in a unique component of $\mathcal{I}_{d,g,R}$, more precisely it fills up an open, dense subset of an irreducible component of $\mathcal{I}_{d,g,R}$ which is generically smooth, superabundant, of dimension $\lambda_{d,g,R} + \sigma_{d,g,R}$. Indeed, by (37), for $[X] \in \mathcal{S}_{d,g,R}$ general we have that

$$\dim T_{[X]}(\mathcal{S}_{d,g,R}) = h^0(X, N_{X/\mathbb{P}^R}) = \dim T_{[X]}(\mathcal{I}_{d,g,R}) = \dim \mathcal{S}_{d,g,R} = \lambda_{d,g,R} + \sigma_{d,g,R}.$$

□

Remark 4 It is clear from the construction that $\mathcal{S}_{d,g,R}$ lies in a component of $\mathcal{I}_{d,g,R}$ which cannot dominate $\mathcal{M}_g$. Indeed, the modular morphism of such a component maps to the *Hurwitz space* $\mathcal{H}_{\gamma,m,g}$ parametrizing isomorphism classes of genus-g curves arising as m-sheeted, ramified covers of irrational curves of genus γ.

Acknowledgements This collaboration has benefitted of funding from the MIUR Excellence Department Project awarded to the Department of Mathematics, University of Rome Tor Vergata (CUP: E83-C18000100006) and from the MIUR Excellence Department Project awarded to the Department of Mathematics and Physics, University Roma Tre. Both authors are members of INdAM–GNSAGA. The authors deeply thank the Referee for the very kind referee report, having enthusiastic words and extremely useful advices and remarks.

References

1. E. Arbarello, M. Cornalba, P.A. Griffiths, J. Harris, Geometry of algebraic curves, vol. I, in *Grundlehren der mathematischen Wissenschaften*, vol. 267, (Springer, New York, 1985)
2. E. Arbarello, M. Cornalba, P.A. Griffiths, Geometry of algebraic curves, vol. II, in *Grundlehren der mathematischen Wissenschaften*, vol. 268, (Springer, New York, 2011)
3. E. Ballico, C. Fontanari, A few remarks about the Hilbert scheme of smooth projective curves. Comm. Algebra **42**, 3895–3901 (2014)
4. F. Bastianelli, C. Ciliberto, F. Flamini, P. Supino, Gonality of curves on general hypersurfaces. J. Math. Pures Appl. **125**, 94–118 (2019)
5. F. Bastianelli, C. Ciliberto, F. Flamini, P. Supino, On complete intersections containing a linear subspace. Geom. Ded. **204**, 231–239 (2020)
6. F. Bastianelli, C. Ciliberto, F. Flamini, P. Supino, On Fano schemes of linear subspaces of general complete intersections. Arch. Math. **115**, 639–645 (2020)
7. A. Beauville, R. Donagi, The varieties of lines of a cubic fourfold. C.R. Acad. Sci. Paris, Ser. 1 **301**, 703–706 (1985)
8. A. Calabri, C. Ciliberto, F. Flamini, R. Miranda, Non-special scrolls with general moduli. Rend. Circ. Mat. Palermo **57**, 1–31 (2008)
9. A. Calabri, C. Ciliberto, F. Flamini, R. Miranda, Special scrolls whose base curve has general moduli. Contemp. Math. **496**, 133–155 (2009)
10. L. Caporaso, A compactification of the universal picard variety over the moduli space of stable curves. J. Amer. Math. Soc. **7**(3), 589–660 (1994)
11. G. Casnati, T. Ekedahl, Covers of algebraic varieties I. A general structure theorem, covers of degree 3, 4 and Enriques surfaces. J. Algebr. Geom. **5**, 439–460 (1996)
12. Y. Choi, H. Iliev, S. Kim, Reducibility of the Hilbert scheme of smooth curves and families of double covers. Taiwan. J. Math. **21**(3), 583–600 (2017)
13. Y. Choi, H. Iliev, S. Kim, Components of the Hilbert Scheme of smooth projective curves using ruled surfaces. Manuscripta math. **164**, 395–408 (2021)
14. C. Ciliberto, On rationally determined line bundles on a family of projective curves with general moduli. Duke Math. J. **55**(4), 909–917 (1987)
15. C. Ciliberto, J. Harris, R. Miranda, On the surjectivity of the Wahl map. Duke Math. J. **57**(3), 829–858 (1988)
16. C. Ciliberto, A. Lopez, R. Miranda, Some remarks on the obstructedness of cones over curves of low genus, in *Higher-Dimensional Complex Varieties (Trento, 1994)* (de Gruyter, Berlin, 1996), pp. 167–182
17. C. Ciliberto, E. Sernesi, Families of varieties and the Hilbert scheme, in *Lectures on Riemann Surfaces*. Proceedings of the College on Riemann Surfaces (Trieste, 1987). (World Scientific, Singapore, 1989), pp. 428–499
18. H. Clemens, P. Griffiths, The intermediate Jacobian of the cubic threefold. Ann. Math. **95**, 281–356 (1972)
19. P. Deligne, D. Mumford, The irreducibility of the space of curves of given genus. Publications mathématiques I.H.E.S. **36**, 75–109 (1969)
20. A. Deopurkar, A. Patel, The Picard rank conjecture for the Hurwitz spaces of degree up to five. Algebra Number Theory **9**(2), 459–492 (2015)
21. A. Deopurkar, A. Patel, Vector bundles and finite covers, pp. 1–30 (2019). arXiv:1608.01711v3 [math.AG]

22. R. Donagi, Group law on the intersection of two quadrics. Annali Sc. N. Sup. Pisa **7**, 217–239 (1980)
23. R. Donagi, Generic Torelli for projective hypersurfaces. Compositio Math. **50**, 325–353 (1983)
24. L. Ein, Hilbert scheme of smooth space curves. Ann. Sci. École Norm. Sup. **19**(4), 469–478 (1986)
25. L. Ein, The irreducibility of the Hilbert scheme of smooth space curves, in *Algebraic geometry, Bowdoin 1985 (Brunswick, Maine, 1985)*, vol. 46, Sympos. Pure Math. (Amer. Math. Soc., 1987), pp. 83–87
26. G. Fano, Sul sistema ∞^2 di rette contenute in una varietà cubica generale dello spazio a quattro dimensioni. Atti Accad. Sc. Torino **39**, 778–792 (1904)
27. F. Flamini, E. Sernesi, The curve of lines on a prime Fano threefold of genus 8. Internat. J. Math. **21**, 1561–1584 (2010)
28. L. Fuentes, M. Pedreira, The projective theory of ruled surfaces. Note Mat. **24**(1), 25–63 (2005)
29. M. Green, R. Lazarsfeld, On the projective normality of complete linear series on an algebraic curve. Invent. Math. **83**(1), 73–90 (1986)
30. J. Harris, Curves in projective space, with the collaboration of D. Eisenbud, in *Séminaire de Mathématiques Supérieures*, **85**, Presses de l'Université de Montréal, Montréal, Quebec (1982)
31. J. Harris, Curves and their moduli. Algebraic geometry, Bowdoin, 1985 (Brunswick, Maine, 1985). Proc. Sympos. Pure Math. **46**(1), 99–143, Amer. Math. Soc., Providence, RI (1987)
32. J. Harris, I. Morrison, *Moduli of Curves*, Graduate Texts in Mathematics, vol. 187. (Springer, New York, 1998)
33. R. Hartshorne, Algebraic geometry. Graduate Texts in Math., vol. 52 (Springer, New York, 1977)
34. A. Iliev, D. Markushevich, The Abel–Jacobi map for a cubic threefold and periods of Fano threefolds of degree 14. Doc. Math. **5**, 23–47 (2000)
35. C. Keem, Reducible Hilbert scheme of smooth curves with positive Brill-Noether number. Proc. Amer. Math. Soc. **122**, 349–354 (1994)
36. M. Maruyama, On automorphism groups of ruled surfaces. J. Math. Kyoto. Univ. **11**, 89–112 (1971)
37. E. Mezzetti, G. Sacchiero, Gonality and Hilbert schemes of smooth curves, in *Algebraic curves and projective geometry (Trento, 1988)*, vol. 1389, Lecture Notes in Math. (Springer, 1989), pp. 183–194
38. E. Sernesi, On the existence of certain families of curves. Invent. Math. **75**, 25–57 (1984)
39. E. Sernesi, Deformation of Algebraic schemes, in *Grundlehren der mathematischen Wissenschaften*, vol. 334, (Springer, New York, 2006)
40. F. Severi, *Vorlesungen über algebraische Geometrie* (Teubner, Leibzig, 1921)
41. C. Voisin, Théorème de Torelli pour le cubiques de $\mathbb{P}^5$. Invent. Math. **86**, 577–601 (1986)
42. J. Wahl, Gaussian maps on algebraic curves. J. Differ. Geom. **32**, 77–98 (1990)

Siegel Modular Forms of Degree Two and Three and Invariant Theory

Gerard van der Geer

Dedicated to Ciro Ciliberto on the occasion of his 70th birthday

Abstract This is a survey based on the construction of Siegel modular forms of degree 2 and 3 using invariant theory in joint work with Fabien Cléry and Carel Faber.

Keywords Modular form · Siegel modular form · Invariant theory · Algebraic curves

1991 Mathematics Subject Classification. 10D · 11F46 · 14H10 · 14H45 · 14J15 · 14K10

1 Introduction

Modular forms are sections of naturally defined vector bundles on arithmetic quotients of bounded symmetric domains. Often such quotients can be interpreted as moduli spaces and sometimes this moduli interpretation allows a description as a stack quotient under the action of an algebraic group like GL_n. In such cases, classical invariant theory can be used for describing modular forms.

In the 1960s, Igusa used the close connection between the moduli of principally polarized complex abelian surfaces and the moduli of algebraic curves of genus two to describe the ring of scalar-valued Siegel modular forms of degree two (and level 1) in terms of invariants of the action of GL_2 acting on binary sextics, see [20, 21]. Igusa

G. van der Geer (✉)
Korteweg-de Vries Instituut, Universiteit van Amsterdam, Postbus 94248, 1090 GE, Amsterdam, The Netherlands
e-mail: geer@science.uva.nl

T. Dedieu et al. (eds.), *The Art of Doing Algebraic Geometry*, Trends in Mathematics,
https://doi.org/10.1007/978-3-031-11938-5_9

used theta functions and a crucial step in Igusa's approach was provided by Thomae's formulas from the nineteenth century that link theta constants for hyperelliptic curves to the cross ratios of the branch points of the canonical map of the hyperelliptic curve to $\mathbb{P}^1$.

In the 1980s Tsuyumine, continuing the work of Igusa, used the connection between the moduli of abelian threefolds and curves of genus 3 to describe generators for the ring of scalar-valued Siegel modular forms of degree 3 (and level 1). He used the moduli of hyperelliptic curves of genus 3 as an intermediate step and used theta functions and the invariant theory of binary octics as developed by Shioda, see [26, 30].

The description of the moduli of curves of genus 2 (resp. 3) in terms of a stack quotient of GL_2 acting on binary sextics (resp. of GL_3 acting on ternary quartics) makes it possible to construct the modular forms directly from the stack quotient without the recourse to theta functions or cross ratios. This applies not only to scalar-valued modular forms, but to vector-valued modular forms as well. Covariants (or concomitants) yield explicit modular forms in an efficient way. This is in contrast to earlier and more laborious methods of constructing vector-valued Siegel modular forms of degree 2 and 3 that use theta functions.

In joint work with Fabien Cléry and Carel Faber [7–9] we exploited this for the construction of Siegel modular forms of degree 2 and 3. In degree 2 the universal binary sextic, the most basic covariant, defines a meromorphic Siegel modular form $\chi_{6,-2}$ of weight $(6, -2)$. Substituting the coordinates of $\chi_{6,-2}$ in a covariant produces a meromorphic modular form that becomes holomorphic after multiplication by an appropriate power of χ_{10}, a cusp form of weight 10 associated to the discriminant. For degree 3 we can play a similar game, now involving the universal ternary quartic and a meromorphic Teichmüller modular form $\chi_{4,0,-1}$ of weight $(4, 0, -1)$ that becomes a holomorphic Siegel modular form $\chi_{4,0,8}$ of weight $(4, 0, 8)$ after multiplication with χ_9, a Teichmüller form of weight 9 related to the discriminant.

With this approach it is easy to retrieve Igusa's result on the ring of scalar modular forms of degree 2. Another advantage of this direct approach is that one can treat modular forms in positive characteristic as well. Thus it enabled the determination of the rings of scalar-valued modular forms of degree 2 in characteristic 2 and 3, two cases that were unaccounted for so far, see [6, 33].

In this survey we sketch the approach and indicate how one constructs Siegel modular forms of degree 2 and 3. We show how to derive the results on the rings of scalar-valued modular forms of degree 2.

2 Siegel Modular Forms

Classically, Siegel modular forms are described as functions on the Siegel upper half space. We recall the definition.

For $g \in \mathbb{Z}_{\geq 0}$ we set $\mathcal{L} = \mathbb{Z}^{2g}$ with generators $e_1, \ldots, e_g, f_1, \ldots, f_g$ and define a symplectic form $\langle\, , \,\rangle$ on $\mathcal{L}$ via $\langle e_i, f_j \rangle = \delta_{ij}$. The Siegel modular group $\Gamma_g =$

$\mathrm{Aut}(\mathcal{L}, \langle\, ,\, \rangle)$ of degree g is the automorphism group of this symplectic lattice. Here we write an element $\gamma \in \Gamma_g$ as a matrix $\begin{pmatrix} a & b \\ c & d \end{pmatrix}$ of four $g \times g$ blocks using the basis e_i and f_i; we often abbreviate this as $\gamma = (a, b; c, d)$. The group Γ_g acts on the Siegel upper half space

$$\mathfrak{H}_g = \{\tau \in \mathrm{Mat}(g \times g, \mathbb{C}) : \tau^t = \tau,\ \mathrm{Im}(\tau) > 0\}$$

via $\tau \mapsto \gamma(\tau) = (a\tau + b)(c\tau + d)^{-1}$.

A scalar-valued Siegel modular form of weight k and degree $g > 1$ is a holomorphic function $f : \mathfrak{H}_g \to \mathbb{C}$ satisfying $f(\gamma(\tau)) = \det(c\tau + d)^k f(\tau)$ for all $\gamma = (a, b; c, d) \in \Gamma_g$. If $\rho : \mathrm{GL}(g) \to \mathrm{GL}(V)$ is a complex representation of GL_g then a vector-valued Siegel modular form of weight ρ and degree $g > 1$ is a holomorphic map $f : \mathfrak{H}_g \to V$ satisfying

$$f(\gamma(\tau)) = \rho(c\tau + d) f(\tau) \qquad \text{for all } \gamma = (a, b; c, d) \in \Gamma_g \tag{1}$$

We may restrict to irreducible representations ρ. For $g = 1$ we have to require an additional growth condition for $y = \mathrm{Im}(\tau) \to \infty$.

However, for an algebraic geometer modular forms are sections of vector bundles. Let $\mathcal{A}_g$ be the moduli space of principally polarized abelian varieties of dimension g. This is a Deligne-Mumford stack of relative dimension $g(g + 1)/2$ over $\mathbb{Z}$. It carries a universal principally polarized abelian variety $\pi : \mathcal{X}_g \to \mathcal{A}_g$. This provides $\mathcal{A}_g$ with a natural vector bundle $\mathbb{E} = \mathbb{E}^{(g)}$, the Hodge bundle, defined as

$$\mathbb{E} = \pi_*(\Omega^1_{\mathcal{X}_g/\mathcal{A}_g}) .$$

Starting from $\mathbb{E}$ we can create new vector bundles. Each irreducible representation ρ of GL_g defines a vector bundle $\mathbb{E}_\rho$ by applying a Schur functor (or just by applying ρ to the transition functions of $\mathbb{E}$). In particular, we have the determinant line bundle $L = \det(\mathbb{E})$. Scalar-valued modular forms of weight k are sections of $L^{\otimes k}$ and these form a graded ring. In fact, for $g \geq 2$ and each commutative ring F we have the ring

$$R_g(F) = \oplus_k H^0(\mathcal{A}_g \otimes F, L^{\otimes k}) .$$

The moduli space $\mathcal{A}_g$ can be compactified. There is the Satake compactification, in some sense a minimal compactification, based on the fact that L is an ample line bundle on $\mathcal{A}_g$. This compactification $\mathcal{A}_g^*$ is defined as $\mathrm{Proj}(R_g)$ and satisfies the inductive property

$$\mathcal{A}_g^* = \mathcal{A}_g \sqcup \mathcal{A}_{g-1}^* .$$

Restriction to the 'boundary' $\mathcal{A}_{g-1}^*$ induces a map called the Siegel operator

$$\Phi : R_g(F) \to R_{g-1}(F) .$$

We will also use (smooth) Faltings-Chai type compactifications $\tilde{\mathcal{A}}_g$ and over these the Hodge bundle extends [14]. We will denote the extension also by $\mathbb{E}$.

For $g > 1$ the Koecher principle holds: sections of $\mathbb{E}_\rho$ over $\mathcal{A}_g$ extend to regular sections of the extension of $\mathbb{E}_\rho$ over $\tilde{\mathcal{A}}_g$, see [14, Prop 1.5, p. 140]. For $g = 1$ this does not hold since the boundary in $\mathcal{A}_1^*$ is a divisor, and we define modular forms of weight k as sections of $L^{\otimes k}$ over $\tilde{\mathcal{A}}_1$. If D denote the divisor added to $\tilde{\mathcal{A}}_g$ to compactify $\mathcal{A}_g$, then elements of $H^0(\tilde{\mathcal{A}}_g, \mathbb{E}_\rho \otimes \mathcal{O}(-D))$ are called cusp forms.

We will write $M_\rho(\Gamma_g)(F)$ for $H^0(\tilde{\mathcal{A}}_g \otimes F, \mathbb{E}_\rho)$ or simply $M_\rho(\Gamma_g)$ when F is clear. The space of cusp forms is denoted by $S_\rho(\Gamma_g)$. By the Koecher principle the spaces $M_\rho(\Gamma_g)(F)$ and $S_\rho(\Gamma_g)(F)$ do not depend on the choice of a Faltings-Chai compactification.

Over the complex numbers if $\rho : \mathrm{GL}(g) \to \mathrm{GL}(V)$ is an irreducible representation, elements of $H^0(\tilde{\mathcal{A}}_g \otimes \mathbb{C}, \mathbb{E}_\rho)$ correspond to holomorphic functions $f : \mathfrak{H}_g \to V$ satisfying (1). Such a function allows a Fourier expansion

$$f(\tau) = \sum_{n \geq 0} a(n)\, q^n \,,$$

where the sum is over symmetric $g \times g$ half-integral matrices (meaning $2n$ is integral and even on the diagonal) which are positive semi-definite, $a(n) \in V$ and q^n is shorthand for $e^{2\pi i \mathrm{Tr}(n\tau)}$.

The definition

$$R_g(F) = \oplus_k H^0(\tilde{\mathcal{A}}_g \otimes F, L^{\otimes k})$$

for a commutative ring F allows one speak of modular forms in positive characteristic by taking $F = \mathbb{F}_p$. One cannot define such modular forms by Fourier series.

We summarize what is known about the rings $R_g(F)$. It is a classical result that the ring $R_1(\mathbb{C})$ is freely generated by two Eisenstein series E_4 and E_6 of weights 4 and 6. Deligne determined in [10] the ring $R_1(\mathbb{Z})$ and the rings $R_1(\mathbb{F}_p)$. He showed that

$$R_1(\mathbb{Z}) = \mathbb{Z}[c_4, c_6, \Delta]/(c_4^3 - c_6^2 - 1728\,\Delta)\,,$$

where Δ is a cusp form of weight 12 and c_4 and c_6 are of weight 4 and 6. Reduction modulo p gives a surjection of $R_1(\mathbb{Z})$ to $R_1(\mathbb{F}_p)$ for $p \geq 5$. Moreover, Deligne showed that $R_1(\mathbb{F}_p)$ in characteristic 2 and 3 is given by

$$R_1(\mathbb{F}_2) = \mathbb{F}_2[a_1, \Delta], \quad R_1(\mathbb{F}_3) = \mathbb{F}_3[b_2, \Delta],$$

where in each case Δ is a cusp form of weight 12 and a_1 (resp. b_2) is a modular form of weight 1 (resp. 2).

In [20] Igusa determined the ring $R_2(\mathbb{C})$. He showed that the subring $R_2^{\mathrm{ev}}(\mathbb{C})$ of even weight modular forms is generated freely by modular forms of weight 4, 6, 10, and 12 and $R_2(\mathbb{C})$ is generated over $R_2^{\mathrm{ev}}(\mathbb{C})$ by a cusp form of weight 35 whose square lies in $R_2^{\rm ev}(\mathbb{C})$; see also [21]. Later [22] he also determined the ring $R_2(\mathbb{Z})$; it has 15 generators of weights ranging from 4 to 48.

In characteristic $p \geq 5$ the structure of the rings $R_2(\mathbb{F}_p)$ is similar to that of $R_2(\mathbb{C})$, see [2, 24]; these are generated by forms of weight 4, 6, 10, 12, and 35. The structure of $R_2(\mathbb{F}_p)$ for $p = 2$ and 3 was determined recently in [6, 33]. All these cases can be dealt with easily using the approach with invariant theory.

In degree 2 one can provide the R_2-module

$$M = \oplus_{j,k} M_{j,k}(\Gamma_2) \qquad \text{with } M_{j,k}(\Gamma_2) = H^0(\tilde{\mathcal{A}}_2, \mathrm{Sym}^j(\mathbb{E}) \otimes \det(\mathbb{E})^k)$$

with the structure of a ring using the projection of GL_2-representations $\mathrm{Sym}^m(V) \otimes \mathrm{Sym}^n(V) \to \mathrm{Sym}^{m+n}(V)$ with V the standard representation by interpreting $\mathrm{Sym}^j(V)$ as the space of homogeneous polynomials of degree j in two variables, say x_1, x_2 and performing multiplication of polynomials. The ring M is not finitely generated as Grundh showed, see [3, p. 234]. The dimensions of the spaces $S_{j,k}(\Gamma_2)(\mathbb{C})$ are known by Tsushima [29] for $k \geq 4$; for $k = 3$ they were obtained independently by Petersen and Taïbi [25, 28].

For fixed j the $R_2^{\rm ev}(\mathbb{C})$-modules

$$\oplus_k M_{j,2k}(\Gamma_2)(\mathbb{C}) \quad \text{and} \quad \oplus_k M_{j,2k+1}(\Gamma_2)(\mathbb{C})$$

are finitely generated modules and their structure has been determined in a number of cases by Satoh, Ibukiyama and others, see the references in [8]. Invariant theory makes it easier to obtain such results.

For $g = 3$ the results are less complete. Tsuyumine showed in 1985 [30] that the ring $R_3(\mathbb{C})$ is generated by 34 generators. Recently Lercier and Ritzenthaler showed in [23] that 19 generators suffice.

3 Moduli of Curves of Genus Two as a Stack Quotient

We start with $g = 2$. Let F be a field of characteristic $\neq 2$ and $V = \langle x_1, x_2 \rangle$ the F-vector space with basis x_1, x_2. The algebraic group GL_2 acts on V via $(x_1, x_2) \mapsto (ax_1 + bx_2, cx_1 + dx_2)$ for $(a, b; c, d) \in \mathrm{GL}_2(F)$. We will write $V_{j,k} = \mathrm{Sym}^j(V) \otimes \det(V)^{\otimes k}$ for $j \in \mathbb{Z}_{\geq 0}$ and $k \in \mathbb{Z}$. This is an irreducible representation of GL_2. The underlying vector space can be identified with the space of homogeneous polynomials of degree j in x_1, x_2. We will denote by $V_{j,k}^0$ the open subspace of polynomials with non-vanishing discriminant.

The moduli space $\mathcal{M}_2$ of smooth projective curves of genus 2 over F allows a presentation as an algebraic stack

$$\mathcal{M}_2 \xrightarrow{\sim} [V_{6,-2}^0/\mathrm{GL}_2]$$

Here the action of $(a, b; c, d) \in \mathrm{GL}_2(F)$ is by $f(x_1, x_2) \mapsto (ad - bc)^{-2} f(ax_1 + bx_2, cx_1 + dx_2)$.

Indeed, if C is a curve of genus 2 the choice of a basis ω_1, ω_2 of $H^0(C, K)$ with $K = \Omega^1_C$ defines a canonical map $C \to \mathbb{P}^1$. Let ι denote the hyperelliptic involution of C. Choosing a non-zero element $\eta \in H^0(C, K^3)^{\iota=-1}$ yields eight elements η^2, $\omega_1^6, \omega_1^5\omega_2, \ldots, \omega_2^6$ in the 7-dimensional space $H^0(C, K^6)^{\iota=1}$ and thus a non-trivial relation.

In inhomogeneous terms, this gives us an equation $y^2 = f$ with $f \in F[x]$ of degree 6 with non-vanishing discriminant. The space $H^0(C, K)$ has a basis $xdx/y, dx/y$. If we let GL_2 act on (x, y) via $(x, y) \mapsto ((ax + b)/(cx + d), y(ad - bc)/(cx + d)^3)$ then this action preserves the form of the equation $y^2 = f$ if we take f in $V_{6,-2}$. Then $\lambda\,\mathrm{Id}_V$ acts via λ^2 on $V_{6,-2}$. Thus the stabilizer of a generic element f is of order 2. Moreover $-\mathrm{Id}_V$ acts by $y \mapsto -y$ on y and the action of GL_2 on the differentials is by the standard representation.

Conclusion 3.1 The pull back of the Hodge bundle $\mathbb{E}$ on $\mathcal{M}_2$ under the composition $V^0_{6,-2} \to [V^0_{6,-2}/\mathrm{GL}_2] \xrightarrow{\sim} \mathcal{M}_2$ is the equivariant bundle V.

The moduli space $\overline{\mathcal{M}}_2$ can be constructed from the projectivized space $\mathbb{P}(V_{6,-2})$ of binary sextics. The discriminant defines a hypersurface $\mathbb{D}$ whose singular locus has codimension 1 in $\mathbb{D}$. The locus of binary sectics with three coinciding roots forms an irreducible component $\mathbb{D}'$ of the singular locus. To illustrate the relation between $\mathbb{P}(V_{6,-2})$ at a general point of $\mathbb{D}'$ and $\overline{\mathcal{M}}_2$ at a point of the locus δ_1 in $\overline{\mathcal{M}}_2$ of stable curves whose Jacobian is a product of two elliptic curves, we reproduce the picture of [12, p. 80].

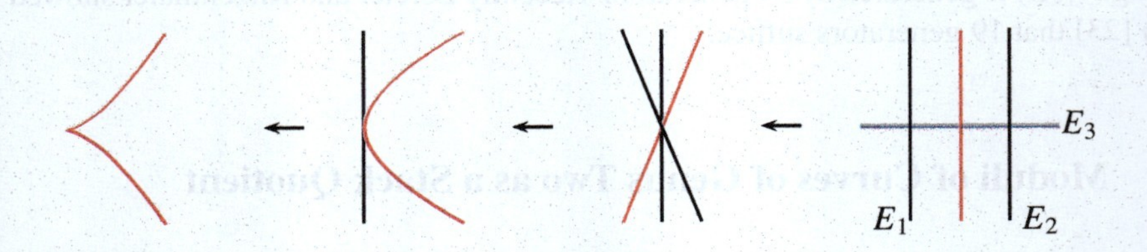

Here we look at a plane Π intersecting $\mathbb{D}$ transversally at a general point of $\mathbb{D}'$. One blows up three times, starting at $\Pi \cap \mathbb{D}'$, and then blows down the exceptional divisors E_1 and E_2; after that E_3 corresponds to the locus δ_1 in $\overline{\mathcal{M}}_2$; in $\mathcal{A}_2$ this corresponds to the locus $\mathcal{A}_{1,1}$ of products of elliptic curves.

4 Invariant Theory of Binary Sextics

We review the invariant theory of GL_2 acting on binary sextics. Let $V = \langle x_1, x_2 \rangle$ be a 2-dimensional vector space over a field F. By definition an invariant for the action of GL_2 acting on the space $\mathrm{Sym}^6(V)$ of binary sextics is an element invariant under $\mathrm{SL}_2(F) \subset \mathrm{GL}_2(F)$. If we write

$$f = \sum_{i=0}^{6} a_i x_1^{6-i} x_2^i \qquad (2)$$

for an element of $\mathrm{Sym}^6(V)$ and thus take $(a_0, \ldots, a_6)$ as coordinates on $\mathrm{Sym}^6(V)$, then an invariant is a polynomial in $a_0, \ldots, a_6$ invariant under $\mathrm{SL}_2(F)$. The discriminant of a binary sextic, a polynomial of degree 10 in the a_i, is an example.

For $F = \mathbb{C}$ the ring of invariants was determined by Clebsch, Bolza and others in the 19th century. It is generated by invariants A, B, C, D, E of degrees 2, 4, 6, 10 and 15 in the a_i. Also for $F = \mathbb{F}_p$ we have generators of these degrees. We refer to [15, 20].

A covariant for the action of GL_2 on binary sextics is an element of $V \oplus \mathrm{Sym}^6(V)$ invariant under the action of SL_2. Such an element is a polynomial in $a_0, \ldots, a_6$ and x_1, x_2. One way to make such covariants is to consider equivariant embeddings of an irreducible GL_2-representation U into $\mathrm{Sym}^d(\mathrm{Sym}^6(V))$. Equivalently, we consider an equivariant embedding

$$\varphi : \mathbb{C} \hookrightarrow \mathrm{Sym}^d(\mathrm{Sym}^6(V)) \otimes U^\vee .$$

Then $\Phi = \varphi(1)$ is a covariant. If U has highest weight $(\lambda_1 \geq \lambda_2)$ then Φ is homogeneous of degree d in $a_0, \ldots, a_6$ and degree $\lambda_1 - \lambda_2$ in x_1, x_2. We say that Φ has degree d and order $\lambda_1 - \lambda_2$.

The simplest example is the universal binary sextic f given by (2); it corresponds to taking $U = \mathrm{Sym}^6(V)$.

Another example is the Hessian of f. Indeed, we decompose in irreducible representations

$$\mathrm{Sym}^2(\mathrm{Sym}^6(V)) = V[12, 0] \oplus V[10, 2] \oplus V[8, 4] \oplus V[6, 6] ,$$

where $V[a, b] = \mathrm{Sym}^{a-b}(V) \otimes \det(V)^b$ is the irreducible representation of highest weight (a, b). By taking $U = V[12, 0]$ we find the covariant $\Phi = f^2$ and by taking $U = V[10, 2]$ we get the Hessian; $U = V[6, 6]$ gives the invariant A.

The covariants form a ring $\mathcal{C}$ and the invariants form a subring $I = I(2, 6)$. The ring of covariants $\mathcal{C}$ was studied intensively at the end of the nineteenth century and the beginning of the twentieth century. The ring $\mathcal{C}$ is finitely generated and Grace and Young presented 26 generators for the ring $\mathcal{C}$, see [16]. These 26 covariants are constructed as transvectants by differentiating in a way similar to the construction of the Hessian. The kth transvectant of two forms $g \in \mathrm{Sym}^m(V)$, $h \in \mathrm{Sym}^n(V)$ is defined as

$$(g,h)_k = \frac{(m-k)!(n-k)!}{m!\,n!} \sum_{j=0}^{k} (-1)^j \binom{k}{j} \frac{\partial^k g}{\partial x_1^{k-j} \partial x_2^j} \frac{\partial^k h}{\partial x_1^j \partial x_2^{k-j}}$$

and the index k is usually omitted if $k = 1$. Examples of the generators are $C_{1,6} = f$, $C_{2,0} = (f, f)_6$, $C_{2,4} = (f, f)_4$, $C_{3,2} = (f, C_{2,4})_4$. We refer to [8] for a list of these 26 generators.

5 Covariants of Binary Sextics and Modular Forms

The Torelli morphism induces an embedding $\mathcal{M}_2 \hookrightarrow \mathcal{A}_2$. The complement of the image is the locus $\mathcal{A}_{1,1}$ of products of elliptic curves. As a compactification we can take $\tilde{\mathcal{A}}_2 = \overline{\mathcal{M}}_2$.

We now fix the field F to be $\mathbb{C}$ or a finite prime field $\mathbb{F}_p$.

In the Chow ring $\mathrm{CH}^*_{\mathbb{Q}}(\tilde{\mathcal{A}}_2) \otimes F$ we have the cycle relation

$$10\lambda_1 = 2[\mathcal{A}_{1,1}] + [D]$$

with $\lambda_1 = c_1(\mathbb{E})$ the first Chern class of $\mathbb{E}$ and D the divisor that compactifies $\mathcal{A}_2 \otimes F$. This implies that there exists a modular form of weight 10 with divisor $2\,\mathcal{A}_{1,1} + D$, hence a cusp form. It is well-defined up to a non-zero multiplicative constant. We will normalize it later. We denote it by $\chi_{10} \in R_2(F)$.

We let V be the F-vector space with basis x_1, x_2. The fact that the pullback of the Hodge bundle $\mathbb{E}$ under

$$V^0_{6,-2} \to [V^0_{6,-2}/\mathrm{GL}_2] \to \mathcal{M}_2 \otimes F \hookrightarrow \mathcal{A}_2 \otimes F \tag{3}$$

is the equivariant bundle V implies that a section of $L^k = \det(\mathbb{E})^k$ pulls back to an invariant of degree k. We thus get an embedding of the ring of scalar-valued modular forms of degree 2 into the ring of invariants

$$R_2(F) \hookrightarrow I(2,6)(F)\,.$$

Conversely, an invariant of degree d defines a section of L^d on $\mathcal{M}_2 \otimes F$, hence a rational (meromorphic) modular form of weight d that is holomorphic outside $\mathcal{A}_{1,1} \otimes F$. By multiplying it with an appropriate power of χ_{10} it becomes holomorphic on $\mathcal{A}_2 \otimes F$, hence on all of $\tilde{\mathcal{A}}_2 \otimes F$. We thus get maps

$$R_2(F) \hookrightarrow I(2,6)(F) \xrightarrow{\nu} R_2(F)[1/\chi_{10}] \tag{4}$$

the composition of which is the identity on $R_2(F)$.

From the description of the moduli $\overline{\mathcal{M}}_2$ given above one sees that the image of a cusp form is an invariant divisible by the discriminant D. The image of χ_{10} is a non-zero multiple of the discriminant D. We may fix χ_{10} by requiring that $\nu(D) = \chi_{10}$.

This extends to the case of vector-valued modular forms. Let

$$M(F) = \oplus_{j,k} M_{j,k}(\Gamma_2)(F)$$

denote the ring of vector-valued modular forms of degree 2.

Proposition 5.1 *Pullback via (3) defines homomorphisms*

$$M(F) \hookrightarrow \mathcal{C}(2,6)(F) \xrightarrow{\nu} M(F)[1/\chi_{10}],$$

the composition of which is the identity.

A modular form of weight (j,k) corresponds to a covariant of degree $d = j/2 + k$ and order j. A covariant of degree d and order r gives rise to a meromorphic modular form of weight $(r, d - r/2)$.

The most basic covariant is the universal binary sextic f. By construction $\nu(f)$ is a meromorphic modular form of weight $(6, -2)$. Therefore the central question is: *Which rational modular form is $\nu(f)$?*

Let $\mathcal{A}_{1,1} \subset \mathcal{A}_2$ be the locus of products of elliptic curves. Under the map

$$\mathcal{A}_1 \times \mathcal{A}_1 \to \mathcal{A}_{1,1} \to \mathcal{A}_2$$

the pullback of the Hodge bundle $\mathbb{E} = \mathbb{E}^{(2)}$ is $p_1^*\mathbb{E}^{(1)} \oplus p_2^*\mathbb{E}^{(1)}$ with p_1 and p_2 the projections of $\mathcal{A}_1 \times \mathcal{A}_1$ on its factors. The pullback of an element $h \in M_{j,k}(\Gamma_2)$ thus can be identified with an element of

$$\bigoplus_{i=0}^{j} M_{k+j-i}(\Gamma_1) \otimes M_{k+i}(\Gamma_1).$$

Near a point of $\mathcal{A}_{1,1}$ we can write such an element symbolically as

$$h = \sum_{i=0}^{j} \eta_j X_1^{j-i} X_2^i,$$

where the X_i are dummy variables to indicate the vector coordinates, and such that the coefficient η_j defines the element of $M_{k+j-i}(\Gamma_1) \otimes M_{k+i}(\Gamma_1)$.

In particular, we have

$$\nu(f) = \sum_{i=0}^{6} \alpha_i X_1^{6-i} X_2^i \,,$$

where α_i are rational functions near a point of $\mathcal{A}_{1,1}$. By interchanging x_1 and x_2 (that corresponds to the element $\gamma \in \Gamma_2$ that interchanges e_1 and e_2) we see that $\alpha_{6-i} = \alpha_i$ for $i = 0, \ldots, 3$.

Proposition 5.2 *If* $\mathrm{char}(F) \neq 2$ *and* $\neq 3$, *then* $\dim S_{6,8}(\Gamma_2)(F) = 1$ *and* $\chi_{10}\nu(f)$ *is a generator of* $S_{6,8}(\Gamma_2)(F)$.

Proof We shall use that $\dim S_{6,8}(\Gamma_2)(\mathbb{C}) \geq 1$. Indeed, we know an explicit cusp form of weight $(6, 8)$, see below. (Alternatively, we know the dimensions of $S_{j,k}(\Gamma_2)(\mathbb{C})$ for $k \geq 4$, see [29]; in particular we know $\dim S_{6,8}(\Gamma_2)(\mathbb{C}) = 1$.) By semi-continuity this implies that $\dim S_{6,8}(\Gamma_2)(F) \geq 1$.

The restriction of an element of $S_{6,8}(\Gamma_2)(F)$ to the locus $\mathcal{A}_{1,1} \otimes F$ lands in

$$\bigoplus_{i=0}^{6} S_{8+6-i}(\Gamma_1)(F) \otimes S_{8+i}(\Gamma_1)(F) \,,$$

and as we have $\dim S_k(\Gamma_1)(F) = 0$ for $k < 12$ it vanishes on $\mathcal{A}_{1,1} \otimes F$.

The tangent space to $\mathcal{A}_2$ at a point $[X = X_1 \times X_2]$ of $\mathcal{A}_{1,1}$, with X_i elliptic curves, can be identified with

$$\mathrm{Sym}^2(T_X) = \mathrm{Sym}^2(T_{X_1}) \oplus (T_{X_1} \otimes T_{X_2}) \oplus \mathrm{Sym}^2(T_{X_2})$$

with T_X (resp T_{X_i}) the tangent space at the origin of X (resp. X_i), and with the middle term corresponding to the normal space. Thus we see that the pullback of the conormal bundle of $\mathcal{A}_{1,1}$ to $\mathcal{A}_1 \times \mathcal{A}_1$ is the tensor product of the pullback of the Hodge bundles on the two factors $\mathcal{A}_1$.

Let $h \in S_{6,8}(\Gamma_2)(F)$ and write h as

$$h = \sum_{i=0}^{6} \eta_i \, X_1^{6-i} X_2^i$$

locally at a general point of $\mathcal{A}_{1,1} \otimes F$. If we consider the Taylor development in the normal direction of $\mathcal{A}_{1,1}$ of the form h that vanishes on $\mathcal{A}_{1,1} \otimes F$ then the first non-zero Taylor term of η_i, say the rth term, is an element of

$$S_{14-i+r}(\Gamma_1)(F) \otimes S_{8+i+r}(\Gamma_1)(F) \,.$$

Since $S_k(\Gamma_1)(F) = (0)$ for $k < 12$, a non-zero rth Taylor term of η_i can occur only for $14 - i + r \geq 12$ and $8 + i + r \geq 12$. We thus find:

$$\mathrm{ord}_{\mathcal{A}_{1,1}}(\eta_0, \ldots, \eta_6) \geq (4, 3, 2, 1, 2, 3, 4) \,.$$

Lemma 5.3 *We have* $\mathrm{ord}_{\mathcal{A}_{1,1}}(\eta_3) = 1$.

Proof If $\mathrm{ord}_{\mathcal{A}_{1,1}}(\eta_3) \geq 2$ then h/χ_{10} is a regular form in $S_{6,-2}(\Gamma_2)$ and we write it as $h/\chi_{10} = \sum_{i=0}^{6} \xi_i X_1^{6-i} X_2^i$ with $\xi_i = \eta_i/\chi_{10}$ regular. Then the invariant $A = 120\, a_0 a_6 - 20\, a_1 a_5 + 8\, a_2 a_4 - 3\, a_3^2$ defines a non-zero regular modular form

$$\nu(A) = 120\, \xi_0 \xi_6 - 20\, \xi_1 \xi_5 + 8\, \xi_2 \xi_4 - 3\, \xi_3^2$$

in $M_2(\Gamma_2)(F)$. But restriction to $\mathcal{A}_{1,1}$ gives for even k an exact sequence

$$0 \to M_{k-10}(\Gamma_2)(F) \to M_k(\Gamma_2)(F) \to \mathrm{Sym}^2(M_k(\Gamma_1)(F)) \tag{5}$$

with the second arrow multiplication by χ_{10}. This implies that $\dim M_2(\Gamma_2)(F) = 0$ for $\mathrm{char}(F) \neq 2$ and $\neq 3$. This proves the lemma. □

The image of a non-zero element $\chi_{6,8}$ of $S_{6,8}$ in $\mathcal{C}(2, 6)$ is a covariant of degree 11 and order 6. But since $\chi_{6,8}$ is a cusp form, this covariant is divisible by the discriminant which is of degree 10. Therefore, $\chi_{6,8}/\chi_{10}$ corresponds to a covariant of degree 1, hence is a non-zero multiple of f. This implies that $\dim S_{6,8}(\Gamma_2)(F) = 1$. □

Corollary 5.4 *If we write* $\nu(f) = \sum_{i=0}^{6} \alpha_i X_1^{6-i} X_2^i$ *then*

$$\mathrm{ord}_{\mathcal{A}_{1,1}}(\alpha_0, \ldots, \alpha_6) \geq (2, 1, 0, -1, 0, 1, 2)$$

and $\mathrm{ord}_{\mathcal{A}_{1,1}}(\alpha_3) = -1$.

6 Constructing Vector-Valued Modular Forms of Degree 2

Now that we know $\nu(f)$ by Proposition 5.2 we can describe the map $\nu : \mathcal{C}(2, 6) \to M[1/\chi_{10}]$ explicitly. Recall that a covariant is a polynomial in $a_0, \ldots, a_6$ and x_1, x_2. We arrive at the following conclusion.

Proposition 6.1 *The map* $\nu : \mathcal{C}(2, 6) \to M[1/\chi_{10}]$ *is substitution of* α_i *for* a_i *(and* X_i *for* x_i*).*

In order to efficiently apply the proposition we need to know the coordinates of a generator $\chi_{6,8}$ of $S_{6,8}$ very explicitly.

Remark 6.2 If $F \neq \mathbb{F}_2$ the moduli space $\mathcal{A}_2[2]$ of level 2 is a Galois cover of $\mathcal{A}_2$ with group $\mathrm{Sp}(2, \mathbb{Z}/2\mathbb{Z})$. This group is isomorphic to the symmetric group $\mathfrak{S}_6$. The sign character of $\mathfrak{S}_6$ defines a character ϵ of Γ_2. The pullback of χ_{10} under $\pi : \mathcal{A}_2[2] \to \mathcal{A}_2$ is a square χ_5^2 since the pullback of D under $\tilde{\mathcal{A}}_2[2] \to \tilde{\mathcal{A}}_2$ is divisible by 2 as a divisor. Thus χ_5 is a modular form of weight 5 with character ϵ.

Let now $F = \mathbb{C}$. Recall that $\chi_{6,8}$ vanishes on $\mathcal{A}_{1,1}$. Dividing $\chi_{6,8}$ by χ_5 provides a holomorphic vector-valued modular form $\chi_{6,3} \in M_{6,3}(\Gamma_2, \epsilon)(F)$ with character ϵ. Such a form can be constructed as follows.

We consider the six odd order two theta functions $\vartheta_i(\tau, z)$ with $(\tau, z) \in \mathfrak{H}_2 \times \mathbb{C}^2$. The gradient $G_i = (\partial\vartheta_i/\partial z_1, \partial\vartheta_i/\partial z_2)(\tau, 0)$ is a modular form of weight $(1, 1/2)$ on some congruence subgroup, but the product of the transposes of these six gradients defines a vector-valued modular form of weight $(6, 3)$ on Γ_2 with character ϵ. The product $\chi_{6,8} = \chi_5\chi_{6,3}$ is a cusp form of weight $(6, 8)$ on Γ_2. A non-zero multiple of its Fourier expansion starts with (with $q_1 = e^{2\pi i\tau_{11}}$, $q_2 = e^{2\pi i\tau_{22}}$ and $r = e^{2\pi i\tau_{12}}$)

$$\chi_{6,8}(\tau) = \begin{pmatrix} 0 \\ 0 \\ r^{-1}-2+r \\ 2(r-r^{-1}) \\ r^{-1}-2+r \\ 0 \\ 0 \end{pmatrix} q_1q_2 + \begin{pmatrix} 0 \\ 0 \\ -2(r^{-2}+8r^{-1}-18+8r+r^2) \\ 8(r^{-2}+4r^{-1}-4r-r^2) \\ -2(7r^{-2}-4r^{-1}-6-4r+7r^2) \\ 12(r^{-2}-2r^{-1}+2r-r^2) \\ -4(r^{-2}-4r^{-1}+6-4r+r^2) \end{pmatrix} q_1q_2^2$$

$$+ \begin{pmatrix} -4(r^{-2}-4r^{-1}+6-4r+r^2) \\ 12(r^{-2}-2r^{-1}+2r-r^2) \\ -2(7r^{-2}-4r^{-1}-6-4r+7r^2) \\ 8(r^{-2}+4r^{-1}-4r-r^2) \\ -2(r^{-2}+8r^{-1}-18+8r+r^2) \\ 0 \\ 0 \end{pmatrix} q_1^2q_2 + \begin{pmatrix} 16(r^{-3}-9r^{-1}+16-9r+r^3) \\ -72(r^{-3}-3r^{-1}+3r-r^3) \\ +128(r^{-3}-2+r^3) \\ -144(r^{-3}+5r^{-1}-5r-r^3) \\ +128(r^{-3}-2+r^3) \\ -72(r^{-3}-3r^{-1}+3r-r^3) \\ 16(r^{-3}-9r^{-1}+16-9r+r^3) \end{pmatrix} q_1^2q_2^2 + \dots$$

Proposition 6.1 provides an extremely effective way of constructing complex vector-valued Siegel modular forms of degree 2. Let us give a few examples. In the decomposition

$$\mathrm{Sym}^2(\mathrm{Sym}^6(V)) = V[12, 0] \oplus V[10, 2] \oplus V[8, 4] \oplus V[6, 6]$$

of $\mathrm{Sym}^2(\mathrm{Sym}^6(V))$ the covariant H defined by $V[10, 2]$ is the Hessian and by Corollary 5.4 gives rise to a form $\chi_{8,8} = \nu(H)\chi_{10} \in S_{8,8}(\Gamma_2)$ and using the Fourier expansion of $\chi_{6,8}$ we obtain the Fourier expansion of $\chi_{8,8}$. Similarly, the covariant corresponding to $V[8, 4]$ gives a form $\chi_{4,10}$ after multiplication with χ_{10}. Finally, the covariant defined by $V[6, 6]$ is the invariant A and defines the cusp form $\chi_{12} = \nu(A)\chi_{10}$. We refer to [7] for more details.

As an illustration of this we refer to the website [1] that gives the Fourier series for generators for all cases where $\dim S_{j,k}(\Gamma_2) = 1$.

Another illustration of the efficacity of the construction of modular forms appears when one considers the modules $\oplus_k M_{j,k}(\Gamma_2)$ and $\oplus_k M_{j,k}(\Gamma_2, \epsilon)$. Let R_2^{ev} be the ring of scalar-valued modular forms of even weight. The structure of the R_2^{ev}-modules

$$\oplus_k M_{j,2k}(\Gamma_2), \qquad \oplus_k M_{j,2k+1}(\Gamma_2)$$

has been determined for $j = 2, 4, 6, 8, 10$ by Satoh, Ibukiyama, Kiyuna, van Dorp and Takemori using various methods. Using covariants one can uniformly treat these cases and the cases of modular forms with character for the same values of j

$$\oplus_k M_{j,2k}(\Gamma_2, \epsilon), \qquad \oplus_k M_{j,2k+1}(\Gamma_2, \epsilon)\,.$$

For example, the R_2-module $\oplus_k M_{2,2k+1}(\Gamma_2, \epsilon)$ is free with generators of weight $(2, 9)$, $(2, 11)$, and $(2, 17)$ and the module $\oplus_k M_{10}^{10,2k}(\Gamma_2, \epsilon)$ is free with 10 generators. We refer to [8].

Yet another application of the construction of modular forms via covariants deals with small weights. It is known by Skoruppa [27] that $\dim S_{j,1}(\Gamma_2) = 0$. He proved this using Fourier-Jacobi forms. We conjecture $\dim S_{j,2}(\Gamma_2) = 0$ and proved this for $j \leq 52$ using covariants. We refer to [5].

As a final illustration, for $k = 3$ the smallest j such that $\dim S_{j,3}(\Gamma_2) \neq 0$ is 36. It is not difficult to construct a generator of $S_{36,3}(\Gamma_2)$ using covariants, see [5].

7 Rings of Scalar-Valued Modular Forms

The approach explained in the preceding section makes it easy to find generators for the rings $R_2(F) = \oplus_k M_k(\Gamma_2)(F)$ of modular forms of degree 2 for $F = \mathbb{C}$ or $F = \mathbb{F}_p$. We write ν_F for the map $I(2,6)(F) \to R_2(F)[1/\chi_{10}]$. We denote by $R_2^{\mathrm{ev}}(F)$ the subring of even weight modular forms.

The degree 2 invariant A of a binary sextic $f = \sum_{i=0}^{6} a_i x_1^{6-i} x_2^i$ can be written as

$$120\, a_0 a_6 - 20\, a_1 a_5 + 8\, a_2 a_4 - 3\, a_3^2 \,.$$

Corollary 5.4 implies that $\nu_F(A)$ cannot be regular for $F = \mathbb{C}$ or $\mathbb{F}_p$ with $p \geq 5$, but also that $\nu_F(AD)$ is a cusp form $\chi_{12} \in S_{12}(\Gamma_2)(F)$ of weight 12.

In degree 4 there is the invariant B given by

$$(81\, a_0 a_6 + 9\, a_1 a_5) a_3^2 - 3\,(15\, a_0 a_4 a_5 + 15\, a_1 a_2 a_6 + a_1 a_4^2 + a_2^2 a_5) a_3 + \cdots + a_2^2 a_4^2$$

and Corollary 5.4 implies that it defines a regular modular form $\psi_4 = \nu_F(B)$ of weight 4.

The invariant C of degree 6 is given by

$$18\,(9\, a_0 a_6 + 4\, a_1 a_5) a_3^4 - 6\,(33\, a_0 a_4 a_5 + 33\, a_1 a_2 a_6 + 4\, a_1 a_4^2 + 4 a_2^2 a_5) a_3^3 + \cdots$$

and in a similar way one sees that $AB - 3\,C$ starts with

$$1458\, a_0 a_6 a_3^4 - 486\,(a_0 a_4 a_5 + a_1 a_2 a_6) a_3^3 + \cdots$$

and defines a regular modular form $\psi_6 = \nu_F(AB - 3C)$ of weight 6.

The discriminant D starts as

$$729\, a_0^2 a_6^2\, a_3^6 - 54(9\, a_0^2 a_4 a_5 a_6 - 2\, a_0^2 a_5^3 + 9\, a_0 a_1 a_2 a_6^2 - 2\, a_1^3 a_6^2) a_3^5 + \cdots$$

and is seen to have order 2 along $\mathcal{A}_{1,1}$. It defines a cusp form that is a non-zero multiple of χ_{10}.

Proposition 7.1 *For $F = \mathbb{C}$ or $F = \mathbb{F}_p$ with $p \geq 5$ the modular forms ψ_4, ψ_6, χ_{10} and χ_{12} generate $R_2^{\mathrm{ev}}(F)$.*

Proof The algebraic independence of A, B, C, D shows that the generators are algebraically independent. Therefore $\psi_4, \psi_6, \chi_{10}, \chi_{12}$ generate a graded subring $T(F) \subseteq R_2^{\mathrm{ev}}(F)$ such that for even k we have

$$\dim T_k(F) = \frac{k^3}{17280} + O(k^2)\,.$$

Now by Riemann-Roch we have for even k

$$\dim M_k(\Gamma_2)(F) = \frac{c_1(L)^3}{3!}k^3 + O(k^2)$$

since $c_1(L)^3 = 1/2880$, [32, p. 72]. Therefore there cannot be more generators. Note that $4 \cdot 6 \cdot 10 \cdot 12 = 2880$. □

Remark 7.2 Restriction to $\mathcal{A}_{1,1}$ shows that $\psi_4, \psi_6, \chi_{10}, \chi_{12}$ generate $M_k(\Gamma_2)(F)$ for $k \leq 12$. Let $d(k) = \dim_F M_k(\Gamma_2)(F)$ and $t(k) = \dim_F T_k(F)$. Then $t(k) \leq d(k)$ and for even k the exact sequence (5) yields

$$d(k) \leq d(k-10) + \frac{c(k)(c(k)+1)}{2}$$

with $c(k) = \dim_F M_k(\Gamma_1)(F)$. Now one easily sees $t(k) - t(k-10) = c(k)(c(k)+1)/2$. Thus if we assume $d(k-10) = t(k-10)$ we get

$$t(k) \leq d(k) \leq d(k-10) + \frac{c(k)(c(k)+1)}{2} = t(k)$$

and this provides via induction another proof that $\psi_4, \psi_6, \chi_{10}$ and χ_{12} generate $R_2^{\mathrm{ev}}(F)$ for $F = \mathbb{C}$ or $\mathbb{F}_p$ with $p \geq 5$.

The odd degree invariant E (of degree 15) of binary sextics starts with

$$-729(a_0^2a_5^3 - a_1^3a_6)a_3^{10} + \ldots$$

and one checks that it has order -3 along $\mathcal{A}_{1,1}$. So $\chi_{10}^2\nu_F(E)$ defines a regular cusp form $\chi_{35} \in S_{35}(\Gamma_2)(F)$ with order 1 along $\mathcal{A}_{1,1}$.

Let now $\mathrm{char}(F) \neq 2$. The locus in $\mathcal{A}_2 \otimes F$ of principally polarized abelian surfaces X with $\mathrm{Aut}(X)$ containing $\mathbb{Z}/2\mathbb{Z} \times \mathbb{Z}/2\mathbb{Z}$ consists of two irreducible divisors $H_1 = \mathcal{A}_{1,1}$ and H_4, the Humbert surface of degree 4 of abelian surfaces isogenous with a product by an isogeny of degree 4. In terms of moduli of curves, H_4 is the locus of curves that are double covers of elliptic curves. We know that the cycle class of $H_1 + H_4$ is $35\lambda_1$ in $\mathrm{Pic}_{\mathbb{Q}}(\mathcal{A}_2)$, see [31, p. 218].

Lemma 7.3 *Suppose that* $\mathrm{char}(F) \neq 2$. *A modular form* $f \in M_k(\Gamma_2)(F)$ *with* k *odd vanishes on* H_1 *and* H_4.

Proof An abelian surface $[X] \in H_1$ or $[X] \in H_4$ possesses an involution that acts by -1 on $H^0(X, \Omega_X^2)$. □

Corollary 7.4 *The form* χ_{35} *as a section of* $L^{\otimes 35}$ *has as divisor* $H_1 + H_4 + D$ *with* D *the divisor at infinity.*

We can now easily derive the results of Igusa and Nagaoka (see [20, 24], and also [19]).

Theorem 7.5 *Let* $F = \mathbb{C}$ *or* $F = \mathbb{F}_p$ *with* $p \geq 5$. *Then the ring* $R_2(\mathbb{F}_p)$ *is generated over* $R_2^{\mathrm{ev}}(\mathbb{F}_p) = F[\psi_4, \psi_6, \chi_{10}, \chi_{12}]$ *by the cusp form* χ_{35} *of weight* 35 *with* $\chi_{35}^2 \in R_2^{\mathrm{ev}}(F)$.

Proof Any odd weight modular form vanishes on H_1 and H_4, hence is divisible by χ_{35}. □

Remark 7.6 The same argument proves Theorem 7.5 for any commutative ring F in which 6 is invertible. It can also be used to obtain Igusa's result on the ring $R_2(\mathbb{Z})$.

Now positive characteristic sometimes allows more modular forms than characteristic zero. We know that the locus in $\mathcal{A}_g \otimes \mathbb{F}_p$ of abelian varieties of p-rank $< g$ has cycle class $(p-1)\lambda_1$, [13, 32]. This implies that there is a non-zero modular form of weight $p-1$ in characteristic p. This modular form is called the Hasse invariant of degree g and weight $p-1$. The image of the Hasse invariant of degree g under the Siegel operator is the Hasse invariant of degree $g-1$.

The Hasse invariants for degree 1 and characteristic 2 and 3 appear as the generators a_1 and b_2 in

$$R_1(\mathbb{F}_2) = \mathbb{F}_2[a_1, \Delta], \quad R_1(\mathbb{F}_3) = \mathbb{F}_3[b_2, \Delta].$$

The degree 2 invariant A of binary sextics reduces to $a_1a_5 - a_2a_4$ modulo 3 and in view of Conclusion 5.4 defines a form $\nu_{\mathbb{F}_3}(A) \in M_2(\Gamma_2)(\mathbb{F}_3)$ and it must agree with the Hasse invariant (up to a non-zero multiplicative scalar) as there is only one invariant of degree 2 (up to multiplicative scalars). A careful analysis of the invariants in characteristic 3 leads to the description of the ring $R_2(\mathbb{F}_3)$ given in [33].

Theorem 7.7 *The subring* $\mathcal{R}_2^{\mathrm{ev}}(\mathbb{F}_3)$ *of modular forms of even weight is generated by forms of weights* 2, 10, 12, 14, *and* 36 *and has the form*

$$\mathcal{R}_2^{\text{ev}}(\mathbb{F}_3) = \mathbb{F}_3[\psi_2, \chi_{10}, \psi_{12}, \chi_{14}, \chi_{36}]/J$$

with J the ideal generated by the relation $\psi_2^3\chi_{36} - \chi_{10}^3\psi_{12} - \psi_2^2\chi_{10}\chi_{14}^2 + \chi_{14}^3$. Moreover, $\mathcal{R}_2(\mathbb{F}_3)$ is generated over $\mathcal{R}_2^{\text{ev}}$ by a form χ_{35} of weight 35 whose square lies in $R_2^{\text{ev}}(\mathbb{F}_3)$. The ideal of cusp forms is generated by $\chi_{10}, \chi_{14}, \chi_{35}, \chi_{36}$.

The case of characteristic 2 was treated in joint work with Cléry in [5]. In the case of characteristic 2 a curve of genus 2 is not described by a binary sextic. Instead we find an equation

$$y^2 + a\,y + b = 0$$

with a (resp. b) in $k[x]$ of degree ≤ 3 (resp. ≤ 6) and the hyperelliptic involution is $y \mapsto y + a$. It comes with a basis $xdx/a, dx/a$ of regular differentials. In this case we look at pairs $(a, b) \in V_{3,-1} \times V_{6,-2}$ with $V_{n,m} = \text{Sym}^n(V) \otimes \det(V)^m$. Let $\mathcal{V}^0 \subset V_{3,-1} \times V_{6,-2}$ be the open subset defining smooth hyperelliptic curves. Now we have an action of GL_2 and an action of $\text{Sym}^3(V)$ via

$$(a, b) \mapsto (a, b + v^2 + va)$$

Together this defines a stack quotient

$$[\mathcal{V}^0/\text{GL}_2 \ltimes V_{3,-1}]$$

Now by an invariant we mean a polynomial in the coefficients $a_0, \ldots, a_3$ and $b_0, \ldots, b_6$ that is invariant under $\text{SL}(V) \ltimes \text{Sym}^3(V)$. Let $\mathcal{K}$ be the ring of invariants. A first example is the square root of the discriminant of a:

$$K_1 = a_0a_3 + a_1a_2\,.$$

As an analog of (5) we now get homomorphisms

$$R_2(\mathbb{F}_2) \hookrightarrow \mathcal{K} \xrightarrow{\nu} R_2(\mathbb{F}_2)[1/\chi_{10}]$$

the composition of which is the identity.

In order to construct characteristic 2 invariants one can still use binary sextics as Igusa suggested in [20]. Indeed, one lifts the curve given by $y^2 + ay + b = 0$ to the Witt ring, say defined by $y^2 + \tilde{a}y + \tilde{b} = 0$ and takes an invariant of the binary sextic given by $\tilde{a}^2 + 4\tilde{b}$, then divides these by the appropriate power of 2 and reduces modulo 2.

For example, the degree 2 invariant of binary sextics yields in this way an invariant K_2 that equals K_1^2. A degree 4 invariant yields an invariant K_4 that turns out to be divisible by K_1. We thus find an invariant K_3 of degree 3.

The Hasse invariant ψ_1 must map to K_1. As in characteristic 3 a careful analysis gives the orders of a_i and b_i along $\mathcal{A}_{1,1}$ and we can deduce for an invariant K the order of $\nu(K)$ along $\mathcal{A}_{1,1}$. The ring $R_2(\mathbb{F}_2)$ was described in [5].

Theorem 7.8 *The ring $\mathcal{R}_2(\mathbb{F}_2)$ is generated by modular forms of weights* 1, 10, 12, 13, *and* 48 *satisfying one relation of weight* 52*:*

$$\mathcal{R}_2(\mathbb{F}_2) = \mathbb{F}_2[\psi_1, \chi_{10}, \psi_{12}, \chi_{13}, \chi_{48}]/(R)$$

with $R = \chi_{13}^4 + \psi_1^3 \chi_{10} \chi_{13}^3 + \psi_1^4 \chi_{48} + \chi_{10}^4 \psi_{12}$. *The ideal of cusp forms is generated by* χ_{10}, χ_{13} *and* χ_{48}.

8 Moduli of Curves of Genus Three and Invariant Theory of Ternary Quartics

Now we turn to genus 3 treated in [9] and consider the moduli space $\mathcal{M}_3^{\rm nh}$ of non-hyperelliptic curves of genus 3 over a field F. This is an open part of the moduli space $\mathcal{M}_3$ with as complement the divisor $\mathcal{H}_3$ of hyperelliptic curves. Let now $V = \langle x_0, x_1, x_2 \rangle$ be the 3-dimensional F-vector space with basis x_0, x_1, x_2. We let $V_{4,0,-1}$ be the irreducible representation $\mathrm{Sym}^4(V) \otimes \det(V)^{-1}$. The underlying space is the space of ternary quartics. It contains the open subset $V_{4,0,-1}^0$ of ternary quartics with non-vanishing discriminant; that is, the ternary quartics that define smooth plane quartic curves.

It is known that $\mathcal{M}_3^{\rm nh}$ has a description as stack quotient

$$\mathcal{M}_3^{\rm nh} \xrightarrow{\sim} [V_{4,0,-1}^0/\mathrm{GL}_3]$$

Indeed, if C is a non-hyperelliptic curve of genus 3 then a choice of basis of $H^0(C, K)$ defines an embedding of C into $\mathbb{P}^2$ and the image satisfies an equation $f(x_0, x_1, x_2) = 0$ with f homogeneous of degree 4. In order that the action on the space of differentials with basis

$$x_i\,(x_0 dx_1 - x_1 dx_0)/(\partial f/\partial x_2), \qquad i = 0, 1, 2$$

is the standard representation V we need to twist $\mathrm{Sym}^4(V)$ by $\det(V)^{-1}$. Then $\lambda \mathrm{Id} \in \mathrm{GL}_3(F)$ acts by λ on $V_{4,0,-1}$ and we arrive at the familiar stack quotient $[Q/\mathrm{PGL}_3]$ with Q the space of smooth projective curves of degree 4 in $\mathbb{P}^2$ by first dividing by the multiplicative group of multiples of the diagonal.

Conclusion 8.1 The pull back of the Hodge bundle $\mathbb{E}$ on $\mathcal{M}_3^{\rm nh}$ under

$$V_{4,0,-1}^0 \to [V_{4,0,-1}^0]/\mathrm{GL}_2] \xrightarrow{\sim} \mathcal{M}_3^{\rm nh}$$

is the equivariant bundle V.

Therefore we now look at the invariant theory of GL_3 acting on ternary quartics $\mathrm{Sym}^4(V)$ with $V = \langle x, y, z \rangle$ the standard representation of $\mathrm{GL}_3(V)$. We write the universal ternary quartic f as

$$f = a_0x^4 + a_1x^3y + \cdots + a_{14}z^4$$

in a lexicographic way. We fix coordinates for $\wedge^2 V$

$$\hat{x} = y \wedge z,\ \hat{y} = z \wedge x,\ \hat{z} = x \wedge y\,.$$

Recall that an irreducible representation ρ of GL_3 is determined by its highest weight $(\rho_1 \geq \rho_2 \geq \rho_3)$. This representation appears in

$$\mathrm{Sym}^{\rho_1-\rho_2}(V) \otimes \mathrm{Sym}^{\rho_2-\rho_3}(\wedge^2 V) \otimes \det(V)^3$$

An invariant for the action of GL_3 on $\mathrm{Sym}^4(V)$ is a polynomial in $a_0, \ldots, a_{14}$ invariant under SL_3. Instead of the notion of covariant we consider here the notion of a concomitant. A concomitant is a polynomial in $a_0, \ldots, a_{14}$ and in x, y, z and $\hat{x}, \hat{y}, \hat{z}$ that is invariant under the action of SL_3. The most basic example is the universal ternary quartic f.

Concomitants can be obtained as follows. One takes an equivariant map of GL_3-representations

$$U \hookrightarrow \mathrm{Sym}^d(\mathrm{Sym}^4(V))$$

or equivalently the equivariant embedding

$$\varphi : \mathbb{C} \longrightarrow \mathrm{Sym}^d(\mathrm{Sym}^4(V)) \otimes U^\vee$$

Then $\Phi = \varphi(1)$ is a concomitant. If U is an irreducible representation of highest weight $\rho_1 \geq \rho_2 \geq \rho_3$ then Φ is of degree d in $a_0, \ldots, a_{14}$, of degree $\rho_1 - \rho_2$ in x, y, z and degree $\rho_2 - \rho_3$ in $\hat{x}, \hat{y}, \hat{z}$.

The invariants form a ring $I(3, 4)$ and the concomitants $\mathcal{C}(3, 4)$ form a module over $I(3, 4)$. For more on the ring $I(3, 4)$ see [11].

9 Concomitants of Ternary Quartics and Modular Forms of Degree 3

The starting point for the construction of modular forms of degree 3 is the Torelli morphism

$$t : \mathcal{M}_3 \to \mathcal{A}_3$$

defined by associating to a curve of genus 3 its Jacobian. This is a morphism of Deligne-Mumford stacks of degree 2 ramified along the hyperelliptic locus $\mathcal{H}_3$. Indeed, every abelian variety has an automorphism of order 2, but a generic curve of genus 3 does not have non-trivial automorphisms. Hyperelliptic curves have an automorphism of order 2 that induces -1_{Jac} on the Jacobian.

There is a Siegel modular form $\chi_{18} \in S_{18}(\Gamma_3)$ constructed by Igusa [21]. It is defined as the product of the 36 even theta constants of order 2. The divisor of χ_{18} in the standard compactification (defined by the second Voronoi fan) $\tilde{\mathcal{A}}_3$ is

$$\mathcal{H}_3 + 2D$$

with D the divisor at infinity.

The pullback under the Torelli morphism of the Hodge bundle $\mathbb{E}$ on $\mathcal{A}_3$ is the Hodge bundle of $\mathcal{M}_3$. The Hodge bundle on $\mathcal{M}_3$ extends to the Hodge bundle over $\overline{\mathcal{M}}_3$, denoted again by $\mathbb{E}$. For each irreducible representation ρ of GL_3 have a bundle $\mathbb{E}_\rho$ on $\overline{\mathcal{M}}_3$ constructed by applying a Schur functor. We thus can consider

$$T_\rho = H^0(\overline{\mathcal{M}}_3, \mathbb{E}_\rho)$$

and elements of it are called Teichmüller modular forms of weight ρ and genus (or degree) 3. There is an involution ι acting on the stack $\mathcal{M}_3$ associated to the double cover $\mathcal{M}_3 \to \mathcal{A}_3$. If the characteristic is not 2 we can thus split T_ρ into ± 1-eigenspaces under ι

$$T_\rho = T_\rho^+ \oplus T_\rho^- \, .$$

We can identify the invariants under ι with Siegel modular forms

$$T_\rho^+ = M_\rho(\Gamma_3) \tag{6}$$

while the space T_ρ^- consists of the genuine Teichmüller modular forms.

The pullback of χ_{18} to $\mathcal{M}_3$ is a square χ_9^2 with χ_9 a Teichmüller modular form of weight 9 constructed by Ichikawa [17, 18].

Using the identification (6) we have

$$\chi_9 \, T_\rho^- \subset S_{\rho'}(\Gamma_3) \quad \text{with } \rho' = \rho \otimes \det^9 .$$

We will now use the invariant theory of ternary quartics Conclusion 8.1 implies that the pullback of a scalar-valued Teichmüller modular form of weight k is an invariant of weight $3k$ in $I(3,4)$. An invariant of degree $3d$ defines a meromorphic Teichmüller modular form of weight d on $\overline{\mathcal{M}}_3$ that becomes holomorphic after multiplication by an appropriate power of χ_9. Indeed, an invariant of degree $3d$ is defined by an equivariant embedding $\det(V)^{4d} \hookrightarrow \mathrm{Sym}^{3d}(\mathrm{Sym}^4(V))$ or taking care of the necessary twisting by

$$\det(V)^d \hookrightarrow \mathrm{Sym}^{3d}(\mathrm{Sym}^4(V)) \otimes \det(V)^{-3d} \, .$$

We thus get

$$T \longrightarrow I(3,4) \longrightarrow T[1/\chi_9] \, ,$$

where the composition of the arrows is the identity. In particular, the Teichmüller modular form χ_9 maps to an invariant of degree 27 and since it is a cusp form one can check that it must be divisible by the discriminant, hence is a multiple of the discriminant.

We can extend this to vector-valued Teichmüller modular forms

$$\Sigma \longrightarrow \mathcal{C}(3,4) \xrightarrow{\nu} \Sigma[1/\chi_9]$$

with the T-module Σ defined as

$$\Sigma = \oplus_\rho T_\rho$$

with ρ running through the irreducible representations of GL_3.

We can ask what the image $\nu(f)$ of the universal ternary quartic is. By construction it is a meromorphic modular form of weight $(4, 0, -1)$. Here the weight refers to the irreducible representation $\mathrm{Sym}^4(V) \otimes \det(V)^{-1}$ of GL_3.

We know that there exists a holomorphic modular cusp form $\chi_{4,0,8}$ of weight $(4, 0, 8)$, see [4] and below.

Proposition 9.1 *Over $\mathbb{C}$ the Siegel modular modular form $\chi_9\, \nu(f)$ is a generator of $S_{4,0,8}(\Gamma_3)(\mathbb{C})$.*

Proof The cusp form $\chi_{4,0,8}$ maps to a concomitant of degree 28 that is divisible by the discriminant. Therefore, $\chi_{4,0,-1} = \chi_{4,0,8}/\chi_9$ corresponds to a concomitant of degree 1. This must be a non-zero multiple of f. □

If we write the universal ternary quartic lexicographically as

$$f = a_0 x^4 + a_1 x^3 y + \cdots + a_{14} z^4$$

and we write the meromorphic Teichmüller form $\chi_{4,0,-1}$ similarly lexicographically as

$$\chi_{4,0,-1} = \alpha_0 X^4 + \alpha_1 X^3 Y + \cdots + \alpha_{14} Z^4$$

with dummy variables X, Y, Z to indicate the coordinates of $\chi_{4,0,-1}$, we arrive at the analog for degree 3:

Proposition 9.2 *The map $\nu : \mathcal{C}(3,4) \to T[1/\chi_9]$ is given by substituting α_i for a_i (and X, Y, Z for x, y, z and $\hat{X}, \hat{Y}, \hat{Z}$ for $\hat{x}, \hat{y}, \hat{z}$).*

In the following, we restrict to $F = \mathbb{C}$. One way to construct a generator of $S_{4,0,8}(\Gamma_3)(\mathbb{C})$ is to take the Schottky form of degree 4 and weight 8 that vanishes on the Torelli locus. We can develop it along $\mathcal{A}_{3,1}$, the locus in $\mathcal{A}_4$ of products of abelian threefolds and elliptic curves. It restriction to $\mathcal{A}_{3,1}$ is a form in $S_8(\Gamma_3) \otimes S_8(\Gamma_1)$ and thus vanishes. The first non-zero term in the Taylor expansion along $\mathcal{A}_{3,1}$ is

$$\chi_{4,0,8} \otimes \Delta \in S_{4,0,8}(\Gamma_3) \otimes S_{12}(\Gamma_1)$$

Since the Schottky form can be constructed explicitly with theta functions we can easily obtain the beginning of the Fourier expansion. We refer to [4] for the details.

In [9] we formulated a criterion that tells us which elements of $\mathcal{C}(3, 4)$ will give holomorphic modular forms. We can associate to a concomitant its order along the locus of double conics by looking at its order in t when we evaluate it on the ternary quartic $t\,f + q^2$ where q is a sufficiently general quadratic form in x, y, z. Then the result is the following, see [9].

Theorem 9.3 *Let c be a concomitant of degree d and $v(c)$ its order along the locus of double conics. If d is odd then $\nu(c)\chi_9$ is a Siegel modular form with order $v(c) - (d-1)/2$ along the hyperelliptic locus. If d is even, then the order of $\nu(c)$ is $v(c) - d/2$.*

We formulate a corollary. Let $M_{i,j,k}(\Gamma_3)^{(m)}$ be the space of Siegel modular forms of weight (i, j, k) vanishing with multiplicity $\geq m$ at infinity. (The weight (i, j, k) corresponds to the irreducible representation of GL_3 of highest weight $(i + j + k, j + k, k)$.) Moreover, let $\mathcal{C}_{d,\rho}(-m\,DC)$ be the vector space of concomitants of type (d, ρ) that have order $\geq m$ along the locus of double conics. (Type (d, ρ) means belonging to an irreducible representation U of highest weight ρ occurring in $\mathrm{Sym}^d(\mathrm{Sym}^4(V))$.)

Corollary 9.4 *The exists an isomorphism*

$$\mathcal{C}_{d,\rho}(-m\,DC) \xrightarrow{\sim} M^{(d-2m)}_{\rho_1-\rho_2,\rho_2-\rho_3,\rho_3+9(d-2m)}$$

given by $c \mapsto \nu(c)\chi_9^{d-2m}$.

This allows now the construction of Siegel modular forms and Teichmüller modular forms of degree 3. In fact, in principle, all of them. As a simple example, we decompose

$$\mathrm{Sym}^2(\mathrm{Sym}^4(V)) = V[8, 0, 0] + V[6, 2, 0] + V[4, 4, 0]\,.$$

The concomitant corresponding to $U = V[8, 0, 0]$ yields via ν the symmetric square of $\nu(f)$. The concomitant corresponding to $V[6, 2, 0]$ yields a form in that after multiplication by χ_{18} becomes a holomorphic form in $S_{4,2,16}$ vanishing with multiplicity 2 at infinity. Similarly, the concomitant c corresponding to $U = V[4, 4, 0]$ yields a cusp form $\nu(c)\chi_{18} \in S_{0,4,16}$ vanishing with multiplicity 2 at infinity. We refer for more examples to [9].

The method also allows to treat the positive characteristic case. We hope to come back to it on another occasion.

References

1. J. Bergström, F. Cléry, C. Faber, G. van der Geer, Siegel Modular Forms of Degree Two and Three (2017). http://smf.compositio.nl
2. S. Böcherer, S. Nagaoka, On mod p properties of Siegel modular forms. Math. Ann. **338**, 421–433 (2007)
3. J. Bruinier, G. van der Geer, G. Harder, D. Zagier, *The 1-2-3 of modular forms* (Universitext, Springer, 2007)
4. F. Cléry, G. van der Geer, Constructing vector-valued Siegel modular forms from scalar-valued Siegel modular forms. Pure Appl. Math. Q. **11**(1), 21–47 (2015)
5. F. Cléry, G. van der Geer, On vector-valued Siegel modular forms of degree 2 and weight $(j, 2)$. Documenta Mathematica **23**, 1129–1156 (2018)
6. F. Cléry, G. van der Geer, Modular forms of degree two and curves of genus two in characteristic two. IMRN **2022**(7), 5204–5218. arXiv:2003.00249
7. F. Cléry, C. Faber, G. van der Geer, Covariants of binary sextics and vector-valued Siegel modular forms of genus 2. Math. Ann. **369**(3–4), 1649–1669 (2017)
8. F. Cléry, C. Faber, G. van der Geer, Covariants of binary sextics and modular forms of degree 2 with character. Math. Comp. **88**(319), 2423–2441 (2019)
9. F. Cléry, C. Faber, G. van der Geer, Concomitants of ternary quartics and vector- valued modular forms of genus three. Selecta Math. **26**, 55 (2020)
10. P. Deligne, Courbes elliptiques: Formulaire (d'après J. Tate), in *Modular Functions IV*, Lecture Notes in Mathematics, vol. 476 (Springer, Berlin, 1975), pp. 53–73
11. J. Dixmier, On the projective invariants of quartic plane curves. Adv. Math. **64**, 279–304 (1987)
12. R. Donagi, R. Smith, The structure of the Prym map. Acta Math. **146**, 25–102 (1981)
13. T. Ekedahl, G. van der Geer, Cycle classes of the E-O stratification on the moduli of abelian varieties, in *Algebra, Arithmetic, and Geometry*, ed. by Y. Tschinkel, Y. Zarhin. Progress in Mathematics, vol. 269 (Birkhäuser, Basel, 2010)
14. G. Faltings, C.-L. Chai, *Degeneration of abelian varieties*. Ergebnisse der Mathematik und ihrer Grenzgebiete (3), vol. 22. (Springer, Berlin, 1990)
15. W. Geyer, Invarianten binärer Formen, in *Classification of Algebraic Varieties and Compact Complex Manifolds*, Lecture Notes in Math, vol. 412 (Springer, Berlin, 1974), pp. 36–69
16. J.H. Grace, A. Young, *The Algebra of Invariants* (Cambridge University Press, Cambridge, 1903)
17. T. Ichikawa, On Teichmüller modular forms. Math. Ann. **299**, 731–740 (1994)
18. T. Ichikawa, Teichmüller modular forms of degree 3. Amer. J. Math. **117**, 1057–1061 (1995)
19. T. Ichikawa, Siegel modular forms of degree 2 over rings. J. Number Theory **129**, 818–823 (2009)
20. J-I. Igusa, Arithmetic variety of moduli for genus two. Ann. Math. **72**, 612–649 (1960)
21. J-I. Igusa, Modular forms and projective invariants. Am. J. Math. **89**, 817–855 (1967)
22. J-I. Igusa, On the ring of modular forms of degree two over $\mathbb{Z}$. Am. J. Math. **101**, 149–183 (1979)
23. R. Lercier, C. Ritzenthaler, Siegel modular forms of degree three and invariants of ternary quartics. Preprint (2019). arXiv:1907.07431
24. S. Nagaoka, Note on mod p Siegel modular forms. I, II. Math. Zeitschrift **235**, 405–420 (2000), Math. Zeitschrift **251**, 821–826 (2005)
25. D. Petersen, Cohomology of local systems on the moduli of principally polarized abelian surfaces. Pac. J. Math. **275**, 39–61 (2015)
26. T. Shioda, On the graded ring of invariants of binary octavics. Am. J. Math. **89**, 1022–1046 (1967)
27. N.-P. Skoruppa, *Über den Zusammenhang zwischen Jacobiformen und Modulformen halbganzen Gewichts*. Inaugural-Dissertation, Bonner Mathematische Schriften 159, Bonn (1984)
28. O. Taïbi, Dimensions of spaces of level one automorphic forms for split classical groups using the trace formula. Ann. Sci. Éc. Norm. Supér. (4) **50**(2), 269–344 (2017)

29. R. Tsushima, An explicit dimension formula for the spaces of generalized automorphic forms with respect to Sp(2, $\mathbb{Z}$). Proc. Japan Acad. Ser. A Math. Sci. **59**(4), 139–142 (1983)
30. S. Tsuyumine, On Siegel modular forms of degree three. Amer. J. Math. **108**, 755–862 (1986). Addendum to "On Siegel modular forms of degree three." Amer. J. Math. **108**, 1001–1003 (1986)
31. G. van der Geer, *Hilbert Modular Surfaces* (Springer, 1987)
32. G. van der Geer, Cycles on the moduli space of abelian varieties, in *Moduli of Curves and abelian varieties*, Aspects Math., vol. E33 (Vieweg, Braunschweig, 1999)
33. G. van der Geer, The ring of modular forms of degree two in characteristic three. Mathematische Zeitschrift. https://doi.org/10.1007/s00209-020-02621-6

On Intrinsic Negative Curves

Antonio Laface and Luca Ugaglia

Abstract Let $\mathbb{K}$ be an algebraically closed field of characteristic 0. A curve of $(\mathbb{K}^*)^2$ arising from a Laurent polynomial in two variables is *intrinsic negative* if its tropical compactification has negative self-intersection. The aim of this note is to start a systematic study of these curves and to relate them with the problem of computing Seshadri constants of toric surfaces.

Keywords Toric surfaces · Seshadri constants

2010 Mathematics Subject Classification: 14M25 · 14C20

1 Introduction

Following the work of González Anaya, González, Karu [16], Kurano [21] and Kurano Matsuoka [22], we define a class of curves on the blowing-up of toric surfaces at a general point. Let f be a Laurent polynomial in two variables and let $\Gamma \subseteq (\mathbb{K}^*)^2$ be its zero locus. The normal fan to the Newton polygon Δ of f defines a toric variety $\mathbb{P}$ such that the compactification of Γ is contained in the smooth locus of $\mathbb{P}$. Such a compactification is called *tropical*; see [28]. Denote by $X := \mathrm{Bl}_e\mathbb{P}$ the blowing-up of $\mathbb{P}$ at the image e of $(1, 1)$ and let C be the strict transform of the compactified curve. We say that C is an *intrinsic negative curve* (resp. *non-positive*) if $C^2 < 0$ (resp. $C^2 \leq 0$); cfr. [21, Definition 3.1]. Our first result is the construction of infinite

Both authors have been partially supported by Proyecto FONDECYT Regular n. 1190777. The second author is member of INdAM—GNSAGA.

A. Laface (✉)
Departamento de Matemática, Universidad de Concepción, Casilla 160-C, Concepción, Chile
e-mail: alaface@udec.cl

L. Ugaglia
Dipartimento di Matematica e Informatica, Università degli studi di Palermo, Via Archirafi 34, 90123 Palermo, Italy
e-mail: luca.ugaglia@unipa.it

T. Dedieu et al. (eds.), *The Art of Doing Algebraic Geometry*, Trends in Mathematics,
https://doi.org/10.1007/978-3-031-11938-5_10

families of intrinsic non-positive curves. In the following table, $\mathrm{lw}(\Delta)$ is the lattice width of Δ, defined in Sect. 2, while $g(C)$ is the genus of the curve C.

Theorem 1 *There exist infinite families of non-positive intrinsic curves, whose Newton polygons are listed in the following table:*

	vertices of Δ	$\mathrm{lw}(\Delta)$	C^2	$g(C)$
(i)	$\begin{bmatrix} 0 & m & 1 \\ 0 & 1 & m \end{bmatrix}$	$m \geq 2$	-1	0
(ii)	$\begin{bmatrix} 0 & m-3 & m & m-1 & m-2 \\ 0 & 0 & 1 & m & m-1 \end{bmatrix}$	$m \geq 4$	-1	0
(iii)	$\begin{bmatrix} 0 & 0 & 2 & m-4 & m-1 & m & m-1 \\ 0 & 1 & m & m & m-1 & m-2 & m-3 \end{bmatrix}$	$m = 2k \geq 8$	-2	0
(iv)	$\begin{bmatrix} 0 & m-2 & m & m-1 & m-2 \\ 0 & 0 & 1 & m & m-1 \end{bmatrix}$	$m \geq 4$	0	0
(v)	$\begin{bmatrix} 0 & m-4 & m & m-2 & m-3 \\ 0 & 0 & 1 & m & m-1 \end{bmatrix}$	$m = 2k+4 \geq 6$	0	1

Before stating the next result, we recall that given a projective variety $\mathbb{P}$, an ample class H and a point $x \in \mathbb{P}$, the *Seshadri constant* of H at x can be defined as

$$\varepsilon(H, x) := \inf_{x \in C} \frac{H \cdot C}{\mathrm{mult}_x(C)}$$

where the infimum is taken over all irreducible curves through x. The problem of finding Seshadri constants of algebraic surfaces has been widely studied (see, for instance, [4, 14, 15, 26] and the references therein). When $\mathbb{P}$ is a toric surface, there are three possibilities for $x \in \mathbb{P}$: either the point is torus-invariant, or it lies on a torus-invariant curve or it is general. In the first two cases, since the blowing-up $\mathrm{Bl}_x\mathbb{P}$ admits the action of a torus of dimension two and one, respectively, it is possible to describe the effective cone (see [2, 11, 27, Sect. 5.4] for a description of the Cox ring), and hence to compute the Seshadri constant (see [5, Sect. 4] and [20, Sect. 3.2]). Concerning a general point, in [20, Theorem 1.3]) a lower bound for the Seshadri constant is given.

In this note, we focus on the case of a general point. In particular, we prove some relations between the geometry of a lattice polygon Δ and the Seshadri constant $\varepsilon(H_\Delta, e)$, where $(\mathbb{P}_\Delta, H_\Delta)$ is the *toric pair* defined by Δ (see Sect. 2 for the definition) and $e \in \mathbb{P}_\Delta$ is a general point. The recent interest in these Seshadri constants and more generally in the Cox ring of blow-ups of toric varieties at a general point has been motivated by the work of Castravet and Tevelev [9] where the authors prove that the finite generation of the Cox ring of $\overline{M}_{0,n}$ implies that of certain blow-ups of toric varieties at a general point.

In order to state our result, given a non-negative integer m denote by $\mathcal{L}_\Delta(m)$ the linear system of Laurent polynomials whose exponents are integer points of Δ and such that all the partial derivatives up to order $m-1$ vanish at $(1, 1)$. If we denote by $\mathrm{vol}(\Delta)$ the normalized volume of Δ (that is twice its Euclidean area), we have the following (the first inequality is well-known [1, Theorem 0.1] and [20], but we state it anyway for the sake of completeness).

Theorem 2 *Let $\Delta \subseteq \mathbb{Q}^2$ be a lattice polygon, $(\mathbb{P}_\Delta, H_\Delta)$ be the corresponding toric pair and $\varepsilon := \varepsilon(H_\Delta, e)$ be the Seshadri constant at $e \in \mathbb{P}_\Delta$. Then the following hold.*

(i) $\varepsilon \leq \mathrm{lw}(\Delta)$.
(ii) *If* $\mathrm{vol}(\Delta) > \mathrm{lw}(\Delta)^2$ *then* $\varepsilon \in \mathbb{Q}$.
(iii) *If there exists $m \in \mathbb{N}$ such that* $\mathrm{vol}(\Delta) \leq m^2$ *and $\mathcal{L}_\Delta(m) \neq \emptyset$, then $\varepsilon \in \mathbb{Q}$ and* $\varepsilon \leq \mathrm{vol}(\Delta)/m$.
(iv) *If moreover $\mathcal{L}_\Delta(m)$ contains an irreducible curve, then* $\varepsilon = \mathrm{vol}(\Delta)/m$.

We remark that Theorem 2 provides in some cases (like [18, Example 5.7]) an alternative proof for the rationality of the Seshadri constant of a toric surface at a general point. Moreover, it allows to compute the exact value of the Seshadri constant $\varepsilon(H_\Delta, e)$ when (X_Δ, H_Δ) is the toric pair associated with the Newton polygon of an intrinsic non-positive curve.

Corollary 1 *Let $f \in \mathbb{K}[u^{\pm 1}, v^{\pm 1}]$ be a Laurent polynomial with Newton polygon Δ and multiplicity m at $(1, 1)$, such that the corresponding intrinsic curve $C \subseteq X_\Delta$ is non-positive, i.e. $C^2 \leq 0$. Then the Seshadri constant of the ample divisor H_Δ of the toric surface $\mathbb{P}_\Delta$ at a general point $e \in \mathbb{P}_\Delta$ is*

$$\varepsilon = \frac{\mathrm{vol}(\Delta)}{m}.$$

In particular the polygons of the infinite families appearing in Theorem 1 have Seshadri constant $\varepsilon = \mathrm{vol}(\Delta)/\mathrm{lw}(\Delta) \in \mathbb{Q}$.

In order to prove the last statement, we are going to apply Theorem 2(iv), showing that in each case there exists an irreducible curve in $\mathcal{L}_\Delta(m)$, where $m = \mathrm{lw}(\Delta)$ and $\mathrm{vol}(\Delta) \leq m^2$. We remark that for the triangles of type (i) in Theorem 1, the upper bound of Theorem 2(iii) coincides with the lower bound given by [20, Sect. 3.2], so that it is also possible to deduce the exact value of the Seshadri constant without producing the irreducible curve, but for all the other families of polygons appearing in Theorem 1, the two bounds are different (see also Remark 4.2).

The paper is structured as follows. In Sect. 1, after recalling some definitions and results about toric varieties and lattice polytopes we introduce intrinsic curves and prove some preliminary results. In Sect. 2, we consider infinite families of intrinsic curves: we first prove Theorem 1, and then we construct an infinite family of intrinsic negative curves on a given toric surface (Example 3.4). Finally, Sect. 3 is devoted to Seshadri constants on toric surfaces. We first prove Theorem 2 and Corollary 1, and then we discuss some possible applications to the study of the blowing-up of weighted projective planes at a general point.

2 Intrinsic Curves

Let us first recall some definitions and set some notations we are going to use throughout this note.

Let $\Delta \subseteq \mathbb{Q}^n$ be a *lattice polytope*, i.e. a polytope whose vertices have integer coordinates. We recall that given a non-zero primitive vector $v \in \mathbb{Z}^n$, the *lattice width of* Δ *in the direction* v is $\mathrm{lw}_v(\Delta) := \max(\Delta, v) - \min(\Delta, v)$ and the *lattice width of* Δ is $\mathrm{lw}(\Delta) := \min\{\mathrm{lw}_v(\Delta) : v \in \mathbb{Z}^n\}$; see [24, Definition 1.8].

Given a lattice polytope $\Delta \subseteq \mathbb{Q}^n$, we can define a pair $(\mathbb{P}_\Delta, H_\Delta)$ consisting of a toric variety $\mathbb{P}_\Delta$ together with a very ample divisor H_Δ. The toric variety is the normalization of the closure of the image of the following monomial morphism:

$$g_\Delta \colon (\mathbb{K}^*)^n \to \mathbb{P}^{|\Delta\cap\mathbb{Z}^n|-1}, \qquad u \mapsto [u^w : w \in \Delta \cap \mathbb{Z}^n],$$

where $u = (u_1, \dots, u_n) \in (\mathbb{K}^*)^n$. It is possible to show that the action of the torus $(\mathbb{K}^*)^n$ on itself extends to an action on $\mathbb{P}_\Delta$ and that the subset of prime torus-invariant divisors is finite and in bijection with the set of facets of Δ. Let $D_1, \dots, D_r$ be such divisors and $v_1, \dots, v_r$ be the inward normal vectors to the facets of Δ. Each v_i defines a linear form $\mathbb{Q}^n \to \mathbb{Q}$ by $w \mapsto w \cdot v_i$ and the very ample divisor is [18, Proposition 3.1]:

$$H_\Delta := -\sum_{i=1}^{r} \min_{w\in\Delta}\{w \cdot v_i\} D_i. \tag{1}$$

On the other hand, any divisor D on $\mathbb{P}_\Delta$ is equivalent to a combination $\sum_i a_i D_i$, so that we can associate with it the *Riemann-Roch polytope*

$$\Delta_D := \{w \in \mathbb{Q}^n : w \cdot v_i \geq -a_i, \ \forall i = 1, \dots, r\}$$

(see [11, Sect. 4.3]). We remark that if D is very ample, then the toric variety associated with Δ_D coincides with $\mathbb{P}_\Delta$. We recall that if Δ is a very ample polytope [11, Definition 2.2.17], then the closure of the image of g_Δ is a normal variety by [11, Theorem 2.3.1] and thus it coincides with $\mathbb{P}_\Delta$. Moreover, by [11, Corollary 2.2.19], Δ is very ample if $n = 2$.

From now on, we restrict to the case $n = 2$, i.e. $\Delta \subseteq \mathbb{Q}^2$ is a lattice polygon, so that $\mathbb{P}_\Delta$ is a normal toric surface. We will denote by $e \in \mathbb{P}_\Delta$ the image via g_Δ of the neutral element of the torus, by $\pi : X_\Delta \to \mathbb{P}_\Delta$ the blowing-up of $\mathbb{P}_\Delta$ at e and by E the exceptional divisor. Given an $m \in \mathbb{N}$, we will denote by $\mathcal{L}_\Delta(m)$ the sublinear system of $|H_\Delta|$ consisting of sections having multiplicity at least m at e.

Definition 2.1 Let $f \in \mathbb{K}[u^{\pm 1}, v^{\pm 1}]$ be an irreducible Laurent polynomial, $\Delta \subseteq \mathbb{Q}^2$ be the Newton polygon of f, i.e. the convex hull of its exponents and $\Gamma \subseteq \mathbb{P}_\Delta$ be the closure of $V(f) \subseteq (\mathbb{K}^*)^2$. We say that the strict transform $C \subseteq X_\Delta$ of Γ is the *intrinsic curve* defined by f and that C is

- an *intrinsic negative* (resp. *non-positive*) curve if $C^2 < 0$ (resp. $C^2 \leq 0$);

- an *intrinsic* $(-n)$*-curve* if $C^2 = -n < 0$ and $p_a(C) = 0$;
- *expected* in $X_{\Delta'}$, with $\Delta \subseteq \Delta'$ if $|\Delta' \cap \mathbb{Z}^2| > \binom{m+1}{2}$.

In what follows, we will often use the notation $\mathbb{P}$, X and H, omitting the subscript when it is clear from the context.

We remark that, with the notation above, $\Gamma \subseteq \mathbb{P}$ is an element of the very ample linear series $|H|$ and $\Gamma \in \mathcal{L}_\Delta(m)$ if f has a multiplicity of at least m at $(1, 1)$, that is, all the partial derivatives of f up to order $m - 1$ vanish at $(1, 1)$. Moreover, if the multiplicity is m, then the strict transform $C \subseteq X$ of Γ is a Cartier divisor such that

$$C^2 = \mathrm{vol}(\Delta) - m^2 \qquad p_a(C) = \frac{1}{2}\left(\mathrm{vol}(\Delta) - |\partial\Delta \cap \mathbb{Z}^2| + m - m^2\right) + 1. \quad (2)$$

See, for instance, [8, Sect. 4]. By abuse of notation, we will sometimes refer to the Newton polygon Δ as the *Newton polygon of* C, and we will simply say that C is *expected* if it is expected in X_Δ, that is, the linear system $\mathcal{L}_\Delta(m)$ has a non-negative expected dimension.

Our first result is about the characterization of Newton polygons of expected non-positive curves.

Proposition 2.2 *Let C be an intrinsic non-positive expected curve with Newton polygon Δ and multiplicity m at e. Then one of the following holds:*

- $\mathrm{vol}(\Delta) = m^2$ *and* $|\partial\Delta \cap \mathbb{Z}^2| = m$;
- $\mathrm{vol}(\Delta) = m^2$ *and* $|\partial\Delta \cap \mathbb{Z}^2| = m + 2$;
- $\mathrm{vol}(\Delta) = m^2 - 1$ *and* $|\partial\Delta \cap \mathbb{Z}^2| = m + 1$.

In particular, $C^2 \in \{-1, 0\}$ *and* $p_a(C) \in \{0, 1\}$.

Proof Let us denote by $b := |\partial\Delta \cap \mathbb{Z}^2|$ the number of boundary lattice points of Δ and by $i := |\Delta \cap \mathbb{Z}^2| - b$ the number of interior lattice points. Recall that by Pick's formula, $\mathrm{vol}(\Delta) = 2i + b - 2$. Since C is expected and non-positive, we have $|\Delta \cap \mathbb{Z}^2| \geq \binom{m+1}{2} + 1$ and $\mathrm{vol}(\Delta) \leq m^2$. By (2), the non-negativity of the arithmetic genus of C gives $\frac{1}{2}(\mathrm{vol}(\Delta) - b + m - m^2) + 1 \geq 0$. The three inequalities in terms of i and b are $2i + 2b \geq m(m + 1), 2i + b - 2 \leq m^2, 2i - m^2 + m \geq 0$. From these, one deduces that one of the following holds:

$$\begin{cases} b = m \\ i = \frac{m^2-m}{2} + 1 \end{cases} \begin{cases} b = m + 2 \\ i = \frac{m^2-m}{2} \end{cases} \begin{cases} b = m + 1 \\ i = \frac{m^2-m}{2}. \end{cases}$$

Since $C^2 = \mathrm{vol}(\Delta) - m^2$ and $C \cdot K = m - b$, in the first two cases we have $C^2 = 0$ and $p_a(C) = 1$ and 0, respectively, while in the last one $C^2 = -1$ and $p_a(C) = 0$.□

Proposition 2.3 *All the non-equivalent polygons for intrinsic non-positive curves of multiplicity* ≤ 4 *are the following.*

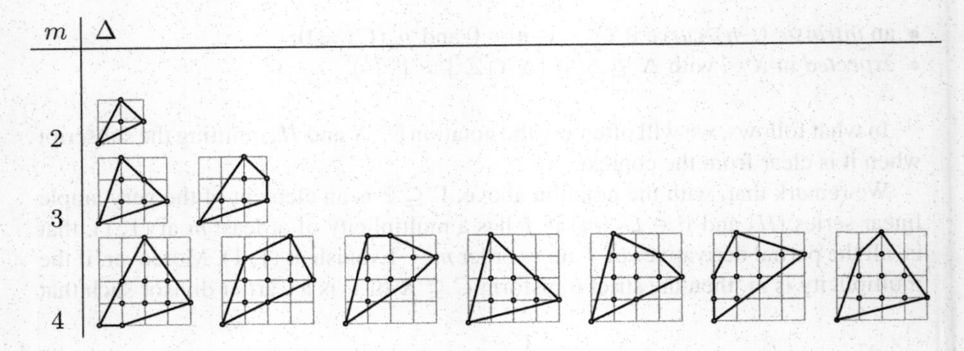

Proof We use the database of polygons with small volume [3] to analyze all the polygons with volume at most 15 and such that $\mathrm{vol}(\Delta) - m^2 \leq 0$. For each such polygon Δ, we compute $\mathcal{L}_\Delta(m)$, where $m \leq 4$, with the aid of the function `FindCurves` of the Magma library:

https://github.com/alaface/non-polyhedral/blob/master/lib.m

We take only the pairs (Δ, m) such that $\mathcal{L}_\Delta(m)$ contains exactly one element, which is irreducible. Finally, we check that in each of these cases the Newton polygon coincides with Δ. □

Remark 2.4 The above intrinsic curves are all expected. In all but the last case, they are intrinsic (-1)-curves; in the last case, the curve has self-intersection 0 and genus 1. The smallest value of m for an unexpected intrinsic negative curve is 5. The lattice polygon Δ is the following:

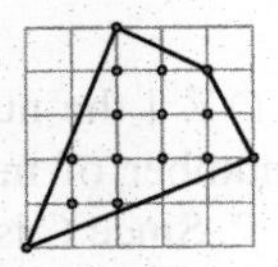

One has $|\partial\Delta \cap \mathbb{Z}^2| = m - 1$ and $|\Delta \cap \mathbb{Z}^2| = \binom{m+1}{2}$ which imply that the corresponding curve has arithmetic genus 1. The curve is defined by the Laurent polynomial

$$\begin{gathered}1 - 8uv + 3uv^2 + 6u^2v^4 - u^2v^5 + 3u^2v + 20u^2v^2 \\ -18u^2v^3 - 18u^3v^2 + 8u^3v^3 + 6u^4v^2 - u^4v^4 - u^5v^2,\end{gathered}$$

which is the unique one whose Newton polygon is contained in Δ and has multiplicity 5 at $(1, 1)$. Its strict transform in X_Δ is smooth of genus 1 and self-intersection -1.

The proof of Proposition 2.3 suggests the following definitions for a pair (Δ, m).

Definition 2.5 Let Δ be a lattice polygon, m a positive integer and set $p_a := \frac{1}{2}(\mathrm{vol}(\Delta) - |\partial\Delta \cap \mathbb{Z}^2| + m - m^2) + 1$. We say that (Δ, m) is

- *numerically negative* (resp. *non-positive*) if $\mathrm{vol}(\Delta) - m^2 < 0$ (resp. $\le$);
- a *$(-n)$-pair* if $\mathrm{vol}(\Delta) - m^2 = -n < 0$ and $p_a = 0$;
- *expected* if $|\Delta \cap \mathbb{Z}^2| > \binom{m+1}{2}$.

Remark 2.6 Clearly, if C is an intrinsic negative curve, then the pair (Δ, m), consisting of the Newton polygon of C and the multiplicity of $\Gamma = \pi(C)$ at e, is numerically negative. On the other hand, if a pair (Δ, m) is numerically negative, in general there does not exist an intrinsic negative curve associated with it. Indeed, first of all it can happen that $\mathcal{L}_\Delta(m)$ is empty (see Example 2.7). Furthermore, even if (Δ, m) is expected (so that $\mathcal{L}_\Delta(m)$ is not empty), in some cases it contains only reducible curves (see Example 2.8).

Example 2.7 Consider the following polygon Δ:

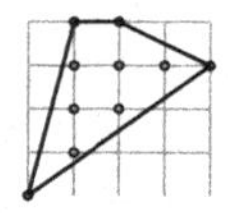

We have that $\mathrm{vol}(\Delta) = 14$ and $|\partial\Delta \cap \mathbb{Z}^2| = 4$, so that the pair $(\Delta, 4)$ is numerically a (-2)-pair. A direct computation shows that $\mathcal{L}_\Delta(4) = \emptyset$, so that there does not exist an intrinsic (-2)-curve with Newton polygon Δ and multiplicity 4.

Example 2.8 The polygon Δ, whose Minkowski decomposition $\Delta = \Delta_1 + \Delta_2$ is given in the below picture

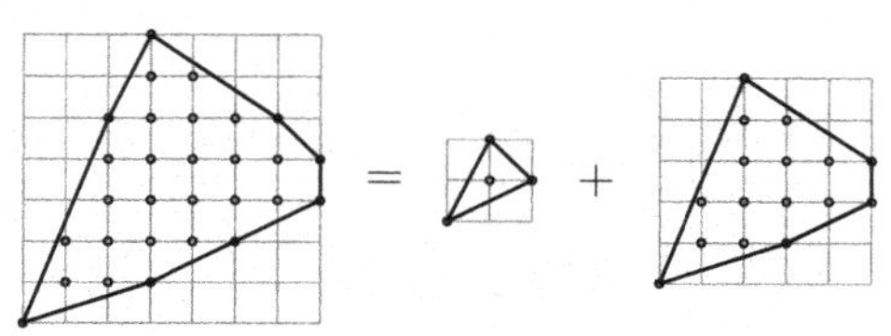

has $\mathrm{vol}(\Delta) = 48$, $|\partial\Delta \cap \mathbb{Z}^2| = 8$ and $|\Delta \cap \mathbb{Z}^2| = \binom{7+1}{2} + 1$, so that $(\Delta, 7)$ is an expected (-1)-pair. The only element in $\mathcal{L}_\Delta(7)$ is defined by

$$(u^2v + uv^2 - 3uv + 1) \cdot (u^5v^3 - 2u^5v^2 - 6u^4v^3 + 11u^4v^2 - 2u^3v^4 + 17u^3v^3 - \\ 24u^3v^2 - u^3v - u^2v^5 + 7u^2v^4 - 22u^2v^3 + 21u^2v^2 + 5u^2v + 4uv^2 - 9uv + 1).$$

The factorization implies the Minkowski decomposition $\Delta = \Delta_1 + \Delta_2$. The polygon Δ_1 corresponds to an intrinsic (-1)-curve C_1, while Δ_2 corresponds to an intrinsic curve C_2 of self-intersection 0 and genus 1. By Proposition 2.12 it follows that $C_1 \cdot C_2 = 0$.

Remark 2.9 Finally, we observe that if the pair (Δ, m) is numerically non-positive and $\mathcal{L}_\Delta(m)$ contains an irreducible curve Γ, then a Laurent polynomial f of $\Gamma \cap (\mathbb{K}^*)^2$ defines an intrinsic non-positive curve. Indeed, by definition the strict transform

$C \subseteq X_\Delta$ of Γ satisfies $C^2 = \mathrm{vol}(\Delta) - m^2 \leq 0$. Moreover, since the Newton polygon Δ' of f is contained in Δ, we also have that $\mathrm{vol}(\Delta') - m^2 \leq \mathrm{vol}(\Delta) - m^2 \leq 0$. We remark that if Δ' is strictly contained in Δ, then the self-intersection of the intrinsic curve is strictly smaller than C^2 (see Example 2.10).

Example 2.10 Let Δ and Δ' be the following polygons, respectively, from left to right.

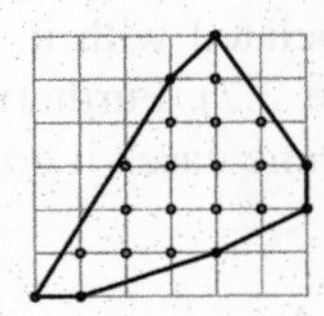 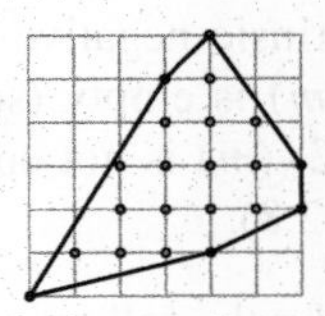

One has $\mathrm{vol}(\Delta) = 35$, $|\partial\Delta \cap \mathbb{Z}^2| = 7$ and $|\Delta \cap \mathbb{Z}^2| = \binom{6+1}{2} + 1$, so that $(\Delta, 6)$ is an expected (-1)-pair. The linear system $\mathcal{L}_\Delta(6)$ contains a unique irreducible curve defined by the following polynomial

$$f := -4u^6v^3 + 3u^6v^2 - 6u^5v^4 + 30u^5v^3 - 18u^5v^2 - u^4v^6 + 2u^4v^5 + 17u^4v^4 - 62u^4v^3 + 25u^4v^2 \\ +4u^4v + 4u^3v^5 - 26u^3v^4 + 50u^3v^3 + 2u^3v^2 - 10u^3v + 6u^2v^3 - 27u^2v^2 + 6u^2v + 6uv - 1.$$

The Newton polygon of f is Δ', since u is the only monomial (corresponding to a lattice point of Δ) that does not appear in f. In particular, $\mathrm{vol}(\Delta') = 34$ and $|\partial\Delta \cap \mathbb{Z}^2| = 6$, so that f defines an intrinsic (unexpected) (-2)-curve.

Definition 2.11 Given two polygons Δ_1, Δ_2 their *mixed volume* is

$$\mathrm{vol}(\Delta_1, \Delta_2) := \frac{1}{2}(\mathrm{vol}(\Delta_1 + \Delta_2) - \mathrm{vol}(\Delta_1) - \mathrm{vol}(\Delta_2)).$$

We conclude the section by showing how the mixed volume of two lattice polygons relates to the intersection product of curves on a toric variety whose fan refines the normal fans of the two polygons.

Proposition 2.12 *Let (Δ_1, m_1), (Δ_2, m_2) be two pairs, each of which consists of a lattice polygon together with a positive integer. Assume that $\mathcal{L}_{\Delta_i}(m_i)$ is non-empty and let $C_i \subseteq X_{\Delta_i}$ be the strict transform of a curve in the linear system. Let X be a surface which admits birational morphisms $\phi_i : X \to X_{\Delta_i}$ for $i = 1, 2$. Then*

$$\phi_1^*(C_1) \cdot \phi_2^*(C_2) = \mathrm{vol}(\Delta_1, \Delta_2) - m_1 m_2.$$

Proof Let $\pi_i : X_{\Delta_i} \to \mathbb{P}_{\Delta_i}$ be the blowing-up at $e \in \mathbb{P}_{\Delta_i}$ with exceptional divisor E_i. Since $C_i \sim \pi_i^* H_i - m_i E_i$, where H_i is a very ample divisor on $\mathbb{P}_{\Delta_i}$, we can assume that the support of the divisor $\pi_i^* H_i - m_i E_i$ does not contain any singular point of X_{Δ_i}. Thus, C_i is a Cartier divisor of X_{Δ_i} and the pullback ϕ_i^* is defined on C_i. The intersection product $\phi_1^*(C_1) \cdot \phi_2^*(C_2)$ does not depend on the surface X because all such surfaces differ by exceptional divisors, which have zero intersection product with the pullbacks of C_1 and C_2. We can then choose $X := X_\Delta$ to be the blowing-up of $\mathbb{P}_\Delta$ at the general point e, where $\Delta := \Delta_1 + \Delta_2$. Since H_i is very ample on $\mathbb{P}_{\Delta_i}$, its pullback is base point free in $\mathbb{P}_\Delta$. By Bertini's theorem, the general elements of these two linear systems intersect transversely at distinct points which, without loss of generality, we can assume to be contained in $(\mathbb{K}^*)^2$. By the Bernstein-Kushnirenko theorem, the number of these intersections is $\mathrm{vol}(\Delta_1, \Delta_2)$, so that the statement follows after taking into account the intersections of the two curves at $e \in \mathbb{P}_\Delta$. □

3 Infinite Families

In this section, we construct infinite families of intrinsic negative curves. First of all, we produce infinite families of toric surfaces, each of which corresponds to an intrinsic negative curve. Then, in Example 3.4, we construct an infinite family of intrinsic negative curves on a given toric surface.

Lemma 3.1 *Let $f_1, f_2, f_3, f_4 \in \mathbb{K}[t]$ and let m be the maximal degree of the four polynomials. Assume that $f_1 - f_2 = f_4 - f_3$, the polynomials f_1, f_2, f_3 and f_4 are coprime and* $\deg(f_1 - f_2) = m$. *Then the image of the following rational map*

$$\mathbb{P}^1 \dashrightarrow (\mathbb{K}^*)^2, \qquad t \mapsto (f_1/f_2, f_3/f_4)$$

has multiplicity m at $(1, 1)$.

Proof Moreover, $\deg(f_1 - f_2) = m$ and $\mathbb{K}$ algebraically closed imply that $f_1 - f_2$ has m roots. Any root of $f_1 - f_2$ is also a root of $f_3 - f_4$, so that we conclude that there are m values of t (counting multiplicities) whose image is the point $(1, 1)$. □

Proof of Theorem 1 First of all, each polygon Δ of type (v) satisfies $\mathrm{lw}(\Delta) = m$, $\mathrm{vol}(\Delta) = m^2$ and $|\partial\Delta \cap \mathbb{Z}^2| = m$, so that the pair (Δ, m) is numerically 0 and expected (in particular it has arithmetic genus 1). Moreover, in [8, Sect. 6] it has been shown that if we set $m = 2k + 4$, for each $k \geq 1$ there exists an irreducible curve of genus 1 whose Newton polygon is Δ. Therefore, we are left with cases (i)–(iv), for which the arithmetic genus of the pair is 0. In these cases, consider the polynomial functions $f_1, f_2, f_3, f_4 := f_1 - f_2 + f_3$ given in the following table.

	f_1	f_2	f_3
(i)	-1	$\sum_{i=1}^{m} t^i$	t^m
(ii)	$(m-1)t-(m-2)$	$-(t-1)t^{m-1}$	$-(t-1)^3\left(t^{m-3}+\sum_{i=0}^{m-4}(m-2-i)t^i\right)$
(iii)	$a^{2k-2}(t-1)$	$\frac{t^{2k-3}(t-a^2)(t^2-a^2)}{a^2}$	$\frac{t^{2k-1}(t-a^2)}{a^2}$, $a := \frac{k-1}{k-2}$
(iv)	$2t-1$	$(1-t)t^{m-1}$	$-(t-1)^2\left(\sum_{i=1}^{m-2} t^i - 1\right)$

These polynomials satisfy the hypotheses of Lemma 3.1, so that the image of the map $\varphi(t) = (f_1/f_2, f_3/f_4)$ has a point of multiplicity m at e. In order to conclude, we have to show that in each case the Newton polygon is the one given in the first column of the table within the proposition. To this aim, we will use [13, Theorem 1.1] which, given a parametric curve $\Gamma \subseteq (\mathbb{K}^*)^2$, provides a description of the normal fan of the Newton polygon of Γ together with the length of the edges, in terms of the zeroes of the four polynomials $f_1, \ldots, f_4$. For the sake of completeness, we explain in detail case (i). In this case, the map φ is defined by

$$\varphi(t) = \left(-\frac{1}{t\sum_{i=0}^{m-1} t^i}, -\frac{t^m}{\sum_{i=0}^{m-1} t^i}\right).$$

Since φ satisfies $\mathrm{ord}_0(\varphi) = (-1, m)$, $\mathrm{ord}_\infty(\varphi) = (m, -1)$ and $\mathrm{ord}_{q_i}(\varphi) = (-1, -1)$, for all the $m-1$ roots $q_1, \ldots, q_{m-1}$ of $\sum_{i=0}^{m-1} t^i$, and these are the only values of t for which $\mathrm{ord}(\varphi)$ does not vanish, by [13, Theorem 1.1], the rays of the normal fan of the Newton polygon of $\overline{\varphi(\mathbb{P}^1)}$ are $(-1, m)$, $(m, -1)$ and $(-1, -1)$. Moreover, the first two rays correspond to two edges of lattice length 1 while the third one has length $m-1$. We conclude that the Newton polygon has vertices $(0, 0)$, $(m, 1)$, $(1, m)$.

Remark 3.2 The triangles of type (i) in Theorem 1 are indeed equivalent to the ones with vertices $(0, 0)$, $(m-1, 0)$, $(m, m+1)$, i.e. $IT(m-1, 1)$ in the notation of [16, Theorem 1.1.A]. Therefore, as a byproduct of Theorem 1 we obtain an alternative (short) proof of [17, Theorem 1.1].

Observe that for each infinite family of Theorem 1, the slopes of (some of) the edges change with m, so that also the toric surfaces change. We are now going to give an example of an infinite family of negative curves lying on the blowing-up of a fixed toric surface. First of all, we recall a construction from [8]. Given an expected lattice polygon $\Delta \subseteq \mathbb{Q}^2$ of width $m := \mathrm{lw}(\Delta)$, with $\mathrm{vol}(\Delta) = m^2$ and $|\partial\Delta \cap \mathbb{Z}^2| = m$, if $\mathcal{L}_\Delta(m)$ contains a unique irreducible element, then its strict transform $C \subseteq X := X_\Delta$ is a curve of arithmetic genus one with $C^2 = 0$. Whenever C is smooth, we denote by

$$\mathrm{res}\colon \mathrm{Pic}(X) \to \mathrm{Pic}(C)$$

the pullback induced by the inclusion. It is not difficult to show that the image of the above map is contained in $\mathrm{Pic}(C)(\mathbb{Q})$. If $\mathrm{res}(C) \in \mathrm{Pic}^0(C)$ is non-torsion then, by [8, Sec. 3], the divisor $K_X + C$ is linearly equivalent to an effective divisor whose support can be contracted by a birational morphism $\phi\colon X \to Y$. The surface Y has at most Du Val singularities and nef anticanonical divisor $-K_Y \sim C$ (here with abuse of notation, we denote by the same letter C a curve which lives in different birational surfaces and is disjoint from the exceptional locus).

Lemma 3.3 *If* $\mathrm{Pic}(Y)$ *has rank three, then* $K_Y^\perp \cap \overline{\mathrm{Eff}}(Y) = \mathbb{Q}_{\geq 0} \cdot [C]$.

Proof The class $[C]$ spans an extremal ray of $\overline{\mathrm{Eff}}(Y)$ because the curves contracted by ϕ are disjoint from C and thus $\mathrm{res}(C)$ is non-torsion also on Y. As a consequence, $\overline{\mathrm{Eff}}(Y)$ is non-polyhedral by [8, Lemma 3.3]. By [8, Lemma 3.14], the minimal resolution of singularities $\pi : Z \to Y$ is a smooth rational surface Z of Picard rank 10, nef anticanonical class $-K_Z$ and non-polyhedral effective cone $\overline{\mathrm{Eff}}(Z)$. Observe that the root sublattice of $\mathrm{Pic}(Z)$ spanned by classes of (-2)-curves over singularities of Y has rank $R = 7$. Assume now that D is an effective divisor such that $D \cdot K_Y = 0$. Then D is pushforward of an effective divisor D' of Z with $D' \cdot K_Z = 0$. Since $-K_Z$ is nef, by adjunction $D' = \sum_i a_i C_i + nC$, where each C_i is a (-2)-curve and $a_i, n \geq 0$. By [8, Corollary 3.17], the fact that $\mathrm{res}(C_i) = 0$ for any i and the fact that $\overline{\mathrm{Eff}}(Y)$ is non-polyhedral we conclude that all the C_i are contracted by π. Thus, D is linearly equivalent to a positive multiple of C. □

We are now going to consider a particular lattice polygon Δ satisfying the above conditions (it is number 24 in [8, Table 3]), and we are going to show that the blowing-up of the corresponding toric surface contains infinitely many negative curves (see Remark 3.5).

Proposition 3.4 *Let* X *be the blowing-up at a general point of the toric surface* $\mathbb{P}$, *defined by the following lattice polygon* $\Delta \subseteq \mathbb{Q}^2$:

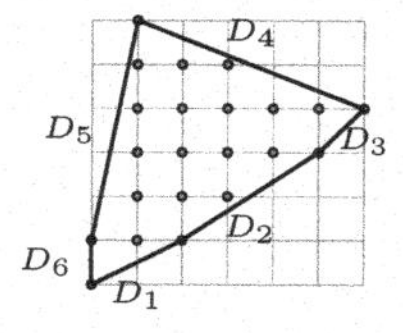

Then there is a birational morphism $\phi\colon X \to Y$ *onto a rational surface* Y *of Picard rank three with only Du Val singularities. Moreover, if* $D_1, \ldots, D_6$ *are the pullbacks of the prime invariant divisors of* $\mathbb{P}$, *ordered according to the picture, the pushforward on* Y *of the divisor*

$$E_k := (7k^2 - 1)D_3 + (161k^2 - 49k - 9)D_4 + (70k^2 - 53k + 9)D_5$$
$$+(14k^2 - 12k + 2)D_6 - (42k^2 - 19k)E$$

is linearly equivalent to a (-1)*-curve for any integer* $k \neq 0$.

Proof Let C be the curve of X defined by the unique element in $\mathcal{L}_\Delta(6)$, which has equation

$$f := -1 + 2v + 7uv - 3u^2v - 23uv^2 + 6u^2v^2 + 2u^3v^2 + 18uv^3 + 20u^2v^3 - 26u^3v^3 + 10u^4v^3 - 2u^5v^3 - 12uv^4 - 11u^2v^4 + 6u^3v^4 + 5u^4v^4 - 4u^5v^4 + u^6v^4 + 5uv^5 + 3u^2v^5 - 2u^3v^5 - uv^6.$$

Let us denote by $v_1, \ldots, v_6$ the primitive generators of the rays of the normal fan to Δ, which are the columns of the following matrix:

$$P := \begin{pmatrix} -1 & -2 & -1 & -2 & 5 & 1 \\ 2 & 3 & 1 & -5 & -1 & 0 \end{pmatrix}.$$

By (1), the divisor $\pi(C)$ is linearly equivalent to $-\sum_{i=1}^{6} \min_{w\in\Delta}\{w \cdot v_i\}\pi(D_i)$ so that

$$C \sim D_2 + 2D_3 + 32D_4 + D_5 - 6E.$$

By (2), the curve C has self-intersection $C^2 = 0$ and it is smooth of genus 1. It is isomorphic, over the rational numbers, to the elliptic curve of equation $y^2 + y = x^3 + x^2$, labeled 43.a1 in the LMFDB database. Its Mordell-Weil group $\mathrm{Pic}^0(C)(\mathbb{Q})$ is free of rank one so that $\mathrm{res}(C)$ is either trivial or non-torsion. The first possibility is ruled out by the fact that $\dim |C| = \dim \mathcal{L}_\Delta(6) = 0$ and the exact sequence of the ideal sheaf of C in X (see [8, Lemma 3.2] for details). Thus, $\mathrm{res}(C)$ is non-torsion, which implies that $h^0(X, nC) = 1$, for any positive integer n, so that $[C]$ spans an extremal ray of $\overline{\mathrm{Eff}}(X)$. By [8, Corollary 3.12], the divisor $K_X + C$ is linearly equivalent to an effective divisor whose support can be contracted. This contraction is the morphism ϕ in the statement. We claim that

$$K_X + C \sim 3C_1 + 2C_2,$$

where $C_1 \sim 5D_4 + D_5 - E$ and $C_2 \sim 5D_5 + D_6 - E$ are the strict transforms of the two one-parameter subgroups corresponding to the width directions $(1, 0)$ and $(0, 1)$ of Δ. To prove the claim, it suffices to observe that the divisor $K_X + C - 3C_1 - 2C_2 \sim -D_1 + D_3 + 16D_4 - 13D_5 - 3D_6$ is principal, being a linear combination of the rows of the above matrix P. Using the intersection matrix of $D_1, \ldots, D_6, E$

$$\begin{pmatrix} -3/2 & 1 & 0 & 0 & 0 & 1/2 & 0 \\ 1 & -1 & 1 & 0 & 0 & 0 & 0 \\ 0 & 1 & -16/7 & 1/7 & 0 & 0 & 0 \\ 0 & 0 & 1/7 & 4/189 & 1/27 & 0 & 0 \\ 0 & 0 & 0 & 1/27 & -5/27 & 1 & 0 \\ 1/2 & 0 & 0 & 0 & 1 & -9/2 & 0 \\ 0 & 0 & 0 & 0 & 0 & 0 & -1 \end{pmatrix},$$

we can see that $C_i \cdot C = C_1 \cdot C_2 = 0$ for $i = 1, 2$ and E_k has integer intersection product with all the D_i, in particular it is a Cartier divisor. Moreover,

$$E_k \cdot C_1 = E_k \cdot C_2 = 0 \qquad E_k^2 = E_k \cdot K = -1, \tag{3}$$

so that by Riemann-Roch E_k is effective. To prove that $\phi_* E_k$ is irreducible, we proceed as follows. Since X has Picard rank 5 and $\phi\colon X \to Y$ contracts $C_1 \cup C_2$, the surface Y has Picard rank 3. The pushforward $\phi_* E_k$ is an effective and Cartier divisor, because we are contracting curves which have intersection product zero with E_k. Moreover, since $-K_Y$ is nef and

$$-K_Y \cdot \phi_* E_k = 1,$$

we deduce that $\phi_* E_k$ is either irreducible or it can be written as $D + D'$, with D irreducible and reduced and D' orthogonal to K_Y. By Lemma 3.3, D' must be equivalent to a multiple of C, so that we can write $\phi_* E_k = D + nC$, for some $n > 0$. Since D is Cartier, both D^2 and $D \cdot K_Y$ are integers, and moreover, $-K_Y$ being nef, by the genus formula $D^2 \geq -2$. The case $D^2 = -2$ can be ruled out since otherwise D would be in $K_Y^{\perp}$. Thus, $D^2 = D \cdot K_Y = -1$ and $-1 = \phi_* E_k^2 = D^2 + 2nD \cdot C = -1 + 2n > 0$ gives again a contradiction. It follows that $\phi_* E_k$ is linearly equivalent to a (-1)-curve. □

Remark 3.5 The way we determined the divisors E_k has been by solving the diophantine Eqs. (3). We also remark that a priori the curve E_k could be reducible of the form $E_k = \Gamma_k + a_1(k)C_1 + a_2(k)C_2$, with Γ_k irreducible and $a_1(k), a_2(k) \geq 0$. So we are only showing the existence of infinitely many negative curves Γ_k on X, which are not necessarily (-1)-curves. Moreover, even if $E_k = \Gamma_k$, the Newton polygon of E_k does not necessarily coincide with the Riemann-Roch polygon Δ_k of the curve E_k, so that E_k could be a (-1)-curve but not an intrinsic one. The Riemann-Roch polygon has vertices corresponding to the columns of the following matrix:

$$\begin{pmatrix} 0 & 0 & 7k^2-4k & 14k^2-12k+2 & 35k^2-12k-1 & 42k^2-19k \\ 0 & 7k^2+k & 42k^2-19k & 7k^2-6k+1 & 21k^2-6k-1 & 28k^2-13k \end{pmatrix}.$$

In particular, if we set $m = 42k^2 - 19k$, we have that $\mathrm{vol}(\Delta_k) = m^2 - 1$, $|\partial\Delta_k \cap \mathbb{Z}^2| = m + 1$ and $\mathrm{lw}(\Delta_k) = m$, so that (Δ_k, m) is numerically a (-1)-pair. For small values of k it is possible to check, with the help of the software Magma [6], that the Newton polygon of the (-1)-curve is indeed Δ_k, but for general k we are not able to prove it. Therefore, Γ_k is not necessarily an intrinsic (-1)-curve, but it is anyway an intrinsic negative curve (see Remark 2.6).

4 Seshadri Constants

In this section, we first prove Theorem 2 and Corollary 1, and then we discuss some consequences of the study of the effective cone of the blowing-up of weighted projective planes.

We will need the following preliminary result about Seshadri constants on projective surfaces.

Lemma 4.1 *Let Y be a projective surface, H an ample divisor of Y and $\pi : X \to Y$ be the blowing-up of Y at a smooth point $p \in Y$ with exceptional divisor E.*

(i) If there is a positive integer m such that $\pi^*H - mE$ is the class of an effective curve $C = \sum_{i=1}^r a_i C_i$ with $C^2 \leq 0$, then

$$\varepsilon(H, p) = \min_i \left\{ \frac{\pi^*H \cdot C_i}{E \cdot C_i} \right\} \leq m + \frac{C^2}{m}.$$

(ii) If furthermore C is irreducible, then $\varepsilon(H, p) = m + \frac{C^2}{m}$.

Proof We prove (i). Let $\varepsilon := \varepsilon(H, p)$. Observe that we can write $C^2 + (m - \varepsilon)E \cdot C = (\pi^*H - mE + (m - \varepsilon)E) \cdot C = (\pi^*H - \varepsilon E) \cdot C \geq 0$, and, since $E \cdot C = m$, we get

$$\varepsilon \leq m + \frac{C^2}{m} \leq m.$$

If C is nef then $\varepsilon \geq m$, so that $\varepsilon = m$ and $C^2 = 0$. This implies $C \cdot C_i = 0$, and hence $\varepsilon = \pi^*H \cdot C_i / E \cdot C_i$ for any i, proving the statement in this case. If C is not nef then $\varepsilon < m$. If α is such that $\varepsilon < \alpha < m$ then $\pi^*H - \alpha E$ is effective and non-nef. Let C' be an irreducible curve such that $(\pi^*H - \alpha E) \cdot C' < 0$, then $(\pi^*H - mE) \cdot C' < 0$ as well. Thus $C' = C_i$ for some i. Since α can be chosen arbitrarily close to ε, we conclude that $(\pi^*H - \varepsilon E) \cdot C_j = 0$ for some j, and the statement follows. Statement (ii) is an immediate consequence of (i) and the equality $\pi^*H \cdot C = C^2 + m^2$. □

Proof of Theorem 2. We prove (i). Let $v \in N$ be a width direction, that is, $\mathrm{lw}_v(\Delta) = \mathrm{lw}(\Delta)$ and let $C_v \subseteq X_\Delta$ be the strict transform of the one-parameter subgroup of the torus defined by v. If $\mu > \mathrm{lw}(\Delta)$, then $(\pi^*H - \mu E) \cdot C_v < 0$, so that $\pi^*H - \mu E$ is not nef. This proves the statement.

We prove (ii). Observe that $(\pi^*H - \varepsilon E)^2 \geq (\pi^*H - \mathrm{lw}(\Delta)E)^2 = \mathrm{vol}(\Delta) - \mathrm{lw}(\Delta)^2 > 0$, where the first inequality is by (i). By the Riemann-Roch theorem, the class of $\pi^*H - \varepsilon E$ is in the interior of the effective cone $\mathrm{Eff}(X_\Delta)$. It follows that the Seshadri constant is computed by a curve $C \subseteq X_\Delta$. From $(\pi^*H - \varepsilon E) \cdot C = 0$ one concludes that ε is a rational number (see also [23, Remark 2.3]).

Statements (iii) and (iv) are consequence of Lemma 4.1 and the fact that $C^2 = \mathrm{vol}(\Delta) - m^2$.

Proof of Corollary 1. Observe that by hypothesis $\mathrm{vol}(\Delta) - m^2 = C^2 \leq 0$, so that the hypothesis (iii) of Theorem 2 is satisfied. Moreover, C being irreducible we have $\mathrm{lw}(\Delta) - m = C \cdot C_v \geq 0$, where C_v is the strict transform of the one-parameter subgroup of the torus defined by the width direction v. Thus, also hypothesis (iv) of Theorem 2 is satisfied and the statement follows.

Remark 4.2 Observe that the best lower bound for ε we can get from [20, Theorem 1.3] in the case of a toric surface is either the width $\mathrm{lw}(\Delta)$, or the biggest length of a segment inside Δ. For instance, if Δ is a triangle of type (i) in Theorem 1, consider the projection $\pi : \mathbb{Q}^2 \to \mathbb{Q}$ onto the second factor. If we take the fiber $\Delta \cap \pi^{-1}(1)$, by [20, Theorem 1.3] we have the inequality

$$\varepsilon(H, e) \geq \min\{m, m - 1/m\} = m - 1/m.$$

Since $\mathrm{vol}(\Delta) = m^2 - 1$, by Theorem 2 (iii), we also have the inequality $\varepsilon(H, e) \leq (m^2 - 1)/m$, so that we can conclude that $\varepsilon(H, e)$ is indeed equal to $m - 1/m$, no need of showing that there exists an irreducible curve $C \in \mathcal{L}_\Delta(m)$.

We also remark that in the remaining cases of Theorem 1, the bound given by [20, Theorem 1.3] is not sharp.

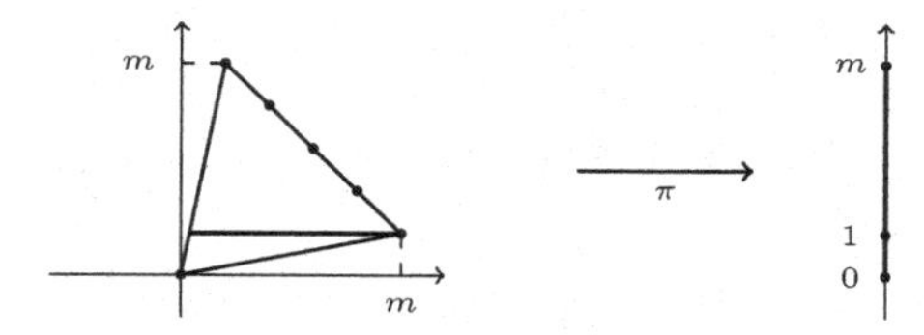

In the same vein, if the lattice polygon Δ contains a segment of lattice length $\mathrm{lw}(\Delta)$, then [20, Theorem 1.3] gives the bound $\varepsilon(H, e) \geq \mathrm{lw}(\Delta)$. But since by Theorem 2 (i) we also have the opposite inequality, we can immediately conclude that the Seshadri constant $\varepsilon(H, e)$ is indeed equal to $\mathrm{lw}(\Delta)$.

For instance, this shows that for any $m \geq 4$, the polygon with vertices $(0, 0)$, $(0, 1)$, $(m, 1)$, $(1, m)$ corresponds to a toric surface with Seshadri constant $\varepsilon(H, e) = m$ (even if it is not hard to find a parametrization as we did with the families of Theorem 1).

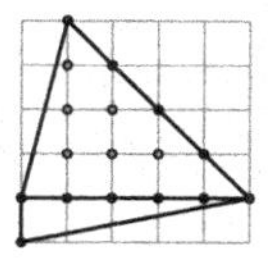

4.1 Weighted Projective Planes

We briefly recall an open problem about the existence of certain irreducible curves in weighted projective planes, and its relation with intrinsic curves. For a comprehensive reference on known facts and open problems on blow-ups of weighted projective planes, see [7, Sect. 6].

Let a, b, c be three positive pairwise coprime integers, $\mathbb{P}(a, b, c)$ be the corresponding weighted projective plane and $\pi : X(a, b, c) \to \mathbb{P}(a, b, c)$ be the blowing-up at the general point $e := (1, 1, 1)$ with exceptional divisor E. The divisor class group of $X := X(a, b, c)$ is free of rank 2 and the effective cone $\mathrm{Eff}(X)$ is in general unknown. By the Riemann-Roch theorem, $\mathrm{Eff}(X)$ contains the positive light cone Q (shaded region) with extremal rays generated by $R_\pm = \pi^*H \pm \frac{1}{\sqrt{abc}}E$.

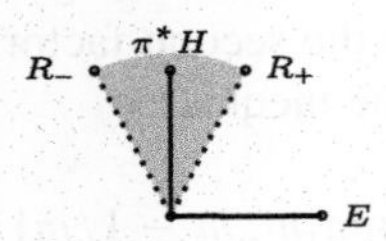

The question is whether $\mathrm{Eff}(X)$ is bounded by the $\mathbb{R}$-divisor R_-, so that $\varepsilon(H, e) = 1/\sqrt{abc}$, or by the class of a negative curve (lying below the ray R_-). In many examples (see, for instance, [16, 19]), the existence of the negative curve has been proved, but in general the question is still open, and in fact it is conjectured that for some triples a, b, c (such as 9, 10, 13) that negative curve does not exist. We remark that proving this conjecture would not only give an example of a surface having non-rational Seshadri constant, but it would also imply the following well-known conjecture (see [25]), in some particular cases, as explained in Proposition 4.4.

Conjecture 4.3 (*Nagata's Conjecture*) Let $\pi : X_r \to \mathbb{P}^2$ be the blowing-up at $r \geq 10$ points in very general position and let $E_1, \dots, E_r$ be the exceptional divisors. Then the class $\pi^*\mathcal{O}(1) - \frac{1}{\sqrt{r}} \sum_i E_i$ is nef on X_r.

So far, Nagata's conjecture has been proved only when r is a perfect square [25], but the following result shows that finding a triple a, b, c, such that the Seshadri constant at the general point $e \in X(a, b, c)$ is $1/\sqrt{abc}$, would imply Nagata's conjecture for $r = abc$ (it can be found, for instance, in [12, Proposition 5.2], but we give anyway a brief proof for the sake of completeness).

Proposition 4.4 *If* $\varepsilon(H, e) = 1/\sqrt{abc}$ *then Nagata's conjecture holds for* abc *points in the plane.*

Proof Let $f : \mathbb{P}^2 \to \mathbb{P}(a, b, c)$ be the morphism defined by $(x, y, z) \mapsto (x^a, y^b, z^c)$ and let Y_r be the blowing-up of $\mathbb{P}^2$ at the $r := abc$ points of $f^{-1}(e)$. Since R_- is nef, also

$$f^*R_- = L - \frac{1}{\sqrt{abc}} \sum_{i=1}^{abc} E_i$$

is nef on Y_r. If we denote by X_r the blowing-up of $\mathbb{P}^2$ at r points in a very general position, then $\overline{\mathrm{Eff}}(X_r) \subseteq \overline{\mathrm{Eff}}(Y_r)$ by semicontinuity of the dimension of cohomology. Thus, $\overline{\mathrm{Nef}}(X_r) \supseteq \overline{\mathrm{Nef}}(Y_r)$ so that f^*R_- is nef also on X_r. □

Remark 4.5 If C is a very general smooth irreducible curve of positive genus g, then the Néron-Severi group, over the rational numbers, of the symmetric product $\mathrm{Sym}^2(C)$ has rank two. In [10, Proposition 3.1], the authors show that if Nagata's conjecture is true and $g \geq 9$, then the effective cone of $\mathrm{Sym}^2(C)$ is open on one side.

The equality $\varepsilon(H, e) = 1/\sqrt{abc}$ holds if and only if there does not exist a negative curve in $X(a, b, c)$ having class $d\pi^*H - mE$, with $d/m < \sqrt{abc}$. A partial result in this direction is given by [22, Theorem 5.4], where the authors show that if the negative curve is expected, then d is bounded from above and this result allows them to conclude that there are no such curves on certain $X(a, b, c)$, like $X(9, 10, 13)$.

On the other hand, we can also say that the equality $\varepsilon(H, e) = 1/\sqrt{abc}$ holds if and only if there exists a sequence $\pi^*d_nH - m_nE$ of classes of positive irreducible curves in $X(a, b, c)$ such that $d_n/m_n \to \sqrt{abc}$, that is, these classes approach the ray R_- from the inside of the light cone. In this direction, observe that an intrinsic negative curve can appear as a positive curve in $X(a, b, c)$. We discuss this approach for $X(9, 10, 13)$ by producing many intrinsic (-1)-curves which are positive curves on the surface. We proceed using the fact that the Cox ring of $X(a, b, c)$ is isomorphic to the extended saturated Rees algebra (see [2, Proposition 4.1.3.8] or [19]):

$$R[I]^{\mathrm{sat}} := \bigoplus_{m\in\mathbb{Z}} (I^m : J^\infty)t^{-m} \subseteq R[t^{\pm 1}],$$

where $R = \mathbb{K}[x, y, z]$ is the Cox ring of the weighted projective plane, I is the ideal of $(1, 1, 1)$ in Cox coordinates, $J = \langle x, y, z\rangle$ is the irrelevant ideal and $I^{-m} = R$ for any $m \geq 0$. Using this, we can compute a minimal generating set consisting of homogeneous elements of given bounded multiplicity at $(1, 1, 1)$. In the case of $X(9, 10, 13)$, fixing the maximum of the multiplicity to be 30, we found 52 generators. In the following table we display the degrees of these generators together with the self-intersection of the corresponding intrinsic curve and its genus (while the self-intersection of the curve on $X(9, 10, 13)$ is $d^2/abc - m^2$).

d	m	C^2	p_a	d	m	C^2	p_a	d	m	C^2	p_a	d	m	C^2	p_a
36	1	0	0	213	6	−2	0	516	15	0	1	823	24	−1	0
39	1	0	0	243	7	0	1	549	16	−1	0	824	24	−1	0
40	1	0	0	309	9	−1	0	550	16	−1	0	858	25	−1	0
83	2	−1	0	310	9	−1	0	551	16	−1	0	891	26	−1	0
109	3	−1	0	312	9	−1	0	585	17	−1	0	892	26	−1	0
110	3	−1	0	313	9	−1	0	652	19	−1	0	893	26	0	1
113	3	−1	0	378	11	−1	0	653	19	−1	0	893	26	3	3
139	4	−1	0	379	11	−1	0	686	20	0	1	926	27	−1	0
140	4	−1	0	380	11	−1	0	720	21	1	2	959	28	0	1
143	4	−1	0	413	12	−1	0	721	21	−1	0	960	28	0	1
208	6	−1	0	481	14	−1	0	755	22	−1	0	994	29	−1	0
209	6	−1	0	482	14	−1	0	789	23	0	1	1028	30	0	1
210	6	−1	0	483	14	−1	0	790	23	1	2	1029	30	−1	0

The slope d/m which best approximates $\sqrt{9 \cdot 10 \cdot 13} \sim 34.20526$ is $959/28 = 34.25$, realized by the last curve. The best approximation given by an intrinsic (-1)-curve of the above list is $891/26 \sim 34.26923$. The following question naturally arises:

Question 4.6 Is it possible to construct an infinite family of intrinsic (-1)-curves appearing as positive curves in $X(9, 10, 13)$, and whose slopes approach $\sqrt{9 \cdot 10 \cdot 13}$?

References

1. F. Ambro, A. Ito, Successive minima of line bundles. Adv. Math. **365**(38), 107045 (2020)
2. I. Arzhantsev, U. Derenthal, J. Hausen, A. Laface, *Cox Rings*, Cambridge Studies in Advanced Mathematics, vol. 144 (Cambridge University Press, Cambridge, 2015)
3. G. Balletti, Enumeration of lattice polytopes by their volume. Discret. Comput. Geom. (2020)
4. T. Bauer, Seshadri constants on algebraic surfaces. Math. Ann. **313**(3), 547–583 (1999)
5. T. Bauer, S. Di Rocco, B. Harbourne, M. Kapustka, A. Knutsen, W. Syzdek, T. Szemberg, A primer on Seshadri constants, Interactions of classical and numerical algebraic geometry. Contemporary Mathematics, vol. 496 (American Mathematical Society, Providence, RI, 2009), pp. 33–70
6. W. Bosma, J. Cannon, C. Playoust, The Magma algebra system. I. The user language. J. Symb. Comput. **24**(3–4), 235–265 (1997). Computational algebra and number theory (London, 1993)
7. A.-M. Castravet, Mori dream spaces and blowing-ups, Algebraic geometry: salt Lake City 2015 Proceedings of Symposia in Pure Mathematics, vol. 97 (American Mathematical Society, Providence, RI, 2018), pp. 143–167
8. A.-M. Castravet, A. Laface, J. Tevelev, L. Ugaglia, Blown-up toric surfaces with non-polyhedral effective cone (2020). arXiv:2009.14298
9. A.-M.T. Castravet, Jenia, $\overline{M}_{0,n}$ is not a Mori dream space. Duke Math. J. **164**(8), 1641–1667 (2015)
10. C. Ciliberto, A. Kouvidakis, On the symmetric product of a curve with general moduli. Geom. Dedicata **78**(3), 327–343 (1999)
11. D.A. Cox, J.B. Little, H.K. Schenck, *Toric Varieties*, Graduate Studies in Mathematics, vol. 124 (American Mathematical Society, Providence, RI, 2011)
12. S.D. Cutkosky, K. Kurano, Asymptotic regularity of powers of ideals of points in a weighted projective plane. Kyoto J. Math. **51**(1), 25–45 (2011)
13. C. D'Andrea, M. Sombra, The Newton polygon of a rational plane curve. Math. Comput. Sci. **4**(1), 3–24 (2010)
14. L. Ein, R. Lazarsfeld, Seshadri constants on smooth surfaces. Algébrique **218**, 177–186 (1993). Journées de Géométrie d'Orsay (Orsay, 1992)
15. Ł Farnik, T. Szemberg, J. Szpond, H. Tutaj-Gasińska, Restrictions on Seshadri constants on surfaces. Taiwanese J. Math. **21**(1), 27–41 (2017)
16. J. González-Anaya, J.L. González, K. Karu, Constructing non-Mori dream spaces from negative curves. J. Algebra **539**, 118–137 (2019)
17. J. González-Anaya, J.L. González, K. Karu, On a family of negative curves. J. Pure Appl. Algebra **223**(11), 4871–4887 (2019)
18. C. Haase, A. Küronya, L. Walter, Toric Newton-Okounkov functions with an application to the rationality of certain Seshadri constants on surfaces (2020). arXiv:2008.04018
19. J. Hausen, S. Keicher, A. Laface, On blowing up the weighted projective plane. Math. Z. **290**(3–4), 1339–1358 (2018)
20. A. Ito, Seshadri constants via toric degenerations. J. Reine Angew. Math. **695**, 151–174 (2018)

21. K. Kurano, Equations of negative curves of blow-ups of Ehrhart rings of rational convex polygons (2021). arXiv:2101.02448
22. K. Kurano, N. Matsuoka, On finite generation of symbolic Rees rings of space monomial curves and existence of negative curves. J. Algebra **322**(9), 3268–3290 (2009)
23. A. Küronya, V. Lozovanu, Convex bodies appearing as Okounkov bodies of divisors. Adv. Math. **229**(5), 2622–2639 (2012)
24. A. Laface, L. Ugaglia, On base loci of higher fundamental forms of toric varieties. J. Pure Appl. Algebra **224**(12), 106447, 18 (2020)
25. M. Nagata, On the 14-th problem of Hilbert. Amer. J. Math. **81**, 766–772 (1959)
26. M. Nakamaye, Seshadri constants and the geometry of surfaces. J. Reine Angew. Math. **564**, 205–214 (2003)
27. P. Orlik, P. Wagreich, Algebraic surfaces with k^*-action. Acta Math. **138**(1–2), 43–81 (1977)
28. J. Tevelev, Compactifications of subvarieties of tori. Am. J. Math. **129**(4), 1087–1104 (2007)

On the Extendability of Projective Varieties: A Survey

Angelo Felice Lopez

Abstract We give a survey of the incredibly beautiful amount of geometry involved with the problem of realizing a projective variety as hyperplane section of another variety.

Keywords Extendability · Canonical curves · K3 surfaces · Fano threefolds

Mathematics Subject Classification: Primary 14N05 · 14J40 · Secondary 14J28 · 14H51

1 Introduction

At the beginning of the 90's, I attended a seminar talk by Ciro Ciliberto at the University of Milan. The main upshot, beautifully conveyed by the speaker, was to make a connection between two apparently unrelated theorems that had appeared recently and to investigate the consequences of this awareness.

The story that followed in the subsequent years, and the research still going on today, will be recollected in this survey, through the magnifying lens of my views on the subject, very much influenced by that talk.

To state the theorems in question, we start with some definitions and notation.

Appendix by Thomas Dedieu: Institut de Mathématiques de Toulouse UMR5219 Université de Toulouse CNRS. UPS IMT, 31062 Toulouse Cedex 9, France
e-mail: thomas.dedieu@math.univ-toulouse.fr

Research partially supported by PRIN "Advances in Moduli Theory and Birational Classification" and GNSAGA-INdAM

Dedicated to Ciro Ciliberto on the occasion of his 70th birthday. What I learned from him, both from a mathematical and a human point of view, is invaluable.

A. F. Lopez (✉)
Dipartimento di Matematica e Fisica, Università di Roma Tre, Largo San Leonardo Murialdo 1, 00146 Roma, Italy
e-mail: lopez@mat.uniroma3.it

T. Dedieu et al. (eds.), *The Art of Doing Algebraic Geometry*, Trends in Mathematics,
https://doi.org/10.1007/978-3-031-11938-5_11

Definition 1.1 Let $X \subset \mathbb{P}^r$ be an irreducible nondegenerate variety of codimension at least 1. Let $k \geq 1$ be an integer. We say that X is *k-extendable* if there exists a variety $Y \subset \mathbb{P}^{r+k}$ different from a cone over X, with $\dim Y = \dim X + k$ and having X as a section by an r-dimensional linear space. We say that X is *precisely k-extendable* if it is k-extendable but not $(k+1)$-extendable. The variety Y is called a *k-extension of X*. We say that X is *extendable* if it is 1-extendable.

It is, of course, a basic question in projective geometry to understand when a variety is extendable and if so, how much. Even when a variety is extendable, it can be extendable in different ways and with chains of extensions of different lengths (see the example in [105]).

On the other hand, unless one has a very good knowledge of the variety and its very ample line bundles, it is usually a difficult but fascinating problem.

A general remark to be made is that two general approaches have been predominant. Researchers have often taken an "optimistic" or a "pessimistic" approach. The first one: fix X and try to prove that it is not extendable or to classify all of its extensions. The second one: start with Y, study its hyperplane sections and try to prove that such a Y cannot exist. We will give examples of both.

To give a measure in this direction, let us define

Definition 1.2 Let $X \subset \mathbb{P}^r$ be a smooth irreducible nondegenerate variety of codimension at least 1 with normal bundle N_X. We set

$$\alpha(X) = h^0(N_X(-1)) - r - 1.$$

The first result, due to Zak [105] and L'vovsky [69] (see also [10, 17]), is as follows:

Theorem 1.3 (Zak-L'vovsky's theorem) *Let $X \subset \mathbb{P}^r$ be a smooth irreducible nondegenerate variety of codimension at least* 1 *and suppose that X is not a quadric. If $\alpha(X) \leq 0$, then X is not extendable.*

Given an integer $k \geq 2$, suppose that either:

(i) $\alpha(X) < r$ (L'vovsky's version), or
(ii) $H^0(N_X(-2)) = 0$ (Zak's version).

If $\alpha(X) \leq k-1$, then X is not k-extendable.

Now to shift to the second result, let us first give the following

Definition 1.4 Let C be a smooth irreducible curve. The *Wahl map of C* is the map

$$\Phi_{\omega_C} : \Lambda^2 H^0(\omega_C) \to H^0(\omega_C^3)$$

defined by $\Phi_{\omega_C}(f dz \wedge g dz) = (fg' - gf')dz^3$.

In the same years, Wahl [100] introduced this map (so named in the subsequent years) and proved the following seminal

Theorem 1.5 (Wahl's theorem) *Let C be a smooth irreducible curve. If C sits on a K3 surface, then Φ_{ω_C} is not surjective.*

A beautiful geometric proof of this theorem was also given by Beauville and Mérindol [16].

The connection between Zak-L'vovsky's and Wahl's theorem will be made in the next sections.

In this survey, we will focus on the extendability of projective varieties. The question was investigated already by the Italian and British schools, it went on with the school of adjunction theory and received a big push after the theorems of Zak-L'vovsky and Wahl. It is a very active area of research still going strong today, where Ciro Ciliberto stands as one of the main contributors.

We apologize for not treating, due to lack of space, several important aspects very much related to extendability, such as projective duality, deformation theory and graded pieces of the T^1 sheaf.

2 Extendability in General

In order to understand Zak-L'vovsky's theorem, we start with some simple but useful observations. Even though we do not know a precise reference, they have been all well-known for a long time.

Proposition 2.1 *Let $X \subset \mathbb{P}^r = \mathbb{P}V$ be a smooth irreducible nondegenerate variety of dimension $n \geq 1$ and of codimension at least 1. Then:*

(i) $h^0(T_{\mathbb{P}^r|X}(-1)) \geq r + 1$ *and equality holds if* $n \geq 2$;
(ii) $N_X(-1)$ *is globally generated;*
(iii) $\alpha(X) \geq 0$;
(iv) *If* $h^1(\mathcal{O}_X) = 0$ *and either* $n \geq 3$ *or* $n = 2$ *and the multiplication map* $V \otimes H^0(\omega_X) \to H^0(\omega_X(1))$ *is surjective, then* $\alpha(X) = h^1(T_X(-1))$.

Proof From the twisted Euler sequence

$$0 \to \mathcal{O}_X(-1) \to V^\vee \otimes \mathcal{O}_X \to T_{\mathbb{P}^r|X}(-1) \to 0 \tag{1}$$

we get that $T_{\mathbb{P}^r|X}(-1)$ is globally generated, $h^0(T_{\mathbb{P}^r|X}(-1)) \geq r + 1$ and that equality holds if $n \geq 2$ by Kodaira vanishing. Hence, we have (i). Now the twisted normal bundle sequence

$$0 \to T_X(-1) \to T_{\mathbb{P}^r|X}(-1) \to N_X(-1) \to 0 \tag{2}$$

implies (ii). Moreover, since $H^0(T_X(-1)) = 0$ by [72, Theorem 8], [99, Theorem 1] (unless X is a plane conic, but then $\alpha(X) = 0$), we get that

$$\alpha(X) = h^0(N_X(-1)) - r - 1 \geq h^0(T_{\mathbb{P}^r|X}(-1)) - r - 1 \geq 0$$

that is (iii). Now assume that $h^1(\mathcal{O}_X) = 0$. If $n \geq 3$, then $H^2(\mathcal{O}_X(-1)) = 0$ by Kodaira vanishing and we get that $H^1(T_{\mathbb{P}^r|X}(-1)) = 0$ by (1). When $n = 2$ the same is achieved by observing that the map $H^2(\mathcal{O}_X(-1)) \to V^\vee \otimes H^2(\mathcal{O}_X)$ is injective. Then (2) gives (iv). □

Let us take a look at the behaviour of $\alpha(X)$ under hyperplane sections.

Proposition 2.2 *Let $X \subset \mathbb{P}^r$ be a smooth irreducible nondegenerate variety of dimension $n \geq 1$ and of codimension at least* 1. *Let $Y \subset \mathbb{P}^{r+1}$ be a smooth extension of X. If $H^0(N_X(-2)) = 0$, then $H^0(N_Y(-2)) = 0$ and $\alpha(Y) \leq \alpha(X) - 1$.*

Proof Since $(N_{Y/\mathbb{P}^{r+1}})_{|X} \cong N_{X/\mathbb{P}^r}$, for every $i \geq 1$, we have an exact sequence

$$0 \to N_Y(-i-1) \to N_Y(-i) \to N_X(-i) \to 0. \tag{3}$$

Since $H^0(N_X(-2)) = 0$ we get from (3) that $h^0(N_Y(-i)) = h^0(N_Y(-i-1))$ for $i \geq 2$, and therefore, $h^0(N_Y(-2)) = 0$. Then, again from (3) with $i = 1$, we see that

$$\alpha(Y) = h^0(N_Y(-1)) - r - 2 \leq h^0(N_X(-1)) - r - 2 = \alpha(X) - 1.$$

□

We will now give an idea of how to prove a much-simplified version of Zak-L'vovsky's theorem (Zak's version). We will consider only the case of a chain of smooth extensions. For singular versions, see [10, 12]. In Sect. 5, we give an example of Zak of a variety that is extendable but not smoothly extendable.

Proof of Theorem 1.3 *(simplified version).*

Proof Recall that $\alpha(X) \geq 0$ by Proposition 2.1(iii). We first record the following

Claim 2.3 If $\alpha(X) = 0$, then $H^0(N_X(-2)) = 0$.

Proof If X has codimension 1 and degree d, then $d \geq 3$ and $N_X(-1) \cong \mathcal{O}_X(d-1)$. But this gives

$$\alpha(X) \geq h^0(\mathcal{O}_X(2)) - r - 1 > 0$$

a contradiction. Therefore, codim $X \geq 2$. This implies that, at every point $x \in X$, the $\mathcal{O}_{X,x}$-module $N_X(-1)_x$ has rank at least 2.

Assume that $H^0(N_X(-2)) \neq 0$ and let $\sigma \in H^0(N_X(-2))$ be non-zero. Let $\{\tau_0, \ldots, \tau_r\}$ be a basis of $\mathrm{Im}\{H^0(\mathcal{O}_{\mathbb{P}^r}(1)) \to H^0(\mathcal{O}_X(1))\}$. Then $\sigma \otimes \tau_0, \ldots, \sigma \otimes \tau_r \in H^0(N_X(-1))$ are linearly independent, thus they are a basis. On the other hand, $N_X(-1)$ is globally generated by Proposition 2.1(ii), hence the sections $\sigma \otimes \tau_j, 0 \leq j \leq r$ generate $N_X(-1)_x$. But the $\sigma_x \otimes (\tau_j)_x$ generate an $\mathcal{O}_{X,x}$-module of rank 1, a contradiction. □

Therefore, if $\alpha(X) = 0$, the theorem follows by Claim 2.3 and Propositions 2.2 and 2.1(iii).

Next, let $k \geq 2$. Suppose that $H^0(N_X(-2)) = 0$ and that X is k-extendable. We show, by induction on k, that $\alpha(X) \geq k$. This will of course complete the proof of the theorem.

Let $Y \subset \mathbb{P}^{r+1}$ be a smooth extension of X. By Proposition 2.2, we have that $H^0(N_Y(-2)) = 0$ and $\alpha(X) \geq \alpha(Y) + 1$. If $k = 2$, we have that Y is extendable, hence $\alpha(Y) \geq 1$ by the first part of the theorem. Therefore, $\alpha(X) \geq 2$. If $k \geq 3$, since Y is $(k-1)$-extendable, we have by induction that $\alpha(Y) \geq k - 1$, hence $\alpha(X) \geq k$. □

Remark 2.4 It is easy to see that the conditions (i) and (ii) in Zak-L'vovsky's theorem are essential. For example, one can take a hypersurface $X \subset \mathbb{P}^{n+1}$ of degree $d \geq 3$. Then $\alpha(X) > n + 1$ and $H^0(N_X(-2)) \neq 0$. On the other hand, X is infinitely extendable. See Remark 5.24 for an example of canonical curves.

One immediate consequence of Zak-L'vovsky's theorem is the one mentioned below. As a matter of fact, this appeared before Theorem 1.3, at least for smooth or normal extensions, and it actually has stronger consequences [9, 17, 48, 94], but we will not be concerned with them here.

Proposition 2.5 *Let $X \subset \mathbb{P}^r$ be a smooth irreducible nondegenerate variety of dimension $n \geq 2$, of codimension at least 1 and suppose that X is not a quadric. If $H^1(T_X(-1)) = 0$, then X is not extendable.*

Proof By (2) and [72, Theorem 8], [99, Theorem 1] we see that $h^0(N_X(-1)) = h^0(T_{\mathbb{P}^r|X}(-1))$. Then $\alpha(X) = 0$ by Proposition 2.1(i), hence X is not extendable by Theorem 1.3. □

3 How to Estimate $\alpha(X)$: A Fortunate Coincidence

Let $X \subset \mathbb{P}^r$ be a smooth irreducible nondegenerate variety of dimension $n \geq 1$ and of codimension at least 1.

How can we compute $\alpha(X)$?

Unless one has a very good knowledge of the normal bundle N_X (or of T_X in the case of Proposition 2.1(iv)), it turns out that it is quite difficult to compute $\alpha(X)$.

However, in dimension 1, a fortunate coincidence, explained below, comes to help.

All the results that follow are of course not new (see for example [30, 101]), but we include them for the benefit of the reader.

Lemma 3.1 *Let $C \subset \mathbb{P}^r$ be a smooth irreducible nondegenerate linearly normal curve of positive genus. Let $\mu_{\mathcal{O}_C(1),\omega_C} : H^0(\mathcal{O}_C(1)) \otimes H^0(\omega_C) \to H^0(\omega_C(1))$ be the multiplication map on sections.*

Consider the map, called the Gaussian map,

$$\Phi_{\mathcal{O}_C(1),\omega_C} : H^0(\Omega_{\mathbb{P}^r|C} \otimes \omega_C(1)) \to H^0(\omega_C^2(1)).$$

Then:

(i) $H^0(\Omega_{\mathbb{P}^r|C} \otimes \omega_C(1)) \cong \operatorname{Ker} \mu_{\mathcal{O}_C(1),\omega_C}$*;*
(ii) Given the identification in (i), we have that $\Phi_{\mathcal{O}_C(1),\omega_C}(s \otimes t) = sdt - tds$*;*
(iii) $\alpha(C) = \operatorname{cork} \Phi_{\mathcal{O}_C(1),\omega_C}$*.*

Proof The twisted dual Euler sequence

$$0 \to \Omega_{\mathbb{P}^r|C} \otimes \omega_C(1) \to H^0(\mathcal{O}_C(1)) \otimes \omega_C \to \omega_C(1) \to 0$$

gives rise to the following exact cohomology sequence

$$0 \to H^0(\Omega_{\mathbb{P}^r|C} \otimes \omega_C(1)) \to H^0(\mathcal{O}_C(1)) \otimes H^0(\omega_C) \overset{\mu_{\mathcal{O}_C(1),\omega_C}}{\longrightarrow} H^0(\omega_C(1)) \to \qquad (4)$$
$$\to H^1(\Omega_{\mathbb{P}^r|C} \otimes \omega_C(1)) \to H^0(\mathcal{O}_C(1)) \otimes H^1(\omega_C) \to 0.$$

This gives (i). Moreover, as $\mu_{\mathcal{O}_C(1),\omega_C}$ is surjective by [29], [5, Theorem 1.6], we have from (4) that

$$h^1(\Omega_{\mathbb{P}^r|C} \otimes \omega_C(1)) = r + 1. \qquad (5)$$

Now the twisted dual normal bundle sequence

$$0 \to N_C^\vee \otimes \omega_C(1) \to \Omega_{\mathbb{P}^r|C} \otimes \omega_C(1) \to \omega_C^2(1) \to 0$$

gives rise to the following exact cohomology sequence:

$$H^0(\Omega_{\mathbb{P}^r|C} \otimes \omega_C(1)) \overset{\Phi_{\mathcal{O}_C(1),\omega_C}}{\longrightarrow} H^0(\omega_C^2(1)) \to H^1(N_C^\vee \otimes \omega_C(1)) \to H^1(\Omega_{\mathbb{P}^r|C} \otimes \omega_C(1)) \to 0. \qquad (6)$$

Thus, we get (ii). Also, from (6), (5) and Serre duality, we get

$$h^0(N_C(-1)) = r + 1 + \operatorname{cork} \Phi_{\mathcal{O}_C(1),\omega_C}$$

that is (iii). □

Remark 3.2 Both the Gaussian map $\Phi_{\mathcal{O}_C(1),\omega_C}$ and the Wahl map Φ_{ω_C} have an ancestor in the map $\mu_1 : \operatorname{Ker} \mu_{W,\omega_C(-L)} \to H^0(\omega_C^2)$, where $W \subseteq H^0(L)$. This was introduced and studied in 1981 by Arbarello and Cornalba in [4, Sect. 4] (see also [2, Bib. notes to Chap. XXI], [5, Sect. 4]).

By the above lemma, now the connection between Zak-L'vovsky's theorem and Wahl's theorem is clear from the following

Proposition 3.3 *Let* $C \subset \mathbb{P}^{g-1}$ *be a canonically embedded smooth curve of genus* $g \geq 3$*. Then*

$$\alpha(C) = \operatorname{cork} \Phi_{\omega_C}.$$

Moreover, if $S \subset \mathbb{P}^g$ is a smooth surface having C as a hyperplane section, then S is a K3 surface.

Proof First of all, Lemma 3.1(iii) says that

$$\alpha(C) = \operatorname{cork} \Phi_{\omega_C, \omega_C}.$$

Now $\Phi_{\omega_C, \omega_C}$ vanishes on symmetric tensors by Lemma 3.1(ii), hence

$$\alpha(C) = \operatorname{cork} \Phi_{\omega_C}.$$

For the second part, recall that C is projectively normal by M. Noether's theorem. Hence, the commutative diagram

$$\begin{array}{ccc} H^0(\mathcal{O}_{\mathbb{P}^g}(l)) & \twoheadrightarrow & H^0(\mathcal{O}_{\mathbb{P}^{g-1}}(l)) \\ \downarrow & & \downarrow\!\!\downarrow \\ H^0(\mathcal{O}_S(l)) & \longrightarrow & H^0(\mathcal{O}_C(l)) \end{array}$$

implies that the map $H^0(\mathcal{O}_S(l)) \to H^0(\mathcal{O}_C(l))$ is surjective for every $l \geq 0$. From the exact sequence

$$0 \to \mathcal{O}_S(l-1) \to \mathcal{O}_S(l) \to \mathcal{O}_C(l) \to 0,$$

we deduce that $h^1(\mathcal{O}_S(l-1) \leq h^1(\mathcal{O}_S(l))$ for every $l \geq 0$, and therefore, $h^1(\mathcal{O}_S(l)) = 0$ for every $l \geq 0$. In particular $q(S) = 0$. Also, $h^1(\mathcal{O}_S(1)) = 0$ and the exact sequence

$$0 \to \mathcal{O}_S \to \mathcal{O}_S(1) \to \omega_C \to 0$$

implies that $h^0(\omega_S) = h^2(\mathcal{O}_S) \geq h^1(\omega_C) = 1$, hence $K_S \geq 0$. On the other hand

$$\mathcal{O}_C(1) \cong \omega_C \cong \omega_S(1) \otimes \mathcal{O}_C$$

hence $K_S \cdot C = 0$ and this implies that $K_S = 0$. Thus, S is a K3 surface. □

We remark that singular extensions of canonical curves can exist and are studied in [41, 42].

We can now pair Proposition 2.2 and Lemma 3.1 to get the following way to estimate $\alpha(X)$.

Proposition 3.4 *Let $X \subset \mathbb{P}^r$ be a smooth irreducible nondegenerate variety of dimension $n \geq 1$. Let C be a smooth curve section of X and assume that it is linearly normal and $H^0(N_C(-2)) = 0$. Then*

$$\alpha(X) \leq \operatorname{cork}\Phi_{\mathcal{O}_C(1),\omega_C} - n + 1.$$

Proof Apply Lemma 3.1 and an iteration of Proposition 2.2. □

The emerging philosophy is that if one can control the corank of the Gaussian maps of the curve section of X, then one can give information on the extendability of X.

As far as we know, aside from passing to the curve section, there is only one other general result that allows studying the extendability of surfaces without passing to the hyperplane section, but considering suitable linear systems on the surface. This will be discussed in Sect. 8.

However, in [25], another interesting method for calculating $h^1(T_S(-1))$, when S is an Enriques surface, is employed. This method potentially generalizes to other surfaces.

Moreover, in [85, 86], a very nice method of studying extendability of varieties covered by lines is introduced. Some applications of this method are mentioned in the next section.

4 Old Days

Already, in the beginning of the last century, several mathematicians of the Italian and British schools, such as Castelnuovo, Del Pezzo, Scorza, Terracini, Edge, Semple and Roth, just to mention a few, studied the extendability problem.

For example, Scorza [89–91] proved that a Veronese variety of dimension $n \geq 2$ (see also [92] for $n = 2$) or a Segre variety, except a quadric surface, is not extendable. Terracini [96] also proved non-extendability of Veronese varieties. Di Fiore and Freni [39] proved that a Grassmannian in its Plücker embedding, except $\mathbb{G}(1,3) \subset \mathbb{P}^5$, is not extendable.

We can quickly recover these results by applying Proposition 2.5.

Proposition 4.1 *The following varieties $X \subset \mathbb{P}^r$ of dimension $n \geq 2$ and codimension at least* 1*, are not extendable:*

(i) any abelian variety or any finite unramified covering of such;
(ii) a Veronese variety;
(iii) a Segre variety, except a quadric surface;
(iv) a Grassmannian in its Plücker embedding, except $\mathbb{G}(1,3) \subset \mathbb{P}^5$.

Proof For the Veronese surface $X \subset \mathbb{P}^9$, observe that $H^1(\Omega^1_{\mathbb{P}^9|X}) = \mathbb{C}$ by (1). Now the map

$$\mathbb{C} = H^1(\Omega^1_{\mathbb{P}^9}) \to H^1(\Omega^1_{\mathbb{P}^9|X}) \to H^1(\Omega^1_X) = \mathbb{C}$$

is an isomorphism, hence so is $H^1(\Omega^1_{\mathbb{P}^9|X}) \to H^1(\Omega^1_X)$. By dualizing, we get the isomorphism

$$H^1(T_X(-1)) \to H^1(T_{\mathbb{P}^9|X}(-1)).$$

Now using (2) and Proposition 2.1(i) we get that $\alpha(X) = 0$, hence Theorem 1.3 applies.

In all other cases, it is easily seen (for coverings use [48, Proposition 2.7]) that $H^1(T_X(-1)) = 0$, hence Proposition 2.5 applies. □

More generally and recently Russo, using a description of the Hilbert scheme of lines passing through a general point, proves in [85, 86] that irreducible hermitian symmetric spaces in their homogeneous embedding and adjoint homogeneous manifolds, with some obvious exceptions, are not extendable.

A celebrated result that also deserves to be mentioned in the recollection of the old days is the so-called Babylonian tower theorem [13, 14, 36, 46, 97]:

Theorem 4.2 *Let $X \subset \mathbb{P}^r$ be a l.c.i. closed subscheme of pure codimension $c \geq 1$. If X extends to a l.c.i. closed subscheme $Y \subset \mathbb{P}^{r+m}$ of pure codimension c for all $m > 0$, then X is a complete intersection.*

On the other hand, if one drops the l.c.i. condition for all the extensions, then it is easy to get examples of smooth varieties that are infinitely extendable and are not complete intersections, such as arithmetically Cohen–Macaulay space curves (a more general result is in [11]).

We end this section by just mentioning many very nice results obtained by Beltrametti, Sommese, Van de Ven, Fujita, Badescu, Serrano, Lanteri, Palleschi, and in general by the people working in adjunction theory. It would be too long, for the purposes of this survey, to recall them here. We merely recall a short list of some relevant papers here [47, 48, 64–66, 93–95], referring the reader to [17] and references therein.

5 Extendability of Canonical Curves, K3 Surfaces and Fano Threefolds

The most beautiful, in my view, part of the story is the incredible amount of nice mathematics that revolved around the study of extendability of canonical curves, still going strong today. This is very much intertwined with the study of extendability of K3 surfaces, as the second part of Proposition 3.3 shows.

Throughout this section, we will let $C \subset \mathbb{P}^{g-1}$ be a canonically embedded smooth irreducible curve of genus $g \geq 3$ (just called a *canonical curve*).

While for $g = 3, 4$, we know that C is a complete intersection in $\mathbb{P}^{g-1}$, namely a plane quartic or a complete intersection of a quadric and a cubic in $\mathbb{P}^3$, the first case to be considered, in terms of extendability, is $g = 5$. In that genus, the first phenomenon occurs: the curve might be trigonal or not. It is a famous theorem of Enriques and Petri that then C is the intersection of three quadrics (hence extendable) if and only if it is not trigonal. Thus, the Brill–Noether theory of C comes into play, as far as extendability is concerned. As we will see, this is one of the main themes.

We now concentrate on what happens for a general curve, the meaning of general to be clear along the way.

In a series of beautiful papers, the cases of genus $6 \le g \le 11$ were settled by Mori and Mukai.

Theorem 5.1 ([71, 73, 75–77])
Let $C \subset \mathbb{P}^{g-1}$ be a canonical curve of genus g. Then:

(i) If $g = 6$, then C extends to a quadric section of $G(2, 5)$ if and only if C has finitely many g^1_4's;
(ii) If $g = 7$, then C extends to a section of $OG(5, 10)$ if and only if C has no g^1_4's;
(iii) If $g = 8$, then C extends to a section of $G(2, 6)$ if and only if C has no g^2_7's;
(iv) If $g = 9$, then C extends to a section of $SpG(3, 6)$ if and only if C has no g^1_5's;
(v) If $g = 10$, then a general C is not smoothly extendable;
(vi) If $g = 11$, then a general C is smoothly extendable.

In the above theorem, $G(k, n)$ is the usual Grassmannian, $OG(k, 2n)$ is the orthogonal Grassmannian, that is the variety that parametrizes k-dimensional vector subspaces of $\mathbb{C}^{2n}$ that are isotropic with respect to a nondegenerate symmetric bilinear form on $\mathbb{C}^{2n}$ and $SpG(n, 2n)$ is the symplectic Grassmannian, that is, the Grassmannian of Lagrangian subspaces of a $2n$-dimensional symplectic vector space.

Taking into account the second part of Proposition 3.3, the above theorem clearly suggests that one should look more closely at the moduli spaces of curves and K3 surfaces.

Let $\mathcal{M}_g$ be the moduli space of curves of genus g. Let $\mathcal{K}_g$ be the moduli stack of $K3$ surfaces of genus g, that is pairs (S, L) with S a smooth K3 surface, and L an ample, globally generated line bundle on S with $L^2 = 2g - 2$. Let $\mathcal{KC}_g$ be the moduli space of pairs (S, C) such that $(S, \mathcal{O}_S(C)) \in \mathcal{K}_g$. Then one has a forgetful morphism

$$c_g : \mathcal{KC}_g \to \mathcal{M}_g \tag{7}$$

and clearly the question is when is c_g dominant. Since $\dim \mathcal{KC}_g = g + 19$ and $\dim \mathcal{M}_g = 3g - 3$ one gets that this is possible if $g \le 11$. Now, on the one hand, Theorem 5.1 shows that c_g is dominant if $g \le 9$ or $g = 11$. On the other hand, a clean connection with Wahl's theorem was made by Ciliberto, Harris and Miranda (later proved also by Voisin).

Theorem 5.2 ([32, 98])
Let C be a general curve of genus $g = 10$ or $g \ge 12$. Then Φ_{ω_C} is surjective. In particular C, in its canonical embedding, is not extendable.

The second part (see also [50, Corollary p. 26]) follows by Proposition 3.3 and Zak-L'vovsky's theorem.

Once the dominance of c_g is settled, the next question in order is the study of its nonempty fibres, or, in other words, the study of how many ways can a hyperplane section of a K3 surface extend. A speculation about this was already made in [28], where some evidence was presented to the idea that the corank of the Wahl map, in that case, for $g = 11$ or $g \geq 13$, should be one. This speculation turned out to be true.

We will use the following.

Definition 5.3 A *prime K3 surface of genus g* is a smooth K3 surface $S \subset \mathbb{P}^g$ of degree $2g-2$ such that $\mathrm{Pic}(S)$ is generated by the hyperplane bundle. We denote by $\mathcal{H}_g$ the unique component of the Hilbert scheme of prime K3 surfaces of genus g.

The idea of [33] is to degenerate a prime K3 surface to the union of two rational normal scrolls and then to degenerate the latter to a union of planes whose hyperplane section is a suitable graph curve having corank one Wahl map. The result obtained is the following

Theorem 5.4 (corank one theorem [33, Theorem 4])
Suppose that $g = 11$ or $g \geq 13$. Let S be a K3 surface represented by a general point in $\mathcal{H}_g$ and let C be a general hyperplane section of S. Then $\mathrm{cork}\Phi_{\omega_C} = 1$.

As a matter of fact, when $g \geq 17$, the same result holds for any smooth section of S (see [34, Theorem 2.15 and Remark 2.23]).

Together with some other computations of coranks (see [33]), there are two immediate consequences.

The first one is about the map c_g (see (7)).

Theorem 5.5 ([33, Theorem 5])

(i) If $3 \leq g \leq 9$ and $g = 11$, then c_g is dominant and the general fibre is irreducible;
(ii) $\mathrm{codim}\,\mathrm{Im} c_{10} = 1$*;*
(iii) $\mathrm{codim}\,\mathrm{Im} c_{12} = 2$*;*
(iv) If $g = 11$ or $g \geq 13$, then c_g is birational onto its image;
(v) If $g = 11$ or $g \geq 13$, then a general canonical curve, that is a hyperplane section of a prime K3 surface, lies on a unique one, up to projective transformations.

The fact that c_g is birational onto its image was later reproved by Mukai [74, 80], Arbarello, Bruno and Sernesi [6] and Feyzbakhsh [43–45]. This is part of Mukai's beautiful program of reconstructing the K3 surface from a moduli space of sheaves on the hyperplane section. We will not be concerned about these matters here.

Returning now to the degeneration method explained after Definition 5.3, it allows also to prove that $H^0(N_C(-2)) = 0$ for a general canonical curve C of genus $g \geq 7$ that is hyperplane section of a prime K3 surface [33, Lemma 4]. Together with the "corank one theorem" (Theorem 5.4), this gives rise to a simple application of Zak-L'vovsky's theorem and Proposition 3.3.

We will use the following.

Definition 5.6 A *prime Fano threefold of genus g* is a smooth anticanonically embedded threefold $X \subset \mathbb{P}^{g+1}$ of degree $2g-2$ such that $\mathrm{Pic}(X)$ is generated by the hyperplane bundle. We denote by $\mathcal{V}_g$ the Hilbert scheme of prime Fano threefolds of genus g.

Then we have the following results, reproving in a quite simple way the important classification results of Fano threefolds [54, 55, 78, 79].

Theorem 5.7 ([33, Theorems 6 and 7])

(i) If $g = 11$ or $g \geq 13$, then there is no prime Fano threefold of genus g;
(ii) If $6 \leq g \leq 10$ or $g = 12$, then $\mathcal{V}_g$ is irreducible and the examples of Fano and Iskovskih fill out all of $\mathcal{V}_g$.

Concerning the last statement, the meaning is that just the knowledge of an example for every g allows concluding that they fill out the Hilbert scheme. The main idea behind (ii) (and also behind Theorem 5.9 below) is that a prime Fano threefold is projectively Cohen–Macaulay, hence (see [82, p. 46]) flatly degenerates to the cone over its hyperplane section S. Now, knowledge of the cohomology of the normal bundle of S allows proving that this cone is a smooth point of the Hilbert scheme. Then to obtain irreducibility one uses the fact that such K3 surfaces S are general in $\mathcal{H}_g$.

Another important observation in [33, Table 2] is that, if $6 \leq g \leq 10$ or $g = 12$, and C is a general hyperplane section of a general $K3$ surface of genus g, then $\operatorname{cork}\Phi_{\omega_C} > 1$, thus suggesting that the curve might be precisely $(\operatorname{cork}\Phi_{\omega_C})$-extendable.

Also, this turned out to be true and related to another milestone in the study of Fano varieties: Fano threefolds of index greater than 1 and Mukai varieties.

We will state, for simplicity of exposition, only the results about Picard rank 1 and $g \geq 3$. The interested reader can consult [26, 34] for further results.

Definition 5.8 For $r \geq 2$, let $\mathcal{H}_{r,g}$ be the component of the Hilbert scheme whose general elements are obtained by embedding prime K3 surfaces of genus g via the r-th multiple of the primitive class.

We denote by $\mathcal{V}_{n,r,g}$ the Hilbert scheme of smooth nondegenerate varieties $X \subset \mathbb{P}^N$ of dimension $n \geq 3$, such that $\rho(X) = 1$ and whose general surface section is a K3 surface represented by a point in $\mathcal{H}_{r,g}$. For $n \geq 4$, a *Mukai variety* is a variety X with $\rho(X) = 1$ represented by a point in $\mathcal{V}_{n,1,g}$.

With similar computations of coranks and cohomology of the normal bundle, the following classification results were obtained in [34] (note also [27], even though the classification results are not affected).

The first one is for Fano threefolds of index $r \geq 2$. Again, as the classification was known before, the novelty is (ii).

Theorem 5.9 ([34, Theorem 3.2])

(i) If $(r, g) \notin \{(2, 3), (2, 4), (2, 5), (2, 6), (3, 4), (4, 3)\}$, then there is no Fano threefold in $\mathcal{V}_{3,r,g}$;
(ii) If $(r, g) \in \{(2, 3), (2, 4), (2, 5), (2, 6), (3, 4), (4, 3)\}$, then $\mathcal{V}_{3,r,g}$ is irreducible and the examples of Fano and Iskovskih form a dense open subset of smooth points of $\mathcal{V}_{3,r,g}$.

Similarly, for Mukai varieties, classified by Mukai himself [79], we have (for the definition of $n(g)$ see [34, Table 3.14]) the following, where again (ii) was new.

Theorem 5.10 ([34, Theorem 3.15])

(i) If $(r, g) \notin \{(1, s), 6 \leq s \leq 10, (2, 5)\}$ or if $(r, g) \in \{(1, s), 6 \leq s \leq 10, (2, 5)\}$ and $n > n(g)$, then there is no Mukai variety in $\mathcal{V}_{n,1,g}$;
(ii) If $(r, g) \in \{(1, s), 6 \leq s \leq 10, (2, 5)\}$ and $4 \leq n \leq n(g)$, then $\mathcal{V}_{n,1,g}$ is irreducible and the examples of Mukai form a dense open subset of smooth points of $\mathcal{V}_{n,1,g}$.

So much for general canonical curves! What about extendability of *any* canonical curve?

One way to measure how far a curve is from being general is via its Clifford index, that we now recall.

Definition 5.11 Let C be a smooth irreducible curve of genus $g \geq 4$. The Clifford index of C is

$$\mathrm{Cliff}(C) = \min\{\deg L - 2r(L), L \text{ a line bundle on } C \text{ such that } h^0(L) \geq 2, h^1(L) \geq 2\}.$$

Recall that $\mathrm{Cliff}(C) = 0$ if and only if C is hyperelliptic; $\mathrm{Cliff}(C) = 1$ if and only if C is trigonal or isomorphic to a plane quintic; $\mathrm{Cliff}(C) = 2$ if and only if C is tetragonal or isomorphic to a plane sextic. A general curve of genus g has Clifford index $\lfloor \frac{g-1}{2} \rfloor$.

Already Beauville and Mérindol, in their proof of Wahl's theorem, observed that if a smooth irreducible curve C sits on a K3 surface S, then the surjectivity of Φ_{ω_C} implies the splitting of the normal bundle sequence

$$0 \to T_C \to T_{S|C} \to N_{C/S} \to 0.$$

This introduced the idea, exploited by Voisin in [98], that the elements of $(\mathrm{Coker}\, \Phi_{\omega_C})^\vee$ should be interpreted as ribbons, or infinitesimal surfaces, embedded in $\mathbb{P}^g$ and extending a canonical curve C. We will return to this later.

In 1996, Wahl proposed a possible converse to his theorem (Theorem 1.5), that, as we will see, is very much connected with Voisin's idea and further developments.

Theorem 5.12 ([103, Theorem 7.1])

Let $C \subset \mathbb{P}^{g-1}$ be a canonical curve of genus $g \geq 8$ and $\mathrm{Cliff}(C) \geq 3$. *Suppose that C satisfies*

$$H^1(\mathcal{I}^2_{C/\mathbb{P}^{g-1}}(k)) = 0 \text{ for every } k \geq 3. \tag{8}$$

Then C is extendable if and only if Φ_{ω_C} is not surjective.

Concerning the condition (8), Wahl proved that it is satisfied by a general canonical curve, while it is not satisfied by a general tetragonal curve of genus $g \geq 8$. He also conjectured that (8) holds for every canonical curve C with $\mathrm{Cliff}(C) \geq 3$. This conjecture was then proved, in almost all cases, in a beautiful paper by Arbarello, Bruno and Sernesi.

Theorem 5.13 ([8, Theorem 1.3])
Let $C \subset \mathbb{P}^{g-1}$ be a canonical curve of genus $g \geq 11$ and $\mathrm{Cliff}(C) \geq 3$. *Then*

$$H^1(\mathcal{I}^2_{C/\mathbb{P}^{g-1}}(k)) = 0 \text{ for every } k \geq 3.$$

Besides playing a crucial role in what follows, this theorem, together with Wahl's theorem and Theorem 5.12, gives

Corollary 5.14 ([8, Corollary 1.4])
Let $C \subset \mathbb{P}^{g-1}$ be a canonical curve of genus $g \geq 11$ and $\mathrm{Cliff}(C) \geq 3$. *Then C is extendable if and only if Φ_{ω_C} is not surjective.*

In the same paper Arbarello, Bruno and Sernesi, also proved another important result. We first recall

Definition 5.15 A smooth irreducible curve C is called a *Brill–Noether–Petri curve* if the multiplication map $H^0(L) \otimes H^0(K_C - L) \to H^0(K_C)$ is injective for every line bundle L on C.

Following an idea of Mukai, Voisin suggested in [98] to study the relation between the extendability of Brill–Noether–Petri curves to K3 surfaces and the non-surjectivity of Φ_{ω_C}. Wahl conjectured in [103] that for $g \geq 8$ this is an equivalence. While this conjecture turned out to be false [3, 7], it was essentially true and this is the other result by Arbarello, Bruno and Sernesi mentioned above.

Theorem 5.16 ([8, Theorem 1.1])
Let C be a Brill–Noether–Petri curve of genus $g \geq 12$. Then C lies on a polarized K3 surface, or on a limit thereof, if and only if its Wahl map is not surjective.

This circle of ideas and results was closed, as of now, by several beautiful theorems proved in [35]. The paper is full of very interesting results, we will describe only some of them.

The first one is that, conversely to Zak-L'vovsky's theorem, the condition $\alpha(X) \geq k$ is sufficient for the k-extendability of most canonical curves and K3 surfaces. This was obtained by a generalization of the method of Voisin and Wahl of integrating ribbons, as follows. Note that the vanishing in Theorem 5.13 is fundamental for this method to work.

Theorem 5.17 ([35, Theorem 2.1])
Let C be a smooth curve of genus g with $\mathrm{Cliff}(C) \geq 3$. *Consider the following two statements:*

(i) $\mathrm{cork}\Phi_{\omega_C} \geq k + 1$;

(ii) *There exists an arithmetically Gorenstein normal variety $Y \subset \mathbb{P}^{g+k}$, not a cone, with $\dim(Y) = k + 2, \omega_Y = \mathcal{O}_Y(-k)$, which has a canonical image of C as a section with a $(g-1)$-dimensional linear subspace of $\mathbb{P}^{g+k}$ (in particular, $C \subset \mathbb{P}^{g-1}$ is $(k+1)$-extendable).*

If $g \geq 11$, then (i) implies (ii). Conversely, if $g \geq 22$ and the canonical image of C is a hyperplane section of a smooth K3 surface in $\mathbb{P}^g$, then (ii) implies (i).

As a matter of fact, the extension Y in the theorem above is universal, in the sense that it gives all surface extensions of C, see [35, Corollary 5.5]. A general criterion for universal extensions of projective varieties was given in [104, Theorem A]. The latter paper contains also nice results about extending a variety to a Calabi–Yau variety.

As for the corank of the Wahl map, the following result is proved by Ciliberto, Dedieu and Sernesi.

Theorem 5.18 ([35, Corollary 2.10])

Let C be a smooth curve of genus $g > 37$ with $\mathrm{Cliff}(C) \geq 3$. If the canonical model of C is a hyperplane section of a K3 surface S, possibly with ADE singularities, then $\mathrm{cork}\,\Phi_{\omega_C} = 1$.

It is an intriguing open question to find examples of higher corank.

Question 1 [35, Question 2.14]

Does there exist any Brill–Noether general curve of genus $g \geq 11, g \neq 12$, such that $\mathrm{cork}\,\Phi_{\omega_C} > 1$?

As explained in [35, Remark 2.13], all canonical curves in smooth Fano threefolds with Picard number greater than 1 are Brill–Noether special.

Before stating the next result, let us define

Definition 5.19 The Clifford index $\mathrm{Cliff}(S, L)$ of a polarized K3 surface (S, L) is the Clifford index of any smooth curve $C \in |L|$.

Note that this does not depend on the choice of C by [49]. In the definition it is assumed that a general curve in $|L|$ is smooth and irreducible, which, being L ample, happens by [87, Proposition 8.1 and Theorem 3.1] unless $L = kE + R$ with $k \geq 3, E^2 = 0, R^2 = -2, E.R = 1$.

Now if C is a smooth hyperplane section of a K3 surface $S \subset \mathbb{P}^g$, then $\mathrm{Cliff}(C) = \mathrm{Cliff}(S, L)$, and in [35, Corollary 2.8], as a consequence of the proof of Theorem 5.22 below, the authors prove that

$$\mathrm{cork}\,\Phi_{\omega_C} = h^1(T_S(-1)) + 1.$$

Then Theorem 5.17 gives

Theorem 5.20 ([35, Theorem 2.18])

Let $(S, L) \in \mathcal{K}_g$ be a polarized K3 surface with $\mathrm{Cliff}(S, L) \geq 3$. Consider the following two statements:

(i) $h^1(T_S \otimes L^\vee) \geq k$;
(ii) *There exists an arithmetically Gorenstein normal variety* $X \subset \mathbb{P}^{g+k}$, *not a cone, with* $\dim(X) = k+2$, $\omega_X = \mathcal{O}_X(-k)$, *having the image of S by the linear system* $|L|$ *as a section with a g-dimensional linear subspace of* $\mathbb{P}^{g+k}$.

If $g \geq 11$, *then (i) implies (ii). Conversely, if* $g \geq 22$, *then (ii) implies (i).*

On the other hand, extendability of K3 surfaces, is possible only for bounded genus, using results of Prokhorov [84] and Karzhemanov [56, 58]:

Theorem 5.21 ([35, 2.9])
Let $S \subset \mathbb{P}^g$ *be a K3 surface possibly with ADE singularities. If* $g = 35$ *or* $g \geq 38$, *then S is not extendable.*

This bound is sharp [84, 105]. Moreover, the example of genus 37 is a K3 surface that is extendable but not smoothly extendable.

As for the map c_g (see (7)), Ciliberto, Dedieu and Sernesi prove that the fibres are smooth over curves with Clifford index at least 3 and $g \geq 11$.

Theorem 5.22 ([35, Theorem 2.6])
Let $(S, C) \in \mathcal{KC}_g$ *with* $g \geq 11$ *and* $\operatorname{Cliff}(C) \geq 3$. *Then*

$$\dim \operatorname{Ker} dc_{g\,(S,C)} = \dim c_g^{-1}(C) = \operatorname{cork} \Phi_{\omega_C} - 1.$$

A similar theorem [35, Theorem 2.19] holds for the analogous map for Fano threefolds and K3 surfaces. See also [24, 26] for the map c_g in the non-primitive case.

Another interesting case on which one can compute the corank of the Wahl map is for curves C lying on a Gorenstein surface S with $h^1(\mathcal{O}_S) = 0$. As a matter of fact, it happens often that $\operatorname{cork} \Phi_{\omega_C} = h^0(-K_S)$ (see [104]). If, in addition, $K_S + C$ is ample and C is not one of a special list (see loc. cit.), then Wahl gives in [104, Theorem 4.5] an explicit construction of the universal extension of C in the canonical embedding. In particular, all extensions of C have bad singularities. See Remark 5.24 for the case of plane curves.

We end this section by recalling what happens, and what is known, for curves C with $\operatorname{Cliff}(C) \leq 2$.

The Wahl map, in these cases, was studied, among others, by Ciliberto–Miranda, Wahl and Brawner. We collect here what is known (see [20, 22] for more details).

Theorem 5.23 ([20–22, 28, 100, 102])
Let C be a smooth irreducible curve of genus g:

(i) *If C is hyperelliptic of genus* $g \geq 2$, *then* $\operatorname{cork} \Phi_{\omega_C} = 3g - 2$ *(in fact this characterizes hyperelliptic curves);*
(ii) *If C is trigonal and* $g \geq 4$, *then* $\operatorname{cork} \Phi_{\omega_C} = g + 5$;
(iii) *If C is a plane curve of degree at least* 5, *then* $\operatorname{cork} \Phi_{\omega_C} = 10$;
(iv) *If C is a general tetragonal curve and* $g \geq 7$, *then* $\operatorname{cork} \Phi_{\omega_C} = 9$.

Remark 5.24 In [35, Example 9.7] (see also [104]), a 10-extension of the canonical embedding of any smooth plane curve of degree $d \geq 4$ is constructed. Together with Zak-L'vovsky's theorem, this shows that, if $d \geq 7$, then the curve is precisely $(\operatorname{cork} \Phi_{\omega_C})$-extendable. On the other hand, if $d = 5, 6$, then Zak-L'vovsky's theorem does not apply since $H^0(N_{C/\mathbb{P}^{g-1}}(-2)) \neq 0$ (see [28]) and $\alpha(C) = \operatorname{cork} \Phi_{\omega_C} = 10 > g - 1$. As it turns out, these curves actually are $(\alpha(C) + 1)$-extendable (see [24, Remark 3.7] and the appendix).

As far as we know, it is not known if trigonal or tetragonal curves are precisely $(\operatorname{cork} \Phi_{\omega_C})$-extendable.

6 Extendability of Enriques Surfaces and Enriques-Fano Threefolds

For Enriques surfaces, we also have that they are not extendable if the sectional genus is large enough. It is an application of Theorem 8.1.

Theorem 6.1 ([62])
Let $S \subset \mathbb{P}^r$ be an Enriques surface of sectional genus $g \geq 18$. Then S is not extendable.

A more precise result for $g = 15$ and 17 is proved in [62, Proposition 12.1]. Actually, a complete list of line bundles associated to possibly extendable very ample linear systems on an Enriques surface is available to the authors. See also below for more precise results.

As for the extendability of Enriques surfaces, let us define

Definition 6.2 An *Enriques–Fano threefold* is an irreducible three-dimensional variety $X \subset \mathbb{P}^N$ having a hyperplane section S that is a smooth Enriques surface, and such that X is not a cone over S. We will say that X has genus g if g is the genus of its general curve section.

In analogy with Fano threefolds, we obtain a genus bound. The bound was also proved, with completely different methods, by Prokhorov [83] (see also [56, 57]). Moreover, he produced an example of genus 17. For possible genera, see also [15, 88] for cyclic quotient terminal singularities and [62, Proposition 13.1], [83] for examples with other singularities. A comprehensive list of all known Enriques–Fano threefolds and their properties can be found in Martello's thesis [70].

Theorem 6.3 ([62, Theorem 1.5])
Any Enriques–Fano threefold has genus $g \leq 17$.

Both the study of extendability of Enriques surfaces and the moduli map analogous to the one on K3 surfaces (see (7)) were recently vastly extended in [25]. We will recall only some of the many results contained in that paper, referring the reader to [25] for a more detailed version.

Definition 6.4 Let S be an Enriques surface and let L be a line bundle on S such that $L^2 > 0$. We set

$$\phi(L) = \min\{E \cdot L : E \in NS(S), E^2 = 0, E > 0\}.$$

Let $\mathcal{E}_{g,\phi}$ be the moduli space of pairs (S, L) where S is an Enriques surface, L is an ample line bundle with $L^2 = 2g - 2$ and $\phi(L) = \phi$. Let $\mathcal{EC}_{g,\phi}$ be the moduli space of triples (S, L, C) where $(S, L) \in \mathcal{E}_{g,\phi}$ and $C \in |L|$ is a smooth irreducible curve. Let $\mathcal{R}_g$ be the moduli space of Prym curves, that is of pairs (C, η) with C a smooth irreducible curve of genus g and η a non-zero 2-torsion element of $\mathrm{Pic}^0(C)$.

Note that $\mathcal{E}_{g,\phi}$ has in general many irreducible components. There is a classification in low genus in [25, Sect. 2] and more generally in [59].

We have the diagram

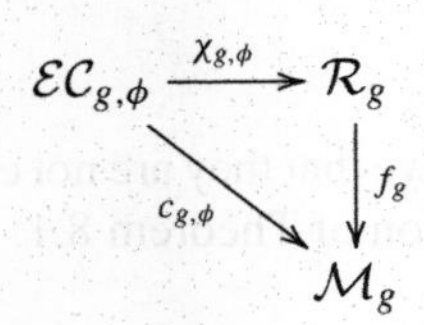

where $\chi_{g,\phi}(S, L, C) = (C, K_{S|C})$, f_g is the degree $2^{2g} - 1$ forgetful covering map and $c_{g,\phi} = f_g \circ \chi_{g,\phi}$.

Then we have

Theorem 6.5 ([25, Theorem 1, 2, 3])

(i) If $\phi \geq 3$, then $\chi_{g,\phi}$ is generically injective on any irreducible component of $\mathcal{EC}_{g,\phi}$, with the exception of 10 components where the dimension of the general fibre is given in the list in [25, Theorem 1];
(ii) $\chi_{g,2}$ is generically finite on all irreducible components of $\mathcal{EC}_{g,2}$ when $g \geq 10$. For $g \leq 9$ the dimension of a general fibre of $\chi_{g,2}$ on the various irreducible components of $\mathcal{EC}_{g,2}$ is given in the list in [25, Theorem 2];
(iii) The dimension of a general fibre of $\chi_{g,1}$ and of $c_{g,1}$ is $\max\{10 - g, 0\}$. Hence $c_{g,1}$ dominates the hyperelliptic locus if $g \leq 10$ and is generically finite if $g \geq 10$.

This has the following nice consequence

Corollary 6.6 ([25, Corollary 1.1])

A general curve of genus 2, 3, 4 and 6 lies on an Enriques surface, whereas a general curve of genus 5 or ≥ 7 does not. A general hyperelliptic curve of genus g lies on an Enriques surface if and only if $g \leq 10$.

The authors then went on to study the extendability of Enriques surfaces (related to the fibre dimension above), proving

Theorem 6.7 ([25, Corollary 1.2])

Let $S \subset \mathbb{P}^r$ be an Enriques surface not containing any smooth rational curve. If S is 1-extendable, then $(S, \mathcal{O}_S(1))$ belongs to the following list (see [25] for precise definitions)

$$\mathcal{E}_{17,4}^{(IV)^+}, \mathcal{E}_{13,4}^{(II)^+}, \mathcal{E}_{13,3}^{(II)}, \mathcal{E}_{10,3}^{(II)}, \mathcal{E}_{9,4}^{+}, \mathcal{E}_{9,3}^{(II)}, \mathcal{E}_{7,3}.$$

Furthermore, the members of this list are all at most 1-extendable, except for members of $\mathcal{E}_{10,3}^{(II)}$, which are at most 2-extendable, and of $\mathcal{E}_{9,4}^{+}$, which are at most 3-extendable.

See also [25, Remark 6.1] for nodal Enriques surfaces.

7 Extendability of Curves in Other Embeddings

We will collect a sample of the non-extendability results for curves. We will not treat, in this section, the case of canonical curves, as it was the object of a specific section.

Perhaps the first significant one is the non-extendability of elliptic normal curves of degree at least 10, as followed by Del Pezzo [37] and Nagata's [81] classification of surfaces of degree d in $\mathbb{P}^d$. Another important result is Castelnuovo–Enriques's theorem [23, 40], that proves that for a curve of degree large enough with respect to the genus, the only smooth extensions can be ruled surfaces (see [31, 51] for a modern proof; see also [50]).

There are of course many other results in this direction. Our choice is to mention the ones that are relevant to the approach via Zak-L'vovsky's theorem and Gaussian maps.

One important distinction to be made is between the case of *any* curve, and the case of a *general* curve, which is a curve with general moduli.

For a given curve, we have the following results according to the Clifford index of C.

Theorem 7.1 ([61, Corollary 2.10], [19, Theorem 2])

Let $C \subset \mathbb{P}^r$ be a smooth irreducible nondegenerate linearly normal curve of genus $g \geq 4$ and degree d. Then C is not extendable if:

(i) *C is trigonal, $g \geq 5$ and $d \geq \max\{4g-6, 3g+7\}$;*
(ii) *C is a plane quintic and $d \geq 26$;*
(iii) $\mathrm{Cliff}(C) = 2$ *and* $d \geq 4g-3$;
(iv) $\mathrm{Cliff}(C) \geq 3$ *and* $d \geq 4g+1-3\,\mathrm{Cliff}(C)$.

For curves with general moduli, we have

Theorem 7.2 ([67, Corollary 1.7])

Let $C \subset \mathbb{P}^r$ be a curve of genus g with general moduli and degree d. Then C is not extendable if:

(i) $d \geq 2g+15$ *for* $3 \leq g \leq 4$;

(ii) $d \geq 2g + 13$ *for* $5 \leq g \leq 8$;
(iii) $d \geq 2g + 10$ *for* $g \geq 9$.

For general embeddings of curves with general moduli, we have

Theorem 7.3 ([67, Corollary 1.7])
Let $C \subset \mathbb{P}^r$ *be a general linearly normal degree* d *embedding a curve of genus* g *with general moduli. Then* C *is not extendable if:*

(i) $d \geq g + 14$ *for* $3 \leq g \leq 4$;
(ii) $d \geq g + 12$ *for* $5 \leq g \leq 8$;
(iii) $d \geq g + 9$ *for* $g \geq 9$.

Another possibility is to study the extendability of curves in terms of general Brill–Noether embeddings.

Theorem 7.4 ([67, Theorem 1.9])
Let d, g, r *be integers such that* $\rho(d, g, r) \geq 0, d < g + r, r \geq 11$ *or* $r = 9$. *If* C *is a curve with general moduli and* L *is a general line bundle in* $W^r_d(C)$, *then* $C \subset \mathbb{P}^r = \mathbb{P}H^0(L)$ *is not extendable.*

8 Extendability of Other Surfaces

In this section, we will outline some results about the extendability of surfaces that are neither $K3$ nor Enriques surfaces.

Most results that we will mention are applications of the following:

Theorem 8.1 ([62, Theorem 1.1])
Let $S \subset \mathbb{P}^r$ *be a smooth irreducible linearly normal surface and let* H *be its hyperplane bundle. Assume there is a base-point free and big line bundle* D_0 *on* S *with* $H^1(H - D_0) = 0$ *and such that the general element* $D \in |D_0|$ *is not rational and satisfies:*

(i) the Gaussian maps $\Phi_{H_D, \omega_D(D_0)}$ *is surjective;*
(ii) the multiplication maps μ_{V_D, ω_D} *and* $\mu_{V_D, \omega_D(D_0)}$ *are surjective*

where $V_D := \mathrm{Im}\{H^0(S, H - D_0) \to H^0(D, (H - D_0)_{|D})\}$. *Then*

$$\alpha(S) \leq \operatorname{cork} \Phi_{H_D, \omega_D}.$$

Now define

Definition 8.2 Let S be a smooth surface and let L be an effective line bundle on S such that the general divisor $D \in |L|$ is smooth and irreducible. We say that L is *trigonal*, if D is such. We denote by $\mathrm{Cliff}(L)$ the Clifford index of D. Moreover, when $L^2 > 0$, we set

$\varepsilon(L) = 3$ if L is trigonal; $\varepsilon(L) = 5$ if $\mathrm{Cliff}(L) \geq 3$, $\varepsilon(L) = 0$ if $\mathrm{Cliff}(L) = 2$;

$$m(L) = \begin{cases} \frac{16}{L^2} & \text{if } L.(L+K_Y) = 4; \\ \frac{25}{L^2} & \text{if } L.(L+K_Y) = 10 \text{ and the general} \\ & \text{divisor in } |L| \text{ is a plane quintic;} \\ \frac{3L.K_Y+18}{2L^2} + \frac{3}{2} & \text{if } 6 \leq L.(L+K_Y) \leq 22 \text{ and } L \text{ is trigonal;} \\ \frac{2\,L.K_Y - \varepsilon(L)}{L^2} + 2 & \text{otherwise.} \end{cases}$$

We give first a general result, application of Theorem 8.1, that holds on any surface, once given a suitable embedding.

Theorem 8.3 ([62, Corollary 3.3])
Let $S \subset \mathbb{P}V$ be a smooth irreducible surface with $V \subseteq H^0(mL + D)$, where L is a base-point free, big, nonhyperelliptic line bundle on S with $L \cdot (L + K_S) \geq 4$ and $D \geq 0$ is a divisor. Suppose that S is regular or linearly normal and that m is such that $H^1((m-2)L + D) = 0$ and $m > \max\{m(L) - (L \cdot D)/L^2, (L \cdot K_S + 2 - L \cdot D)/L^2 + 1\}$. Then S is not extendable.

See also [62, Corollary 3.5] for a better result in the case of adjoint embeddings.
We now list what is known in terms of the Kodaira dimension. We will concentrate on minimal surfaces.

8.1 Surfaces of Kodaira Dimension $-\infty$

For embeddings of $\mathbb{P}^2$, the non-extendability was settled by Scorza [90]. See also Proposition 4.1.

As explained in [17, Sect. 5.5], if a variety X is a $\mathbb{P}^d$-bundle over a smooth variety and X is an ample divisor on a locally complete intersection variety Y, then in almost all cases, including the case of surfaces, Y is a $\mathbb{P}^{d+1}$-bundle over the same base and X belongs to the tautological linear system. This shows that l.c.i. extensions of $\mathbb{P}^d$-bundles are $\mathbb{P}^{d+1}$-bundles.

On the other hand, as far as we know, not so many results are known, in general, about the extendability of ruled surfaces.

A result was proved in [38, Theorem 3.3.22] for rational ruled surfaces, as an application of Theorem 8.1. We believe that this can be extended to other ruled surfaces.

8.2 Surfaces of Kodaira Dimension 0

The cases of K3 and Enriques surfaces have been treated above. Abelian and bielliptic surfaces are not extendable by Proposition 4.1.

8.3 Surfaces of Kodaira Dimension 1

The only results that we are aware of are the ones in [68], which hold especially for Weierstrass fibrations.

8.4 Surfaces of General Type

We are not aware of results, except the one below, that applies, for example, for pluricanonical embeddings.

Theorem 8.4 ([62, Corollary 1.2]) *Let $S \subset \mathbb{P}V$ be a minimal surface of general type with base-point free and nonhyperelliptic canonical bundle and $V \subset H^0(mK_S + D)$), where $D \geq 0$ and either D is nef or D is reduced and K_S is ample. Suppose that S is regular or linearly normal and that*

$$m \geq \begin{cases} 9 & \text{if } K_Y^2 = 2; \\ 7 & \text{if } K_Y^2 = 3; \\ 6 & \text{if } K_Y^2 = 4 \text{ and the general curve in } |K_Y| \text{ is trigonal or if} \\ & K_Y^2 = 5 \text{ and the general curve in } |K_Y| \text{ is a plane quintic;} \\ 5 & \text{if either the general curve in } |K_Y| \text{ has Clifford index 2 or} \\ & 5 \leq K_Y^2 \leq 9 \text{ and the general curve in } |K_Y| \text{ is trigonal;} \\ 4 & \text{otherwise.} \end{cases}$$

Then S is not extendable.

9 Appendix: Extendability of Canonical Models of Plane Quintics

by THOMAS DEDIEU

Let C be a smooth plane curve of degree $d \geq 5$, which we most often consider in its canonical embedding in $\mathbf{P}^{g-1}$, $g = \frac{1}{2}(d-1)(d-2)$. By [102, Remark 4.9] one has $\alpha(C) = 10$. If $d \geq 7$ then C has Clifford index strictly larger than 2, hence $h^0(N_{C/\mathbf{P}^{g-1}}(-k)) = 0$ for all $k > 1$, and Theorems 1.3 and 5.17 apply together, to the effect that C is precisely 10-extendable; a universal extension of C is described in [35, Sect. 9], which shows that all surface extensions of C are rational. If $d = 5$ (resp. 6), then C has Clifford index 1 (resp. 2), $h^0(N_{C/\mathbf{P}^{g-1}}(-2)) = 3$ (resp. 1), and $h^0(N_{C/\mathbf{P}^{g-1}}(-k)) = 0$ for all $k > 2$, see [60] and the references therein. The expectation in this case is that there should exist a 12 (resp. 10) dimensional family of non-isomorphic surface extensions of C, and inside this a 2 (resp. 0) dimensional family of surfaces in which the first infinitesimal neighbourhood of C is trivial, see [26, Sect. 1]. The case $d = 6$ is studied in detail in [26] and [24, Sect. 3.2]: the above expectations are met, and a suitable embedding of the double cover of $\mathbf{P}^2$ branched over C is a $K3$ surface extending C; this $K3$ surface is an anticanonical divisor of the weighted projective space $\mathbf{P}(1^3, 3)$, and so are all $K3$ extensions of C. In this appendix I show that a similar situation holds in the case $d = 5$.

I am grateful to the anonymous referee for pointing out [87, Theorem 7.2], which triggered a substantial improvement of the results presented in this text.

Proposition 9.1 *Let C be a smooth plane quintic, in its canonical embedding $C \subset \mathbf{P}^5$. The moduli space of $K3$ extensions of C has dimension* 12. *Inside the moduli space of $K3$ extensions of C, there is a 2-dimensional family of extensions in which the first infinitesimal neighbourhood of C is trivial. There exists a* 14*-extension of C, which is an arithmetically Gorenstein normal Fano variety of dimension* 15 *and index* 13.

The precise meaning of 9.1 is that the fibre of the moduli map $c_6 : \mathcal{KC}_6 \to \mathcal{M}_6$ (introduced in (7)) over a general plane quintic has dimension 12. By way of comparison, note that c_6 is surjective and has general fibre of dimension 10.

Our story begins with the following construction of Ide [53, Sect. 4.2].

Construction 9.2 *Let f be a degree* 5 *homogeneous polynomial in $\mathbf{x} = (x_0, x_1, x_2)$ defining $C \subset \mathbf{P}^2$, and ℓ a linear functional in $\mathbf{x}$ defining a line Γ intersecting C transversely. Then the degree* 5 *weighted hypersurface S in $Y = \mathbf{P}(1^3, 2)$ defined by*

$$\ell(\mathbf{x})y^2 + f(\mathbf{x}) = 0,$$

in homogeneous coordinates $(\mathbf{x}, y)$ so that the x_i have weight 1 *and y has weight* 2*, is a $K3$ surface (with an ADE singularity), and C is the complete intersection of S and the degree* 2 *hypersurface Π defined by $y = 0$.*

A geometric construction of S is as follows. Let $\varepsilon : P' \to \mathbf{P}^2$ be the blow-up at the five points of $\Gamma \cap C$, and Γ' and C' be the proper transforms of Γ and C respectively. Let $\pi : S' \to P'$ be the double cover branched over the smooth curve $\Gamma' + C'$. The pull-back $\pi^*\Gamma'$ is a (-2)-curve, and S is obtained from S' by contracting it to an A_1 double point.

The latter contraction is realized by the map given by the complete linear system $|-\pi^*K_{P'}|$. Arguing as in [26, Proposition 2.2], one sees that the latter has dimension 6, has $\pi^*\Gamma'$ as a fixed part, and the general member of its moving part is mapped birationally by $\varepsilon \circ \pi$ to a smooth quintic C_1 passing through the 5 points of $\Gamma \cap C$ and otherwise everywhere tangent to C. Note that $|-K_{P'}|$ has Γ' as fixed part and the pull-back by ε of the linear system of conics as moving part, hence for all $D \in |C'|$ its restriction to D is the canonical divisor K_D. This implies that the image of the map given by $|-\pi^*K_{P'}|$ is a $K3$ surface $S \subset \mathbf{P}^6$ (with one ordinary double point), having C as a hyperplane section. As a sideremark, note that $|-\pi^* K_{P'}|$ maps the pull-backs by π of the 5 exceptional curves of ε to 5 lines passing through the node of S.

The above description is conveniently complemented by adopting the following equation based point of view.

Weighted projective geometry 9.3 *For useful background on the subject, I recommend [63, Exercises V.1.3] and [52]. The weighted projective space* $Y = \mathbf{P}(1^3, 2)$ *has only one singular point, namely the coordinate point* $(0 : 0 : 0 : 1)$ *at which it has a quotient singularity of type* $\frac{1}{2}(1, 1, 1)$*, that is an ordinary quadruple point (locally isomorphic to the cone over the Veronese surface* $v_2(\mathbf{P}^2)$*). It has dualizing sheaf* $\mathcal{O}(5)$*, which is not invertible. All (weighted) quintic hypersurfaces in* Y *pass through the singular point* $(0 : 0 : 0 : 1)$*, and the general such has an ordinary double point there.*

By adjunction, the general quintic surface in Y has trivial (invertible) canonical sheaf, hence is a $K3$ surface (with canonical singularities). Moreover, the general complete intersection of hypersurfaces of degrees 5 and 2 is smooth, with (invertible) canonical sheaf $\mathcal{O}(2)$.

Maintaining the notation of 9.2, the curve C may be embedded in Y as the complete intersection defined by

$$y = f(\mathbf{x}) = 0. \tag{9}$$

On the other hand, the automorphisms of Y are given in homogeneous coordinates by

$$(\mathbf{x} : y) \mapsto (A\mathbf{x} : ay + Q(\mathbf{x})) \tag{10}$$

with $A \in \mathrm{GL}(3)$, $a \in \mathbf{C}^*$, and $Q \in H^0(\mathbf{P}(1^3), \mathcal{O}(2))$, see for instance [1, Sect. 8]. One may therefore put every complete intersection of hypersurfaces of degrees 2 and 5 in the form (9) by acting with an automorphism of Y, which shows that it is isomorphic to a plane quintic curve.

The sheaf $\mathcal{O}(2)$ is invertible on Y, and the associated complete linear system induces an embedding $\phi : Y \to \mathbf{P}^6$ with image a cone over a Veronese surface $v_2(\mathbf{P}^2) \subset \mathbf{P}^5$, with vertex a point. The map ϕ sends S to a $K3$ surface of degree 10 in $\mathbf{P}^6$ passing through the vertex, and having C (in its canonical embedding) as a hyperplane section: in other words $\phi(S)$ is an extension of the canonical curve C. It coincides with the model of S in $\mathbf{P}^6$ given by $|-\pi^* K_{P'}|$ and described in 9.2. The sheaf $\mathcal{O}_Y(1)$ is not invertible on Y and neither is its restriction to S; the associated complete linear series induces the rational map $S \dashrightarrow \mathbf{P}^2$ coinciding with $\varepsilon \circ \pi$ off the node.

Proof (Proof of Proposition 9.1) As we have seen, we may consider C as the complete intersection in Y of two hypersurfaces of degrees 2 and 5 as in (9). The linear system of quintics containing C has dimension

$$h^0(Y, \mathcal{I}_{C/Y}(5)) - 1 = h^0(Y, \mathcal{O}_Y(3)) = 13,$$

and its general element gives a $K3$ surface extending $C \subset \mathbf{P}^5$. By the description in (10), the automorphisms of Y fixing the hypersurface $y = 0$ are all of the form $(\mathbf{x} : y) \mapsto (\mathbf{x} : ay)$, hence they form a 1-dimensional group. From this we conclude that there is a 12-dimensional family of mutually non-isomorphic $K3$ extensions of C.

On the other hand, by [87, Remark 7.13] every $K3$ surface containing C is contained in a cone over the Veronese surface, that is in $\mathbf{P}(1^3, 2)$ in its model given by $\mathcal{O}(2)$. By the computation of the divisor class group of weighted projective spaces [18, Theorem 7.1] and adjunction, such a surface is necessarily of the same kind as those considered above, i.e., it is a weighted quintic in $\mathbf{P}(1^3, 2)$ in which C is cut out by a quadric. Therefore the 12-dimensional family found above is the full fibre of c_6 over C, and 9.1 is proved.

Inside this family, there is a 2-dimensional family of surfaces in which the first infinitesimal neighbourhood of C is trivial, namely those surfaces constructed as in 9.2 by taking a double cover branched over the disjoint union of C itself and a line: that the infinitesimal neighbourhood is indeed trivial in this case follows from the argument given in [16, p. 875]. We get a 2-dimensional family by letting the line Γ move freely in $\mathbf{P}^2$: the double cover of the blow-up remembers the 5 points of $\Gamma \cap C$, hence two general lines give two non-isomorphic surfaces. This proves 9.1.

Eventually, to prove 9.1 we use Totaro's construction as in [24, Sect. 3.2]. Namely we take X the quintic hypersurface in $\mathbf{P}(1^3, 2, 2^{13})$ given in homogeneous coordinates $(\mathbf{x} : y : \mathbf{z})$ by

$$f(\mathbf{x}) + G_1(\mathbf{x}, y)z_1 + \cdots + G_{13}(\mathbf{x}, y)z_{13} = 0,$$

where $G_1, \ldots, G_{13}$ form a basis of $H^0(Y, \mathcal{O}(3))$, and embed it in $\mathbf{P}^{19}$ with the complete linear system $|\mathcal{O}_X(2)|$, which is $\frac{1}{13}$ of the anticanonical series. That $X \subset \mathbf{P}^{19}$ is arithmetically Gorenstein and normal follows from the fact that it has canonical curves as linear sections as in [35, Sect. 5]. $\square$

Remark 9.4 Not all the $K3$ extensions of a plane quintic C may be constructed as in 9.2, as a dimension count shows. In fact, by [87, Theorem 7.2] if S' is a smooth $K3$ surface containing C, then $C \sim 2L + \Gamma'$ in S', where L and Γ' are irreducible curves of respective genera 2 and 0, and $L \cdot \Gamma' = 1$. The linear system $|L|$ induces a double cover $\pi : S' \to \mathbf{P}^2$ branched over a sextic $B \subset \mathbf{P}^2$ mapping Γ' to a line Γ, and there are the following two possibilities.

(a) If Γ' is contained in the ramification locus of π, then B contains Γ, and in this case S' and its image S by $|C|$ are as in the Construction 9.2.

(b) If Γ' is not contained in the ramification locus of π, then Γ is a tri-tangent line to B and $\pi^*\Gamma = \Gamma' + \Gamma''$, with Γ'' another irreducible rational curve such that $L \cdot \Gamma'' = 1$, and $\Gamma' \cdot \Gamma'' = 3$. Then $L \sim \Gamma' + \Gamma''$ as it is linearly equivalent to the pull-back of a line by π. One thus has

$$C \sim 2L + \Gamma' \sim 3L - \Gamma'',$$

and the general member of $|C|$ is mapped birationally by π to a smooth plane quintic everywhere tangent to B. The map induced by $|C|$ contracts Γ' and maps Γ'' to a degree 5 curve spanning a $\mathbf{P}^3$ and with a space triple point at the image of Γ'.

Type (b) extensions of C are thus double covers of the plane branched over a sextic B which is everywhere tangent to the sextic $C + \Gamma$ for some line Γ. By [26, Theorem 4.1] there is a 10-dimensional family of such $K3$ extensions for each choice of Γ, hence a 12-dimensional family if we let Γ move. The general extension of C is thus of type (b).

This may also be seen directly from the equations. Let C be given in Y by equations $y = f(\mathbf{x}) = 0$ as in (9). Then the general quintic S in Y containing C has equation

$$\ell(\mathbf{x})y^2 + 2g(\mathbf{x})y + f(\mathbf{x}) = 0 \tag{11}$$

with g a non-zero degree 3 homogeneous polynomial, and ℓ as before a non-zero linear form. To see this surface birationally as a double cover of the plane, we put the equation (11) in the form

$$\ell\Big(y + \frac{g}{\ell}\Big)^2 + \Big(f - \frac{g^2}{\ell}\Big) = 0,$$

which shows that S is birational to the double cover branched over the plane sextic $(\ell f - g^2 = 0)$, which is everywhere tangent to both C and the line $(\ell = 0)$.

Remark 9.5 As has been observed above in 9.3 and 9.4, all $K3$ extensions of C are singular. This contrasts with the fact, pointed out to me by Edoardo Sernesi, that the rational surface extensions of canonical plane curves constructed in [35, Sect. 9] have one elliptic singularity which in the case of quintics is smoothable. These rational surface extensions may only be partially smoothed, to $K3$ surfaces with one ordinary double point, meaning that there is a global obstruction to a total smoothing.

Remark 9.6 As was the case for plane sextics [24, Remk 3.7], canonical plane quintics provide curves which are not complete intersections, for which the assumptions of either form of Theorem 1.3 don't hold, and which are extendable more than α times.

Moreover, canonical plane quintics show that Zak's claim [105, p. 278], that if an n-dimensional $X \subset \mathbf{P}^N$ has $\alpha(X) < (N - n - 1)(N + 1)$ then it is at most $\alpha(X)$-extendable, cannot hold without an additional assumption.

Remark 9.7 We can play the same game with plane quartics and cubics as with sextics and quintics. Let $\varepsilon : P' \to \mathbf{P}^2$ be the blow-up at the 8 (resp. 9) points of the transverse intersection $\Gamma \cap C$, with Γ a conic and C a smooth quartic (resp. $E_1 \cap E_2$, with E_1 and E_2 two smooth cubics), and $\pi : S' \to P'$ be the double cover branched over the proper transforms of Γ and C (resp. E_1 and E_2). In the former case, the linear system $| -\pi^* K_{P'}|$ maps S' to a quartic surface S in $\mathbf{P}(1^3, 1) = \mathbf{P}^3$ with an ordinary double point. In the latter case we get what I would describe as a virtual cubic surface in $\mathbf{P}(1^3, 0)$; the linear system $|-\pi^* K_{P'}|$ is then of the form $|2F|$ with F an elliptic curve, hence gives a map to $\mathbf{P}^2$ contracting S' onto a smooth conic.

References

1. A. Al Amrani, Classes d'idéaux et groupe de Picard des fibrés projectifs tordus, in *Proceedings of Research Symposium on K-Theory and its Applications (Ibadan, 1987)*, vol. 2 (1989), pp. 559–578
2. E. Arbarello, M. Cornalba, P.A. Griffiths. *Geometry of Algebraic Curves. Volume II. With a Contribution by Joseph Daniel Harris*. Grundlehren der Mathematischen Wissenschaften [Fundamental Principles of Mathematical Sciences], vol. 268 (Springer, Heidelberg, 2011). xxx+963 pp
3. E. Arbarello, A. Bruno, Rank-two vector bundles on Halphen surfaces and the Gauss-Wahl map for Du Val curves. J. Éc. polytech. Math. **4**, 257–285 (2017)
4. E. Arbarello, M. Cornalba, On a conjecture of Petri. Comment. Math. Helv. **56**(1), 1–38 (1981)
5. E. Arbarello, E. Sernesi, Petri's approach to the study of the ideal associated to a special divisor. Invent. Math. **49**(2), 99–119 (1978)
6. E. Arbarello, A. Bruno, E. Sernesi, Mukai's program for curves on a K3 surface. Algebr. Geom. **1**(5), 532–557 (2014)
7. E. Arbarello, A. Bruno, G. Farkas, G. Saccà, Explicit Brill-Noether-Petri general curves. Comment. Math. Helv. **91**(3), 477–491 (2016)
8. E. Arbarello, A. Bruno, E. Sernesi, On hyperplane sections of K3 surfaces. Algebr. Geom. **4**(5), 562–596 (2017)
9. L. Bădescu, Infinitesimal deformations of negative weights and hyperplane sections, in *Algebraic Geometry* (L'Aquila, 1988), 1–22, Lecture Notes in Mathematics, vol. 1417 (Springer, Berlin, 1990)
10. L. Bădescu, On a result of Zak-L'vovsky, in *Projective Geometry with Applications*, 57–73, *Lecture Notes in Pure and Applied Mathematics*, vol. 166 (Dekker, New York, 1994)
11. E. Ballico, Extending infinitely many times arithmetically Cohen-Macaulay and Gorenstein subvarieties of projective spaces. arXiv:2102.06457
12. E. Ballico, C. Fontanari, Gaussian maps, the Zak map and projective extensions of singular varieties. Res. Math. **44**(1–2), 29–34 (2003)
13. W. Barth, Submanifolds of low codimension in projective space, in *Proceedings of the International Congress of Mathematicians*, vol. 1 (Vancouver, B.C., 1974), pp. 409-413. Canad. Math. Congress, Montreal, Que., 1975
14. W. Barth, A. Van de Ven, A decomposability criterion for algebraic 2-bundles on projective spaces. Invent. Math. **25**, 91–106 (1974)
15. L. Bayle, Classification des variétés complexes projectives de dimension trois dont une section hyperplane générale est une surface d'Enriques. J. Reine Angew. Math. **449**, 9–63 (1994)
16. A. Beauville, J.Y. Mérindol, Sections hyperplanes des surfaces K3. Duke Math. J. **55**(4), 873–878 (1987)
17. M.C. Beltrametti, A.J. Sommese, The adjunction theory of complex projective varieties. De Gruyter Expositions in Mathematics, vol. 16 (Walter de Gruyter & Co., Berlin, 1995)
18. M. Beltrametti, L. Robbiano, Introduction to the theory of weighted projective spaces. Expo. Math. **4**, 111–162 (1986)
19. A. Bertram, L. Ein, R. Lazarsfeld, Surjectivity of Gaussian maps for line bundles of large degree on curves, in *Algebraic Geometry* (Chicago, IL, 1989), 15–25, *Lecture Notes in Mathematics*, vol. 1479 (Springer, Berlin, 1991)

20. J.N. Brawner, The Gaussian map Φ_K for curves with special linear series. Thesis (Ph.D.)-The University of North Carolina at Chapel Hill (1992), 73 p
21. J.N. Brawner, The Gaussian-Wahl map for trigonal curves. Proc. Am. Math. Soc. **123**(5), 1357–1361 (1995)
22. J.N. Brawner, Tetragonal curves, scrolls, and K3 surfaces. Trans. Am. Math. Soc. **349**(8), 3075–3091 (1997)
23. G. Castelnuovo, Massima dimensione dei sistemi lineari di curve piane di dato genere. Ann. Mat. **18**(2), 119–128 (1890)
24. C. Ciliberto, T. Dedieu, K3 curves with index $k>1$. Boll. dell'Unione Mat. Ital. **15** 87–115 (2022). arXiv:2012.10642
25. C. Ciliberto, T. Dedieu, C. Galati, A.L. Knutsen, Moduli of curves on Enriques surfaces. Adv. Math. **365**, 107010, 42 (2020)
26. C. Ciliberto, T. Dedieu. *Double covers and extensions*, to appear in Kyoto J. Math. arXiv:2008.03109
27. C. Ciliberto, A. Lopez, R. Miranda, Corrigendum to: classification of varieties with canonical curve section via Gaussian maps on canonical curves. Am. J. Math. **120**(1), 1–21 (1998). Am. J. Math. **143**(6), 1661–1663 (2021)
28. C. Ciliberto, R. Miranda, Gaussian maps for certain families of canonical curves, in *Complex Projective Geometry* (Trieste, 1989/Bergen, 1989), 106–127 *London Mathematical Society Lecture Note series*, vol. 179 (Cambridge University Press, Cambridge, 1992)
29. C. Ciliberto, Sul grado dei generatori dell'anello canonico di una superficie di tipo generale. Rend. Sem. Mat. Univ. Politec. Torino **41**(3), 83–111 (1983)
30. C. Ciliberto, R. Miranda, On the Gaussian map for canonical curves of low genus. Duke Math. J. **61**(2), 417–443 (1990)
31. C. Ciliberto, F. Russo, Varieties with minimal secant degree and linear systems of maximal dimension on surfaces. Adv. Math. **200**(1), 1–50 (2006)
32. C. Ciliberto, J. Harris, R. Miranda, On the surjectivity of the Wahl map. Duke Math. J. **57**(3), 829–858 (1988)
33. C. Ciliberto, A. Lopez, R. Miranda, Projective degenerations of K3 surfaces, Gaussian maps, and Fano threefolds. Invent. Math. **114**(3), 641–667 (1993)
34. C. Ciliberto, A. Lopez, R. Miranda, Classification of varieties with canonical curve section via Gaussian maps on canonical curves. Am. J. Math. **120**(1), 1–21 (1998)
35. C. Ciliberto, T. Dedieu, E. Sernesi, Wahl maps and extensions of canonical curves and K3 surfaces. J. Reine Angew. Math. **761**, 219–245 (2020)
36. I. Coandă, A simple proof of Tyurin's Babylonian tower theorem. Comm. Algebra **40**(12), 4668–4672 (2012)
37. P. del Pezzo, Sulle superficie dell'n^{mo} ordine immerse nello spazio di n dimensioni. Rend. del Circolo Matematico di Palermo **1**, 241–271 (1887)
38. L. Di Biagio, Some issues about the extendability of projective surfaces. Roma Tre Undergraduate Thesis (2004). http://ricerca.mat.uniroma3.it/users/lopez/Tesi-DiBiagio.pdf
39. L. Di Fiore, S. Freni, On varieties cut out by hyperplanes into Grassmann varieties of arbitrary indexes. Rend. Istit. Mat. Univ. Trieste **13**(1–2), 51–57 (1981)
40. F. Enriques, Sulla massima dimensione dei sistemi lineari di dato genere appartenenti a una superficie algebrica. Atti Reale Acc. Sci. Torino **29**, 275–296 (1894)
41. D.H.J. Epema, Surfaces with canonical hyperplane sections. CWI Tract 1. Stichting Mathematisch Centrum, Centrum voor Wiskunde en Informatica, Amsterdam (1984)
42. D.H.J. Epema, Surfaces with canonical hyperplane sections. Nederl. Akad. Wetensch. Indag. Math. **45**, 173–184 (1983)
43. S. Feyzbakhsh, Erratum to Mukai's program (reconstructing a K3 surface from a curve) via wall-crossing. J. Reine Angew. Math. **765**, 101–137. J. Reine Angew. Math. **768**(183) (2020)
44. S. Feyzbakhsh, Mukai's Program (reconstructing a K3 surface from a curve) via wall-crossing, II. arXiv:2006.08410
45. S. Feyzbakhsh, Mukai's program (reconstructing a K3 surface from a curve) via wall-crossing. J. Reine Angew. Math. **765**, 101–137 (2020)

46. H. Flenner, Babylonian tower theorems on the punctured spectrum. Math. Ann. **271**(1), 153–160 (1985)
47. T. Fujita, Classification theories of polarized varieties, in *London Mathematical Society Lecture Note Series*, vol. 155 *Cambridge University Press, Cambridge, 1990*, xiv+205 pp
48. T. Fujita, Impossibility criterion of being an ample divisor. J. Math. Soc. Japan **34**(2), 355–363 (1982)
49. M. Green, R. Lazarsfeld, Special divisors on curves on a K3 surface. Invent. Math. **89**(2), 357–370 (1987)
50. J. Harris, D. Mumford, On the Kodaira dimension of the moduli space of curves. With an appendix by William Fulton. Invent. Math. **67**(1), 23–88 (1982)
51. R.C. Hartshorne, Curves with high self-intersection on algebraic surfaces. Inst. Hautes Études Sci. Publ. Math. No. **36**, 111–125 (1969)
52. A.R. Iano-Fletcher, Working with weighted complete intersections, in *Explicit Birational Geometry of 3-folds*, London Mathematical Society Lecture Note Series, vol. 281 (Cambridge University Press, Cambridge, 2000), pp. 101–173
53. M. Ide, Every curve of genus not greater than eight lies on a $K3$ surface. Nagoya Math. J. **190**, 183–197 (2008)
54. V.A. Iskovskih, Fano threefolds. II. Izv. Akad. Nauk SSSR Ser. Mat. **42**(3), 506–549 (1978)
55. V.A. Iskovskih, Fano threefolds. I. Izv. Akad. Nauk SSSR Ser. Mat. **41**(3), 516–562, 717 (1977)
56. I.V. Karzhemanov, On Fano threefolds with canonical Gorenstein singularities. Mat. Sb. **200**(8), 111–146 (2009); translation in Sb. Math. **200**(7–8), 1215–1246 (2009)
57. I.V. Karzhemanov, On some Fano-Enriques threefolds. Adv. Geom. **11**(1), 117–129 (2011)
58. I.V. Karzhemanov, Fano threefolds with canonical Gorenstein singularities and big degree. Math. Ann. **362**(3–4), 1107–1142 (2015)
59. A.L. Knutsen, On moduli spaces of polarized Enriques surfaces. J. Math. Pures Appl. **144**(9), 106–136 (2020)
60. A.L. Knutsen, Global sections of twisted normal bundles of $K3$ surfaces and their hyperplane sections, Atti Accad. Naz. Lincei Rend. Lincei Mat. Appl. **31**(1), 57–79 (2020)
61. A.L. Knutsen, A.F. Lopez, Surjectivity of Gaussian maps for curves on Enriques surfaces. Adv. Geom. **7**(2), 215–247 (2007)
62. A.L. Knutsen, A.F. Lopez, R. Muñoz, On the extendability of projective surfaces and a genus bound for Enriques-Fano threefolds. J. Diff. Geom. **88**(3), 485–518 (2011)
63. J. Kollár, Rational curves on algebraic varieties, in *Ergebnisse der Mathematik und ihrer Grenzgebiete*. 3. Folge. A Series of Modern Surveys in Mathematics, vol. 32 (Springer, Berlin, 1996)
64. A. Lanteri, M. Palleschi, A.J. Sommese, On triple covers of $\mathbb{P}^n$ as very ample divisors, in *Classification of Algebraic Varieties* (L'Aquila, 1992), 277–292, Contemp. Math. **162**, American Mathematical Society, Providence, RI (1994)
65. A. Lanteri, M. Palleschi, A.J. Sommese, Double covers of $\mathbb{P}^n$ as very ample divisors. Nagoya Math. J. **137**, 1–32 (1995)
66. A. Lanteri, M. Palleschi, A.J. Sommese, Del Pezzo surfaces as hyperplane sections. J. Math. Soc. Japan **49**(3), 501–529 (1997)
67. A.F. Lopez, Surjectivity of Gaussian maps on curves in $\mathbb{P}^r$ with general moduli. J. Algebr. Geom. **5**(4), 609–631 (1996)
68. A.F. Lopez, R. Muñoz, J.C. Sierra, On the extendability of elliptic surfaces of rank two and higher. Ann. Inst. Fourier (Grenoble) **59**(1), 311–346 (2009)
69. S. L'vovsky, Extensions of projective varieties and deformations. I, II. Michigan Math. J. **39**(1), 41–51, 65–70 (1992)
70. V. Martello, On Enriques-Fano threefolds. Ph.D. Thesis, Università della Calabria (2021)
71. S. Mori, S. Mukai, The uniruledness of the moduli space of curves of genus 11, in *Algebraic Geometry* (Tokyo/Kyoto, 1982), 334–353, *Lecture Notes in Mathematics*, vol. 1016 (Springer, Berlin, 1983)

72. S. Mori, H. Sumihiro, On Hartshorne's conjecture. J. Math. Kyoto Univ. **18**(3), 523–533 (1978)
73. S. Mukai, Curves and Grassmannians, in *Algebraic Geometry and Related Topics* (Inchon, 1992), 19–40, Conference Proceedings Lecture Notes Algebraic Geometry, I (International Press, Cambridge, MA 1993)
74. S. Mukai, Curves and K3 surfaces of genus eleven, in *Moduli of Vector Bundles* (Sanda, 1994; Kyoto, 1994), 189–197, *Lecture Notes in Pure and Applied Mathematics*, vol. 179 (Dekker, New York, 1996)
75. S. Mukai, Curves and symmetric spaces, II. Ann. Math. **172**(2, 3), 1539–1558 (2010)
76. S. Mukai, Curves and symmetric spaces. I. Am. J. Math. **117**(6), 1627–1644 (1995)
77. S. Mukai, Curves, K3 surfaces and Fano 3-folds of genus ≤ 10, in *Algebraic Geometry and Commutative Algebra*, vol. I (Kinokuniya, Tokyo, 1988), pp. 357–377
78. S. Mukai, Fano 3-folds, in *Complex Projective Geometry* (Trieste, 1989/Bergen, 1989), 255–263, *London Mathematical Society Lecture Note Series*, vol. 179 (Cambridge University Press, Cambridge, 1992)
79. S. Mukai, Biregular classification of Fano 3-folds and Fano manifolds of coindex 3. Proc. Nat. Acad. Sci. U.S.A. **86**(9), 3000–3002 (1989)
80. S. Mukai, Non-abelian Brill-Noether theory and Fano 3-folds. Sugaku Expo. **14**(2), 125–153 (2001)
81. M. Nagata, On rational surfaces. I. Irreducible curves of arithmetic genus 0 or 1. Mem. Coll. Sci. Univ. Kyoto Ser. A. Math. **32**, 351–370 (1960)
82. H.C. Pinkham, Deformations of algebraic varieties with G_m action. Astérisque, No. **20**. Société Mathématique de France, Paris (1974), i+131 pp
83. Y.G. Prokhorov, On Fano-Enriques varieties. Mat. Sb. **198**(4), 117–134 (2007); translation in Sb. Math. **198**(3–4), 559–574 (2007)
84. Y.G. Prokhorov, The degree of Fano threefolds with canonical Gorenstein singularities. Mat. Sb. **196**(1), 81–122 (2005); translation in Sb. Math. **196**(1–2), 77–114 (2005)
85. F. Russo, Lines on projective varieties and applications. Rend. Circ. Mat. Palermo **61**(2, 1), 47–64 (2012)
86. F. Russo, On the geometry of some special projective varieties, in *Lecture Notes of the Unione Matematica Italiana*, vol. 18 (Springer, Cham; Unione Matematica Italiana, Bologna, 2016)
87. B. Saint-Donat, Projective models of K-3 surfaces. Am. J. Math. **96**, 602–639 (1974)
88. T. Sano, On classifications of non-Gorenstein $\mathbb{Q}$ -Fano 3-folds of Fano index 1. J. Math. Soc. Japan **47**(2), 369–380 (1995)
89. G. Scorza, *Opere Scelte*, vol. I (Edizioni Cremonese, Napoli, 1960)), pp. 376–386
90. G. Scorza, Sopra una certa classe di varietà razionali. Rend. Circ. Mat. Palermo **28**, 400–401 (1909)
91. G. Scorza, Sulle varietà di Segre. Atti Accad. Sci. Torino **45**, 119–31 (1910)
92. C. Segre, Sulle varietà normali a tre dimensioni composte da serie semplici di piani. Atti Accad. Sci. Torino **21**, 95–115 (1885)
93. F. Serrano, The adjunction mappings and hyperelliptic divisors on a surface. J. Reine Angew. Math. **381**, 90–109 (1987)
94. A.J. Sommese, On manifolds that cannot be ample divisors. Math. Ann. **221**(1), 55–72 (1976)
95. A.J. Sommese, A. Van de Ven, On the adjunction mapping. Math. Ann. **278**(1–4), 593–603 (1987)
96. A. Terracini, Alcune questioni sugli spazi tangenti e osculatori ad una varietà. Nota I. Atti Accad. Torino **49** (1913)
97. A.N. Tjurin, Finite-dimensional bundles on infinite varieties. Izv. Akad. Nauk SSSR Ser. Mat. **40**(6), 1248–1268, 1439 (1976)
98. C. Voisin, Sur l'application de Wahl des courbes satisfaisant la condition de Brill-Noether-Petri. Acta Math. **168**(3–4), 249–272 (1992)
99. J.M. Wahl, A cohomological characterization of $\mathbb{P}^n$. Invent. Math. **72**(2), 315–322 (1983)
100. J.M. Wahl, The Jacobian algebra of a graded Gorenstein singularity. Duke Math. J. **55**(4), 843–871 (1987)

101. J.M. Wahl, Deformations of quasihomogeneous surface singularities. Math. Ann. **280**(1), 105–128 (1988)
102. J.M. Wahl, Gaussian maps on algebraic curves. J. Diff. Geom. **32**(1), 77–98 (1990)
103. J.M. Wahl, On cohomology of the square of an ideal sheaf. J. Algebr. Geom. **6**(3), 481–511 (1997)
104. J.M. Wahl, Hyperplane sections of Calabi-Yau varieties. J. Reine Angew. Math. **544**, 39–59 (2002)
105. F.L. Zak, Some properties of dual varieties and their applications in projective geometry, in *Algebraic Geometry* (Chicago, IL, 1989), 273–280, *Lecture Notes in Mathematics*, vol. 1479 (Springer, Berlin, 1991)

The Minimal Cremona Degree of Quartic Surfaces

Massimiliano Mella

Abstract Two birational projective varieties in $\mathbb{P}^n$ are Cremona Equivalent if there is a birational modification of $\mathbb{P}^n$ mapping one onto the other. The minimal Cremona degree of $X \subset \mathbb{P}^n$ is the minimal integer among all degrees of varieties that are Cremona Equivalent to X. The Cremona Equivalence and the minimal Cremona degree is well understood for subvarieties of codimension at least 2 while both are in general very subtle questions for divisors. In this note, I compute the minimal Cremona degree of quartic surfaces in $\mathbb{P}^3$. This allows me to show that any quartic surface of elliptic ruled type has nontrivial stabilizers in the Cremona group.

Keywords Birational maps · Cremona equivalence · Embeddings · Hypersurfaces

2010 Mathematics Subject Classification. 14E25 · 14E05 · 14N05 · 14E07

1 Introduction

Birational geometry and birational maps are one of the most peculiar aspects of Algebraic Geometry. Among the many interests of Ciro in all realms of Algebraic Geometry, and actually Mathematics, an important spot has to be reserved for birational arguments, not only for their intrinsic interests, but also for their link to the Italian school of geometry and projective geometry.

The most studied birational object is certainly the Cremona Group, which is the group of birational self-maps of the projective space

$$Cr_n := \{f : \mathbb{P}^n \dashrightarrow \mathbb{P}^n \mid \text{birational map}\}.$$

Dedicated to Ciro Ciliberto with admiration

M. Mella (✉)
Dipartimento di Matematica e Informatica, Università di Ferrara, Via Machiavelli 30, 44121 Ferrara, Italy
e-mail: mll@unife.it

T. Dedieu et al. (eds.), *The Art of Doing Algebraic Geometry*, Trends in Mathematics,
https://doi.org/10.1007/978-3-031-11938-5_12

This group is wild, from almost all points of view, see [7] for a nice introduction. In this note, I will focus my attention on a problem that is related to the wildness of Cr_n, the so-called Cremona Equivalence.

Let $X, Y \subset \mathbb{P}^N$ be irreducible and reduced birational varieties. The subvariety X and Y are said to be Cremona Equivalent if there is a birational modification $\varphi : \mathbb{P}^N \dashrightarrow \mathbb{P}^N$ such that $\varphi(X) = Y$ and φ is an isomorphism of the generic point of X. The Cremona Equivalence problem has an old history that I will resume in Sect. 2 and a quite recent evolution thanks to the modern tools of birational geometry inherited from Minimal Model Program and Sarkisov program.

As a matter of fact, any pair of birational projective varieties is Cremona equivalent as long as their codimension is at least two. This is a quite surprising result proved in [28] and improved in [11]. Note that this forces Cr_n to contain, as a set, all groups of birational modifications of its subvarieties.

The divisorial case is quite intricate. It is easy to give examples of non Cremona Equivalent divisors, [28], but it is quite hard to understand the divisors that are Cremona Equivalent to a given one. For instance, rational divisor in $\mathbb{P}^n$ Cremona Equivalent to a hyperplane is only known for $\mathbb{P}^2$ and in a less precise way $\mathbb{P}^3$.

A natural notion arising from Cremona Equivalence is that of minimal Cremona degree; see Definition 2.1. The complete classification of minimal Cremona degree plane curves is known, [5, 29], and Ciro together with Alberto Calabri completed also the classification of minimal Cremona degree linear system of plane curves, [5].

Recently, in a series of papers, [26, 27, 29], I tried to shed some light on the surface case and here I present the classification of minimal Cremona degree surfaces of degree at most 4; see Theorem 3.13. This is done following the main ideas in [27] and plugging in the detailed description of singularities of non rational quartic surfaces obtained in a series of papers by Umezu and Urabe, [34–36]. This is the real bottleneck of my methods: the need for a complete understanding of the singularities of divisors I am considering. This prevents me to extend this classification to surfaces of higher degrees.

I want to finish the introduction by thanking Ciro for all he taught me during our long friendship and for all the nice moments we shared both in life and in mathematics. I am in debt, more than you ever thought.

2 History and Background

Let $C \subset \mathbb{P}^2$ be an irreducible and reduced plane curve. It is natural to ask what is the minimal degree of curves that are equivalent to C via a Cremona modification. This is a classical problem studied since the XIXth century by Cremona and Noether. More generally, one can introduce the notion of minimal Cremona degree as follows.

Definition 2.1 Let $X \subset \mathbb{P}^n$ be an irreducible and reduced hypersurface. The minimal Cremona degree of X is

$$\min\{d | X \text{is Cremona Equivalent to a hypersurface of degree} d\}.$$

The divisor X is of minimal Cremona degree if its degree is equal to its minimal Cremona degree. That is it not possible to lower its degree with a Cremona modification in Cr_n.

As I said, the case of plane curves has been widely treated in the old times, [8, 9, 20]; see also the beautiful books of Coolidge [13]and Conforto [12] for a complete account of the result proved by that time. More recently, the subject has been studied with the theory of log pairs, [18, 23, 31], and finally with a mixture of old and new techniques, a complete classification of minimal Cremona degree irreducible plane curves has been achieved in [5, 29].

As a matter of fact, the Cremona equivalence of a plane curve is dictated by its singularities but, unfortunately, its minimal Cremona degree cannot be guessed without a partial resolution of those, [29, Example 3.18]. Due to this, it is quite hard even in the plane curve case to determine the minimal Cremona degree of a fixed curve simply by its equation. The main tool developed in the XXth century and improved by Ciro and Alberto is the theory of adjoint linear systems. Let $D \subset \mathbb{P}^N$ be a divisor and $f : X \to \mathbb{P}^N$ a log resolution of $(\mathbb{P}^N, D)$, with D_X the strict transform divisor. The adjoint linear system, with $m \geq n$, is

$$adj_{n,m}(D) = f_*(|nD_X + mK_X|).$$

S. Kantor first noticed that the dimension of adjoint linear systems is invariant under Cremona modifications. It is easy to see that $adj_{n,m}(D)$ is independent of the resolution f, as long as $m \geq n$, therefore, a divisor of minimal Cremona degree 1 has all adjoint linear systems empty.

It is quite natural to ask whether the opposite is true, and actually, for plane curves, this is one of the main results obtained at the beginning of XXth century.

Theorem 2.2 ([10]) *An irreducible and reduced curve $C \subset \mathbb{P}^2$ is Cremona Equivalent to a line if and only if all of its adjoints vanish.*

In modern terms, also related to the Abhyankar–Moh problem [1], we may rephrase this result by saying that a plane curve C is Cremona Equivalent to a line if and only if its log Kodaira dimension is negative. Pushing the theory of adjoint linear systems Ciro and Alberto were able to classify minimal Cremona degree curves, minimal Cremona degree linear systems and the contractibility of configurations of lines, [5, 6].

It is then natural to investigate surfaces in $\mathbb{P}^3$, keeping in mind that, quite often, the numerical invariants related to the canonical class and its log variants are not subtle enough in higher dimensions. Think of the beautiful Castelnuovo's rationality Theorem and the wild rationality behavior of Fano 3-folds.

For the Cremona Equivalence of surfaces, it is useful to adopt the ♮-Minimal Model Program, developed in [25] or minimal model program with scaling [3]. In this way, a criterion for detecting surfaces Cremona equivalent to a plane has been given in [29]. The criterion, inspired by the previous work of Coolidge on curves Cremona equivalent to lines [13], allows to determine all rational surfaces that are Cremona equivalent to a plane, [29, Theorem 4.15]. Unfortunately, worse than in the plane curve case, the criterion requires not only the resolution of singularities, but also a control on different log varieties attached to the pair $(\mathbb{P}^3, S)$.

Let us start to enter the Cremona Equivalence problem for surfaces with some notations and definitions.

Definition 2.3 Let (T, H) be a $\mathbb{Q}$-factorial uniruled threefold and H an irreducible and reduced Weil divisor on T. Let

$$\rho = \rho_H = \rho(T, H) =: \sup\{m \in \mathbb{Q} | H + mK_T \text{is an effective} \mathbb{Q}\text{-divisor }\} \geq 0,$$

be the (effective) threshold of the pair (T, H).

Remark 2.4 The threshold is not a birational invariant of pairs and it is not preserved by blowing up. Consider a plane $H \subset \mathbb{P}^3$ and let $Y \to \mathbb{P}^3$ be the blow up of a point in H then $\rho(Y, H_Y) = 0$, while $\rho(\mathbb{P}^3, H) = 1/4$. For future reference, note that both are less than one.

In [29], to overcome this problem, it was introduced the notion of good models and of sup threshold.

Definition 2.5 Let (Y, S_Y) be a threefold pair. The pair (Y, S_Y) is a birational model of the pair (T, S) if there is a birational map $\varphi : T \dashrightarrow Y$ such that φ is well defined on the generic point of S and $\varphi(S) = S_Y$. A good model, [29], is a pair (Y, S_Y) with S_Y smooth and Y terminal and $\mathbb{Q}$-factorial.

Remark 2.6 Let (T, S) be a pair, to produce a good model it is enough to consider a log resolution of (T, S). Clearly, there are infinitely many good models for any pair, and running a directed MMP, one can find the one that is more suitable for the needs of the moment.

The threshold allowed to produce an equivalent condition to being Cremona Equivalent to a plane, [29, Theorem 4.15], but unfortunately it is almost impossible to check this condition on specific examples. More recently, [27], a numerical trick allowed to simplify the criteria and provided an effective test for a large class of rational surfaces.

Lemma 2.7 ([27]) *Let (T, S) and (T_1, S_1) be birational models of a pair. Assume that (T, S) has canonical singularities. If $\rho(T, S) = a \geq 1$ then $\rho(T_1, S_1) \geq a$.*

As a direct consequence of Lemma 2.7, one can reformulate the condition of being Cremona Equivalent to a plane as follows.

Corollary 2.8 ([27]) *A rational surface $S \subset \mathbb{P}^3$ is Cremona equivalent to a plane if and only if there is a good model (T, S_T) of $(\mathbb{P}^3, S)$ with $0 < \rho(T, S_T) < 1$.*

There is a class of divisors that are always Cremona Equivalent to a hyperplane.

Remark 2.9 Let $S \subset \mathbb{P}^3$ be a monoid, that is an irreducible and reduced surface of degree d with a point, say p, of multiplicity $d-1$. Then $S = (x_3 F_{d-1} + F_d = 0)$, consider the linear system

$$\mathcal{L} := \{(F_{d-1}x_0 = 0), (F_{d-1}x_1 = 0), (F_{d-1}x_2 = 0), S\}.$$

Then $\varphi_{\mathcal{L}} : \mathbb{P}^3 \dashrightarrow \mathbb{P}^3$ is a birational modification and $\varphi_{\mathcal{L}}(S)$ is a plane. That is any monoid is Cremona Equivalent to a plane.

As a warm-up, I apply Corollary 2.8 and Remark 2.9 to determine the minimal Cremona degree of all surfaces of degree at most 3.

Proposition 2.10 *Let $S \subset \mathbb{P}^3$ be an irreducible and reduced surface of degree at most 3 and σ its minimal Cremona degree. Then $\sigma \in \{1, 3\}$ and*

$\sigma = 1$ *if and only if S is rational*
$\sigma = 3$ *if and only if S is not rational, i.e., S is a cone over an elliptic curve.*

Proof The statement is immediate in degree 2 by Remark 2.9. Let S be a rational cubic. If S is smooth, then $(\mathbb{P}^3, S)$ is a good model with $\rho(\mathbb{P}^3, S) = 3/4$; hence we conclude by Corollary 2.8. If S has a double point, then it is a monoid, and Remark 2.9 allows to conclude. If S is a cone, I conclude by [26].

My aim is to improve this result by determining the minimal Cremona degree of quartic surfaces in $\mathbb{P}^3$.

The case of quartics is, as usual, more subtle due to their own intrinsic complexity. Smooth quartic surfaces are the only smooth hypersurfaces with automorphisms not coming from linear automorphisms of $\mathbb{P}^n$, [24]. In a recent paper, K. Oguiso produced examples of isomorphic smooth quartic surfaces that are not Cremona Equivalent, [32]. It is a long-standing problem to determine which quartic surfaces are stabilized by nontrivial subgroups in Cr_3, that is for which quartic surface $S \subset \mathbb{P}^3$ there is a Cremona modification $\omega : \mathbb{P}^3 \dashrightarrow \mathbb{P}^3$ such that ω is not an isomorphism and $\omega(S) = S$. The above problem has been studied by Enriques [16] and Fano [17] and also by Sharpe and coauthors in a series of papers, [30, 33], at the beginning of the XXth century. More recently, Araujo–Corti–Massarenti continued the study of mildly singular quartic surfaces admitting a nontrivial stabilizers in the Cremona Group, [2], in the context of Calabi–Yau pairs preserving symplectic forms.

On the other hand, the singularities of quartic surfaces are completely classified, [15], and there are a few hundreds of non isomorphic rational quartic surfaces, [19]. This allowed, quite surprisingly, to prove the following result.

Theorem 2.11 ([27]) *Let $S \subset \mathbb{P}^3$ be a rational quartic surface then S is Cremona Equivalent to a plane.*

This shows that any rational quartic has a huge stabilizer in the Cremona group disregarding the type of singularity it may have. Indeed it is amazing that, even if there are hundreds of non isomorphic families of rational quartics, the Cremona group of $\mathbb{P}^3$ is playable enough to smooth any of them to a plane. In the next section, I determine the minimal Cremona degree of an arbitrary quartic surface, adapting the techniques used to prove Theorem 2.11 to an arbitrary quartic surface.

3 Minimal Cremona Degree of Quartics

My aim is to study the minimal Cremona degree of an arbitrary quartic. The main tool I use, besides the $\sharp$-Minimal Model techniques, is the complete classification of singularities of quartic surfaces, see [15, 19] and, in particular, the detailed analysis of the singular locus given in [34–36].

Let us start treating quartic cones.

Lemma 3.1 *Let S be a quartic cone in $\mathbb{P}^3$. Then S is of minimal degree if and only if its sectional genus is at least* 2.

Proof A surface of degree less than 4 is either rational or an elliptic cone. By [26, Corollary 2.7], two surface cones are Cremona Equivalent if and only if their hyperplane sections are birational.

Next, I study non normal quartics.

Lemma 3.2 *Let $S \subset \mathbb{P}^3$ be a non normal quartic, which is not a cone, then S is not of minimal Cremona degree.*

Proof If S is rational, I apply Theorem 2.11. Then I assume that S is not rational. Let $Y \subset S$ be the singular locus of S. Then by the classification of [36], I have that either Y is a pair of skew lines or Y is a line and the general hyperplane section, say H, has an A_3 singularity in $H \cap Y$.

Here I mimic part of the proof in [27, Proposition 2.4] Let $L \subset Y$ be a line and $x \in S$ a general point. Consider the linear system Λ of quadrics through L and x. Let $\varphi : \mathbb{P}^3 \dashrightarrow \mathbb{P}^5$ be the map associated with the linear system Λ. I have $\varphi(\mathbb{P}^3) = Z \cong \mathbb{P}^1 \times \mathbb{P}^2$, embedded via the Segre map, and $\varphi(S) = \tilde{S}$ is a divisor of type $(3, 2)$ in $\mathbb{P}^1 \times \mathbb{P}^2$. Note that divisors of type $(1, 0)$ are planes and divisors of type $(0, 1)$ are quadrics, then I have $\deg \tilde{S} = 3 + 4 = 7$.

Claim 3.3 The surface $\tilde{S}$ is singular along a smooth conic.

Proof (Proof of the Claim) If $Y = L \cup R$ is a pair of lines then $L \cap R = \emptyset$ and clearly $f(R)$ is a smooth conic, singular for $\tilde{S}$. Assume that $Y = L$ and let $\nu : T \to \mathbb{P}^3$ be the blow-up of L, with exceptional divisor E. Then $E \cong \mathbb{P}^1 \times \mathbb{P}^1$ and since the general hyperplane section of S has a singular point of type A_3 I have that $\nu_*^{-1}(S) \cap E$ is a conic and $\nu_*^{-1}(S)$ is singular along this conic. This is enough to conclude.

Let $y \in \mathrm{Sing}(\tilde{S})$ be a general point and $\pi : \mathbb{P}^5 \dashrightarrow \mathbb{P}^4$ the projection from y. Then $\pi_{|Z}$ is a birational map, $Y := \pi(Z) \subset \mathbb{P}^4$ is a quadric of rank 4, and $S_Q := \pi(\tilde{S})$ is a surface of degree $7 - 2 = 5$.

Claim 3.4 The vertex of the quadric is a smooth point of S_Q.

Proof The surface $\tilde{S}$ is a divisor of type $(3, 2)$ in Z and it is singular in y. Let l and P, respectively, be the line and the plane passing through x in Z. The general choice of $x \in S$ yields $l \not\subset \tilde{S}$. The line l is mapped to the vertex of the quadric and $\tilde{S}_{|l} = 2x + p$ for some point p. This shows that S_Q contains the vertex of the quadric and it is smooth there.

The threefold Q is a quadric cone and S_Q is singular along a line. Let $z \in \mathrm{Sing}(S_Q)$ be a point. By the Claim 3.4 z is not the vertex of Q. Thus, the projection from z produces a birational model of (Q, S_Q), say $(\mathbb{P}^3, Z)$, with Z a cubic surface. Therefore, S is not of minimal degree.

Remark 3.5 Incidentally, note that Lemma 3.2 gives a different proof of [36, Proposition 2.6], where it is proven that a non normal quartic birational to a ruled surface over a curve of genus 2 is a cone.

Finally, I treat the case of normal quartics. Let us first recall the following well known result, [34, Proposition 8].

Proposition 3.6 *The minimal resolution of a normal quartic surface $S \subset \mathbb{P}^3$ is one of the following:*

(i) a K3 surface,
(ii) a rational surface,
(iii) birationally equivalent to an elliptic ruled surface,
(iv) a ruled surface of genus 3.

It is immediate that quartic surfaces in (i) and (iv) are of minimal Cremona degree, see the proof of Theorem 3.13 for the details. Theorem 2.11 treats surfaces in (ii). Then I am left to study surfaces in (iii). That is surfaces of elliptic ruled type. The main tool I use for this type of surface is the detailed description of their singularities contained in [35]. I summarize what I need in the following Theorem.

Theorem 3.7 ([35, Corollary pg 134]) *Let $S \subset \mathbb{P}^3$ be a normal quartic of elliptic ruled type. Then the set of irreducible components of a minimal resolution of its singular locus contains either two disjoint elliptic curves or an elliptic curve, say E, and one rational curve intersecting E. In particular, a minimal resolution has always at least two irreducible components with at least one elliptic curve.*

I am ready to complete the analysis.

Lemma 3.8 *Let $S \subset \mathbb{P}^3$ be a normal quartic of the elliptic ruled type. Then S is not of minimal Cremona degree.*

Proof The surface S is not rational and not a cone. In particular, S has only singular points of multiplicity 2 and by hypothesis S has at least an irrational point. By [15, 19] classification, the irrational singularities are of the following type, in brackets the corresponding equation of S:

(1) a double point with an infinitely near double line
$[x_0^2x_1^2 + x_0x_1Q_2(x_2, x_3) + F_4(x_1, x_2, x_3) = 0]$,
(2) a tachnode with an infinitely near double line
$[x_0^2x_1^2 + x_0(x_2^3 + x_1Q_2(x_2, x_3)) + F_4(x_1, x_2, x_3) = 0]$.

Let S be a quartic with a singular point of type (a), $a \in \{1, 2\}$, and let $\Lambda_a \subset |\mathcal{O}(2)|$ be the linear system of quadrics having multiplicity $a + 1$ on the valuation associated to the double line. Then it is easy to check that the map

$$\varphi_{\Lambda_a} : \mathbb{P}^3 \dashrightarrow X_a \subset \mathbb{P}^{7-a}$$

is birational.

As observed in [27, Proposition 2.4], we have two cases:

(S_1) $X_1 \subset \mathbb{P}^6$ is the cone over the Veronese surface
(S_2) $X_2 \subset \mathbb{P}^5$ is the cone over the cubic surface $C \subset \mathbb{P}^4$, where C is the projection of the Veronese surface, say V, from a point $z \in V$.

The main point here is that, in both cases, I have $S_a := \varphi_{\Lambda_a}(S) \subset |\mathcal{O}_{\mathbb{P}^{7-a}}(2)|$, in particular, $\deg X_a = 5 - a$ and $\deg S_a = 10 - 2a$.

Case 3.9 (S_2) Assume that S has a point of type (2). Then the pair $(\mathbb{P}^3, S)$ is birational to (X_2, S_2). The surface $S_2 \subset X_2 \subset \mathbb{P}^5$ has degree 6 and X_2 has degree 3. Let $x \in S_2$ be a general point and $\pi : \mathbb{P}^5 \dashrightarrow \mathbb{P}^4$ the projection from x. Then $\pi(X_2) = Q$ is a quadric cone and $S_x := \pi(S_2)$ is a surface of degree 5. Hence, there is a cubic hypersurface $D \subset \mathbb{P}^4$ such that

$$D_{|Q} = S_x + H,$$

for some plane H. Let $y \in S_x$ be a general point and $\pi_y : \mathbb{P}^4 \dashrightarrow \mathbb{P}^3$ the projection from y.

Claim 3.10 $\tilde{S} := \pi_y(S_x)$ is a quartic surface singular along a line.

Proof The point y is general, therefore, $\deg \tilde{S} = 4$. The map $\pi_{y|Q}$ is birational and it contracts the embedded tangent cone $\mathbb{T}_yQ \cap Q = \Pi_1 \cup \Pi_2$ to a pair of lines $l_1 \cup l_2$. Up to reordering, I may assume that $H \cap \Pi_1$ is the vertex of the cone. Therefore, $\tilde{S} \cap \Pi_1$ is a cubic passing through y. Hence, $\tilde{S}$ has multiplicity 2 along l_1.

In particular, the surface S is Cremona Equivalent to a non normal quartic.

Case 3.11 (S_1) Assume that S has a point of type (1). Then $(\mathbb{P}^3, S)$ is birational to (X_1, S_1). First, I prove that S_1 is always singular and on the smooth locus of X_1.

Claim 3.12 S_1 is in the smooth locus of X_1 and S_1 is singular

Proof I need to describe deeper the map $\varphi := \varphi_{\Lambda_1} : \mathbb{P}^3 \dashrightarrow X_1$, following [27, Proposition 2.4].

Let $S \subset \mathbb{P}^3$ be the quartic, I may assume that there is an irrational singular point of type (1) in $p \equiv [1, 0, 0, 0] \in S$ and the equation of S is

$$(x_0^2 x_1^2 + x_0 x_1 Q + F_4 = 0) \subset \mathbb{P}^3.$$

Let $\epsilon : Y \to \mathbb{P}^3$ be the weighted blow-up of p, with weights $(2, 1, 1)$ on the coordinates (x_1, x_2, x_3), and exceptional divisor $E \cong \mathbb{P}(1, 1, 2)$. Then, I have:

- $\epsilon^*(x_1 = 0) = H + 2E$, $\epsilon_{|H} : H \to (x_1 = 0)$ is an ordinary blow-up and $H_{|E}$ is a smooth rational curve;
- $\epsilon^*(S) = S_Y + 4E$,
- $S_{Y|E}$ has at most two irreducible components and in this case both curves are rational,
- $S_{Y|H}$ is a union of four smooth disjoint rational curves.

In particular:

- both H and S_Y are on the smooth locus of E and hence on the smooth locus of Y;
- The surface H is ruled by, the strict transforms of, the lines in the plane $(x_1 = 0)$ passing through the point p.
- S_Y has no further singularities along H
- by Theorem 3.7 the surface S_Y is not a resolution of singularities of S. That is S_Y is singular.

Let l_Y be a general curve in the ruling and $\Lambda_Y = \epsilon_*^{-1}(\Lambda_1)$ the strict transform linear system. Then $E \cdot l_Y = 1$ and by a direct computation I have

- $\Lambda_Y \cdot l_Y = (\epsilon^*(\mathcal{O}(2)) - 2E) \cdot l_Y = 0$
- $S_Y \cdot l_Y = (\epsilon^*(\mathcal{O}(4)) - 4E) \cdot l_Y = 0$
- $K_Y \cdot l_Y = (\epsilon^*(\mathcal{O}(-4)) + 3E) \cdot l_Y = -1.$
- $H \cdot l_Y = (\epsilon^*(\mathcal{O}(1)) - 2E) \cdot l_Y = -1.$

Then H can be blown down to a smooth rational curve with a birational map $\mu : Y \to X_1$ and by construction $S_Y = \mu^* S_1$. This shows that the unique singularity of X_1 is the singular point in E and the surface S_1 is singular.

Let $x \in \mathrm{Sing}(S_1)$ be a singular point. Set $\pi : \mathbb{P}^6 \dashrightarrow \mathbb{P}^5$ be the projection from x. Then $\pi_{|X_1} : X_1 \dashrightarrow X_2$ is birational and $\pi(S_1) \in |\mathcal{O}_{X_2}(2)|$. I am, therefore, back to case (S_2). This shows that, also in this case, $(\mathbb{P}^3, S)$ is Cremona Equivalent to a non normal quartic.

To conclude, observe that S is of elliptic ruled type and it is birational to a non normal quartic, say V. If V is a cone, I conclude by Lemma 3.1 If V is not a cone, I apply Lemma 3.2.

I am ready to compute the minimal Cremona degree of quartic surfaces.

Theorem 3.13 *Let $S \subset \mathbb{P}^3$ be a quartic surface and σ its minimal Cremona degree. Then $\sigma \in \{1, 3, 4\}$ and*

$\sigma = 1$ if and only if S is rational
$\sigma = 3$ if and only if it is of elliptic ruled type, i.e., it is birational to a ruled surface over an elliptic curve
$\sigma = 4$ in all other cases, i.e., S has at most rational double points or it is a cone of sectional genus at least 2.

Proof If S is a cone, I conclude by Lemma 3.1. If S is rational, I conclude by Theorem 2.11. By [28, Lemma 3.1], if S is not of minimal degree, the pair $(\mathbb{P}^3, S)$ has worse than canonical singularities. Therefore, I may assume that if S is not rational, it is not a cone, and $(\mathbb{P}^3, S)$ has worse than canonical singularities. If S is not normal, by Lemma 3.2, it is not of minimal Cremona degree, and being not rational, it has $\sigma = 3$ and it is of elliptic ruled type. If S is normal, by Proposition 3.6, it is of elliptic ruled type. By Lemma 3.8, the surface S is not of minimal degree, and being not rational, it has $\sigma = 3$. On the other hand, all surfaces of degree at most 2 are rational and non rational surfaces of degree 3 are elliptic cones. Therefore, $\sigma = 3$ if and only if S is of elliptic ruled type.

Thanks to the detailed classification of singularities, I am able to easily characterize the minimal Cremona degree of quartic surfaces with isolated singularities.

Corollary 3.14 *Let $S \subset \mathbb{P}^3$ be a quartic surface with isolated singularities and σ its minimal degree. Then*

$\sigma = 1$ if and only if there is a unique elliptic singularity,
$\sigma = 3$ if and only if there are either two elliptic singularities or one singular point of genus 2,
$\sigma = 4$ if and only if it is a cone or has only rational double points.

Proof Immediate by classification in [15].

Remark 3.15 A similar result for non-isolated singularities is possible, thanks to [36], but it is not as neat as the one in Corollary 3.14.

It is hopeless to look for a similar statement in higher degrees. The singularities of surfaces of degrees greater than 4 are not classified. Even the Cremona Equivalence of rational surfaces is not easy to tackle due to the lack of classification of rational surfaces of degrees greater than 4. The most intriguing problem is to determine whether the vanishing of adjoints is equivalent to the Cremona Equivalence to a plane, like in the plane curve case.

Thanks to Theorem 3.13, I am able to prove that any quartic surface whose minimal Cremona degree is less than 4 has a nontrivial stabilizer in the Cremona Group. I start with the following probably known results that I prove for lack of an adequate reference.

Lemma 3.16 *Let $X \subset \mathbb{P}^n$ be a cubic hypersurface, then its stabilizer in Cr_n is nontrivial.*

Proof Let $T \subset \mathbb{P}^{n+1}$ be a cubic hypersurface with a double point in $[0, \ldots, 0, 1]$ and containing X has the hyperplane section $x_{n+1} = 0$. If $(F_3 = 0) = X \subset \mathbb{P}^n$, it is enough to consider

$$T = (x_{n+1}Q + F_3 = 0),$$

for $Q \in \mathbb{C}[x_0, \ldots, x_n]$ general. Then the projection

$$\pi : T \dashrightarrow \mathbb{P}^n = (x_{n+1} = 0) \subset \mathbb{P}^{n+1}$$

from the point $[0, \ldots, 0, 1]$ is a birational map such that $\pi(X) = X$. Fix a general point $p \in X$ and let $\tau_p : T \dashrightarrow T$ be the involution induced by p. That is, $\tau_p(q)$ is the third point of intersection of the line spanned by p and q. By construction $\tau(X) = X$. Hence, $\pi \circ \tau \circ \pi^{-1}$ is a non trivial element in Cr_n that stabilizes X.

Corollary 3.17 *Let $S \subset \mathbb{P}^3$ be a quartic surface of minimal Cremona degree different from* 4. *Then S has a nontrivial stabilizer in Cr_3.*

Proof Let σ be the minimal Cremona degree of S, then $\sigma \in \{1, 3\}$ by Theorem 3.13. If $\sigma = 1$, the result is immediate. If $\sigma = 3$, then S is Cremona equivalent to a cubic cone and I apply Lemma 3.16 to conclude.

Remark 3.18 Note that for quartic surfaces with minimal Cremona degree 4 the situation is completely different and not much is known; see [2] for a modern reference.

References

1. S.S. Abhyankar, T.T. Moh, Embeddings of the line in the plane. J. Reine Angew. Math. **276**, 148–166 (1975)
2. Araujo, C. Corti, A. Massarenti, A. *Pliability of Calabi-Yau Pairs and Subgroups of the Cremona Group of $I\!P^3$* in preparation
3. Birkar, C., Cascini, P., Hacon, C., J. McKernan, Existence of minimal models for varieties of log general type. J. A.M.S. **23**, 405–468 (2010)
4. A. Bruno, K. Matsuki, Log Sarkisov program. Internat. J. Math. **8**(4), 451–494 (1997)
5. A. Calabri, C. Ciliberto, Birational classification of curves on rational surfaces. Nagoya Math. J. **199**, 43–93 (2010)
6. A. Calabri, C. Ciliberto, On the Cremona contractibility of unions of lines in the plane Kyoto. J. Math. **57**, 55–78 (2017)

7. S. Cantat, Cremona group Algebraic geometry: Salt Lake City. Proc. Sympos. Pure Math. **97.1**. Am. Math. Soc. **2018**, 101–142 (2015)
8. G. Castelnuovo, Massima dimensione dei sistemi lineari di curve piane di dato genere. Ann. Mat. **18**(2), 119–128 (1890)
9. G. Castelnuovo, Ricerche generali sopra i sistemi lineari di curve piane. Mem. R. Accad. Sci. Torino Cl. Sci. Fis. Mat. Nat. **42**(2), 137–188 (1890–1891)
10. G. Castelnuovo, F. Enriques, Sulle condizioni di razionalit'a dei piani doppi. Rend. Circ. Mat. Palermo **14**, 290–302 (1900)
11. C. Ciliberto, M.A. Cueto, M. Mella, K. Ranestad, P. Zwiernik, *Cremona Linearizations of Some Classical Varieties from Classical to Modern Algebraic Geometry* (Trends in the History of Science, Birkhäuser/Springer, Cham, 2016), pp.375–407
12. F. Conforto, Le, *superficie razionali* (Zanichelli, Bologna, 1939)
13. J.L. Coolidge, *A Treatise of Algebraic Plane Curves* (Oxford University Press, Oxford, 1928)
14. A. Corti, M. Mella, Birational geometry of terminal quartic 3-folds I. Am. J. Math. **126**, 739–761 (2004)
15. A.I. Degtyarev, Classification of surfaces of degree four having a non-simple singular point Math. USSR Izvestiya **35**, 607–627 (1990)
16. Enriques, F., *Sulle superficie algebriche, che ammettono una serie discontinua di trasformazioni birazionali*. Rendiconti della Reale Accademia dei Lincei, ser. 5, vol. 15, pp. 665–669 (1906)
17. Fano, G. *Sopra alcune superficie del 4-ordine rappresentabili sul piano doppio*. Rendiconti del Reale Istituto Lombardo, ser. 2, vol. 39, pp. 1071–1086 (1906)
18. S. Iitaka, Birational geometry of plane curves. Tokyo J. Math. **22**(2), 289–321 (1999)
19. C.M. Jessop, *Quartic Surfaces with Singular Points* (Cambridge [Eng.] University Press, 1916). https://archive.org/details/cu31924062545383
20. G. Jung, Ricerche sui sistemi lineari di genere qualunque e sulla loro riduzione all'ordine minimo. Annali di Mat. **16**(2), 291–327 (1888)
21. S. Kantor, Sur une théorie des courbes et des surfaces admettant des correspondances univoques. C. R. Acad. Sci. Paris **100**, 343–345 (1885)
22. S. Kantor, Premiers fondements pour une th'eorie des transformations p'eriodiques univoques. Atti Accad. Sci. Fis. Mat. Napoli **4**(2), 1–335 (1891)
23. N.M. Kumar, M.P. Murthy, Curves with negative self intersection on rational surfaces. J. Math. Kyoto Univ. **22**, 767–777 (1983)
24. H. Matsumura, P. Monsky, On the automorphisms of hypersurfaces. J. Math. Kyoto Univ. **3**(3), 347–361 (1963)
25. M. Mella, #-Minimal Model of uniruled 3-folds. Mat. Zeit. **242**, 187–207 (2002)
26. M. Mella, Equivalent birational embeddings III: Cones Rend. Semin. Mat. Univ. Politec. Torino **71**(3–4), 463–472 (2013)
27. M. Mella, Birational geometry of rational quartic surfaces. J. Math. Pures Appl. **141**(9), 89–98 (2020)
28. M. Mella, E. Polastri, Equivalent birational embeddings bull. London Math. Soc. **41**(1), 89–93 (2009)
29. M. Mella, E. Polastri, Equivalent Birational Embeddings II: divisors. Mat. Zeit. **270**, Numbers 3–4, 1141–1161 (2012). https://doi.org/10.1007/s00209-011-0845-3
30. F.M. Morgan, F.R. Sharpe, Quartic surfaces invariant under periodic transformations. Ann. Math. **15**(1–4), 84–92 (1913/14)
31. M. Nagata, On rational surfaces. I. Irreducible curves of arithmetic genus 0 or 1. Mem. Coll. Sci. Univ. Kyoto Ser. A Math. **32**, 351–370 (1960)
32. K. Oguiso, Isomorphic quartic K3 surfaces in the view of Cremona and projective transformations Taiwanese. J. Math. **21**(3), 671–688 (2017)
33. F.R. Sharpe, V. Snyder, Birational transformations of certain quartic surfaces Trans. Am. Math. Soc. **15**(3), 266–276 (1914)
34. Y. Umezu, On normal projective surfaces with trivial dualizing sheaf Tokyo. J. Math. **4**(2), 343–354 (1981)
35. Y. Umezu, Quartic surfaces of elliptic ruled type. Trans. Am. Math. Soc. **283**, 127–143 (1984)
36. T. Urabe, Classification of non-normal quartic surfaces Tokyo. J. Math. **9**, 265–295 (1986)

On the Degree of the Canonical Map of a Surface of General Type

Margarida Mendes Lopes and Rita Pardini

A Ciro, maestro e amico.

Abstract Let X be a minimal complex surface of general type such that its image Σ via the canonical map φ is a surface; we denote by d the degree of φ. In this expository work, first of all we recall the known possibilities for Σ and d when φ is not birational, which are quite a few, and then we consider the question of producing concrete examples for all of them. We present the two main methods of construction of such examples and we give several instances of their application. We end the paper by outlining the state of the art on this topic and raising several questions.

Keywords Surface of general type · Canonical map · Canonical degree

2020 Mathematics Subject Classification: 14J29

1 Introduction

In his epochal paper [1] Beauville undertook the study of the canonical map of surfaces of general type, bringing to light the great variety of possible behaviors of this map. In spite of some later refinements (see for instance [31, 32]) and of the

Research partially supported by FCT/Portugal through UID/MAT/04459/2020 and by project PRIN 2017SSNZAW_004 "Moduli Theory and Birational Classification" of Italian MIUR. The first author is a member of Centro de Análise Matemática, Geometria e Sistemas Dinâmicos of Técnico/ Universidade de Lisboa. The second author is a member of GNSAGA of INDAM

M. Mendes Lopes (✉)
Departamento de Matemática, Instituto Superior Técnico Universidade de Lisboa, Av. Rovisco Pais, 1049-001 Lisboa, Portugal
e-mail: mmendeslopes@tecnico.ulisboa.pt

R. Pardini
Dipartimento di Matematica, Università di Pisa, Largo B. Pontecorvo, 5, 56127 Pisa, Italy
e-mail: rita.pardini@unipi.it

T. Dedieu et al. (eds.), *The Art of Doing Algebraic Geometry*, Trends in Mathematics,
https://doi.org/10.1007/978-3-031-11938-5_13

great many examples in the literature (too many to be listed here), several questions still remain open.

In this paper we focus on the case when the canonical image is a surface and the canonical map is not birational, with the aim to offer the reader a quick guide to the topic. In Sect. 2, we summarize the known general results, and in Sect. 3, we describe the two main methods of construction of examples found in the literature, that is, generating pairs and abelian covers. In Sect. 4, we apply the methods of Sect. 3 to construct an assortment of examples, with the aim of giving a taste of how the methods work. In Sect. 5, we collect some final remarks and several open questions.

We have tried to keep the exposition reader's friendly, hoping that the paper may serve as an introduction to this fascinating topic, besides being a useful reference for surface experts.

Notations and Conventions: We work over the complex numbers. All varieties are assumed to be projective and irreducible, unless otherwise specified. For a product variety $X_1 \times X_2$ we denote by $p_i \colon X_1 \times X_2 \to X_i$, $i = 1, 2$, the two projections and, given $L_i \in Pic(X_i)$, we denote by $L_1 \boxtimes L_2$ the line bundle $p_1^* L_1 \otimes p_2^* L_2$ on $X_1 \times X_2$. We denote linear equivalence of divisors by $\equiv$.

For a smooth surface X we write as usual $p_g(X) := h^0(K_X) = h^2(\mathcal{O}_X)$ and $q(X) := h^0(\Omega^1_X) = h^1(\mathcal{O}_X)$; in case X is singular we set $p_g(X) = p_g(Y)$ and $q(X) = q(Y)$ for Y any desingularization of X. As usual an effective divisor on a surface is said to be *normal crossings* if its only singularities are ordinary double points and it is said to be *simple normal crossings* if it is normal crossings and in addition all its irreducible components are smooth.

2 The Canonical Map

In this section, we denote by X a surface of general type with $p_g := p_g(X) \geq 3$ and by $\varphi \colon X \to \mathbb{P}^{p_g - 1}$ the canonical map; we assume that the image Σ of φ is a surface and we set $d := \deg \varphi$.

2.1 The Canonical Image

The main result on the canonical image in case it is a surface is due to Beauville and is the following:

Theorem 2.1 ([1, Theorem 3.1]) *There are the following possibilities:*

(A) $p_g(\Sigma) = 0$;
(B) $p_g(\Sigma) = p_g(X)$ *and* Σ *is a canonical surface, namely it is the canonical image of a surface of general type* Z *with birational canonical map.*

As explained in [1] (see Example 4.7) for any surface T with $p_g = 0$ one can find minimal surfaces of general type X such that the image of the canonical map of X is

T, by taking a suitable $\mathbb{Z}_2$-cover. On the other hand, surfaces in case (B) with $d \geq 2$ were thought for a long time not to exist. The first example of such a phenomenon is a surface with $d = 2$ whose canonical image is a quintic surface in $\mathbb{P}^3$ with 20 double points (see [1, 3]).

2.2 Bounds on the Canonical Degree

Here we look at the bounds on the degree d of the canonical map. Some of the examples referred to below can be found in Sect. 4.

Recall that writing $K_X \equiv M + Z$, where Z is the fixed part of the canonical linear system $|K_X|$, one has

$$M^2 \geq d \deg \Sigma.$$

Minimal surfaces S of general type with birational canonical map satisfy $K_S^2 \geq 3p_g(S) + q(S) - 7$; this inequality is due to Jongmans ([14, p. 425], see also [9, Theorem. 3.2]). So in case (B) one has $\deg \Sigma \geq 3p_g(X) + q(\Sigma) - 7$ while in case (A) one has $\deg \Sigma \geq p_g(X) - 2$, since Σ generates $\mathbb{P}^{p_g-1}$.

Assume now that X is minimal. Then one has $K_X^2 \geq M^2$ and so in case (A),

$$K_X^2 \geq d(p_g(X) - 2)$$

and in case (B)

$$K_X^2 \geq d(3p_g(X) + q(\Sigma) - 7).$$

Combining the above inequalities with the Bogomolov-Miyaoka-Yau bound $K_X^2 \leq 9\chi(\mathcal{O}_X)$ one obtains for case (A)

$$27 - 9q(X) \geq (d - 9)(p_g(X) - 2) \tag{1}$$

and for case (B)

$$30 - 9q(X) - dq(\Sigma) \geq (d - 3)(3p_g(X) - 7) \tag{2}$$

Thus, as already pointed out in [1], in case (A) $d \leq 9$ if $p_g \geq 30$ and in case (B) $d \leq 3$ if $p_g(X) \geq 13$. Later on Xiao Gang in [32] showed that in fact in case (A) if $p_g(X) > 132$ then $\deg \varphi \leq 8$. For degree 8 there exist examples with unbounded p_g as shown by Beauville (see Example 4.8). More recently in [20] Nguyen Bin produced several other such examples.

Surfaces in case (B) with $d \geq 2$ are much more difficult to find. Note that minimal surfaces in case (B) with $d = 2$ must satisfy

$$K_X^2 \geq 6p_g(X) + 2q(\Sigma) - 14 \tag{3}$$

while if $d = 3$ they must satisfy

$$K_X^2 \geq 9p_g(X) + 3q(\Sigma) - 21 \tag{4}$$

and as such are very near the Bogomolov-Miyaoka-Yau bound. Families with unbounded p_g and $d = 2$ are known (see Sect. 4) but for $d \geq 3$ one knows only sporadic examples.

Many other consequences of the above formulas can be drawn. To point a few:

(1) $d \leq 9$ whenever $q(X) \geq 3$;
(2) as noted first in [24], the maximum possible degree is 36 and can be attained only if $p_g = 3, q = 0$;
(3) if $q(X) > 0$ the maximum possible degree is 27 and can be attained only if $p_g = 3, q = 1$;
(4) in case (B) the maximum possible degree is 9 and can be attained only if $p_g = 4$, $q = 0$ and Σ is a quintic of $\mathbb{P}^3$;
(5) in case (B) if $d = 3$, $q(X) \leq 3$.

Rito in [29] uses the Borisov-Keum equations of a fake projective plane and the Borisov-Yeung equations of the Cartwright-Steger surface (the unique known minimal surface of general type with $K^2 = 9$, $p_g = q = 1$) to show the existence of a surface with $p_g = 3, q = 0$ and canonical map of degree 36, and of a surface with $p_g = 3, q = 1$ and canonical map of degree 27 (as in (2) and (3) above). The first surface is an étale $\mathbb{Z}_2^2$-cover of the fake projective plane and the second an étale $\mathbb{Z}_3$-cover of the Cartwright-Steger surface.

Let us also remark that (again as explained in [1]) that if $p_g \geq 25$ and Σ has Kodaira dimension ≥ 0 then $d \leq 4$. This is an immediate consequence of the fact that a surface in $\mathbb{P}^n$ of degree less than $2n - 2$ has Kodaira dimension $-\infty$ and if the degree is $2n - 2$ either it has Kodaira dimension $-\infty$ or it is a K3-surface that, having $p_g = 1$, cannot be the image of a canonical map by Theorem 2.1.

3 Two Constructions

We describe here the two main constructions used in the literature to produce examples of surfaces with non-birational canonical map.

3.1 Generating Pairs

The first instance of this construction is due to Beauville (cf. [4, Sect. 2.9]), who used it to produce the first known unbounded sequence of surfaces falling in case (B) of Theorem 2.1. Later the construction was studied systematically in [7] and [8], producing more unbounded sequences of such examples.

The idea is to start with a finite map $h: V \to W$ of surfaces and a line bundle L on W satisfying certain conditions (cf. Definition 3.1 below) and use it to construct a sequence of surfaces X_n, $n \geq 3$ such that the canonical image Σ_n of X_n is a surface and:

- $p_g(\Sigma_n) = 0$ and the canonical map of X_n has degree twice the degree of h if the general curve $C \in |L|$ is hyperelliptic;
- $p_g(\Sigma_n) = p_g(X_n)$ and the canonical map of X_n has degree equal to the degree of h if the general curve $C \in |L|$ is not hyperelliptic;
- $\lim_{n\to\infty} p_g(X_n) = +\infty$.

More precisely, we formulate the following:

Definition 3.1 A *generating pair of degree* ν is a pair $(h: V \to W, L)$ where V and W are surfaces with at most canonical singularities, $h: V \to W$ is a finite morphism of degree $\nu \geq 2$ and L is a line bundle on W such that the following conditions are satisfied:

- $q(W) = 0$, $p_g(W) = p_g(V)$;
- $L^2 > 0$, $h^0(L) \geq 2$;
- the general curve $C \in |L|$ is smooth and irreducible of genus $g \geq 2$ and h^*C is also smooth;
- the natural pull back map $H^0(K_W + L) \to H^0(K_V + h^*L)$ is an isomorphism.

Remark 3.2 In the original construction by Beauville (Example 4.1) the surface V is an abelian surface with an irreducible principal polarization, $h: V \to W$ is the quotient map to the Kummer surface and L is the line bundle induced on W by twice the principal polarization of V.

Remark 3.3 The definition of generating pair given here is slightly different from the one in [7]: we do not require V to be smooth nor the map h to be unramified in codimension 1. On the other hand, for the sake of simplicity, we require $q(W) = 0$ instead of deducing this property from the other conditions. Very few examples of generating pairs $(h: V \to W, L)$ are known (cf. [7, Sect. 3]), and they fit both definitions. In addition, they all have $p_g(V) > 0$. In view of the bounds given in Sect. 2.2, Corollary 3.5 below implies that the degree ν of a generating pair is at most 3 if C is not hyperelliptic and at most 4 otherwise. In addition, in [7, Sect. 7] a detailed analysis is carried out, showing that also the numerical possibilities for L^2, $h^0(W, L)$ and the genus g of C are very limited if h^*C is not hyperelliptic. Note that if h^*C is hyperelliptic, the construction below produces surfaces with canonical map of degree 2ν onto a rational surface.

Consider now a generating pair $(h: V \to W, L)$ of degree ν. Let $n \geq 3$ be an integer and consider the linear system $|M_n| := |L \boxtimes \mathcal{O}_{\mathbb{P}^1}(n)|$ on $W \times \mathbb{P}^1$, let $\Sigma_n \in |M_n|$ be a general element, let $X_n \subset V \times \mathbb{P}^1$ be the preimage of Σ_n and let $f: X_n \to \Sigma_n$ the induced map. We have the following (cf. [7]):

Proposition 3.4 *In the above assumptions and notations, one has*

(i) Σ_n *and* X_n *are surfaces of general type with at most canonical singularities;*
(ii) Σ_n *is minimal if* $K_W + L$ *is nef and* X_n *is minimal if* $K_V + h^*L$ *is nef;*
(iii) $p_g(X_n) = p_g(\Sigma_n)$*, hence the canonical map of* X *is composed with* f*;*
(iv) *the canonical image of* Σ_n *is a surface;*
(v) *the canonical map of* Σ_n *is birational if* C *is not hyperelliptic and it has degree 2 onto a rational surface otherwise.*

Proof (*Sketch of proof*) We just outline the proof and refer the reader to Sect. 2 of [7] for full details.

The statement in (i) on the singularities follows by Bertini's theorem because of the assumptions on $|L|$; the fact that X_n and Σ_n are of general type follows from (iv).

By adjunction, we have that K_{Σ_n} is the restriction of $(K_W + L) \boxtimes \mathcal{O}_{\mathbb{P}^1}(n-2)$ and K_{X_n} is the restriction of $(K_V + h^*L) \boxtimes \mathcal{O}_{\mathbb{P}^1}(n-2)$, so (ii) follows immediately. Standard computations and Kawamata-Viehweg's vanishing give $p_g(\Sigma_n) = (n-1)h^0(K_W + L) + p_g(W)$ and $p_g(X_n) = (n-1)h^0(K_V + h^*L) + p_g(V)$, so (iii) follows immediately from Definition 3.1.

Denote by $f_n \colon \Sigma_n \to \mathbb{P}^1$ the fibration induced by the second projection $W \times \mathbb{P}^1 \to \mathbb{P}^1$: since $n \geq 3$ it is easy to see that $|K_{\Sigma_n}|$ separates the fibers of f_n. In addition, the general fiber F_n of f_n is isomorphic to the general curve $C \in |L|$ and the restriction map $H^0((K_W + L) \boxtimes \mathcal{O}_{\mathbb{P}^1}(n-2)) \to H^0(K_F)$ corresponds to the restriction $H^0(K_W + L) \to H^0(K_C)$. The latter map is surjective, since $q(W) = 0$, hence the canonical map of Σ_n restricts to the canonical map of F and statements (iv) and (v) follow immediately. □

The main consequence of Proposition 3.4 is that, as explained at the beginning of the section, a generating pair gives rise to an unbounded sequence of surfaces whose canonical map has a given behaviour:

Corollary 3.5 *Let* $(h \colon V \to W, L)$ *be a generating pair of degree* ν *and let* C *be the general curve of* $|L|$*. Then:*

(i) *if* C *is hyperelliptic, then for all* $n \geq 3$ *the surface* X_n *is an example of case (A) of Theorem 2.1 with degree* $d = 2\nu$ *and the canonical image of* X_n *is a rational surface;*
(ii) *if* C *is not hyperelliptic, then* X_n *is an example of case (B) of Theorem 2.1 with degree* $d = \nu$ *and* Σ_n *is the canonical image of* X_n*.*

Moreover, one has $\nu \leq 4$ *in case* (i) *and* $\nu = 2$ *in case* (ii).

Proof Claims (i) and (ii) follow directly by Proposition 3.4. In case (i), we have $d \leq 8$ (see Sect. 2.2), hence $\nu \leq 4$.

In case (ii), we have $d = \nu \leq 3$ and $q \leq 3$ if $d = 3$ (see again Sect. 2.2). On the other hand, the general curve of $|L|$ has genus $g \geq 3$ since it is not hyperelliptic, so for $d = 3$ by Lemma 3.6 below the examples would have irregularity $\bar{g} - g \geq 2(g-1) \geq 4$, a contradiction. □

For later reference, we record the following:

Lemma 3.6 *Let g be the genus of C and $\bar{g}$ the genus of h^*C; then the surfaces X_n and Σ_n have the following invariants:*

$$\begin{aligned} p_g(X_n) &= p_g(\Sigma_n) = np_g(W) + (n-1)g, \\ q(\Sigma_n) &= 0,\ q(X_n) = q(V) = \bar{g} - g > 0, \\ K^2_{\Sigma_n} &= n(K^2_W - L^2) + 8(n-1)(g-1) \\ K^2_{X_n} &= n(K^2_V - \nu L^2) + 8(n-1)(\bar{g}-1) \end{aligned}$$

Proof Definition 3.1 implies that L is nef and big, so $L \boxtimes \mathcal{O}_{\mathbb{P}^1}(n)$ is also nef and big, and using Kawamata-Viehweg's vanishing one obtains easily $q(\Sigma_n) = q(W) = 0$, $q(X_n) = q(V)$, $h^0(K_W + L) = \chi(K_W + L)$ and $h^0(K_V + h^*L) = \chi(K_V + h^*L)$. Riemann-Roch and the condition $h^0(K_W + L) = h^0(K_V + h^*L)$ in Definition 3.1 now give $q(V) = \bar{g} - g$. Then the remaining formulae are easily computed, recalling from the proof of Proposition 3.4 that $p_g(\Sigma_n) = (n-1)h^0(K_W + L) + p_g(W)$ and applying the adjunction formula to compute $K^2_{\Sigma_n}$ and $K^2_{X_n}$.

Remark 3.7 Lemma 3.6 shows that all the examples of surfaces with non-birational canonical map arising from a generating pair are irregular. However, in [8] unbounded sequences Y_n of surfaces as in case (B) of Theorem 2.1 with $d = 2$ and $q(Y_n) = 0$ are constructed: one starts with a carefully chosen sequence X_n of surfaces arising from a generating pair and admitting a $\mathbb{Z}_3$-action, and sets $Y_n := X_n/\mathbb{Z}_3$. The action of $\mathbb{Z}_3$ kills the irregularity but preserves the features of the canonical map.

3.2 *Abelian Covers*

The standard reference for abelian covers is [22]; in this section, we summarize the properties we need and analyze in detail the base locus of the canonical system of a cover.

Let G be a finite abelian group; we denote by G^* the group $Hom(G, \mathbb{C}^*)$ of characters of G. Here by a *G-cover* we mean a finite map $f\colon X \to Y$ that is the quotient map for a faithful G-action on X, with Y irreducible and X connected. If we assume X normal and Y smooth, then f is automatically flat and the G-action induces a decomposition

$$f_*\mathcal{O}_X = \mathcal{O}_Y \oplus \bigoplus_{\chi \neq 1} L_\chi^{-1}, \tag{5}$$

where the L_χ are line bundles. Note that L_χ is non-trivial for $\chi \neq 1$ by the assumption that X be connected. Since f is flat, the (reduced) branch locus of f is a divisor D. For a component Δ of D we define the *inertia subgroup* of Δ as the subgroup $H < G$ consisting of the elements fixing $f^{-1}(\Delta)$ pointwise. The group H is cyclic and the natural representation of H on the normal bundle to $f^{-1}(\Delta)$ at a general point defines a generator ψ of H^*. So we can group together the components of D with same inertia group and character and write $D = \sum_{(H,\psi)} D_{(H,\psi)}$, where (H, ψ) runs over all the pairs (cyclic subgroup $H < G$, generator of H^*). The line bundles L_χ and the effective divisors $D_{(H,\psi)}$ are the *building data* of the cover and satisfy the following *fundamental relations*:

$$L_\chi + L_{\chi'} \equiv L_{\chi\chi'} + \sum \varepsilon_{\chi,\chi'}^{(H,\psi)} D_{(H,\psi)}, \quad \forall \chi, \chi' \in G^*, \tag{6}$$

where the coefficient $\varepsilon_{\chi,\chi'}^{(H,\psi)}$ is defined as follows. Denote by m_H the order of H. Let $r_\chi^{(H,\psi)}$ be the smallest non-negative integer such that

$$\chi|_H = \psi^{r_\chi^{(H,\psi)}}.$$

Then we have $\varepsilon_{\chi,\chi'}^{(H,\psi)} = 1$ if $r_\chi^{(H,\psi)} + r_{\chi'}^{(H,\psi)} \geq m_H$ and $\varepsilon_{\chi,\chi'}^{(H,\psi)} = 0$ otherwise.

The building data encode all the information on a G-cover, as explained in the following structure theorem:

Theorem 3.8 ([22, Theorem 2.1]) (Notation as above) *Let Y be a smooth irreducible projective variety, let $\{L_\chi\}_{\chi \neq 1}$ be line bundles of Y such that $L_\chi \not\equiv \mathcal{O}_Y$ for every χ and let $D_{(H,\psi)}$ be effective divisors such that $D := \sum_{(H,\psi)} D_{(H,\psi)}$ is reduced. Then,*

(i) $\{L_\chi, D_{(H,\psi)}\}$ are the building data of an abelian cover $f : X \to Y$ with X normal if and only if they satisfy the fundamental relations (6)*;*
(ii) the building data determine $f : X \to Y$ up to G−equivariant isomorphism.

Theorem 3.8 is satisfactory from the theoretical point of view, however it is not so easy to apply, since the number of equations in (6) grows quadratically with the order of G. In fact, it is possible to reduce the number of equations to the rank of the group G; in particular, cyclic covers are described by one relation. The price that one has to pay for this is that the formulation of the theorem is not intrinsic anymore. So, choose characters $\chi_1, \dots \chi_k \in G^*$ such that $G^* =< \chi_1 > \oplus \dots \oplus < \chi_k >$. Denote by d_i the order of χ_i, write $L_i := L_{\chi_i}$, $i = 1, \dots k$ and $r_i^{(H,\psi)} := r_{\chi_i}^{(H,\psi)}$. We call L_i, $D_{(H,\psi)}$ the *reduced building data* of the cover and we have the following *reduced fundamental relations*:

$$d_i L_i \equiv \sum_{(H,\psi)} \frac{d_i r_i^{(H,\psi)}}{m_H} D_{(H,\psi)}, \quad i = 1, \dots k \tag{7}$$

We have the following "reduced version" of Theorem 3.8:

Theorem 3.9 ([22, Proposition 2.1]) (Notation as above) *Let Y be a smooth projective variety, let $\{L_i\}_{i=1,\dots,k}$ be line bundles of Y and let $D_{(H,\psi)}$ be effective divisors such that $D := \sum D_{(H,\psi)}$ is reduced.*

Then,

(i) $\{L_i, D_{(H,\psi)}\}$ *can be extended to a set of building data* $\{L_\chi, D_{(H,\psi)}\}$ *satisfying* (6) *if and only if* $\{L_i, D_{(H,\psi)}\}$ *satisfy* (7)*;*
(ii) $\{L_i, D_{(H,\psi)}\}$ *uniquely determine the* L_χ.

Remark 3.10 Let $f\colon X \to Y$ be a G-cover with reduced building data $\{L_i, D_{(H,\psi)}\}$. Given a character $\chi \in G^*$, it is possible to compute explicitly L_χ from the reduced building data as follows. Write $\chi = \chi_1^{\alpha_1} \cdots \chi_k^{\alpha_k}$ with $0 \le \alpha_k < d_i$, $i = 1, \dots k$ and for every pair (H, ψ) set $\beta_{(H,\psi)} := \lfloor \frac{\sum \alpha_i r_i^{(H,\psi)}}{m_H} \rfloor$; then,

$$L_\chi \equiv \sum_{i=1}^{k} \alpha_i L_i - \sum_{(H,\psi)} \beta_{(H,\psi)} D_{(H,\psi)}. \tag{8}$$

If χ has order d and $Pic(Y)$ has no d-torsion (e.g. Y is simply connected), then L_χ can also be computed as the only solution in $Pic(Y)$ of the equation

$$dL_\chi \equiv \sum_{(H,\psi)} \frac{dr_\chi^{(H,\psi)}}{m_H} D_{(H,\psi)}. \tag{9}$$

Remark 3.11 (*Cyclic covers*) Let $G = \mathbb{Z}_d$, let $\chi \in G^*$ be a generator and let $L = L_\chi$. By Theorem 3.9 in this case the reduced building data have to satisfy just one relation. Choose a primitive d–th root ζ of 1 and let $g \in G$ be the generator such that $\chi(g) = \zeta$. Given a pair (H, ψ), there is a unique integer $0 < \alpha < d$ such that g^α generates H and $\psi(g^\alpha) = \zeta^{d/m_H}$. Conversely, α determines the pair (H, ψ) completely. Hence we can decompose $D = \sum_\alpha D_\alpha$.

Here there is only one reduced fundamental relation:

$$dL \equiv \sum_\alpha \alpha D_\alpha \tag{10}$$

If $D = D_\alpha$ for some α such that $(\alpha, d) = 1$, then it is possible to choose χ in such a way that the reduced fundamental relation is

$$dL \equiv D.$$

This is called a *simple cyclic cover*. Simple cyclic covers are especially easy to describe: they can be realized as hypersurfaces in the total space $V(L)$ of the line bundle L and the adjunction formula (15) below simplifies to

$$K_X \equiv f^*(K_Y + (d-1)L) \tag{11}$$

Remark 3.12 ($\mathbb{Z}_p^k$-*covers*) Take $G = \mathbb{Z}_p^k$, with p a prime number. If we fix a p-th root $\zeta \neq 1$ of 1 we can take the "logarithm" of any character χ of G and regard

G^* as the dual vector space of G. In addition, there is a bijection of $G \setminus \{0\}$ with the set of pairs (character, subgroup) that associates with $v \in G \setminus \{0\}$ the subgroup H generated by v and the character $\psi \in H^*$ such that $\psi(v) = 1$ (recall that we are viewing characters as $\mathbb{Z}_p$-linear functionals). In fact a similar correspondence holds for any finite abelian group G, although it is slightly more complicated (see [10, Sect. 2]).

If we denote by $\widehat{m}$ the smallest non-negative representative of a class $m \in \mathbb{Z}_p$, the fundamental relations (6) read

$$L_\chi + L_{\chi'} \equiv L_{\chi+\chi'} + \sum_{v \neq 1} \left\lfloor \frac{\widehat{\chi(v)} + \widehat{\chi'(v)}}{p} \right\rfloor D_v, \quad \forall \chi, \chi' \in G^*, \tag{12}$$

and the reduced fundamental relations (7) read

$$pL_i \equiv \sum_{v \neq 1} \widehat{\chi_i(v)} D_v, \quad i = 1, \ldots k. \tag{13}$$

Similarly, (9) becomes

$$pL_\chi \equiv \sum_{v \neq 1} \widehat{\chi(v)} D_v. \tag{14}$$

As one would expect in view of Theorem 3.8, geometrical properties of a G-cover can be read off the building data. For instance the cover is *totally ramified*, namely it does not factor through a non-trivial étale cover of Y, iff the inertia subgroups of the components of D generate G. Here is the criterion for smoothness:

Proposition 3.13 ([22, Proposition 3.1]) *Let $f\colon X \to Y$ be an abelian cover with branch divisor $D = \sum_{(H,\psi)} D_{(H,\psi)}$, let $y \in Y$ be a point, let $D_1, \ldots D_s$ be the irreducible components of D that contain y and let (H_i, ψ_i) be the pair subgroup-character associated with D_i, $i = 1, \ldots s$. Then X is smooth above y if and only if*

(1) D_i is smooth at y for every i;
(2) D is normal crossings at y;
(3) the natural map $H_1 \oplus \cdots \oplus H_s \to G$ is injective.

One can compute the numerical invariants of X in terms of the corresponding invariants of Y and of the building data. Let d denote the exponent of G, namely the least common multiple of the orders of the elements of G. Then dK_X is Cartier and one has

$$dK_X \equiv f^* \left(dK_Y + \sum \frac{d(m_H - 1)}{m_H} D_{(H,\psi)} \right) \tag{15}$$

As a consequence, it makes sense to compute

$$K_X^n = |G| \left(K_Y + \sum_{(H,\psi)} \frac{(m_H - 1)}{m_H} D_{(H,\psi)} \right)^n . \tag{16}$$

In addition one has

$$h^i(\mathcal{O}_X) = h^i(\mathcal{O}_Y) + \sum_{\chi \neq 1} h^i({L_\chi}^{-1}), \tag{17}$$
$$\chi(\mathcal{O}_X) = \chi(\mathcal{O}_Y) + \sum_{\chi \neq 1} \chi({L_\chi}^{-1})$$

The group G acts also on the canonical sheaf $\omega_X = \mathcal{O}_X(K_X)$ and the sheaf $f_*\omega_X$ decomposes under the action of G as follows:

$$f_*\omega_X = \omega_Y \oplus \bigoplus_{\chi \neq 1} (\omega_Y \otimes L_\chi) \tag{18}$$

where G acts trivially on ω_Y and acts on $\omega_Y \otimes L_\chi$ via the character χ^{-1}.

Remark 3.14 Equation (18) gives the following G-equivariant decomposition:

$$H^0(X, K_X) = H^0(Y, K_Y) \oplus \bigoplus_{\chi \neq 1} H^0(Y, K_Y + L_{\chi^{-1}}) \tag{19}$$

where G acts on $H^0(Y, K_Y + L_{\chi^{-1}})$ via the character χ. So if we have a G-cover of smooth varieties $f: X \to Y$ such that $h^0(K_Y + L_\chi) \neq 0$ for precisely one character χ, then the canonical system $|K_X|$ is pulled back from Y and the canonical map of X factorizes through f. If in addition the system $|K_Y + L_\chi|$ is birational, then f is birationally equivalent to the canonical map of X. This observation can be used to construct examples of case (A) of Theorem 2.1 (see Sect. 4.2).

Assume now instead that Γ is a subgroup of G such that $h^0(K_Y + L_\chi) = 0$ for every $\chi \notin \Gamma^\perp$. Then the canonical map of X factorizes via the quotient map $X \to Z := X/\Gamma$. The surface Z thus constructed is in general singular, but it has rational singularities and $p_g(Z) = p_g(X)$. So the canonical map of X is the composition of $X \to Z$ with the canonical map of (a smooth model of) Z; in particular, if the canonical map of Z is birational then $X \to Z$ is essentially the canonical map of X. This observation can be used to construct examples of case (B) of Theorem 2.1 (see Sect. 4.2). The difficulty in applying this method is that one needs L_χ to be "very small" for $\chi \notin \Gamma^\perp$, in order to satisfy $h^0(Y, K_Y + L_\chi) = 0$, for$\chi \notin \Gamma^\perp$, and at the same time the building data of the cover must be sufficiently positive for X to be a surface of general type.

We close this section by recalling a result from [15, Sect. 3.4] (cf. also [5] for the case of $\mathbb{Z}_2^2$-covers) that is useful in determining the base locus of the canonical system of a G-cover:

Proposition 3.15 *Let $f: X \to Y$ be a G-cover of smooth varieties with building data $\{L_\chi, D_{(H,\psi)}\}$ and write $f^*D_{(H,\psi)} = m_H R_{(H,\psi)}$. Then the zero locus of the map $f^*(\omega_Y \otimes L_\chi) \to \omega_X$ induced by the decomposition* (18) *is equal to*

$$\sum_{(H,\psi)} (m_H - 1 - r_\chi^{(H,\psi)}) R_{(H,\psi)}.$$

Remark 3.16 By Proposition 3.15, $|K_X|$ is generated by all the effective divisors in $|f^*(\omega_Y \otimes L_\chi)| + \sum_{(H,\psi)}(m_H - 1 - r_\chi^{(H,\psi)})R_{(H,\psi)}$ for χ such that $h^0(\omega_Y \otimes L_\chi) \neq 0$. This gives an explicit way of understanding the base locus of the canonical map of X (for an application see Examples 4.5 and 4.9).

4 Examples

In this section, we present a sampling of examples of the various possibilities for the degree and the canonical image of a surface of general type. We do not aim at completeness, but rather at giving the reader a feeling for how the methods of Sects. 3.1 and 3.2 work. For this reason, we have chosen to list separately the examples obtained via generating pairs and via abelian covers.

4.1 Examples Arising from Generating Pairs

Here we list examples obtained as in Sect. 3.1, using freely the notation introduced there. We refer the reader to [7, Sect. 3] for a detailed description of the various generating pairs.

Example 4.1 (*Case (B), $d = 2$ - Beauville's example*) We take V an abelian surface with an irreducible principal polarization, $h: V \to W$ the quotient map to the Kummer surface and L the line bundle on W induced by twice the principal polarization. This gives a generating pair with $L^2 = 4$, $\nu = 2$ and C a non-hyperelliptic curve of genus 3.

By Proposition 3.4, the construction produces a sequence of minimal surfaces X_n whose canonical map is 2-to-1 onto a canonically embedded minimal surface Σ_n. By Lemma 3.6 we have

$$K^2_{\Sigma_n} = 3p_g(\Sigma_n) - 7; \quad K^2_{X_n} = 6p_g(X_n) - 14, \quad q(X_n) = 2.$$

As seen in inequality (3) of Sect. 2.2, a surface as in case (B) of Theorem 2.1 satisfies $K_X^2 \geq 6p_g(X) - 14$, so this is a limit case. In [16], it is proven that these are essentially the only surfaces as in case (B) of Theorem 2.1 with $K^2 = 6p_g - 14$ and $q \geq 2$.

Example 4.2 (*Case (B)*, $d = 2$) We consider a generating pair constructed as follows. Take A an abelian surface with an irreducible principal polarization, let M be a symmetric theta divisor and let $V \to A$ be the double cover given by the relation $2M \equiv B$, where $B \in |2M|$ is general. The surface V is smooth of general type with $K_V^2 = 4$, $p_g(V) = q(V) = 2$; using the fact that all curves in $|2M|$ are symmetric one shows the existence of an involution ι of V that lifts multiplication by -1 in A. The involution ι has 20 isolated fixed points and acts trivially on $|K_V|$. We let $h\colon V \to W := V/\iota$ be the quotient map: the surface W has canonical singularities and is minimal of general type with $K_W^2 = 2$, $p_g(W) = 2$ and $q(W) = 0$. So we have $h^0(2K_W) = h^0(2K_V) = 5$ and setting $L := K_W$ one obtains a generating pair. Using the fact that W is by construction a double cover of the Kummer surface of A one can show that the general $C \in |L|$ is non-hyperelliptic.

By Proposition 3.4, the construction produces a sequence of minimal surfaces X_n whose canonical map is 2-to-1 onto a canonically embedded minimal surface Σ_n. By Lemma 3.6, we have

$$5K_{\Sigma_n}^2 = 16p_g(\Sigma_n) - 32; \quad 5K_{X_n}^2 = 32p_g(X_n) - 64, \quad q(X_n) = 2.$$

Example 4.3 (*Case (B)*, $d = 2$) This example is similar to the previous one, in that the map $h\colon V \to W$ is the bicanonical map of V and $L = K_W$. Here we consider a non-hyperelliptic curve Γ of genus 3 and take as V the symmetric product $S^2\Gamma$. The surface V is smooth minimal of general type with $K_V^2 = 6$ and $p_g(V) = q(V) = 3$; the bicanonical map $h\colon V \to W$ is the quotient map for the involution ι of V that sends $p + q \in V$ to the only effective divisor linearly equivalent to $K_\Gamma - p - q$. The surface W is minimal, has canonical singularities and invariants $K_W^2 = 3$, $p_g(W) = 3$, $q(W) = 0$.

By Proposition 3.4, the construction produces a sequence of minimal surfaces X_n whose canonical map is 2-to-1 onto a canonically embedded minimal surface Σ_n. By Lemma 3.6 we have

$$7K_{\Sigma_n}^2 = 24p_g(\Sigma_n) - 72; \quad 7K_{X_n}^2 = 48p_g(X_n) - 144, \quad q(X_n) = 3.$$

Example 4.4 (*Case (A)*, $d = 6$) For a detailed description of this generating pair see [2]. Let $B \subset \mathbb{P}^2$ be a sextic curve obtained as the dual curve of a smooth plane cubic and let $W \to \mathbb{P}^2$ be the double cover branched on B. The singularities of B are 9 cusps, that give 9 singularities of type A_2 on W, so W is a K3 surface with canonical singularities. We denote by L the pull back of $\mathcal{O}_{\mathbb{P}^2}(1)$ on W, so we have $L^2 = 2$ and the general curve C of $|L|$ is smooth of genus 2. There exists a $\mathbb{Z}_3$-cover $h\colon V \to W$ branched precisely over the singularities of V. The surface V is abelian and h^*L is a polarization of type $(1, 3)$, so $h^0(K_V + h^*L) = h^0(h^*L) = 3 = h^0(L) = h^0(K_W + L)$ and we have a generating pair with C hyperelliptic.

By Proposition 3.4, the construction produces a sequence of minimal surfaces X_n whose canonical map is 6-to-1 onto a rational surface Σ_n. By Lemma 3.6 we have

$$K_{\Sigma_n}^2 = 2p_g(\Sigma_n) - 4; \quad K_{X_n}^2 = 6p_g(X_n) - 12, \quad q(X_n) = 2.$$

4.2 Examples Constructed as Abelian Covers

While all the constructions of Sect. 4.1 give sequences of examples with unbounded invariants, some of the constructions in this section give sporadic examples.

In all these examples we have $G = \mathbb{Z}_p^k$ with p a prime and we use the additive notation for characters as explained in Remark 3.12; the group of characters is identified with $\mathbb{Z}_p^k$ via the dual basis of the canonical basis of $G = \mathbb{Z}_p^k$.

Example 4.5 (*Case (A),* $p_g = 3, d = 3, \dots 9$) We take $G = \mathbb{Z}_2^2$ and Y a del Pezzo surface of degree $d \geq 3$. For every $0 \neq v \in G$ we choose a curve D_v in $|-K_Y|$ in such a way that $D := \sum_v D_v$ is a simple normal crossings divisor. The reduced fundamental relations are $2L_{10} \equiv 2L_{01} \equiv -2K_Y$, so that $L_{10} \equiv L_{01} \equiv -K_Y$ is the only solution. By Remark 3.10 we get $L_{11} \equiv -K_Y$, too. The corresponding cover $f: X \to Y$ is smooth by Proposition 3.13 and using Eqs. (17) and (16) we get $K_X^2 = d$, $q(X) = 0$ and $p_g(X) = 3$. By (15) we have that $2K_X \equiv f^*(-K_Y)$ is ample and thus X is minimal of general type. Finally, by Proposition 3.15 we see that the system $|K_X|$ is spanned by the three curves $R_v := f^{-1}D_v$, $0 \neq v \in G$. So $|K_X|$ is free and maps X d-to-1 to $\mathbb{P}^2$.

Example 4.6 (*Case (A),* $p_g = 3, d = 16$) This example is due to Persson, [24]. Let $Y = \mathbb{P}^2$ and denote by h the class of a line. Take $G = \mathbb{Z}_2^4$, let $\chi_0 \in G^*$ be the character that maps $v \in G$ to the sum of its coordinates and consider the following building data:

- D_v is a line if $\chi_0(v) = 1$, and $D_v = 0$ otherwise;
- $L_{\chi_0} \equiv 4h$ and $L_\chi \equiv 2h$ if $\chi \neq \chi_0$.

We also assume that the lines D_v are in general position. The surface X is smooth by Proposition 3.13; by (15) we have $2K_X \equiv f^*(2h)$ and therefore X is minimal of general type with $K_X^2 = 16$. Using (17) one computes $q(X) = 0$ and $p_g(X) = 3$. One has $h^0(K_Y + L_\chi) = 0$ for every $\chi \neq \chi_0$, so the canonical map of X coincides with f (see Remark 3.14).

Example 4.7 (*Case (A),* $d = 2$, *unbounded* p_g) Any surface Y with $p_g(Y) = 0$ occurs as the canonical image of a minimal surface X. This fact was noted for the first time in [1].

Choose a line bundle L of Y such that $|2L|$ is base point free and $K_Y + L$ is very ample and let $f: X \to Y$ be the double cover given by the relation $2L \equiv D$, where D is a general element of $|2L|$. The surface X is smooth by Proposition 3.13 and $K_X \equiv f^*(K_Y + L)$ is ample, so X is minimal of general type. In addition we have $|K_X| = f^*|K_Y + L|$ (cf. Remark 3.14) and therefore the canonical map of Y is just f followed by the embedding defined by $|K_Y + L|$.

Example 4.8 (*Case (A),* $d = 8$, *unbounded* p_g) This example is due to Beauville ([1, Exemple 4.3]) and the construction is similar to the previous one. Take $Y = C \times \mathbb{P}^1$, where C is a non-hyperelliptic curve of genus 3. Let $\eta \in Pic(C)$ be a non-zero 2-torsion element and set $L = \eta \boxtimes \mathcal{O}_{\mathbb{P}^1}(m)$, $m \geq 3$; pick a smooth divisor $D \in |2L| =$

$|2mF|$, where F is a fiber of the projection $Y \to \mathbb{P}^1$. We denote by $f: X \to Y$ the double cover given by the relation $2L \equiv D$. As in the previous example, one checks that X is a smooth minimal surface of general type and that the canonical map of X is the composition of f with the map given by the system $|K_Y + L| = |(K_C + \eta) \boxtimes \mathcal{O}_{\mathbb{P}^1}(m-2)|$. Since C is not hyperelliptic, the system $|K_C + \eta|$ is free of degree 4. So $|(K_C + \eta) \boxtimes \mathcal{O}_{\mathbb{P}^1}(m-2)|$ gives a degree 4 map to $\mathbb{P}^1 \times \mathbb{P}^1$, and the canonical map of X has degree 8.

The invariants of X are: $K_X^2 = 16(m-2)$, $p_g(X) = 2m-2$, $q(X) = 3$.

Example 4.9 (*Case (A), $d = 6$, unbounded p_g*) We take $G = \mathbb{Z}_3^2$ and $Y = \mathbb{P}^1 \times \mathbb{P}^1$; we denote by $|F_1|$ and $|F_2|$ the two rulings of $\mathbb{P}^1 \times \mathbb{P}^1$.

We consider the G-cover $f: X \to Y$ given by the following choice of building data:

- $D_{10}, D_{20}, D_{01}, D_{02}$ are distinct fibers of $|F_1|$;
- D_{11} and D_{22} are distinct fibers of $|F_2|$;
- D_{12} and D_{21} are general elements of $|mF_2|$, $m \geq 3$;
- $L_{10} \equiv L_{01} \equiv L_{20} \equiv L_{02} \equiv F_1 + (m+1)F_2$;
- $L_{11} \equiv L_{22} \equiv 2F_1 + F_2$;
- $L_{12} \equiv L_{21} \equiv 2F_1 + mF_2$.

The surface X is smooth by Proposition 3.13; by (15) we have $3K_X \equiv f^*(2F_1 + (4m-2)F_2)$ and therefore X is minimal of general type with $K_X^2 = 16m - 8$. Using (17) one computes $q(X) = 0$ and $p_g(X) = 2m - 2$.

Denote by $\Gamma < G$ the cyclic subgroup generated by $(1, 1)$; one has $h^0(K_Y + L_\chi) = 0$ for every $\chi \notin \Gamma^\perp$, so the canonical map of X is the composition of the quotient map $X \to Z := X/\Gamma$ with the canonical map of a smooth model of Z (see Remark 3.14). The surface Z is a $\mathbb{Z}_3$-cover of Y with building data $D_1 = D_{10} + D_{02} + D_{21}$ and $D_2 = D_{20} + D_{01} + D_{12}$, $L_1 = L_{12}$, $L_2 = L_{21}$. The singularities of Z are either of type A_2, occurring over the singular points of D_1 and D_2, or of type $\frac{1}{3}(1,1)$, occurring over the intersection points of D_1 and D_2; the minimal desingularization $\tilde{Z}$ satisfies $K_{\tilde{Z}}^2 = 4m - 8$, $q(\tilde{Z}) = 0$, $p_g(\tilde{Z}) = 2m - 2$. In addition, it is not hard to check that $K_{\tilde{Z}}$ is nef, hence $\tilde{Z}$ is minimal with $K^2 = 2p_g - 4$, namely it is a Horikawa surface. By [13, Lemma 1.1] the canonical map of $\tilde{Z}$ is 2-to-1 onto a minimal ruled surface of $\mathbb{P}^{2m-3}$ (note that the fibration $|F_2|$ of Y induces a genus 2 fibration on $\tilde{Z}$). So the canonical map of X has degree 6.

For $v \in G \setminus \{0\}$ we write as usual $f^*D_v = 3R_v$. By Proposition 3.15 the canonical system $|K_X|$ is generated by the following linear subsystems:

$$f^*|K_Y + L_{12}| + R_{10} + R_{02} + R_{21} + 2R_{11} + 2R_{22}$$

and

$$f^*|K_Y + L_{21}| + R_{20} + R_{01} + R_{12} + 2R_{11} + 2R_{22}.$$

Since the systems $|K_Y + L_{12}|$ and $|K_Y + L_{21}|$ are base point free, it follows that the fixed locus of $|K_X|$ consists of the divisor $2R_{11} + 2R_{22}$ and of $4m$ simple base points which are the inverse images of the intersection points of D_1 and D_2.

Example 4.10 (*Case (B), $d = 3$*) This example is due to Tan ([30]). We take as Y the del Pezzo surface of degree 6, that is, the blow up of $\mathbb{P}^2$ at three non-collinear points; we denote by $|F_i|, i = 1, 2, 3$ the three pencils of rational curves of Y induced by the pencils of lines through the three blown up points. Recall that $K_Y \equiv -(F_1 + F_2 + F_3)$. We take $G = \mathbb{Z}_3^2$ and consider the G-cover $f : X \to Y$ given by the following choice of building data:

- $D_{10} \in |3F_1|$, $D_{01} \in |3F_2|$, $D_{22} \in |3F_3|$, and $D_v = 0$ for the remaining $v \in G$;
- $L_{10} \equiv F_1 + 2F_3$, $L_{01} \equiv F_2 + 2F_3$.

In addition we assume that $D = D_{10} + D_{01} + D_{22}$ is a simple normal crossings divisor. It is easy to check (cf. Remarks 3.10 and 3.12) that $L_{11} \equiv F_1 + F_2 + F_3$, $L_{22} \equiv 2(F_1 + F_2 + F_3)$ and for the remaining χ one has $L_\chi \equiv F_i + 2F_j$ for some $i \neq j \in \{1, 2, 3\}$. The surface X is smooth by Proposition 3.13; by (15) we have $3K_X \equiv f^*(3(F_1 + F_2 + F_3))$ and therefore X is minimal of general type with $K_X^2 = 54$. Using (17) one computes $q(X) = 0$ and $p_g(X) = 8$.

Denote by $\Gamma < G$ the cyclic subgroup generated by $(1, 2)$; one has $h^0(K_Y + L_\chi) = 0$ for every $\chi \notin \Gamma^\perp$, so the canonical map of X is the composition of the quotient map $X \to Z := X/\Gamma$ with the canonical map of a smooth model of Z (see Remark 3.14). The surface Z is a simple $\mathbb{Z}_3$-cover of Y with branch locus $D = D_{10} + D_{01} + D_{22}$ with canonical singularities (it has A_2 points over the singular points of D). We claim that the canonical map φ_Z of Z is birational. Indeed, by (11) the canonical bundle K_Z is the pullback of $K_Y + 2L_{11} \equiv F_1 + F_2 + F_3$, so the covering map $Z \to Y$ is composed with φ_Z. Since $p_g(Z) = 8 > 7 = h^0(F_1 + F_2 + F_3)$, we conclude that φ_Z is not equal to the covering map and therefore it has degree 1.

Example 4.11 (*Case (B), $d = 3$*) This example is taken from [23]. We take $G = \mathbb{Z}_3^3$ and $Y = \mathbb{P}^2$; we denote by χ_1, χ_2, χ_3 the canonical basis of the space of characters and by h the class of a line in Y. We consider the G-cover $f : X \to Y$ given by the following choice of reduced building data:

- D_v is a line if $\chi_1(v) = 1, \chi_2(v) = 0$ or $\chi_1(v) = 0, \chi_2(v) = 1$, and $D_v = 0$ otherwise;
- $L_{100} \equiv L_{010} \equiv h$, $L_{001} \equiv 2h$;

In addition, we assume that the lines D_v are in general position. The surface X is smooth by Proposition 3.13; by (15) we have $3K_X \equiv f^*(3h)$ and therefore X is minimal of general type with $K_X^2 = 27$.

Let Γ be the subgroup of G generated by $(0, 0, 1)$: it is not hard to check (cf. Remarks 3.10 and 3.12) that for all $\chi \notin \Gamma^\perp$ one has $L_\chi \equiv 2h$ and therefore $h^0(K_Y + L_\chi) = 0$. Using (17) one computes $q(X) = 0$ and $p_g(X) = 5$. By Remark 3.14 the canonical map of X is composed with the quotient map $X \to Z := X/\Gamma$. The surface Z is a $\mathbb{Z}_3^2$-cover with reduced building data:

- $D_{10} = D_{100} + D_{101} + D_{102}$ and $D_{01} = D_{010} + D_{011} + D_{012}$;
- $L_{10} \equiv L_{01} \equiv h$.

So Z is the fibered product of two simple cyclic $\mathbb{Z}_3$-covers, each branched on the union of three lines; it is easily seen to be isomorphic to the complete intersection of two cubics of $\mathbb{P}^4$. The singularities of Z are A_2-points, occurring over the singular points of D_{10} and D_{01}. So Z is a canonically embedded surface and so the canonical map of X has degree 3.

Example 4.12 (*Case (B),* $d = 5$) This example has been found both in [30] and [23], independently.

The construction is similar to the previous one. We take $G = \mathbb{Z}_5^2$ and $Y = \mathbb{P}^2$; we denote by χ_1, χ_2 the canonical basis of the space of characters and by h the class of a line in Y. We consider the G-cover $f : X \to Y$ given by the following choice of reduced building data:

- D_v is a line if $\chi_1(v) = 1$ and $D_v = 0$ otherwise;
- $L_{10} \equiv h$, $L_{01} \equiv 2h$.

In addition, we assume that the lines D_v are in general position. The surface X is smooth by Proposition 3.13; by (15) we have $5K_X \equiv f^*(5h)$ and therefore X is minimal of general type with $K_X^2 = 25$.

Let Γ be the subgroup of G generated by $(1, 0)$: it is not hard to check (cf. Remarks 3.10 and 3.12) that for all $\chi \notin \Gamma^\perp$ one has one has $L_\chi \equiv 2h$ and therefore $h^0(K_Y + L_\chi) = 0$. Using (17) one computes $q(X) = 0$ and $p_g(X) = 4$. By Remark 3.14 the canonical map of X is composed with the quotient map $X \to Z := X/\Gamma$. The surface Z is a simple $\mathbb{Z}_5$-cover of the plane branched over the union D of the 5 branch lines; its singularities are A_4-points, occurring over the singular points of D. It is immediate to see Z is isomorphic to a quintic in $\mathbb{P}^3$, so it is a canonically embedded surface and the canonical map of X has degree 5.

Example 4.13 (*Case (B),* $d = 2$, *unbounded* p_g) This example is one of the examples constructed by Nguyen Bin in [21].

We take $G = \mathbb{Z}_2^3$ and $Y = \mathbb{P}^1 \times \mathbb{P}^1$; we denote by χ_1, χ_2, χ_3 the canonical basis of the space of characters, by F_1 and F_2 the classes of the two rulings of Y. We consider the G-cover $f : X \to Y$ given by the following choice of reduced building data:

- $D_{100}, D_{101} \in |2F_1 + 2F_2|$, $D_{110} \in |2mF_1|$, $D_{111} \in |2nF_2|$ with $m, n \geq 2$, and $D_v = 0$ otherwise;
- $L_{100} \equiv (m + 2)F_1 + (n + 2)F_2$, $L_{010} \equiv mF_1 + nF_2$, $L_{001} \equiv F_1 + (n + 1)F_2$.

In addition, we assume that the divisor $D = \sum_v D_v$ is a simple normal crossings divisor. The surface X is smooth by Proposition 3.13; by (15) we have $2K_X \equiv f^*(2mF_1 + 2nF_2)$ and therefore X is minimal of general type with $K_X^2 = 16mn$. Denote by Γ the subgroup generated by (001). Writing out (14) for all the remaining characters of G, one sees that for every χ one has: (1) $L_\chi \equiv aF_1 + bF_2$ with $a, b > 0$

for every $0 \neq \chi \in G^*$; (2) if $\chi \notin \Gamma^\perp$ then either a or b is equal to 1. Condition (1) gives $q(X) = 0$ (cf. (17)). Condition (2) implies that $h^0(K_Y + L_\chi) = 0$ for all $\chi \notin \Gamma^\perp$, so by Remark 3.14 the canonical map of X is composed with the quotient map $X \to Z := X/\Gamma$. The surface Z is a $\mathbb{Z}_2^2$-cover with building data:

– $D_{10} = D_{100} + D_{101}$ and $D_{11} = D_{110} + D_{111}$ and $D_{01} = 0$;
– $L_{10} \equiv (m+2)F_1 + (n+2)F_2$, $L_{01} \equiv mF_1 + nF_2$ and $L_{11} \equiv 2F_1 + 2F_2$

The singularities of Z are A_1 points occurring above the singular points of D_{10} and D_{11}. The cover $Z \to Y$ can be factored as a composition of two double covers $Z \to W := Z/\Gamma \to Y$, where Γ is the subgroup generated by (11). The double cover $h\colon W \to Y$ is branched on $D_{10} \equiv 4(F_1 + F_2)$, so W is a K3 surface with singularities of type A_1. The double cover $p\colon Z \to W$ is a flat double cover given by the relation $2L \equiv B$, where $L := h^*L_{01}$ and $B = h^*(D_{11})$. We have $K_Z \equiv p^*(K_W + L) \equiv p^*L$, so $K_Z^2 = 8mn$ and $p_g(Z) = h^0(K_W) + h^0(K_W + L) = 1 + (2 + 2mn) = 3 + 2mn$.

The last step is to prove that the canonical map of Z is birational. One looks first at the system $|L|$ on W: the line bundle L is the pull back of the very ample line bundle $mF_1 + nF_2$ of Y and in addition $h^0(L) = 2mn + 2 > (m+1)(n+1) = h^0(mF_1 + nF_2)$, so $|L|$ gives a birational map. A similar argument applied to $K_Z = p^*L$ shows that $|K_Z|$ is birational.

The invariants of X are: $K_X^2 = 16mn$, $p_g(X) = mn + 3$ and $q(X) = 0$.

5 Remarks and Open Questions

The examples of the previous section show the great variety of possible behaviors of the canonical map. In this final section, we try to give a synthesis of the situation and formulate some questions.

5.1 *Case (A) of Theorem 2.1 ($p_g(\Sigma) = 0$)*

We have seen in Sect. 2.2 that if p_g is large enough (≥ 30) the degree d of the canonical map is at most 8.

Unbounded families of examples are known for all the even values $d \leq 8$. As explained in Sect. 4, Example 4.7, for $d = 2$ there are examples with any p_g obtained by taking suitable double covers of surfaces with $p_g = 0$.

The product of two hyperelliptic curves gives rise to examples with unbounded p_g and $d = 4$.

Examples 4.4, 4.9 and [1, Exemple 4.4] give families with $d = 6$ and unbounded p_g.

As already mentioned in Sect. 2.2, [1] (see Example 4.8) gives a family with $d = 8$, unbounded p_g and $q = 3$, while more recently several such families with

$q = 0$ and $q = 1$ have been constructed in [20] as $\mathbb{Z}_2^3$-covers of $\mathbb{P}^2$ blown-up in one point.

On the other hand, for $d = 3, 5, 7$ only sporadic examples are known so far, all having relatively small p_g. In [17] surfaces with canonical map of degree 3, having a smooth canonical curve and whose image is a surface of minimal degree $n - 1$ in $\mathbb{P}^n$ are completely classified and explicitly constructed. In particular surfaces with $K^2 \leq 3p_g - 5$ and canonical map of degree 3 necessarily have $p_g \leq 5$ (cf. [33]). So we are led to ask

Question 5.1 For $d = 3, 5, 7$, are the invariants of surfaces in case (A) of Theorem 2.1 bounded?

Another interesting situation is $p_g = 3$, since in this case the bound on the degree is very large, i.e. $d \leq 36$ (cf. Sect. 2.2).

Question 5.2 For $p_g = 3$, are there examples for every possible $2 \leq d \leq 36$?

For $d \leq 9$, $q = 0$ the answer is affirmative (see, e.g. Example 4.5). As already mentioned in Sect. 2.1, the limit values 36 for $q = 0$ and 27 for $q = 1$ given by inequality (1) were shown to be effective in [29]. In recent times this question has been the subject of intense activity and, without being exhaustive, we would like to mention the example in [28] with $d = 24$, $K^2 = 24$, $p_g = 3$, $q = 0$ and the examples in [11] with $d = 32$, $K^2 = 32$, $p_g = 3$, $q = 0$ and $d = 24$, $K^2 = 24$, $p_g = 3$, $q = 1$. However we believe that there are many gaps to be filled, for instance many of the cases of odd degree.

We note that if $p_g = q = 3$ the possible degrees allowed by inequality (1) vary between 2 and 9 but by the classification of surfaces with $p_g = q = 3$ (see [12, 25]) only $d = 4$, $d = 6$ and $d = 8$ occur (see [6]).

Many other such questions arise. For instance,

Question 5.3 What pairs (p_g, d) with $d \geq 9$ actually occur?

The bounds on p_g given by inequality (1) for $d \geq 10$ and the one for $d = 9$ by [32] (see Sect. 2.1) may not be the best possible and it would be interesting to know how sharp they are.

As an example, for $d = 16$ we know, by inequality (1), that $p_g \leq 5$ and if $p_g = 5$ then $q = 0$. However although examples with $d = 16$ are known for $p_g = 3$, $q = 0$ (see Example 4.6), $p_g = 3$, $q = 2$ ([27]) and $p_g = 4$, $q = 1$ ([18]) the question of the existence of a surface with $d = 16$ and $p_g = 5$ is still open.

5.2 *Case (B) of Theorem 2.1 ($p_g(\Sigma) = p_g(X)$)*

The case when the canonical map has degree $d > 1$ and its image is a canonically embedded surface is perhaps the most intriguing one. We have seen in Sect. 2.2 that for $p_g \geq 13$ one has $d \leq 3$ and that there are unbounded sequences of such surfaces with $d = 2$ (Examples 4.1–4.3, 4.13).

On the other hand, for $d = 3$ the examples we are aware of (Examples 4.10 and 4.11, [19, 26, 30]) satisfy $p_g \leq 8$. So the first question one is led to ask is

Question 5.4 For $d = 3$, are the invariants of surfaces in case (B) of Theorem 2.1 unbounded?

Note that by Corollary 3.5, one cannot hope to construct a sequence of examples using a generating pair as in Sect. 3.1. Note also that, for $d = 3$ the only possible accumulation point for the slope K^2/χ is 9.

These remarks suggest the following questions:

Question 5.5 For $d = 2$ is there an upper bound for the irregularity of a surface in case (B) of Theorem 2.1?

Question 5.6 For $d = 2$ what are the accumulation points of the slopes K^2/χ of surfaces in case (B) of Theorem 2.1?

We know by inequalities (3) and (4) of Sect. 2.2 that these accumulation points must lie in the interval [7, 10]. The sequences of surfaces constructed in Examples 4.1–4.3 have slopes tending to 6, $\frac{32}{5}$, $\frac{48}{7}$, respectively; the examples in [7] are obtained from these by taking the quotient by a suitable $\mathbb{Z}/3$ action and their slopes have the same accumulation points. Example 4.13 and the other examples of [21] have slopes tending to 8.

For small values of p_g, in principle it is possible to have $3 < d \leq 9$ in case (B); more precisely we have the following bounds (see Sect. 2.2):

- if $d = 4$, then $p_g \leq 9$;
- if $d = 5$, then $p_g \leq 7$;
- if $d = 6$, then $p_g \leq 5$;
- if $d = 7, 8$ or 9, then $p_g = 4$.

However except for Example 4.12 that satisfies $d = 5$ and $p_g = 4$ we do not know other examples with $d > 3$; the obvious question is

Question 5.7 For what pairs (d, p_g), with $d > 3$, are there examples of surfaces in case (B) of Theorem 2.1?

References

1. A. Beauville, L'application canonique pour les surfaces de type général. Inv. Math. **55**, 121–140 (1979)
2. Ch. Birkenhake, H. Lange, A family of abelian surfaces and curves of genus four. Manuscr. Math. **85**, 393–407 (1994)
3. F. Catanese, Babbage's conjecture, contact of surfaces, symmetric determinantal varieties and applications. Invent. Math. **63**, 433–465 (1981)

4. F. Catanese, Canonical rings and "special" surfaces of general type. Proc. Symp. Pure Math. **46**, 175–194 (1987)
5. F. Catanese, Singular bidouble covers and the construction of interesting algebraic surfaces. AMS Contemp. Math. **24**(1), 97–120 (1999)
6. F. Catanese, C. Ciliberto, M. Mendes Lopes, On the classification of irregular surfaces of general type with non birational bicanonical map. Trans. Am. Math. Soc. **350**(1), 275–308 (1998)
7. C. Ciliberto, R. Pardini, F. Tovena, Prym varieties and the canonical map of surfaces of general type (XXIX, Annali della Scuola Normale Superiore di Pisa. Classe di Scienze, 2000), 905–938
8. C. Ciliberto, R. Pardini, F. Tovena, Regular canonical covers. Math. Nachrichten **251**, 19–27 (2003)
9. O. Debarre, Inégalités numériques pour les surfaces de type général, with an appendix by A. Beauville. Bull. Soc. Math. France **110**(3), 319–346 (1982)
10. B. Fantechi, R. Pardini, Automorphisms and moduli spaces of varieties with ample canonical class via deformations of abelian covers. Commun. Algebra **25**, 1413–1441 (1997)
11. C. Gleissner, R. Pignatelli, C. Rito, New surfaces with canonical map of high degree, Commun. Anal. Geom, to appear. arXiv:1807.11854 [math.AG]
12. C. Hacon, R. Pardini, Surfaces with $p_g = q = 3$. Trans. Am. Math. Soc. **354**(7), 2631–2638 (2002)
13. E. Horikawa, Algebraic surfaces of general type with small c_1^2. I. Ann. Math. **104**, 357–387 (1976)
14. F. Jongmans, Sur l'étude des surfaces algébriques caractérisées par la condition $p_g \geq 2(p_a + 2)$. Acad. Roy. Belgique. Bull. Cl. Sci. (5) **36**, 485–494 (1950)
15. C. Liedtke, Singular abelian covers of algebraic surfaces. Manuscr. Math. **112**, 375–390 (2003)
16. M. Mendes Lopes, R. Pardini, Irregular canonical double surfaces. Nagoya J. Math. **152**, 203–230 (1998)
17. M. Mendes Lopes, R. Pardini, Triple canonical surfaces of minimal degree. Int. J. Math. **152**, 203–230 (1998)
18. Bin Nguyen, A new example of an algebraic surface with canonical map of degree 16. Arch. Math. **113**, 385–390 (2019)
19. Bin Nguyen, New examples of canonical covers of degree 3. Math. Nach. **295**(3), 450–467 (2022)
20. B. Nguyen, Some unlimited families of minimal surfaces of general type with the canonical map of degree 8. Manuscr. Math. **163**, 13–25 (2020)
21. Bin Nguyen, Some infinite sequences of canonical covers of degree 2. Adv. Geom. **21**(1), 143–148 (2021)
22. R. Pardini, Abelian covers of algebraic varieties. J. Reine Angew. Math. **417**, 191–213 (1991)
23. R. Pardini, Canonical images of surfaces. J. Reine Angew. Math. **417**, 215–219 (1991)
24. U. Persson, Double coverings and surfaces of general type, Algebraic geometry (Proc. Sympos., Univ. Tromsø, Tromsø, vol. 687 of Lecture Notes in Mathematics, vol. 1978 (Springer, Berlin, 1977), 168–195
25. G.P. Pirola, Surfaces with $p_g = q = 3$. Manuscr. Math. **108**, 163–170 (2002)
26. C. Rito, Cuspidal quintics and surfaces with $p_g = 0$, $K^2 = 3$ and 5-torsion. LMS J. Comput. Math. **19**(1), 42–53 (2016)
27. C. Rito, A surface with q=2 and canonical map of degree 16. Mich. Math. J. **66**(1), 99–105 (2017)
28. C. Rito, A surface with canonical map of degree 24. Int. J. Math. **284**(6), 75–84 (2017)
29. C. Rito, Surfaces with canonical map of maximum degree. J. Algebr. Geom. **31**, 127–135 (2022)
30. S. Tan, Surfaces whose canonical maps are of odd degrees. Math. Ann. **292**, 13—29 (1992)
31. G. Xiao, L'irrégularité des surfaces de type général dont le système canonique est composé d'un pinceau. Compositio Math. **56**(2), 251–257 (1985)
32. G. Xiao, Algebraic surfaces with high canonical degree. Math. Ann. **274**, 473–483 (1986)
33. F. Zucconi, Surfaces with canonical map of degree 3 and $K^2 = 3p_g - 5$. Osaka J. Math. **34**(2), 411–428 (1997)

Hyper-Kähler Varieties with a Motive of Abelian Type

Claudio Pedrini

Abstract According to a well-known Conjecture the Chow motive of a hyper-Kähler variety should be of abelian type. In this paper we consider the case of the hyper-Kähler varieties associated to a cubic fourfold $X \subset \mathbf{P}^5$: the Fano variety of lines $F(X)$, the eightfold Z constructed in [13] and the 10-dimensional projective compactification $\bar{\mathcal{J}}$ of the Jacobian fibration constructed in [12]. We show that the Chow motives of $F(X)$ and Z are of abelian type if X has a motive of abelian type. This in fact the case for a one dimensional family of cubic fourfolds inside the Hassett divisor $\mathcal{C}_d$. If X is a general cubic fourfold than the motive $h(\bar{\mathcal{J}})$ is of abelian type if the surface Σ_2 of lines of second type has a motive of abelian type.

1 Introduction

A hyper-Kähler manifold (HK for short) is a simply connected compact hyper-Kähler manifold X such that $H^0(X, \Omega^2_X) = \mathbf{C}\omega$, where ω is a holomorphic 2-form on X which is nowhere degenerate (as a skew symmetric form on the tangent space).

In dimension 2 a HK manifold is a K3 surface. For every even complex dimension, there are two known deformations types of irreducible holomorphic symplectic varieties: the Hilbert scheme $S^{[n]}$ of n-points on a K3 surface S and the generalized Kummer varieties. A generalized Kummer variety X is of the form $X = K^n(A) = a^{-1}(0)$, where A is an abelian surface and $a : A^{[n+1]} \to A$ is the Albanese map. In addition to these two series of examples, there are two exceptional examples. They occur in dimension 6 and in dimension 10 and were discovered by O'Grady by resolving symplectically two singular moduli space of sheaves on symplectic surfaces. These two examples are usually referred to as OG6 and OG10. The hyper-Kähler varieties of OG10 type are not deformation equivalent to $S^{[5]}$.

We will denote by $\mathcal{M}_{rat}(\mathbf{C})$ the (covariant) category of Chow motives over $\mathbf{C}$ (with $\mathbf{Q}$-coefficients), by $\mathcal{M}_{hom}(\mathbf{C})$ the category of homological motives and by $\mathcal{M}_A(\mathbf{C})$ the category of Andre' motives. The category $\mathcal{M}_A(\mathbf{C})$ is obtained from $\mathcal{M}_{hom}(\mathbf{C})$ by

C. Pedrini (✉)
Dipartimento di Matematica, Università di Genova, Genoa, Italy
e-mail: pedrini@dima.unige.it

T. Dedieu et al. (eds.), *The Art of Doing Algebraic Geometry*, Trends in Mathematics,
https://doi.org/10.1007/978-3-031-11938-5_14

formally adjoining the Lefschetz involutions $*_L$ associated to the Lefschetz isomorphisms $H^i(X) \simeq H^{2d-i}(X)$, where X is a smooth projective variety of dimension d, see [1]. The involutions $*_L$ are given by an algebraic correspondence if and only if the standard conjecture $B(X)$ holds true. Therefore under $B(X)$ the category of Andre' motives coincides with $\mathcal{M}_{hom}(\mathbf{C})$. There are functors

$$\mathcal{M}_{rat}(\mathbf{C}) \xrightarrow{F} \mathcal{M}_{hom}(\mathbf{C}) \xrightarrow{G} \mathcal{M}_A(\mathbf{C})$$

Assuming Kimura's Conjecture on the finite dimensionality of motives the functor F is conservative, i.e. it preserves isomorphism, while, under the standard Conjecture $B(X)$, the functor G is an equivalence of categories. Since the motives of K3 surfaces and of cubic 4-folds are of abelian type in the Andre' category $\mathcal{M}_A(\mathbf{C})$ (see [1]) they are conjecturally of abelian type in $\mathcal{M}_{rat}(\mathbf{C})$, i.e. they lie in the subcategory $\mathcal{M}_{rat}^{Ab}(\mathbf{C})$ generated by abelian varieties. For a K3 surface S this result is known only when Pic S has rank ≥ 19, see [17], for Kummer surfaces and in some other scattered cases. For a cubic fourfold X it is known for a family of special cubic fourfolds in $\mathcal{C}_d$, see [2]. Special families of cubic fourfolds with the same property have also been described by Laterveer in [11]: their equations are defined from cubic forms in $\mathbf{P}^4$. By the results in [6] the same conjecture holds for $h(S^{[n]})$ and hence for a HK variety that is birational to $S^{[n]}$, because HK varieties that are birationally equivalent have isomorphic motives in $\mathcal{M}_{rat}(\mathbf{C})$.

In [21] it is proved that the Andre' motive of a hyper-Kähler variety which is deformation equivalent to $S^{[n]}$ lies in the subcategory $\mathcal{M}_A^{Ab} \subset \mathcal{M}_A$ generated by the motives of abelian varieties. By a result in [22] if X_1 and X_2 are deformation equivalent projective hyper-Kähler manifolds the Andre' motive of X_1 is abelian if and only if the Andre' motive of X_2 is abelian. It follows that the projective deformation of a generalized Kummer variety has a motive lying in $\mathcal{M}_A^{Ab}$.

In [8, Corollary 1.15] it is proved that the Andre' motive of all the known projective hyper-Kähler varieties is of abelian type. All these results suggest the following conjecture.

Conjecture 1.1 The motive of a HK manifold is of Abelian type in $\mathcal{M}_{rat}(\mathbf{C})$.

In this note consider the case of the hyper-Kähler varieties associated to a smooth cubic fourfold $X \subset \mathbf{P}^5_{\mathbf{C}}$. The Fano variety of lines $F(X)$, that is deformation equivalent to $S^{[2]}$, the eightfold Z, constructed in [13] from the space of twisted cubic curves on a cubic fourfold not containing a plane, that is deformation equivalent to $S^{[4]}$, and the smooth projective compactification $\bar{\mathcal{J}}$ of the Jacobian fibration, constructed in [12], that is deformation equivalent to $OG10$. Then we have the following result

Theorem 1.2 *The Chow motives of the HK varieties $F(X)$ and Z are of abelian type if the motive of X is of abelian type. For a general X the motive $h(\bar{\mathcal{J}})$ is of abelian type if $h(\Sigma_2) \in \mathcal{M}_{rat}^{Ab}$. Here Σ_2 is the surface of lines of the second type in $F(X)$.*

For the Fano variety of lines $F(X)$ the result comes from the following equality (see [4, 5])

$$\mathrm{Mot}(F(X) = \mathrm{Mot}(X),$$

where, for a smooth projective variety over $\mathbf{C}$, we denote by $\mathrm{Mot}(Y)$ the full pseudo-abelian tensor subcategory of $\mathcal{M}_{rat}(\mathbf{C})$ generated by $h(Y)$ and the Lefshetz motive $\mathbf{L}$.

Hassett in [7] defined a countably infinite union of divisors $\mathcal{C}_d \subset \mathcal{C}$, where $\mathcal{C}$ is the moduli space of cubic 4-folds, parametrizing *special* cubic fourfolds with labelling of discriminant d. He showed that $\mathcal{C}_d$ is irreducible and nonempty if and only if $d \geq 8$ and $d \equiv 0, 2$ [6].

Using a recent result in [2] we get the following.

Theorem 1.3 *Every Hassett divisor $\mathcal{C}_d$ in the moduli space $\mathcal{C}$ of cubic fourfolds contains a one dimensional family of cubic fourfolds X such that the Chow motives of X and of the hyper-Kähler varieties $F(X)$, Z are all of abelian type.*

My thanks are due to Michele Bolognesi and Robert Laterveer for appreciated conversations about the topics of this paper. I also thank the Referee for valuable suggestions correcting some mistakes and inaccurancies.

2 The Motive of the HK Variety Z Constructed in [13]

Let X be a smooth cubic fourfold not containing a plane and let Z be the eight-fold hyper-Kähler variety constructed in [13]. In this section we recall some results showing that $\mathrm{Mot}(Z) \subset \mathrm{Mot}(X)$.

Voisin in [23, Proposition 4.8] defines a rational map

$$\psi : F \times F \dashrightarrow Z \tag{1}$$

The map ψ is defined as follows. Let (l, l') be a general element in $F \times F$, with $l \cap l' = \emptyset$. The surface $S = S_{(l,l')} = X \cap \mathbf{P}^3_{<l,l'>}$ is a smooth cubic surface containing the lines l, l' and the linear system $|\mathcal{O}_S(l - l')(1)|$ is a 2-dimensional linear system of rational cubics on S. For every point $x \in l$ the plane $< x, l' >$ intersects X along the union of the line l' and a residual conic C_x. Then $l' \cup C_x$ is a rational cubic curve belonging to the linear system and we get a member of this linear system which is the image of (l, l') under the rational map ψ. For each pair (l, l') we get a $\mathbf{P}^1 \simeq l$ of such curves. The map (1) is dominant and has degree 6. Therefore the motive $h(Z)$ is isomorphic to a direct summand of $h(\widetilde{F \times F})$, where $\widetilde{F \times F}$ is the blow-up of $F \times F$ along the incidence sub variety I.

The following result in [9, Theorem 3.1] shows that the motive $h(Z)$ is of abelian type in $\mathcal{M}_{rat}(\mathbf{C})$ if $h(X) \in \mathcal{M}^{Ab}_{rat(}\mathbf{C})$.

Theorem 2.1 *Let X be a smooth cubic fourfold not containing a plane. The motive $h(Z)$ is a direct summand of the motive*

$$\bigoplus_{1\le i\le r} h(X^4)(l_i)$$

with $r \in \mathbf{N}$, $l_i \in \mathbf{Z}$.

Since $h(X^4) \in \mathrm{Mot}(X)$ we get

$$\mathrm{Mot}(Z) \subset \mathrm{Mot}(X)$$

The motive $h(X)$ has a reduced Chow-Künneth decomposition

$$h(X) = \mathbf{1} \oplus \mathbf{L} \oplus (\mathbf{L}^2)^{\oplus \rho_2} \oplus t(X) \oplus \mathbf{L}^3 \oplus \mathbf{L}^4,$$

where ρ_2 is the rank of $CH_2(X)$ and $t(X)$ is the transcendental motive, see [4]. Therefore $\mathrm{Mot}(X)$ is generated by $t(X)$ and the Lefschetz motive $\mathbf{L}$.

From the results in [2] we get the following

Theorem 2.2 *Every Hassett divisor $\mathcal{C}_d$(with $d \neq 8$) contains a one dimensional family $\mathcal{F}$ of cubic fourfolds X whose Chow motive is finite dimensional and Abelian. Therefore all the hyper-Kähler varieties Z associated to the cubic fourfolds in $\mathcal{F}$ have a motive of abelian type.*

3 The Motive of the Compactification $\bar{\mathcal{J}}$ in [12]

Let $X \subset \mathbf{P}^5_{\mathbf{C}}$ be general cubic fourfold and let and let $U \subset (\mathbf{P}^5_{\mathbf{C}})^*$ be the open subset parametrizing hyperplanes $H \subset \mathbf{P}^5_{\mathbf{C}}$ such that the cubic threefold $Y_H = H \cap X$ is smooth. Let $\mathcal{J}_U \to U$ be the Lagrangian fibration, whose fiber over the point $H \in U$ is the intermediate Jacobian $J(Y_H)$. In [12] it is proved that there exists a smooth projective compactification $\bar{\mathcal{J}} \to B$, with $B = (\mathbf{P}^5)^*$, which is a hyper-Kähler variety of dimension 10. G. Sacca in [18] showed that this construction can be extended to all smooth cubic fourfolds.

Remark 3.1 In [14, Theorem 7.6] it is proved that for a smooth cubic fourfold outside the Hasset divisors $\mathcal{C}_8$ and $\mathcal{C}_{12}$ there is a unique irreducible holomorphic symplectic compactification of $\mathcal{J}_U$.

The variety $\bar{\mathcal{J}}$ is deformation equivalent to the HK variety OG10, defined by K.O'Grady. OG10 can be described (see [8, Remark 4.1]) as the symplectic resolution

$$\tilde{M} = \mathrm{Bl}_{\Sigma} \to M$$

Here M is the (singular) moduli space of σ-semistable objects in $D^b(S,\alpha)$, with S a irreducible K3 surface and α a Brauer class, with the same Mukai vector $\mathbf{v}$. The singular locus Σ of M consists of strictly σ-semistable objects . In [8, Corollary 4.5] it is proved that the Chow motive $h(\tilde{M})$ belongs to subcategory of $\mathcal{M}_{rat}(\mathbf{C})$ generated by the motive $h(S)$. If the K3 surface S has Picard rank ≥ 19 then its motive and therefore also $h(\tilde{M})$ are of abelian type, see [8, Corollary 4.7].

Let X be a general cubic fourfold and let

$$\begin{array}{ccc} P & \xrightarrow{q} & X \\ p\downarrow & & \\ F & & \end{array} \tag{2}$$

be the incidence diagram, where $F = F(X)$ is the Fano variety of lines. Then $CH^1(F) =< g > \oplus p_* q^*(CH_2(X)_{prim}$, where p and q are the maps in (2), see [20, 21.4]. Here g is the hyperplane section class of the Plücker embedding of F in Grass(1, 5). Let D be an ample uniruled divisor in $CH^1(F)$ whose class equals ng, with $n > 0$. If D is the uniruled divisor coming from Voisin's rational self-map $\phi : F \dashrightarrow F$, as in [20, Sect. 18], then $n = 60$, see [3, Theorem 8]. The uniruled divisor D is associated to the smooth surface $\Sigma_2 \subset F(X)$ of lines of second type that is the indeterminacy locus of ϕ. In the diagram

$$\begin{array}{ccc} E & \xrightarrow{f} & D \subset F \\ \pi\downarrow & & \\ \Sigma_2 & & \end{array} \tag{3}$$

E is the exceptional divisor of the blow-up $\pi : \tilde{F} \to F$ along the surface Σ_2 and the uniruled divisor $D = f(E)$ is normalization equivalent to E. For every $s \in \Sigma_2$ the fiber $\pi^{-1}(s)$ is the curve $R_{(s)} \simeq \mathbf{P}^1$ of lines l'_t. Here $\mathbf{P}^2_t \cdot X = 2s + l'_t$ with $\{\mathbf{P}^2_t\}_{t\in\mathbf{P}^1}$ the planes tangent to X along s. The divisor $D \subset F$ is the unique uniruled divisor in F swept out by rational curves $f(\pi^{-1}(s)$ in the primitive class in $H_2(F,\mathbf{Z})$, see [16, Corollary 0.2].

A line $l \subset X$ is called *special* with respect to the divisor D if the class $[l] \in CH_0(F)$ is represented by a point on D. Note that if l is special with respect to one uniruled divisor $D \subset F$ then it is special with respect to any uniruled divisor, see [19, Lemma 1.1].

For every line $l \subset X \subset \mathbf{P}^5$ there is a $\mathbf{P}^3 \subset (\mathbf{P}^5)^*$ of hyperplane sections Y_H of X containing l. Therefore we get a $\mathbf{P}^3$-bundle $P \to F$. Let Z be the inverse image of the uniruled divisor D in P.Then Z dominates $B = (\mathbf{P}^5)^*$ because any hyperplane section Y_H of X contains a line residual to a special line in X, see [24, Corollary 3.5]. In the fibration $h : Z \to B$ the fibers of h are the images of the curves $D_H = D \cap F(Y_H)$ Let $Z_U \to U$ be the restriction of $h : Z \to B$ to the open subset U parametrizing smooth cubic 3-folds Y_H. Let

$$\phi_H : CH_1(Y_H)_{hom} \simeq J(Y_H) \simeq \mathrm{Alb}(S_H)$$

be the Abel-Jacobi isomorphism, where $S_H = F(Y_H)$ is the Fano surface of lines on Y_H and $\mathrm{Alb}\, S_H$ is an Abelian variety of dimension 5. For every cycle $\alpha \in CH_1(Y_H)_{hom}$ there exist 5 special lines $(l_1, \ldots, l_5) \in D_H^5$ such that

$$\phi_H(\alpha) = \phi_H(\sum_{1 \leq i \leq 5} [l_i] - 5[l_{x_o}]) \in J(Y_H),$$

where x_0 is a fixed point on D_H and l_{x_0} the corresponding line, see [19, Corollary 2.8]. Therefore, for every $H \in U$ we get a surjective map

$$f_H : D_H^5 \to J(Y_H) \tag{4}$$

defined by $f_H(l_1, \ldots, l_5) = \phi_H(\sum_{1 \leq i \leq 5}[l_i] - 5[l_{x_o}])$.
Let $J_U \to U$ be the intermediate Jacobian fibration as in [12] and let

$$\begin{array}{ccc} Z \times_U \cdots \times_U Z & \longrightarrow & U \\ {\scriptstyle f_U}\downarrow & & \downarrow{\scriptstyle =} \\ J_U & \xrightarrow{j_U} & U \end{array} \tag{5}$$

be a diagram of maps between the fibrations $Z \times_U \cdots \times_U Z \to U$ and $j_U : J_U \to U$, where, for every $H \in U$, the map $f_H : D_H^5 \to J(Y_H)$ is the surjective map in (4).

A similar result appears in [24, Corollary 3.5] where it is proved that there is a dominant rational map

$$Z \times_B \cdots \times_B Z \dashrightarrow \bar{\mathcal{J}} \tag{6}$$

Proposition 3.2 *In the diagram (5) the morphism* $f_U : Z \times_U \cdots \times_U Z \to \mathcal{J}_U$ *is finite.*

Proof Let $y \in J_U$ with $y \in J(Y_H)$ and $y = f_H(\alpha)$, where $\alpha = (l_1, \ldots, l_5) \in D^5$. If $(l'_1, \ldots, l'_5) \in D^5$ is such that $f_H((l'_1, \ldots, l'_5)) = y$ then

$$\sum_i [l_i] - 5[l_{x_0}] = \sum_i [l'] - 5[l_{x_0}]$$

in $CH_1(Y_H)_{hom}$ and hence $\sum_i([l_i] - \sum_i [l'_i]) \in K$. Here K is kernel of the group homomorphism $CH_0(D_H)_{hom} \to CH_1(Y_H)_{hom}$. From the exact sequence

$$0 \to K \to CH_0(D_H)_{hom} \to CH_1(Y_H)_{hom} \to 0$$

we get $K \subset T(S_H)$, with $S_H = F(Y_H)$ and

$$0 \to T(S_H) \to CH_0(S_H)_{hom} \to \mathrm{Alb}(S_H) \to 0.$$

Here $T(S_H)$ is the Albanese kernel. There is an isomorphism

$$T(S_H) \simeq \mathcal{R}_{hom} \tag{7}$$

with $\mathcal{R} \subset CH_0(S_H)$, the subgroup generated by triangles, see [20, Remark 20.8].

We claim that, for every $l \in D_H$ there is only a finite number of lines $l' \in D_H$ such that $[l] - [l'] \in K \subset \mathcal{R}_{hom}$.

We have $I_* \mathcal{R}_{hom} = 0$, where

$$I = \{(l, l') \in S_H \times S_H / l \cap l' \neq \emptyset\}$$

and $I_*([l]) = C_l$ is the curve of lines meeting l. Therefore $C_l = C_{l'}$ if $[l] - [l'] \in \mathcal{R}_{hom}$, hence $[l'] \in C_l \cdot D_H$. For every line $l \in S_H$, we have $3C_l = g|_{S_H} \in CH_1(S_H)$ and $(C_l)^2 = 5$ (see [10, 1.15 and 1.17]). Since the Plücker polarization g on $F(X)$ restricts to the Plücker polarization $g|_{S_H}$ on Y_H we get

$$3nC_l = (ng)|_{S_H} = D_H \; D_H \cdot C_l = (3n)C_l^2 = 15n = N \tag{9}$$

because $D = ng \in CH^1(F)$ and $(ng)|_{S_H} = D_H$. From (9) we get

$$[l'] \in C_l \cdot D_H = \{[l_1], \ldots [l_N]\},$$

Having proved the claim we see that there are only a finite number of classes $[l_i']$ in D_H such that $\sum_{1 \leq i \leq 5} [l_i'] - 5[l_{x_0}]$ belongs to the fiber $f^{-1}(z)$.Therefore the morphism $f_U : \mathcal{D} \to \mathcal{J}_U$ is finite. □

Next step is to pass from the diagram (5) to a similar one between the fibrations $Z \times_B \cdots \times_B Z \to B$ and $\bar{j} : \bar{\mathcal{J}} \to B$, where $\bar{\mathcal{J}} \to B$ is the compactification of $\mathcal{J}_U \to U$. The fiber $\bar{\mathcal{J}}_H$ of $\bar{j}$ over $H \in B$ is the compactified Prym variety $\overline{\text{Prym}}(\tilde{C}_l / C_l)$ of an e'tale double cover $\epsilon : \tilde{C}_l \to C_l$ of irreducible, but possibly singular, curves. Here $l \subset Y_H$ is a *very good line* in the sense of [12, Definition 2.9], C_l is a planar quintic whose singularities are in 1-1 correspondence with the singularities of Y_H. The curve C_l is the discriminant of the conic bundle $\pi_l : \tilde{Y}_H \to \mathbf{P}^2$. The Prym variety $\text{Prym}(\tilde{C}_l / C_l)$ is the identity component of the fixed locus of

$$\tau = -i^* : \text{Pic}^0(\tilde{C}_l) \to \text{Pic}^0(\tilde{C}_l),$$

where $i : \tilde{C}_l \to \tilde{C}_l$ the involution induced by the double cover ϵ, see [12, 4.1]. If $[l'] \in \tilde{C}_l$, then $i^*([l']) = [l"]$, with $P^2_{<l,l'>} \cap Y_H = l \cup l' \cup l"$. The Prym variety $\text{Prym}(\tilde{C}_l / C_l)$ is a a dense open subset of $\overline{\text{Prym}}(\tilde{C}_l / C_l)$, see [12, Remark 4.7]. Over the smooth locus U the relative Prym compactification descends to the intermediate Jacobian fiibration, because, if Y_H is smooth, then C_l is a smooth quintic and, by a result of Mumford, $\text{Prym}(\tilde{C}_l / C_l) \simeq J(Y_H)$.

For every $H \in B$ there is an incidence diagram

$$\begin{array}{ccccc}
\tilde{\mathcal{C}}_l & \longrightarrow & P_{Y_H} & \xrightarrow{q_H} & Y_H \\
\downarrow & & p_H \downarrow & & \\
\tilde{C}_l & \longrightarrow & F(Y_H) & &
\end{array} \tag{10}$$

with $\tilde{\mathcal{C}}_l$ the universal curve.The incidence variety P_{Y_H} is a $\mathbf{P}^1$-bundle over $F(Y_H)$, the variety $F(Y_H)$ is irreducible and the map q_H is finite of degree 6, see [12, 2.10]. From the diagram above we get a map of Chow groups

$$\phi_l : CH_1(Y_H) \to CH_0(\tilde{C}_l) \tag{11}$$

sending the class $[l_1]$ of a line $l_1 \in F(Y_H)$ to the 0-cycle $\tilde{C}_l \cdot C_{l_1}$, where

$$\tilde{C}_l \cdot C_{l_1} = \sum_{1 \le i \le 6} [E_i] - [l] - [l_1] + [l_1'] = c_{F(Y_H)} - [l] - [l_1] + [l_1'].$$

Here E_i are the six lines in $F(Y_H)$ trough the point $l \cap l_1 = y$ and $c_{F(Y_H)}$ is the class of the sum of all lines passing trough a general point of Y_H, see [20, Remark 20.8]. The line l_1' is the residual intersection in $\mathbf{P}^2_{<l,l_1>} \cap Y_H$, i.e. $[l_1'] = i^*([l_1])$. If $\alpha = [l_1] - [l_2] \in CH_1(Y_H)_{hom}$ we get

$$\phi_l(\alpha) = [l_2] - [l_1] + \tilde{\sigma}([l_1]) - \tilde{\sigma}([l_2]),$$

and therefore the involution $\tilde{\sigma}$ acts as -1 on $\phi_l(\alpha)$. It follows that the image of ϕ_l, when restricted to the subgroup $CH_1(Y_H)_{hom}$, is contained in $Pic^0(\tilde{C}_l)^- = Fix(-i^*)$. Let's define, for every $H \in B$

$$\bar{f}_H : D^5_H \xrightarrow{g_H} CH_1(Y_H)_{hom} \xrightarrow{\phi_l} \mathrm{Pic}^0(\tilde{C}_l)^- \subset \bar{\mathcal{J}}_H \tag{12}$$

where l is a general line on Y_H, ϕ_l is the map in (11) and $g_H(l_1, \dots, l_5) = \sum_{1 \le i \le 5} [l_i] - 5[l_{x_0}] \in CH_1(Y_H)_{hom}$.

If $H \in U \subset B$ then the map $\bar{f}_H$ coincides with the map f_H in (4). Therefore diagram in (5) can be extended to a similar one over the whole $B = (\mathbf{P}^5)^*$.

The above argument shows that the rational map in (6) can be in fact extended to a morphism $\mu_5 : Z \times_B \cdots \times_B Z \to \bar{\mathcal{J}}_U$.

In [24, Theorem 3.3 (5)] the map μ_5 is used to prove that the compactification $\bar{\mathcal{J}}$ admits a surface decomposition, i.e. there is a diagram

$$\begin{array}{ccc}
Z \times_B \cdots \times_B Z & \xrightarrow{\mu_5} & \bar{\mathcal{J}} \\
f_5 \downarrow & & \\
(\Sigma_2)^5 & &
\end{array} \tag{13}$$

where $\dim(Z \times_B \cdots \times_B Z) = \dim \bar{\mathcal{J}} = \dim(\Sigma_2^5) = 10$ and μ_5, f_5 are generically finite surjective morphisms.

Corollary 3.3 *The diagram in (13) induces a generically finite morphism* $\mu_5 \circ f_5^{-1} : (\Sigma_2)^5 \to \bar{\mathcal{J}}$. *Therefore* $h(\bar{\mathcal{J}})$ *is a direct summand of* $h(\Sigma_2^5)$ *in* $\mathcal{M}_{rat}(\mathbf{C})$.

Proof Let $\alpha = (s_1, \ldots, s_5) \in (\Sigma_2)^5$ and

$$(l_1^{(i)}, \ldots, l_5^{(i)}) = f_5^{-1}(\alpha) \in Z \times_B \cdots \times_B Z),$$

with $l_j^{(i)} \in R_{(s_j)} \subset D_H$. All the lines in $R_{(s_j)}$ have the same class in $CH_0(D_H)$ and

$$\mu_5(f_5^{-1}(\alpha) = \sum_i \bar{f}_H((l_1^{(i)}, \ldots, l_5^{(i)}) \in \bar{\mathcal{J}}_H$$

where $\bar{f}_H$ is the map in (12). Then the result follows from [15, 2.3(vi)].

Proposition 3.4 *Let X be a general cubic fourfold. Then the motive $h(F)$ is a direct summand of $h(\Sigma_2 \times \Sigma_2)$ in $\mathcal{M}_{rat}(\mathbf{C})$.*

Proof Let $\pi : E \to \Sigma_2$ and D be as in (3). Let $T \subset F \times F \times F$ be the triangle variety, i.e. plane sections of X which are the union of 3 lines. In [24, Theorem 3.3] it is proved that the following diagram

$$\begin{array}{ccc} \Gamma & \xrightarrow{\Phi} & F \\ \Psi \big\downarrow & & \\ \Sigma_2 \times \Sigma_2 & & \end{array} \tag{16}$$

gives a surface decomposition for the Fano variety of lines $F = F(X)$. Here

$$\Gamma = pr_{12}^{-1}(E \times E) \cap T \subset F \times F \times F$$

with $\Phi = pr_{3|\Gamma} : \Gamma \to F$ and $\Psi = (\pi, \pi) \circ p_{12} : \Gamma \to \Sigma_2 \times \Sigma_2$. The maps Φ and Ψ are both generically finite and surjective. For a general $l \in F$ the fiber $\Phi^{-1}(l)$ consists of a finite set of triangles $(l_1^{(i)}, l_2^{(i)}, l) \in \Gamma$,with $i = 1 \cdots N$, where $l_1^{(i)}$ and $l_2^{(i)}$ belong to $D_l \cap \tau_l(D_l)$. Here $D_l = S_l \cap D$, with S_l the surface of lines meeting l and $\tau_l : S_l \to S_l$ the involution sending a line l' to $l"$ such that $(l, l', l")$ is a triangle. Then

$$D_l \cdot \tau_l(D_l) = 24n^2 = N,$$

where ng is the class of D in $A^1(F)$, see [19, Proposition 2.6]. For a general element $(s_1, s_2) \in \Sigma_2 \times \Sigma_2$ the fiber $\Psi^{-1}(s_1, s_2)$ is given by a finite set $(l', l", l) \in \Gamma$ with $l' \in R_{(s_1)}$, $l" \in R_{(s_2)}$ and l belonging to the finite set of secants to l' and $l"$. The set $L^2 = \{(l', l") \in R_{(s_1)} \times R_{(s_2)}\}$ is parametrized by a $\mathbf{P}^2$, hence there is a finite number of elements belonging to the incidence subvariety $I \subset F \times F$, i.e. such that $l' \cap l" \neq \emptyset$. For every $(l', l") \in L^2$ there is a finite set of secants to $(l', l")$.

Let

$$\gamma \in A_4(F \times (\Sigma_2 \times \Sigma_2)) = \mathrm{CH}_4(F \times (\Sigma_2 \times \Sigma_2)) \otimes \mathbf{Q}$$

be the correspondence given by the closure of the graph of the map

$$l \to \sum_{1 \leq N} [(s_i, s_i')] \in \Sigma_2 \times \Sigma_2,$$

with $(s_i, s_i') = \Psi((l_i, l_i', l))$ and $\{(l_i, l_i', l)\}_{1 \leq i \leq N} = \Phi^{-1}(l) \in \Gamma$. Since (l_i, l_i', l) is a triangle there is a plane $\mathbf{P}^2$ such that $\mathbf{P}^2 \cdot X = [l_i] + [l_i'] + [l]$. Therefore

$$[l] + [l_i] + [l_i'] = 3c_F$$

for all $i = 1, \ldots, N$, where c_F is the class of any point on a rational surface in F, see [19, Lemma 1.4].

Let $\epsilon \in A_4((\Sigma_2 \times \Sigma_2) \times F)$ be the correspondence defined as follows

$$\epsilon = \{(s_1, s_2), (\phi_*(s_1) + \phi_*(s_2) - 3c_F)\}.$$

where $\phi_*(s_1) + \phi_*(s_2) - 3c_F \in A_0(F)$. Here $\phi : F \dashrightarrow F$ is Voisin's rational map and $\phi_*(s) = s'$ for any $s \in \Sigma_2$ and $s' \in R_{(s)}$, see [20, Lemma 18.3]. The composition $\epsilon \circ \gamma \in A_4(F \times F)$ is the correspondence

$$\{l, (3Nc_F - N[l] - 3Nc_F)\} \subset F \times F \in A_4(F \times F),$$

hence

$$\epsilon \circ (-1/N)\gamma = \Delta_F \in A_4(F \times F).$$

Therefore the map of motives $\epsilon_* : h(\Sigma_2 \times \Sigma_2) \to h(F)$ has a right inverse $(-1/N)\gamma_* : h(F) \to h(\Sigma_2 \times \Sigma_2)$. Then

$$h(\Sigma_2 \times \Sigma_2) \simeq h(F) \oplus M$$

in $\mathcal{M}_{rat}(\mathbf{C})$ which proves that $h(F)$ is a direct summand of $h(\Sigma_2 \times \Sigma_2)$.

Proposition 3.5 *Let X be a general fourfold such that the motive of the surface Σ_2 is of abelian type. Then X and the hyper-Kähler varieties $F(X)$ and $\bar{\mathcal{J}}$ have a motive of abelian type.*

Proof We have $h(\bar{\mathcal{J}}) \in \mathrm{Mot}(\Sigma_2, h(F) \in \mathrm{Mot}(\Sigma_2)$ and $\mathrm{Mot}(X) = \mathrm{Mot}(F)$.

References

1. Y. André, Pour une theorie inconditionelle des motifs. Publ. Math. IHÉS 1–48 (1996)
2. H. Awada, M. Bolognesi, C. Pedrini, *A Family of Special Cubic Fourfolds with Motive of Abelian Type*. arXiv:2007.07193v1 [math.AG]. Accessed 14 Jul. 2020
3. E. Amerik, A computation of invariant of a rational self-map. Ann. Fac. Sci. Toulouse Math. **18**(3), 445–457 (2009)
4. M. Bolognesi, C. Pedrini, The transcendental motive of a cubic fourfold. J. Pure Appl. Algebra (2020). https://doi.org/10.1016/j.jpaa.2020.10633
5. T.-H. Bülles, Motives of moduli spaces on K3 surfaces and of special cubic fourfolds. Manuscripta Math. (2018). https://doi.org/10.10007/s00229-018-1086-0
6. A. de Cataldo, L. Migliorini, The chow groups and the motive of the hilbert scheme of points on a surface. J. Algebra **251**, 824–848 (2002)
7. B. Hassett, Special cubic fourfolds. Compositio. Math. **120**, 1–23 (2000)
8. S. Fioccari, L. Fu, Z. Zhang, *On the Motive of O'Grady's Ten Dimensional Hyperkähler Varieties*. arXiv:1911.06572 [math.AG]
9. L. Fu, R. Laterveer, C. Vial, *The Generalized Franchetta Conjecture for some Hyper-Kähler Varieties II*. arXiv:2002.05490 [math.AG]
10. D. Huybrechts, *The Geometry of Cubic Hypersurfaces*, notes (2019)
11. R. Laterveer, A family of cubic fourfolds with finite-dimensional motive. J. Math. Soc. Jpn. **70**(4), 1453–1473 (2018)
12. R. Laza, G. Sacca', C. Voisin, A hyper-Kähler compactification of the intermediate Jacobian fibration associated with a cubic 4-fold. Acta Math. **218**(1), 55–135 (2017)
13. C. Lehn, M. Lehn, C. Sorger, D. van Straten, Twisted cubics on cubic fourfolds. J. Reine Angew. Math. **731**, 87–128 (2017)
14. G. Mongardi, C. Onorati, *Birational Geometry of Irreducible Holomorphic Symplectic Tenfolds of O'Grady Type*. arXiv:2010.12511v1 [math.AG]. Accessed 23 Oct. 2020
15. J. Murre, J. Nagel, C. Peters, *Lectures on the Theory of Pure Motives*, vol. 61 (University Lecture Series, A.M.S., 2013)
16. G. Oberdieck, J. Shen, Q. Yin, *Rational Curves in the Fano Variety of Cubic 4-folds and Gromov-Witten Invariants*. arXiv:1805.07001v1 [math.AG]. Accessed 18 May 2018
17. C. Pedrini, On the finite dimensionality of a K3 surface. Manuscripta Math. **138**, 59–72 (2012)
18. G. Sacca', *Birational Geometry of the Intermediate Jacobian Fibration* (with an Appendix by C. Voisin). arXiv:2002.01420v1] [math AG]. Accessed Feb. 2020
19. J. Shen, Q. Yin, *K3 Categories, One-cycles on Cubic Fourfolds and the Beauville-Voisin Filtration*. arXiv:1712.0717 v1 [math.AG]
20. M. Shen, C. Vial, The Fourier transform for certain Hyperkälher fourfolds. Mem. AMS **240**(1139), 1–104 (2014)
21. U. Schlickewei, On the Andre' motive of certain irreducible simplectic varieties. Geom. Dedic. **156**(1), 141–149 (2012)
22. A. Soldatenkov,*Deformation Principle and Andre' Motives of Hyper-Kähler Manifolds*. arXiv:1904.11320v2 [math.AG]. Accessed 23 Sept. 2019
23. C. Voisin, Remarks and questions on isotropic subvarieties and 0-cycles on hyper-Kähler varieties, in K3 Surfaces and Their Moduli, in *Proceedings of the Schiermonnikoog Conference 2014, Progress in Mathematics*, Birkhuser, vol. 315, pp. 365–399 (2016)
24. C. Voisin, *Triangle Varieties and Surface Decomposition of Hyper-Kähler Varieties*. arXiv:1810.11848v1 [math.AG]. Accessed 28 Oct. 2018

Finite Quotients of Surface Braid Groups and Double Kodaira Fibrations

Francesco Polizzi and Pietro Sabatino

Abstract Let Σ_b be a closed Riemann surface of genus b. We give an account of some results obtained in the recent papers [6, 18, 19] and concerning what we call here *pure braid quotients*, namely, non-abelian finite groups appearing as quotients of the pure braid group on two strands $\mathsf{P}_2(\Sigma_b)$. We also explain how these groups can be used in order to provide new constructions of double Kodaira fibrations.

Keywords Surface braid groups · Extra-special p-groups · Kodaira fibrations

1 Introduction

A *Kodaira fibration* is a smooth, connected holomorphic fibration $f_1 \colon S \longrightarrow B_1$, where S is a compact complex surface and B_1 is a compact complex curve, which is not isotrivial (this means that not all its fibers are biholomorphic to each other). The genus $b_1 := g(B_1)$ is called the *base genus* of the fibration, whereas the genus $g := g(F)$, where F is any fiber, is called the *fiber genus*. If a surface S is the total space of a Kodaira fibration, we will call it a *Kodaira fibred surface*. For every Kodaira fibration we have $b_1 \geq 2$ and $g \geq 3$, see [12, Theorem 1.1]. Since the fibration is smooth, the condition on the base genus implies that S contains no rational or elliptic curves; hence it is minimal and, by the sub-additivity of the Kodaira dimension, it is of general type, hence algebraic.

Examples of Kodaira fibrations were originally constructed in [1, 14] in order to show that, unlike the topological Euler characteristic, the signature σ of a real

To Professor Ciro Ciliberto on the occasion of his 70th birthday.

F. Polizzi (✉)
Dipartimento di Matematica e Informatica, Università della Calabria, Ponte Bucci Cubo 30B, 87036 Cosenza, Arcavacata di Rende, Italy
e-mail: francesco.polizzi@unical.it

P. Sabatino
Via Val Sillaro 5, 00141 Roma, Italy
e-mail: pietrsabat@gmail.com

T. Dedieu et al. (eds.), *The Art of Doing Algebraic Geometry*, Trends in Mathematics,
https://doi.org/10.1007/978-3-031-11938-5_15

manifold is not multiplicative for fiber bundles. In fact, every Kodaira fibred surface S satisfies $\sigma(S) > 0$, see, for example, the introduction of [16], whereas $\sigma(B_1) = \sigma(F) = 0$, and so $\sigma(S) \neq \sigma(B_1)\sigma(F)$.

A *double Kodaira surface* is a compact complex surface S, endowed with a *double Kodaira fibration*, namely, a surjective, holomorphic map $f: S \longrightarrow B_1 \times B_2$ yielding, by composition with the natural projections, two Kodaira fibrations $f_i: S \longrightarrow B_i$, $i = 1, 2$.

The purpose of this article is to give an account of recent results, obtained in the series of papers [6, 18, 19], concerning the construction of some double Kodaira fibrations (that we call *diagonal*) by means of group-theoretical methods. Let us start by introducing the needed terminology. Let $b \geq 2$ and $n \geq 2$ be two positive integers, and let $\mathsf{P}_2(\Sigma_b)$ be the pure braid group on two strands on a closed Riemann surface of genus b. We say that a finite group G is a *pure braid quotient* of type (b, n) if there exists a group epimorphism

$$\varphi: \mathsf{P}_2(\Sigma_b) \longrightarrow G \tag{1}$$

such that $\varphi(A_{12})$ has order n, where A_{12} is the braid corresponding, via the isomorphism $\mathsf{P}_2(\Sigma_b) \simeq \pi_1(\Sigma_b \times \Sigma_b - \Delta)$, to the homotopy class in $\Sigma_b \times \Sigma_b - \Delta$ of a loop in $\Sigma_b \times \Sigma_b$ "winding once" around the diagonal Δ. Since A_{12} is a commutator in $\mathsf{P}_2(\Sigma_b)$ and $n \geq 2$, it follows that every pure braid quotient is a non-abelian group, see Remark 2.

By Grauert–Remmert's extension theorem together with Serre's GAGA, the existence of a pure braid quotient as in (1) is equivalent to the existence of a Galois cover $\mathbf{f}: S \longrightarrow \Sigma_b \times \Sigma_b$, branched over Δ with branching order n. After Stein factorization, this yields in turn a diagonal double Kodaira fibration $f: S \longrightarrow \Sigma_{b_1} \times \Sigma_{b_2}$. We have $\mathbf{f} = f$, i.e., no Stein factorization is needed, if and only if G is a *strong* pure braid quotient, an additional condition explained in Definition 3.

We are now in a position to state our first results, see Theorems 1, 2, 3:

- If $b \geq 2$ is an integer and $p \geq 5$ is a prime number, then both extra-special p-groups of order p^{4b+1} are non-strong pure braid quotients of type (b, p).
- If $b \geq 2$ is an integer and p is a prime number dividing $b + 1$, then both extra-special p-groups of order p^{2b+1} are pure braid quotients of type (b, p).
- If a finite group G is a pure braid quotient, then $|G| \geq 32$, with equality holding if and only if G is extra-special. Moreover, in the last case, we can explicitly compute the number of distinct quotients maps of type (1), up to the natural action of $\mathrm{Aut}(G)$.

We believe that such results are significant because, although we know that $\mathsf{P}_2(\Sigma_b)$ is residually p-finite for all $p \geq 2$ (see [2, pp. 1481–1490]), it is usually tricky to explicitly describe its non-abelian finite quotients.

The geometrical counterparts of the above group-theoretical statements allow us to construct infinite families of double Kodaira fibrations with interesting numerical properties, for instance, having slope greater than $2 + 1/3$ or signature equal to 16, see Theorems 4, 5, 6:

- Let $f: S_p \longrightarrow \Sigma_{b'} \times \Sigma_{b'}$ be the diagonal double Kodaira fibration associated with a non-strong pure braid quotient $\varphi: \mathsf{P}_2(\Sigma_2) \longrightarrow G$ of type $(2, p)$, where G is an extra-special p-group G of order p^9 and $b' = p^4 + 1$. Then the maximum slope $\nu(S_p)$ is attained for precisely two values of p, namely,

$$\nu(S_5) = \nu(S_7) = 2 + \frac{12}{35}.$$

Furthermore, $\nu(S_p) > 2 + 1/3$ for all $p \geq 5$. More precisely, if $p \geq 7$ the function $\nu(S_p)$ is strictly decreasing and

$$\lim_{p \to +\infty} \nu(S_p) = 2 + \frac{1}{3}.$$

- Let Σ_b be any closed Riemann surface of genus b. Then there exists a double Kodaira fibration $f: S \longrightarrow \Sigma_b \times \Sigma_b$. Moreover, denoting by $\kappa(b)$ the number of such fibrations, we have

$$\kappa(b) \geq \boldsymbol{\omega}(b + 1),$$

where $\boldsymbol{\omega}: \mathbb{N} \longrightarrow \mathbb{N}$ stands for the arithmetic function counting the number of distinct prime factors of a positive integer. In particular,

$$\limsup_{b \to +\infty} \kappa(b) = +\infty.$$

- Let G be a finite group and $\mathbf{f}: S \longrightarrow \Sigma_b \times \Sigma_b$ be a Galois cover, with Galois group G, branched over the diagonal Δ with branching order n. Then $|G| \geq 32$, and equality holds if and only if G is extra-special. If G is extra-special of order 32 and $(b, n) = (2, 2)$, then $\mathbf{f}: S \longrightarrow \Sigma_2 \times \Sigma_2$ is a diagonal double Kodaira fibration such that

$$b_1 = b_2 = 2, \quad g_1 = g_2 = 41, \quad \sigma(S) = 16.$$

As a consequence of the last result, we obtain a sharp lower bound for the signature of a diagonal double Kodaira fibration, see Theorem 7:

- Let $f: S \longrightarrow \Sigma_{b_1} \times \Sigma_{b_2}$ be a diagonal double Kodaira fibration, associated with a pure braid quotient $\varphi: \mathsf{P}_2(\Sigma_b) \longrightarrow G$ of type (b, n). Then $\sigma(S) \geq 16$, and equality holds precisely when $(b, n) = (2, 2)$ and G is an extra-special group of order 32.

Note that our methods show that *every* curve of genus b (and not only some special curve with extra automorphisms) is the basis of a (double) Kodaira fibration and that, in addition, the number of distinct Kodaira fibrations over a fixed base can be arbitrarily large. Furthermore, *every* curve of genus 2 is the base of a (double) Kodaira fibration with signature 16 and this provides, to our knowledge, the first example of positive-dimensional family of (double) Kodaira fibrations with small signature.

The aforementioned examples with signature 16 also provide new "double solutions" to a problem, posed by G. Mess and included in Kirby's problem list in low-dimensional topology, see [13, Problem 2.18 A], asking what is the smallest number b for which there exists a real surface bundle over a real surface with base genus b and non-zero signature. We actually have $b = 2$, also for double Kodaira fibrations, see Theorem 8:

- Let S be double Kodaira surface, associated with a pure braid quotient $\varphi \colon \mathsf{P}_2(\Sigma_b) \longrightarrow G$ of type (2, 2), where G is an extra-special group of order 32. Then the real manifold X underlying S is a closed, orientable 4-manifold of signature 16 that can be realized as a real surface bundle over a real surface of genus 2, with fiber genus 41, in two different ways.

In fact, it is an interesting question whether 16 and 41 are the minimum possible values for the signature and the fiber genus of a (not necessarily diagonal) double Kodaira surface $f : S \longrightarrow \Sigma_2 \times \Sigma_2$, but we will not develop this point here.

The above results paint a rather clear picture regarding pure braid quotients and the relative diagonal double Kodaira fibrations when $|G| \leq 32$. It is natural then to investigate further this topic for $|G| > 32$, and indeed this paper also contains the following new result, see Theorem 9:

- If G is a finite group with $32 < |G| < 64$, then G is not a pure braid quotient.

We provide only a sketch of the proof, which is based on calculations performed by means of the computer algebra system GAP4, see [8]; the details will appear in a forthcoming paper.

Notation and conventions. The order of a finite group G is denoted by $|G|$. If $x \in G$, the order of x is denoted by $o(x)$. The subgroup generated by $x_1, \ldots, x_n \in G$ is denoted by $\langle x_1, \ldots, x_n \rangle$. The center of G is denoted by $Z(G)$ and the centralizer of an element $x \in G$ by $C_G(x)$. If $x, y \in G$, their commutator is defined as $[x, y] = xyx^{-1}y^{-1}$. We denote both the cyclic group of order p and the field with p elements by $\mathbb{Z}_p$. We use sometimes the `IdSmallGroup(G)` label from GAP4 list of small groups. For instance, $\mathsf{S}_4 = G(24, 12)$ means that S_4 is the twelfth group of order 24 in this list.

2 Pure Surface Braid Groups and Finite Braid Quotients

Let Σ_b be a closed Riemann surface of genus $b \geq 2$, and let $\mathscr{P} = (p_1, p_2)$ be an ordered pair of distinct points on Σ_b. A *pure geometric braid* on Σ_b based at $\mathscr{P}$ is a pair (α_1, α_2) of paths $\alpha_i : [0, 1] \longrightarrow \Sigma_b$ such that

- $\alpha_i(0) = \alpha_i(1) = p_i$ for all $i \in \{1, 2\}$
- the points $\alpha_1(t), \alpha_2(t) \in \Sigma_b$ are pairwise distinct for all $t \in [0, 1]$,

see Fig. 1.

Fig. 1 A pure braid on two strands

Definition 1 The *pure braid group* on two strands on Σ_b is the group $\mathsf{P}_2(\Sigma_b)$ whose elements are the pure braids based at $\mathscr{P}$ and whose operation is the usual concatenation of paths, up to homotopies among braids.

It can be shown that $\mathsf{P}_2(\Sigma_b)$ does not depend on the choice of the set $\mathscr{P} = (p_1, p_2)$, and that there is an isomorphism

$$\mathsf{P}_2(\Sigma_b) \simeq \pi_1(\Sigma_b \times \Sigma_b - \Delta, \mathscr{P}) \tag{2}$$

where $\Delta \subset \Sigma_b \times \Sigma_b$ is the diagonal.

The group $\mathsf{P}_2(\Sigma_b)$ is finitely presented for all b, and explicit presentations can be found in [4, 5, 9, 21]. Here we follow the approach in [9, Sects. 1–3], referring the reader to that paper for further details.

Proposition 1 ([9, Theorem 1]) *Let $p_1, p_2 \in \Sigma_b$, with $b \geq 2$. Then the map of pointed topological spaces given by the projection onto the first component*

$$(\Sigma_b \times \Sigma_b - \Delta, \mathscr{P}) \longrightarrow (\Sigma_b, p_1)$$

induces a split short exact sequence of groups

$$1 \longrightarrow \pi_1(\Sigma_b - \{p_1\}, p_2) \longrightarrow \mathsf{P}_2(\Sigma_b) \longrightarrow \pi_1(\Sigma_b, p_1) \longrightarrow 1. \tag{3}$$

For all $j \in \{1, \dots, b\}$, let us consider now the $2b$ elements

$$\rho_{1j}, \tau_{1j}, \rho_{2j}, \tau_{2j} \tag{4}$$

of $\mathsf{P}_2(\Sigma_b)$ represented by the pure braids shown in Fig. 2. If $\ell \neq i$, the path corresponding to ρ_{ij} and τ_{ij} based at p_ℓ is the constant path. Moreover, let A_{12} be the pure braid shown in Fig. 3. In terms of the isomorphism (2), the generators ρ_{ij}, τ_{ij} correspond to the generators of $\pi_1(\Sigma_b \times \Sigma_b - \Delta, \mathscr{P})$ coming from the usual description of Σ_b as the identification space of a regular $2b$-gon, whereas A_{12} corresponds to the homotopy class in $\Sigma_b \times \Sigma_b - \Delta$ of a topological loop in $\Sigma_b \times \Sigma_b$ that "winds once" around Δ.

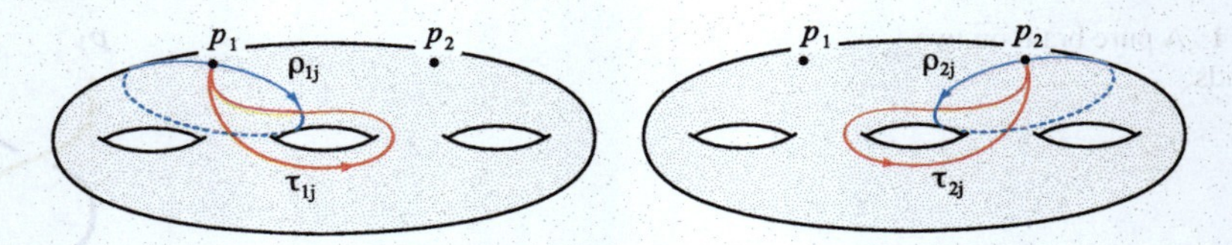

Fig. 2 The pure braids ρ_{1j}, τ_{1j}, ρ_{2j}, τ_{2j} on Σ_b

Fig. 3 The pure braid A_{12} on Σ_b

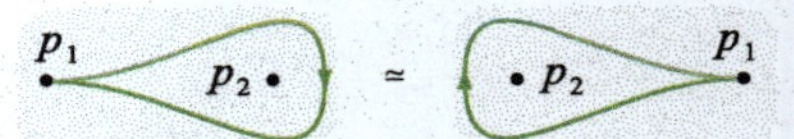

The elements

$$\rho_{21}, \dots, \rho_{2b}, \ \tau_{21}, \dots, \tau_{2b}, \ A_{12} \tag{5}$$

can be seen as generators of the kernel $\pi_1(\Sigma_b - \{p_1\}, \ p_2)$ in (3), whereas the elements

$$\rho_{11}, \dots, \rho_{1b}, \ \tau_{11}, \dots, \tau_{1b} \tag{6}$$

are lifts of a set of generators of $\pi_1(\Sigma_b, \ p_1)$ via the quotient map $\mathsf{P}_2(\Sigma_b) \longrightarrow \pi_1(\Sigma_b, \ p_1)$, namely, they form a complete system of coset representatives for $\pi_1(\Sigma_b, \ p_1)$.

By Proposition 1, the braid group $\mathsf{P}_2(\Sigma_b)$ is a semi-direct product of the two groups $\pi_1(\Sigma_b - \{p_1\}, \ p_2)$ and $\pi_1(\Sigma_b, \ p_1)$, whose presentations are both well-known; then, in order to write down a presentation for $\mathsf{P}_2(\Sigma_b)$, it only remains to specify how the generators in (6) act by conjugation on those in (5). This is provided by the following result, cf. [6, Theorem 1.6], where the conjugacy relations are expressed in the commutator form (i.e., instead of $xyx^{-1} = z$ we write $[x, \ y] = zy^{-1}$).

Proposition 2 ([9, Theorem 7]) *The group $\mathsf{P}_2(\Sigma_b)$ admits the following presentation.*

Generators
$\rho_{1j}, \ \tau_{1j}, \ \rho_{2j}, \ \tau_{2j}, \ A_{12} \quad j = 1, \dots, b.$
Relations

- Surface relations:

$$[\rho_{1b}^{-1}, \ \tau_{1b}^{-1}] \, \tau_{1b}^{-1} \, [\rho_{1\,b-1}^{-1}, \ \tau_{1\,b-1}^{-1}] \, \tau_{1\,b-1}^{-1} \cdots [\rho_{11}^{-1}, \ \tau_{11}^{-1}] \, \tau_{11}^{-1} \, (\tau_{11} \, \tau_{12} \cdots \tau_{1b}) = A_{12}$$

$$[\rho_{21}^{-1}, \ \tau_{21}] \, \tau_{21} \, [\rho_{22}^{-1}, \ \tau_{22}] \, \tau_{22} \cdots [\rho_{2b}^{-1}, \ \tau_{2b}] \, \tau_{2b} \, (\tau_{2b}^{-1} \, \tau_{2\,b-1}^{-1} \cdots \tau_{21}^{-1}) = A_{12}^{-1}$$

- Action of ρ_{1j} :

$$
\begin{aligned}
&[\rho_{1j},\ \rho_{2k}] = 1 && \text{if} \quad j < k \\
&[\rho_{1j},\ \rho_{2j}] = 1 && \\
&[\rho_{1j},\ \rho_{2k}] = A_{12}^{-1}\,\rho_{2k}\,\rho_{2j}^{-1}\,A_{12}\,\rho_{2j}\,\rho_{2k}^{-1} && \text{if} \quad j > k \\
&[\rho_{1j},\ \tau_{2k}] = 1 && \text{if} \quad j < k \\
&[\rho_{1j},\ \tau_{2j}] = A_{12}^{-1} && \\
&[\rho_{1j},\ \tau_{2k}] = [A_{12}^{-1},\ \tau_{2k}] && \text{if} \quad j > k \\
&[\rho_{1j},\ A_{12}] = [\rho_{2j}^{-1},\ A_{12}] &&
\end{aligned}
\tag{7}
$$

- Action of τ_{1j} :

$$
\begin{aligned}
&[\tau_{1j},\ \rho_{2k}] = 1 && \text{if} \quad j < k \\
&[\tau_{1j},\ \rho_{2j}] = \tau_{2j}^{-1}\,A_{12}\,\tau_{2j} && \\
&[\tau_{1j},\ \rho_{2k}] = [\tau_{2j}^{-1},\ A_{12}] && \text{if} \quad j > k
\end{aligned}
$$

$$
\begin{aligned}
&[\tau_{1j},\ \tau_{2k}] = 1 && \text{if} \quad j < k \\
&[\tau_{1j},\ \tau_{2j}] = [\tau_{2j}^{-1},\ A_{12}] && \\
&[\tau_{1j},\ \tau_{2k}] = \tau_{2j}^{-1}\,A_{12}\,\tau_{2j}\,A_{12}^{-1}\,\tau_{2k}\,A_{12}\,\tau_{2j}^{-1}\,A_{12}^{-1}\,\tau_{2j}\,\tau_{2k}^{-1} && \text{if} \quad j > k
\end{aligned}
$$

$$[\tau_{1j},\ A_{12}] = [\tau_{2j}^{-1},\ A_{12}]$$

Remark 1 The inclusion map $\iota\colon \Sigma_b \times \Sigma_b - \Delta \longrightarrow \Sigma_b \times \Sigma_b$ induces a group epimorphism $\iota_*\colon \mathsf{P}_2(\Sigma_b) \longrightarrow \pi_1(\Sigma_b \times \Sigma_b,\ \mathscr{P})$, whose kernel is the normal closure of the subgroup generated by A_{12}. Thus, given any group homomorphism $\varphi\colon \mathsf{P}_2(\Sigma_b) \longrightarrow G$, it factors through $\pi_1(\Sigma_b \times \Sigma_b,\ \mathscr{P})$ if and only if $\varphi(A_{12})$ is trivial.

Tedious but straightforward calculations show that the presentation given in Proposition 2 is invariant under the substitutions

$$A_{12} \longleftrightarrow A_{12}^{-1}, \quad \tau_{1j} \longleftrightarrow \tau_{2\ b+1-j}^{-1}, \quad \rho_{1j} \longleftrightarrow \rho_{2\ b+1-j},$$

where $j \in \{1, \ldots, b\}$. These substitutions correspond to the involution of $\mathsf{P}_2(\Sigma_b)$ induced by a reflection of Σ_b switching the jth handle with the $(b+1-j)$th handle for all j. Hence we can exchange the roles of p_1 and p_2 in (3), and see $\mathsf{P}_2(\Sigma_b)$ as the middle term of a split short exact sequence of the form

$$1 \longrightarrow \pi_1(\Sigma_b - \{p_2\},\ p_1) \longrightarrow \mathsf{P}_2(\Sigma_b) \longrightarrow \pi_1(\Sigma_b,\ p_2) \longrightarrow 1, \tag{8}$$

induced by the projection onto the second component

$$(\Sigma_b \times \Sigma_b - \Delta, \mathscr{P}) \longrightarrow (\Sigma_b, p_2).$$

The elements

$$\rho_{11}, \ldots, \rho_{1b}, \tau_{11}, \ldots, \tau_{1b}, A_{12}$$

can be seen as generators of the kernel $\pi_1(\Sigma_b - \{p_2\}, p_1)$ in (8), whereas the elements

$$\rho_{21}, \ldots, \rho_{2b}, \tau_{21}, \ldots, \tau_{2b}$$

yield a complete system of coset representatives for $\pi_1(\Sigma_b, p_2)$.

We can now define the objects studied in this paper.

Definition 2 Take positive integers $b, n \geq 2$. A finite group G is called a *pure braid quotient of type* (b, n) if there exists a group epimorphism

$$\varphi\colon \mathsf{P}_2(\Sigma_b) \longrightarrow G \tag{9}$$

such that $\varphi(A_{12})$ has order n.

Remark 2 Since we are assuming $n \geq 2$, the element $\varphi(A_{12})$ is non-trivial and so the epimorphism φ does not factor through $\pi_1(\Sigma_b \times \Sigma_b, \mathscr{P})$, see Remark 1. The geometrical relevance of this condition will be explained in Sect. 4. The same condition also shows that a pure braid quotient is necessarily non-abelian, because $\varphi(A_{12})$ is a non-trivial commutator in G, see (7).

Sometimes, we will use the term *pure braid quotient* in order to indicate the full datum of the quotient homomorphism (9), instead of the quotient group G alone.

If G is a pure braid quotient, then the two subgroups

$$\begin{aligned} K_1 &:= \langle \varphi(\rho_{11}), \varphi(\tau_{11}), \ldots, \varphi(\rho_{1b}), \varphi(\tau_{1b}), \varphi(A_{12}) \rangle \\ K_2 &:= \langle \varphi(\rho_{21}), \varphi(\tau_{21}), \ldots, \varphi(\rho_{2b}), \varphi(\tau_{2b}), \varphi(A_{12}) \rangle \end{aligned} \tag{10}$$

are both normal in G, and hence there are two short exact sequences

$$\begin{aligned} 1 \longrightarrow K_1 \longrightarrow G \longrightarrow Q_2 \longrightarrow 1 \\ 1 \longrightarrow K_2 \longrightarrow G \longrightarrow Q_1 \longrightarrow 1, \end{aligned}$$

in which the elements $\varphi(\rho_{21}), \varphi(\tau_{21}), \ldots, \varphi(\rho_{2b}), \varphi(\tau_{2b})$ yield a complete system of coset representatives for Q_2, whereas the elements $\varphi(\rho_{11}), \varphi(\tau_{11}), \ldots, \varphi(\rho_{1b}), \varphi(\tau_{1b})$ yield a complete system of coset representatives for Q_1.

Let us end this section with the following definition, whose geometrical meaning will become clear later, see Remark 5 of Sect. 4.

Definition 3 A pure braid quotient $\varphi\colon \mathsf{P}_2(\Sigma_b) \longrightarrow G$ is called *strong* if $K_1 = K_2 = G$.

3 Extra-Special Groups as Pure Braid Quotients

We know that $\mathsf{P}_2(\Sigma_b)$ is residually p-finite for all prime number $p \geq 2$, see [2, pp. 1481–1490]. This implies that, for every p, we can find a non-abelian finite p-group G that is a pure braid quotient of type $(b,\ q)$, where q is a power of p. However, it can be tricky to explicitly describe some of these quotients.

In this section we will present a number of results in this direction, obtained in the series of articles [6, 18, 19]; our exposition here will closely follow the treatment given in these papers. Let us start by introducing the following classical definition, see for instance [10, p. 183] and [15, p. 123].

Definition 4 Let p be a prime number. A finite p-group G is called *extra-special* if its center $Z(G)$ is cyclic of order p and the quotient $V = G/Z(G)$ is a non-trivial, elementary abelian p-group.

An elementary abelian p-group is a finite-dimensional vector space over the field $\mathbb{Z}_p$, hence it is of the form $V = (\mathbb{Z}_p)^{\dim V}$ and G fits into a short exact sequence

$$1 \longrightarrow \mathbb{Z}_p \longrightarrow G \longrightarrow V \longrightarrow 1. \tag{11}$$

Note that, V being abelian, we must have $[G,\ G] = \mathbb{Z}_p$, namely, the commutator subgroup of G coincides with its center. Furthermore, since the extension (11) is central, it cannot be split, otherwise G would be isomorphic to the direct product of the two abelian groups $\mathbb{Z}_p$ and V, which is impossible because G is non-abelian. It can be also proved that, if G is extra-special, then $\dim V$ is even and so $|G| = p^{\dim V+1}$ is an odd power of p.

For every prime number p, there are precisely two isomorphism classes $M(p)$, $N(p)$ of non-abelian groups of order p^3, namely,

$$M(p) = \langle \mathsf{r},\ \mathsf{t},\ \mathsf{z} \mid \mathsf{r}^p = \mathsf{t}^p = 1,\ \mathsf{z}^p = 1, [\mathsf{r},\ \mathsf{z}] = [\mathsf{t},\ \mathsf{z}] = 1,\ [\mathsf{r},\ \mathsf{t}] = \mathsf{z}^{-1}\rangle$$
$$N(p) = \langle \mathsf{r},\ \mathsf{t},\ \mathsf{z} \mid \mathsf{r}^p = \mathsf{t}^p = \mathsf{z},\ \mathsf{z}^p = 1, [\mathsf{r},\ \mathsf{z}] = [\mathsf{t},\ \mathsf{z}] = 1,\ [\mathsf{r},\ \mathsf{t}] = \mathsf{z}^{-1}\rangle$$

and both of them are in fact extra-special, see [10, Theorem 5.1 of Chap. 5].

If p is odd, then the groups $M(p)$ and $N(p)$ are distinguished by their exponent, which equals p and p^2, respectively. If $p = 2$, the group $M(p)$ is isomorphic to the dihedral group D_8, whereas $N(p)$ is isomorphic to the quaternion group Q_8.

We can now provide the classification of extra-special p-groups, see [10, Sect. 5 of Chap. 5].

Proposition 3 *If $b \geq 2$ is a positive integer and p is a prime number, there are exactly two isomorphism classes of extra-special p-groups of order p^{2b+1}, which can be described as follows.*

- *The central product $\mathsf{H}_{2b+1}(\mathbb{Z}_p)$ of b copies of $M(p)$, having presentation*

$$\begin{aligned}\mathsf{H}_{2b+1}(\mathbb{Z}_p) = \langle\, \mathsf{r}_1, \mathsf{t}_1, \ldots, \mathsf{r}_b, \mathsf{t}_b, \mathsf{z} \mid \mathsf{r}_j^p = \mathsf{t}_j^p = \mathsf{z}^p = 1,\\ [\mathsf{r}_j, \mathsf{z}] = [\mathsf{t}_j, \mathsf{z}] = 1,\\ [\mathsf{r}_j, \mathsf{r}_k] = [\mathsf{t}_j, \mathsf{t}_k] = 1,\\ [\mathsf{r}_j, \mathsf{t}_k] = \mathsf{z}^{-\delta_{jk}} \,\rangle.\end{aligned}$$

If p is odd, this group has exponent p.

- *The central product $\mathsf{G}_{2b+1}(\mathbb{Z}_p)$ of $b-1$ copies of $M(p)$ and one copy of $N(p)$, having presentation*

$$\begin{aligned}\mathsf{G}_{2b+1}(\mathbb{Z}_p) = \langle\, \mathsf{r}_1, \mathsf{t}_1, \ldots, \mathsf{r}_b, \mathsf{t}_b, \mathsf{z} \mid \mathsf{r}_b^p = \mathsf{t}_b^p = \mathsf{z},\\ \mathsf{r}_1^p = \mathsf{t}_1^p = \cdots = \mathsf{r}_{b-1}^p = \mathsf{t}_{b-1}^p = \mathsf{z}^p = 1,\\ [\mathsf{r}_j, \mathsf{z}] = [\mathsf{t}_j, \mathsf{z}] = 1,\\ [\mathsf{r}_j, \mathsf{r}_k] = [\mathsf{t}_j, \mathsf{t}_k] = 1,\\ [\mathsf{r}_j, \mathsf{t}_k] = \mathsf{z}^{-\delta_{jk}} \,\rangle.\end{aligned}$$

If p is odd, this group has exponent p^2.

We are now in a position to state our first two results.

Theorem 1 ([6, Sect. 4], [18, Theorem 3.10]) *If $b \geq 2$ is an integer and $p \geq 5$ is a prime number, then both extra-special p-groups of order p^{4b+1} are pure braid quotients of type $(b,\ p)$. All these quotients are non-strong, in fact K_1 and K_2 have index p^{2b} in G.*

Theorem 2 ([6, Sect. 4], [18, Theorem 3.7]) *If $b \geq 2$ is an integer and p is a prime number dividing $b+1$, then both extra-special p-groups of order p^{2b+1} are strong pure braid quotients of type $(b,\ p)$.*

Theorems 1 and 2 were originally proved by the first author and A. Causin in [6], but only in the case $G = \mathsf{H}_{4b+1}(\mathbb{Z}_p)$ and $G = \mathsf{H}_{2b+1}(\mathbb{Z}_p)$, respectively, by using some group-cohomological results related to the structure of the cohomology algebra $H^*(\Sigma_b \times \Sigma_b - \Delta,\ \mathbb{Z}_p)$. Let us give here a sketch of the argument, referring the reader to the aforementioned paper for full details.

Assuming $p \geq 3$, we identified $\mathsf{H}_{4b+1}(\mathbb{Z}_p)$ with the *symplectic Heisenberg group* $\mathsf{Heis}(V,\omega)$, where

$$V = H_1(\Sigma_b \times \Sigma_b - \Delta,\ \mathbb{Z}_p) \simeq H_1(\Sigma_b \times \Sigma_b,\ \mathbb{Z}_p) \simeq (\mathbb{Z}_p)^{4b}$$

and ω is a symplectic form on V. This group is the central extension

$$1 \longrightarrow \mathbb{Z}_p \longrightarrow \mathsf{Heis}(V,\ \omega) \longrightarrow V \longrightarrow 1 \tag{12}$$

of the additive group V given as follows: the underlying set of $\mathsf{Heis}(V,\ \omega)$ is $V \times \mathbb{Z}_p$, endowed with the group law

$$(v_1, t_1)(v_2, t_2) = \left(v_1 + v_2, \ t_1 + t_2 + \frac{1}{2}\omega(v_1, v_2)\right). \tag{13}$$

By basic linear algebra, all symplectic forms on $(\mathbb{Z}_p)^{4b}$ are equivalent to the standard symplectic form; thus, given two symplectic forms ω_1, ω_2 on V, the two Heisenberg groups $\mathsf{Heis}(V, \omega_1)$, $\mathsf{Heis}(V, \omega_2)$ are isomorphic. Moreover, the center of the Heisenberg group coincides with its commutator subgroup and is isomorphic to $\mathbb{Z}_p$.

Now, let

$$\phi \colon \mathsf{P}_2(\Sigma_b) \longrightarrow V$$

be the group epimorphism given by the composition of the reduction mod p map $H_1(\Sigma_b \times \Sigma_b - \Delta, \mathbb{Z}) \longrightarrow V$ with the abelianization map $\mathsf{P}_2(\Sigma_b) \longrightarrow H_1(\Sigma_b \times \Sigma_b - \Delta, \mathbb{Z})$. We have a commutative diagram

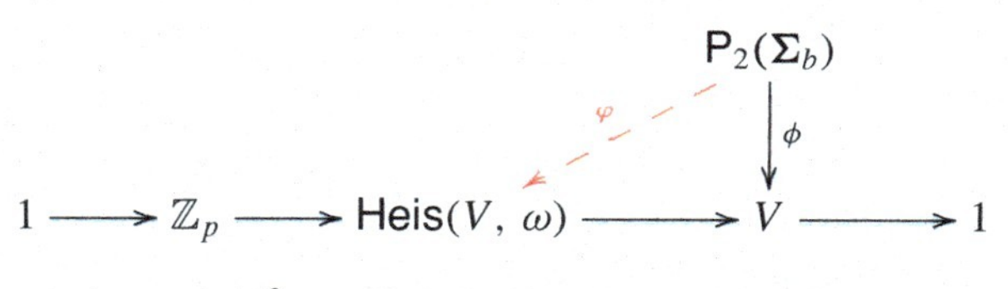

and we denote by $u \in H^2(V, \mathbb{Z}_p)$ the cohomology class corresponding to the bottom Heisenberg extension. Then a lifting $\varphi \colon \mathsf{P}_2(\Sigma_b) \longrightarrow \mathsf{Heis}(V, \omega)$ of ϕ exists if and only if $\phi^* u = 0 \in H^2(\mathsf{P}_2(\Sigma_b), \mathbb{Z}_p)$.

The next step is to provide an interpretation of the cohomological condition $\phi^* u = 0$ in terms of the symplectic form ω, and this is achieved by using the following facts:

- we have a natural identification

$$H^2(V, \mathbb{Z}_p) \simeq \Lambda^2 V^\vee \oplus V^\vee \tag{14}$$

under which the extension class u giving the Heisenberg central extension (12) corresponds to (ω, ϵ). Here $\epsilon \colon V \longrightarrow \mathbb{Z}_p$ stands for the linear functional on V defined by $\epsilon(v) = w^p$, where w is any preimage of v in $\mathsf{Heis}(V, \omega)$;
- we have natural identifications

$$V^\vee \simeq H^1(\Sigma_b \times \Sigma_b - \Delta, \mathbb{Z}_p) \simeq H^1(\Sigma_b \times \Sigma_b, \mathbb{Z}_p)$$

and there is a commutative diagram

$$\begin{array}{ccc} \mathrm{Alt}^2(V) \simeq \wedge^2 V^\vee & \xrightarrow{\ \xi\ } & H^2(\Sigma_b \times \Sigma_b, \mathbb{Z}_p) \\ & \searrow^{\eta} & \downarrow \\ & & H^2(\Sigma_b \times \Sigma_b - \Delta, \mathbb{Z}_p) \end{array}$$

where the vertical map is the quotient by the 1-dimensional vector subspace of $H^2(\Sigma_b \times \Sigma_b, \mathbb{Z}_p)$ generated by the class δ of the diagonal, whereas η and ξ stand for the cup-product maps;

- $\Sigma_b \times \Sigma_b - \Delta$ is an aspherical space, namely, all its higher homotopy group vanish, and so for all $i \geq 1$ there is a natural isomorphism

$$H^i(\Sigma_b \times \Sigma_b - \Delta, \mathbb{Z}_p) \simeq H^i(\mathsf{P}_2(\Sigma_b), \mathbb{Z}_p) \tag{15}$$

where $\mathbb{Z}_p$ is endowed, as an abelian group, with the structure of trivial $\mathsf{P}_2(\Sigma_b)$-module.

Combining all this, we infer that there is a commutative diagram

$$\begin{array}{ccccc} \wedge^2 V^\vee \oplus V^\vee & \xrightarrow{\simeq} & H^2(V, \mathbb{Z}_p) & \xrightarrow{\phi^*} & H^2(\mathsf{P}_2(\Sigma_b), \mathbb{Z}_p) \\ & & \downarrow & & \downarrow \simeq \\ & & \mathrm{Alt}^2(V) \simeq \wedge^2 V^\vee & \xrightarrow{\eta} & H^2(\Sigma_b \times \Sigma_b - \Delta, \mathbb{Z}_p), \end{array}$$

where the isomorphism on the left is (14), the vertical map on the left is the projection onto the first summand and the vertical map on the right is (15). Since the projection of the extension class $u \in H^2(V, \mathbb{Z}_p)$ can be naturally identified with $\omega \in \mathrm{Alt}^2(V)$, we have proved the following

Proposition 4 *The obstruction class $\phi^* u \in H^2(\mathsf{P}_2(\Sigma_b), \mathbb{Z}_p)$ can be naturally interpreted as the image $\eta(\omega) \in H^2(\Sigma_b \times \Sigma_b - \Delta, \mathbb{Z}_p)$ of the symplectic form $\omega \in \mathrm{Alt}^2(V)$ via the cup-product map η.*

As a consequence, we obtain the following lifting criterion, that we believe is of independent interest.

Proposition 5 *A lifting $\varphi\colon \mathsf{P}_2(\Sigma_b) \longrightarrow \mathsf{Heis}(V, \omega)$ of $\phi\colon \mathsf{P}_2(\Sigma_b) \longrightarrow V$ exists if and only if $\eta(\omega) = 0$. Furthermore, if φ exists, then $\varphi(A_{12})$ has order p if and only if $\xi(\omega) \in H^2(\Sigma_b \times \Sigma_b, \mathbb{Z}_p)$ is a non-zero integer multiple of the diagonal class δ. In this case, φ is necessarily surjective.*

Inspired by Proposition 5, we say that a symplectic form $\omega \in \mathrm{Alt}^2(V)$ is *of Heisenberg type* if $\xi(\omega)$ is a non-zero integer multiple of δ; equivalently, ω is of Heisenberg type if $\eta(\omega) = 0$ and $\xi(\omega) \neq 0$. By the previous discussion, it follows that, if ω is of Heisenberg type, $\mathsf{Heis}(V, \omega)$ is a pure braid quotient of type (b, p).

We are therefore left with the task of constructing symplectic forms of Heisenberg type on V. We denote by $\alpha_1, \beta_1, \ldots, \alpha_b, \beta_b$ the images in $H^1(\Sigma_b, \mathbb{Z}_p) = H^1(\Sigma_b, \mathbb{Z}) \otimes \mathbb{Z}_p$ of the elements of a basis of $H^1(\Sigma_b, \mathbb{Z})$ which is symplectic with respect to the cup product; then, we can choose for V the ordered basis

$$r_{11}, t_{11}, \ldots, r_{1b}, t_{1b}, \; r_{21}, t_{21}, \ldots, r_{2b}, t_{2b} \tag{16}$$

where, under the isomorphism $V \simeq H_1(\Sigma_b \times \Sigma_b, \mathbb{Z}_p)$ induced by the inclusion $\iota\colon \Sigma_b \times \Sigma_b - \Delta \longrightarrow \Sigma_b \times \Sigma_b$, the elements $r_{1j}, t_{1j}, r_{2j}, t_{2j} \in V$ are the duals of the elements $\alpha_j \otimes 1, \beta_j \otimes 1, 1 \otimes \alpha_j, 1 \otimes \beta_j \in H^1(\Sigma_b \times \Sigma_b, \mathbb{Z}_p) \simeq H^1(\Sigma_b, \mathbb{Z}_p) \otimes H^1(\Sigma_b, \mathbb{Z}_p)$, respectively.

Since $p \geq 5$, we can find non-zero scalars $\lambda_1, \ldots, \lambda_b, \mu_1, \ldots, \mu_b \in \mathbb{Z}_p$ such that $1 - \lambda_i\mu_1 \neq 0$ for all $i \in \{1, \ldots, b\}$ and

$$\sum_{j=1}^{b} \lambda_j = \sum_{j=1}^{b} \mu_j = 1. \tag{17}$$

Then we consider the alternating form $\omega\colon V \times V \longrightarrow \mathbb{Z}_p$ represented, with respect to the ordered basis (16), by the skew-symmetric matrix

$$\Omega_b = \begin{pmatrix} L_b & J_b \\ J_b & M_b \end{pmatrix} \in \mathrm{Mat}(4b, \mathbb{Z}_p) \tag{18}$$

where the blocks are the elements of $\mathrm{Mat}(2b, \mathbb{Z}_p)$ given by

$$L_b = \begin{pmatrix} \begin{matrix} 0 & \lambda_1 \\ -\lambda_1 & 0 \end{matrix} & & 0 \\ & \ddots & \\ 0 & & \begin{matrix} 0 & \lambda_b \\ -\lambda_b & 0 \end{matrix} \end{pmatrix} \qquad M_b = \begin{pmatrix} \begin{matrix} 0 & \mu_1 \\ -\mu_1 & 0 \end{matrix} & & 0 \\ & \ddots & \\ 0 & & \begin{matrix} 0 & \mu_b \\ -\mu_b & 0 \end{matrix} \end{pmatrix}$$

$$J_b = \begin{pmatrix} \begin{matrix} 0 & -1 \\ 1 & 0 \end{matrix} & & 0 \\ & \ddots & \\ 0 & & \begin{matrix} 0 & -1 \\ 1 & 0 \end{matrix} \end{pmatrix}$$

Standard Gaussian elimination shows that

$$\det \Omega_b = (1 - \lambda_1\mu_1)^2(1 - \lambda_2\mu_2)^2 \cdots (1 - \lambda_b\mu_b)^2 > 0$$

and so ω is non-degenerate. Moreover, a direct computation yields $\xi(\omega) = \delta$, that is, ω is of Heisenberg type. The calculation of the indices of K_1 and K_2 in G is now straightforward, and this completes the proof of Theorem 1 in the case $G = \mathsf{H}_{4b+1}(\mathbb{Z}_p)$.

Now, let us assume that p divides $b + 1$, so that $-b = 1$ holds in $\mathbb{Z}_p$, and take

$$\lambda_1 = \cdots = \lambda_b = \mu_1 = \cdots = \mu_b = -1 \in \mathbb{Z}_p.$$

Therefore relations (17) are satisfied and the same computations as in the previous case show that the corresponding alternating form ω satisfies $\xi(\omega) = \delta$. However, ω is not symplectic, since its associate matrix

$$\Omega_b = \begin{pmatrix} J_b & J_b \\ J_b & J_b \end{pmatrix} \in \mathrm{Mat}(4b, \mathbb{Z}_p)$$

has rank $2b$ and, subsequently, ω has a $2b$-dimensional kernel V_0, namely,

$$V_0 = \langle r_{11} - r_{21}, \, t_{11} - t_{21}, \ldots, r_{1b} - r_{2b}, \, t_{1b} - t_{2b} \rangle. \tag{19}$$

The set $V \times \mathbb{Z}_p$, with the operation (13), is a group whose center equals $V_0 \times \mathbb{Z}_p$ and that, with slight abuse of notation, we denote again by $\mathsf{Heis}(V, \omega)$. Furthermore, the argument in Proposition 5 still applies, providing the existence of a lifting $\mathsf{P}_2(\Sigma_b) \longrightarrow \mathsf{Heis}(V, \omega)$. Setting $W = V/V_0$, the alternating form on V descends to a symplectic form on W, that we denote it again by ω; so $\mathsf{Heis}(W, \omega)$ is a genuine Heisenberg group, endowed with a group epimorphism $\mathsf{Heis}(V, \omega) \longrightarrow \mathsf{Heis}(W, \omega)$. Composing this epimorphism with the lifting $\mathsf{P}_2(\Sigma_b) \longrightarrow \mathsf{Heis}(V, \omega)$, we obtain a group epimorphism $\varphi \colon \mathsf{P}_2(\Sigma_b) \longrightarrow \mathsf{Heis}(W, \omega)$ such that $\varphi(A_{12})$ is non-trivial and central, hence of order p.

Since W is a $\mathbb{Z}_p$-vector space of dimension $2b$, the group $\mathsf{Heis}(W, \omega)$ is isomorphic to $\mathsf{H}_{2b+1}(\mathbb{Z}_p)$; finally, a simple computation based on the expression (19) for $\ker \Omega_b$ yields $K_1 = K_2 = G$, and this shows Theorem 2 in the case $G = \mathsf{H}_{2b+1}(\mathbb{Z}_p)$.

The proof of Theorems 1 and 2 in full generality (i.e., for all extra-special groups) was given in [18], using a completely algebraic technique that avoided the use of symplectic geometry and of group cohomology. It is based on the following

Definition 5 Let G be a finite group. A *diagonal double Kodaira structure* of type (b, n) on G is an ordered set of $4b + 1$ generators

$$\mathfrak{S} = (\mathsf{r}_{11}, \mathsf{t}_{11}, \ldots, \mathsf{r}_{1b}, \mathsf{t}_{1b}, \mathsf{r}_{21}, \mathsf{t}_{21}, \ldots, \mathsf{r}_{2b}, \mathsf{t}_{2b}, \mathsf{z}),$$

with $o(\mathsf{z}) = n$, that are images of the ordered set of generators

$$(\rho_{11}, \tau_{11}, \ldots, \rho_{1b}, \tau_{1b}, \rho_{21}, \tau_{21}, \ldots, \rho_{2b}, \tau_{2b}, A_{12})$$

via a pure braid quotient $\varphi \colon \mathsf{P}_2(\Sigma_b) \longrightarrow G$ of type (b, n). The structure is called *strong* if

$$\langle \mathsf{r}_{11}, \mathsf{t}_{11}, \ldots, \mathsf{r}_{1b}, \mathsf{t}_{1b} \rangle = \langle \mathsf{r}_{21}, \mathsf{t}_{21}, \ldots, \mathsf{r}_{2b}, \mathsf{t}_{2b} \rangle = G.$$

Therefore, checking whether G is a pure braid quotient of type (b, n) is equivalent to checking whether it admits a diagonal double Kodaira structure $\mathfrak{S}$ of type (b, n). Moreover, by definition, $\varphi \colon \mathsf{P}_2(\Sigma_b) \longrightarrow G$ is strong if and only if $\mathfrak{S}$ is.

Let us refer now to the presentations for extra-special p-groups given in Proposition 3. Assuming that p divides $b + 1$, in both cases $G = \mathsf{H}_{2b+1}(\mathbb{Z}_p)$ and $G =$

$\mathsf{G}_{2b+1}(\mathbb{Z}_p)$ we can obtain a strong diagonal double Kodaira structure $\mathfrak{S}$ on G by setting

$$\mathsf{r}_{1j} = \mathsf{r}_{2j} = \mathsf{r}_j, \quad \mathsf{t}_{1j} = \mathsf{t}_{2j} = \mathsf{t}_j$$

for all $j = 1, \dots, b$. The divisibility condition is necessary to ensure that the element of $\mathfrak{S}$ satisfies the two relations coming from the surface relations in Proposition 2. This proves Theorem 2.

In order to prove Theorem 1, it is convenient to consider the following alternative presentation of extra-special p-groups. Consider any non-degenerate, skew-symmetric matrix $A = (a_{jk})$ of order $2b$ over $\mathbb{Z}_p$, and consider the finitely presented groups

$$\begin{aligned} \mathsf{H}(A) = \langle \mathsf{x}_1, \dots, \mathsf{x}_{2b}, \mathsf{z} \mid \mathsf{x}_1^p = \dots = \mathsf{x}_{2b}^p = \mathsf{z}^p = 1, \\ [\mathsf{x}_1, \mathsf{z}] = \dots = [\mathsf{x}_{2b}, \mathsf{z}] = 1, \\ [\mathsf{x}_j, \mathsf{x}_k] = \mathsf{z}^{a_{jk}} \rangle, \end{aligned} \tag{20}$$

$$\begin{aligned} \mathsf{G}(A) = \langle \mathsf{x}_1, \dots, \mathsf{x}_{2b}, \mathsf{z} \mid \mathsf{x}_1^p = \dots = \mathsf{x}_{2b-2}^p = \mathsf{z}^p = 1, \\ \mathsf{x}_{2b-1}^p = \mathsf{x}_{2b}^p = \mathsf{z}, \\ [\mathsf{x}_1, \mathsf{z}] = \dots = [\mathsf{x}_{2b}, \mathsf{z}] = 1, \\ [\mathsf{x}_j, \mathsf{x}_k] = \mathsf{z}^{a_{jk}} \rangle, \end{aligned} \tag{21}$$

where the exponent in $\mathsf{z}^{a_{jk}}$ stands for any representative in $\mathbb{Z}$ of $a_{jk} \in \mathbb{Z}_p$. Standard computations show that $\mathsf{H}(A) \simeq \mathsf{H}_{2b+1}(\mathbb{Z}_p)$ and $\mathsf{G}(A) \simeq \mathsf{G}_{2b+1}(\mathbb{Z}_p)$. Now we can take as A the matrix $\Omega_b \in \mathrm{Mat}(4b, \mathbb{Z}_p)$ given in (18). Setting $G = \mathsf{H}(\Omega_b)$ or $G = \mathsf{G}(\Omega_b)$, the group G is generated by a set of $4b + 1$ elements

$$\mathfrak{S} = \{\mathsf{r}_{11}, \mathsf{t}_{11}, \dots, \mathsf{r}_{1b}, \mathsf{t}_{1b}, \mathsf{r}_{21}, \mathsf{t}_{21}, \dots, \mathsf{r}_{2b}, \mathsf{t}_{2b}, \mathsf{z}\}$$

subject to the relations (20) or (21), respectively. One can check that $\mathfrak{S}$ provides a diagonal double Kodaira structure of type (b, p) on G, and so a diagonal double Kodaira structure of the same type on the isomorphic group $\mathsf{H}_{4b+1}(\mathbb{Z}_p)$ or $\mathsf{G}_{4b+1}(\mathbb{Z}_p)$. This proves Theorem 1.

Remark 3 In particular, the pure braid quotients of smallest order detected by the methods detailed so far are the extra-special groups of order $2^7 = 128$, corresponding to the case $(b, p) = (3, 2)$ in Theorem 2.

Recently, in the paper [19], we were able to significantly lower the value of $|G|$, actually providing a sharp lower bound for the order of a pure braid quotient.

Theorem 3 [19] *Assume that G is a finite group that is a pure braid quotient. Then $|G| \geq 32$, with equality if and only if G is extra-special. In this case, the following holds.*

- *There are precisely* $2211840 = 1152 \cdot 1920$ *distinct group epimorphisms* $\varphi\colon \mathsf{P}_2(\Sigma_2) \longrightarrow G$, *and all of them make* G *a strong pure braid quotient of type* $(2,\ 2)$.
- *if* $G = G(32,\ 49) = \mathsf{H}_5(\mathbb{Z}_2)$, *these epimorphisms form* 1920 *orbits under the natural action of* $\mathrm{Aut}(G)$.
- *if* $G = G(32,\ 50) = \mathsf{G}_5(\mathbb{Z}_2)$, *these epimorphisms form* 1152 *orbits under the natural action of* $\mathrm{Aut}(G)$.

The proof of Theorem 3 is obtained again by looking at the diagonal double Kodaira structures on G, see Definition 5.

Remark 4 A key observation is that if G is a CCT-group, namely, G is not abelian and commutativity is a transitive relation on the set of the non-central elements, then G admits no diagonal double Kodaira structures and, subsequently, it cannot be a pure braid quotient.

A long but straightforward analysis shows that there are precisely eight non-CCT groups with $G \leq 32$, namely, $G = \mathsf{S}_4$ and $G = G(32,\ t)$ with $t \in \{6,\ 7,\ 8,\ 43,\ 44,\ 49,\ 50\}$.

These cases are handled separately, and a refined analysis proves that only $G(32,\ 49)$ and $G(32,\ 50)$, i.e., the two extra-special groups, admit diagonal double Kodaira structures.

Finally, the number of such structures in each case is computed by using the same techniques as in [22]; more precisely, we exploit the fact that $V = G/Z(G)$ can be endowed with a natural structure of 4-dimensional symplectic vector space over $\mathbb{Z}_2$, and that $\mathrm{Out}(G)$ embeds in $\mathrm{Sp}(4,\ \mathbb{Z}_2)$ as the orthogonal group associated with the quadratic form q on V related to the symplectic form $(\cdot\ ,\cdot)$ by $q(\overline{\mathsf{x}}\,\overline{\mathsf{y}}) = q(\overline{\mathsf{x}}) + q(\overline{\mathsf{y}}) + (\overline{\mathsf{x}},\ \overline{\mathsf{y}})$.

4 Geometrical Application: Diagonal Double Kodaira Fibrations

Recall that a *Kodaira fibration* is a smooth, connected holomorphic fibration $f_1\colon S \longrightarrow B_1$, where S is a compact complex surface and B_1 is a compact complex curve, which is not isotrivial (this means that not all fibers are biholomorphic each other). The genus $b_1 := g(B_1)$ is called the *base genus* of the fibration, whereas the genus $g := g(F)$, where F is any fiber, is called the *fiber genus*.

Definition 6 A *double Kodaira surface* is a compact complex surface S, endowed with a *double Kodaira fibration*, namely, a surjective, holomorphic map $f : S \longrightarrow B_1 \times B_2$ yielding, by composition with the natural projections, two Kodaira fibrations $f_i\colon S \longrightarrow B_i$, $i = 1,\ 2$.

With a slight abuse of notation, in the sequel we will use the symbol Σ_b to indicate both a closed Riemann surface of genus b and its underlying real surface. If a finite group G is a pure braid the quotient of type (b, n) then, by using Grauert–Remmert's extension theorem together with Serre's GAGA, the group epimorphism $\varphi\colon \mathsf{P}_2(\Sigma_b) \longrightarrow G$ yields the existence of a smooth, complex, projective surface S endowed with a Galois cover

$$\mathbf{f}\colon S \longrightarrow \Sigma_b \times \Sigma_b$$

with Galois group G and branched precisely over Δ with branching order n, see [6, Proposition 4.4]. Composing the group monomorphisms $\pi_1(\Sigma_b - \{p_i\},\ p_j) \longrightarrow \mathsf{P}_2(\Sigma_b)$ with $\varphi\colon \mathsf{P}_2(\Sigma_b) \longrightarrow G$, we get two homomorphisms

$$\varphi_1\colon \pi_1(\Sigma_b - \{p_2\},\ p_1) \longrightarrow G, \quad \varphi_2\colon \pi_1(\Sigma_b - \{p_1\},\ p_2) \longrightarrow G,$$

whose images are the normal subgroups K_1 and K_2 defined in (10).

By construction, these are the homomorphisms induced by the restrictions $\mathbf{f}_i\colon \Gamma_i \longrightarrow \Sigma_b$ of the Galois cover $\mathbf{f}\colon S \longrightarrow \Sigma_b \times \Sigma_b$ to the fibers of the two natural projections $\pi_i\colon \Sigma_b \times \Sigma_b \longrightarrow \Sigma_b$. Since Δ intersects transversally at a single point all the fibers of the natural projections, it follows that both such restrictions are branched at precisely one point, and the number of connected components of the smooth curve $\Gamma_i \subset S$ equals the index $m_i := [G : K_i]$ of K_i in G.

So, taking the Stein factorizations of the compositions $\pi_i \circ \mathbf{f}\colon S \longrightarrow \Sigma_b$ as in the diagram below

$$\begin{array}{ccc} S & \xrightarrow{\pi_i \circ \mathbf{f}} & \Sigma_b \\ & {\scriptstyle f_i} \searrow \quad \nearrow {\scriptstyle \theta_i} & \\ & \Sigma_{b_i} & \end{array} \tag{22}$$

we obtain two distinct Kodaira fibrations $f_i\colon S \longrightarrow \Sigma_{b_i}$, hence a double Kodaira fibration by considering the product morphism

$$f = f_1 \times f_2\colon S \longrightarrow \Sigma_{b_1} \times \Sigma_{b_2}.$$

Definition 7 We call $f\colon S \longrightarrow \Sigma_{b_1} \times \Sigma_{b_2}$ the *diagonal double Kodaira fibration* associated with the pure braid quotient $\varphi\colon \mathsf{P}(\Sigma_b) \longrightarrow G$. Conversely, we will say that a double Kodaira fibration $f\colon S \longrightarrow \Sigma_{b_1} \times \Sigma_{b_2}$ is *of diagonal type* (b, n) if there exists a pure braid quotient $\varphi\colon \mathsf{P}(\Sigma_b) \longrightarrow G$ of the same type such that f is associated with φ.

Since the morphism $\theta_i : \Sigma_{b_i} \longrightarrow \Sigma_b$ is étale of degree m_i, by using the Hurwitz formula we obtain

$$b_1 - 1 = m_1(b-1), \quad b_2 - 1 = m_2(b-1). \tag{23}$$

Moreover, the fiber genera g_1, g_2 of the Kodaira fibrations $f_1 : S \longrightarrow \Sigma_{b_1}$, $f_2 : S \longrightarrow \Sigma_{b_2}$ are computed by the formulae

$$2g_1 - 2 = \frac{|G|}{m_1}(2b - 2 + \mathfrak{n}), \quad 2g_2 - 2 = \frac{|G|}{m_2}(2b - 2 + \mathfrak{n}),$$

where $\mathfrak{n} := 1 - 1/n$. Finally, the surface S fits into a diagram

$$\begin{array}{ccc} S & \xrightarrow{\quad \mathbf{f} \quad} & \Sigma_b \times \Sigma_b \\ & {\scriptstyle f} \searrow \quad \nearrow {\scriptstyle \theta_1 \times \theta_2} & \\ & \Sigma_{b_1} \times \Sigma_{b_2} & \end{array}$$

so that the double Kodaira fibration $f : S \longrightarrow \Sigma_{b_1} \times \Sigma_{b_2}$ is a finite cover of degree $\frac{|G|}{m_1 m_2}$, branched precisely over the curve

$$(\theta_1 \times \theta_2)^{-1}(\Delta) = \Sigma_{b_1} \times_{\Sigma_b} \Sigma_{b_2}.$$

Such a curve is always smooth, being the preimage of a smooth divisor via an étale morphism. However, it is reducible in general, see [6, Proposition 4.11].

Remark 5 By definition, the pure braid quotient $\varphi \colon \mathsf{P}_2(\Sigma_b) \longrightarrow G$ is strong (see Definition 3) if and only if $m_1 = m_2 = 1$, that in turn implies $b_1 = b_2 = b$, i.e., $f = \mathbf{f}$. In other words, φ is strong if and only if no Stein factorization as in (22) is needed or, equivalently, if and only if the Galois cover $\mathbf{f} \colon S \longrightarrow \Sigma_b \times \Sigma_b$ induced by φ is already a double Kodaira fibration, branched on the diagonal $\Delta \subset \Sigma_b \times \Sigma_b$.

We can now compute the invariants of S as follows, see [6, Proposition 4.8].

Proposition 6 *Let $f : S \longrightarrow \Sigma_{b_1} \times \Sigma_{b_2}$ be a diagonal double Kodaira fibration, associated with a pure braid quotient $\varphi \colon \mathsf{P}_2(\Sigma_b) \longrightarrow G$ of type (b, n). Then we have*

$$c_1^2(S) = |G|\,(2b-2)(4b - 4 + 4\mathfrak{n} - \mathfrak{n}^2)$$
$$c_2(S) = |G|\,(2b-2)(2b - 2 + \mathfrak{n})$$

where $\mathfrak{n} = 1 - 1/n$. As a consequence, the slope and the signature of S can be expressed as

$$\begin{aligned} \nu(S) &= \frac{c_1^2(S)}{c_2(S)} = 2 + \frac{2\mathfrak{n} - \mathfrak{n}^2}{2b - 2 + \mathfrak{n}} \\ \sigma(S) &= \frac{1}{3}\left(c_1^2(S) - 2c_2(S)\right) = \frac{1}{3}\,|G|\,(2b-2)\left(1 - \frac{1}{n^2}\right). \end{aligned} \tag{24}$$

Remark 6 Not all double Kodaira fibrations are of diagonal type. In fact, if S is of diagonal type, then its slope satisfies $\nu(S) = 2 + s$, where $0 < s < 6 - 4\sqrt{2}$, see [18, Proposition 4.11 and Remark 4.12].

We can now specialize these results, by taking as G an extra-special p-group and using what we have proved in Sect. 3.

Fix $b = 2$ and let $p \geq 5$ be a prime number. Then every extra-special p-group G of order $p^{4b+1} = p^9$ is a non-strong pure braid quotient of type $(2,\ p)$ and such that $m_1 = m_2 = p^{2b}$, see Theorem 1. Setting $b' := p^4 + 1$, cf. Eqs. (23), by [6, Proposition 4.11] the associated diagonal double Kodaira fibration $f\colon S \longrightarrow \Sigma_{b'} \times \Sigma_{b'}$ is a cyclic cover of degree p, branched over a reduced, smooth divisor D of the form

$$D = \sum_{c \in (\mathbb{Z}_p)^{2b}} D_c$$

where the D_c are pairwise disjoint graphs of automorphisms of $\Sigma_{b'}$.

By using Proposition 6, we can now construct infinitely many double Kodaira fibrations with slope strictly higher than $2 + 1/3$.

Theorem 4 ([6, Proposition 4.12], [18, Theorem 4.8]) *Let $f\colon S_p \longrightarrow \Sigma_{b'} \times \Sigma_{b'}$ be the diagonal double Kodaira fibration associated with a non-strong pure braid quotient $\varphi\colon \mathsf{P}_2(\Sigma_2) \longrightarrow G$ of type $(2,\ p)$, where G is an extra-special p-group G of order p^9. Then the maximum slope $\nu(S_p)$ is attained for precisely two values of p, namely,*

$$\nu(S_5) = \nu(S_7) = 2 + \frac{12}{35}.$$

Furthermore, $\nu(S_p) > 2 + 1/3$ for all $p \geq 5$. More precisely, if $p \geq 7$ the function $\nu(S_p)$ is strictly decreasing and

$$\lim_{p \to +\infty} \nu(S_p) = 2 + \frac{1}{3}.$$

Remark 7 The original examples by Atiyah, Hirzebruch and Kodaira have slope lying in the interval $(2, 2 + 1/3]$, see [3, p. 221]. Our construction provides an infinite family of Kodaira fibred surfaces such that $2 + 1/3 < \nu(S) \leq 2 + 12/35$, maintaining at the same time a complete control on both the base genus and the signature. By contrast, the ingenious "tautological construction" used in [7] yields a higher slope than ours, namely, $2 + 2/3$, but it involves an étale pullback "of sufficiently large degree", that completely loses control on the other quantities. Note that [16] gives (at least in principle) an effective version of the pullback construction.

If p is a prime number dividing $b+1$, by Theorem 2 every extra-special p-group G of order p^{2b+1} is a strong pure braid quotient of type $(b,\ p)$, and this gives in turn a diagonal double Kodaira fibration $f\colon S \longrightarrow \Sigma_b \times \Sigma_b$, see Remark 5. If $\omega\colon \mathbb{N} \longrightarrow \mathbb{N}$ stands for the arithmetic function counting the number of distinct prime factors of a positive integer, see [11, p. 335], we obtain

Theorem 5 ([6, Corollary 4.18], [18, Theorem 4.5]) *Let Σ_b be any closed Riemann surface of genus b. Then there exists a double Kodaira fibration $f\colon S \longrightarrow \Sigma_b \times \Sigma_b$. Moreover, denoting by $\kappa(b)$ the number of such fibrations, we have*

$$\kappa(b) \geq \omega(b+1).$$

In particular,

$$\limsup_{b \to +\infty} \kappa(b) = +\infty.$$

As far as we know, this is the first construction showing that *all* curves of genus $b \geq 2$ (and not only some special curves with extra automorphisms) appear in the base of at least one double Kodaira fibration $f\colon S \longrightarrow \Sigma_b \times \Sigma_b$. In addition, two Kodaira fibred surfaces corresponding to two distinct prime divisors of $b+1$ are non-homeomorphic, because the corresponding signatures are different: just use (24) with $n = p$ and $|G| = p^{2b+1}$ and note that, for fixed b, the function expressing $\sigma(S)$ is strictly increasing in p. This shows that the number of topological types of S, for a fixed base Σ_b, can be arbitrarily large.

Let us now consider Theorem 3, whose geometrical translation is

Theorem 6 [19] *Let G be a finite group and $\mathbf{f}\colon S \longrightarrow \Sigma_b \times \Sigma_b$ be a Galois cover, with Galois group G, branched over the diagonal Δ with branching order n. Then $|G| \geq 32$, and equality holds if and only if G is extra-special. In this case, the following holds.*

(1) *There exist* $2211840 = 1152 \cdot 1920$ *distinct G-covers* $\mathbf{f}\colon S \longrightarrow \Sigma_2 \times \Sigma_2$, *and all of them are diagonal double Kodaira fibrations such that*

$$b_1 = b_2 = 2, \quad g_1 = g_2 = 41, \quad \sigma(S) = 16. \tag{25}$$

(2) *If* $G = G(32,\ 49) = \mathsf{H}_5(\mathbb{Z}_2)$, *these G-covers form* 1920 *equivalence classes up to cover isomorphisms.*

(3) *If* $G = G(32,\ 50) = \mathsf{H}_5(\mathbb{Z}_2)$, *these G-covers form* 1152 *equivalence classes up to cover isomorphisms.*

As a consequence, we obtain a sharp lower bound for the signature of a diagonal double Kodaira fibration. In fact, the second equality in (24) together with Theorem 6 imply that, for every such fibration, we have

$$\sigma(S) = \frac{1}{3}\,|G|\,(2b-2)\left(1 - \frac{1}{n^2}\right) \geq \frac{1}{3} \cdot 32 \cdot (2 \cdot 2 - 2)\left(1 - \frac{1}{2^2}\right) = 16,$$

and this in turn establishes the following result.

Theorem 7 [19] *Let S be a double Kodaira surface, associated with a pure braid quotient $\varphi \colon \mathsf{P}_2(\Sigma_b) \longrightarrow G$ of type (b, n). Then $\sigma(S) \geq 16$, and equality holds precisely when $(b, n) = (2, 2)$ and G is an extra-special group of order* 32.

Remark 8 If S is a double Kodaira fibration, corresponding to a pure braid quotient $\varphi \colon \mathsf{P}_2(\Sigma_b) \longrightarrow G$ of type (b, n), then, using the terminology in [7], it is *very simple*. Let us denote by $\mathfrak{M}_S$ the connected component of the Gieseker moduli space of surfaces of general type containing the class of S, and by $\mathcal{M}_b$ the moduli space of smooth curves of genus b. Thus, by applying [20, Theorem 1.7] and using the fact that $\Delta \subset \Sigma_b \times \Sigma_b$ is the graph of the identity $\mathrm{id} \colon \Sigma_b \longrightarrow \Sigma_b$, we infer that every surface in $\mathfrak{M}_S$ is still a very simple double Kodaira fibration and that there is a natural map of schemes

$$\mathcal{M}_b \longrightarrow \mathfrak{M}_S,$$

which is an isomorphism on geometric points. Roughly speaking, since the branch locus $\Delta \subset \Sigma_b \times \Sigma_b$ is rigid, all the deformations of S are realized by deformations of $\Sigma_b \times \Sigma_b$ preserving the diagonal, hence by deformations of Σ_b, cf. [6, Proposition 4.22]. In particular, this shows that $\mathfrak{M}_S$ is a connected and irreducible component of the Gieseker moduli space.

Every Kodaira fibred surface S has the structure of a real surface bundle over a real surface, and so $\sigma(S)$ is divisible by 4, see [17]. If, in addition, S has a spin structure, i.e., its canonical class is 2-divisible in $\mathrm{Pic}(S)$, then $\sigma(S)$ is a positive multiple of 16 by Rokhlin's theorem, and examples with $\sigma(S) = 16$ are constructed in [16]. It is not known if there exists a Kodaira fibred surface with $\sigma(S) \leq 12$.

Constructing (double) Kodaira fibrations with small signature is a rather difficult problem. As far as we know, before the present work the only examples with signature 16 were the ones listed in [16, Table 3, Cases 6.2, 6.6, 6.7 (Type 1), 6.9]. The examples in Theorem 7 are new, since both the base genera and the fiber genera are different from the ones in the aforementioned cases. Our results also show that *every* curve of genus 2 is the base of a double Kodaira fibration with signature 16. Thus, we obtain two families of dimension 3 of such fibrations that, to our knowledge, provides the first examples of positive-dimensional families of double Kodaira fibrations with small signature.

Theorem 7 also provide new "double solutions" to a problem, posed by G. Mess and included in Kirby's problem list in low-dimensional topology, see [13, Problem 2.18 A], asking what is the smallest number b for which there exists a real surface bundle over a real surface with base genus b and non-zero signature. We actually have $b = 2$, also for double Kodaira fibrations.

Theorem 8 [19] *Let S be a double Kodaira surface, associated with a pure braid quotient $\varphi \colon \mathsf{P}_2(\Sigma_b) \longrightarrow G$ of type $(2, 2)$, where G is an extra-special group of order* 32. *Then the real manifold X underlying S is a closed, orientable* 4*-manifold of signature* 16 *that can be realized as a real surface bundle over a real surface of genus* 2, *with fiber genus* 41, *in two different ways.*

It is an interesting question whether 16 and 41 are the minimum possible values for the signature and the fiber genus of a (non-necessary diagonal) double Kodaira fibration $f: S \longrightarrow \Sigma_2 \times \Sigma_2$; however, this topic exceeds the scope of this paper.

5 Beyond $|G| = 32$

This last section contains the new result of this article. As we already observed, so far we have detailed a rather clear picture regarding pure braid quotient groups G and the relative diagonal double Kodaira fibrations for $|G| \leq 32$. Indeed, there is no pure braid quotient of order strictly less than 32 and for $|G| = 32$ we have only the two extra-special groups, see Theorem 3. Furthermore, Theorem 2 provides examples of pure braid quotients starting with order equal to 128, see Remark 3, and they are extra-special groups, too. It seems then natural to investigate this matter further for $|G| > 32$, for instance, in order to look for non-extra-special examples. In this direction, we have obtained the following result that highlights the existence of a gap between orders 32 and 64.

Theorem 9 *If G is a finite group with $32 < |G| < 64$, then G is not a pure braid quotient.*

Here we just give a sketch of the proof, while the full details will appear elsewhere. We know that the group G cannot be abelian, see Remark 2, and we also mentioned that CCT-groups cannot be pure braid quotients, see Remark 4. On the other hand, by [19] we know that, if G is a pure braid quotient and admits no proper quotients that are pure braid quotients, then G must be monolithic, i.e., the intersection $\mathrm{soc}(G)$ of its non-trivial normal subgroups is non-trivial. In fact, consider the epimorphism $\varphi\colon \mathsf{P}_2(\Sigma_b) \longrightarrow G$ and assume that there is a non-trivial, normal subgroup N of G such that $\varphi(A_{12}) \notin N$. Then, composing the projection $G \longrightarrow G/N$ with φ, we obtain a pure braid quotient $\bar{\varphi}\colon \mathsf{P}_2(\Sigma_b) \longrightarrow G/N$, which leads to a contradiction. It follows that $\varphi(A_{12}) \in \mathrm{soc}(G)$, in particular G is monolithic.

By Theorem 3 this implies that, if a pure braid quotient G satisfies our assumptions on the order, then G must be monolithic. A straightforward computer calculation with GAP4 now shows that there are precisely two non-abelian groups G with $32 < |G| < 64$ that are both non-CCT and monolithic, namely, $G(54,\ 5)$ and $G(54,\ 6)$; by the remarks above, they are the only possible candidates to be pure braid quotients in that range for $|G|$. Finally, a brute force check (again by using GAP4) shows that these groups admit no diagonal double Kodaira structure, proving our assertion.

Acknowledgements F. Polizzi was partially supported by GNSAGA-INdAM. Both authors thank A. Causin for drawing the figures.

References

1. M.F. Atiyah, The signature of fibre bundles, in *Global Analysis (Papers in honor of K. Kodaira)*, University of Tokyo Press (1969), pp. 73–84
2. V.G. Bardakov, P. Bellingeri, On residual properties of pure braid groups of closed surfaces. Comm. Alg. **37**, 1481–1490 (2009)
3. W. Barth, K. Hulek, C.A.M. Peters, A. Van de Ven, *Compact Complex Surfaces*. Grundlehren der Mathematischen Wissenschaften, vol. 4, Second enlarged edition (Springer, Berlin, 2003)
4. P. Bellingeri, On presentations of surface braid groups. J. Algebra **274**, 543–563 (2004)
5. J. Birman, On braid groups. Comm. Pure Appl. Math. **22**, 41–72 (1969)
6. A. Causin, F. Polizzi, Surface braid groups, finite Heisenberg covers and double Kodaira fibrations, Ann. Sc. Norm. Pisa Cl. Sci. **XXII**(5), 1309–1352 (2021)
7. F. Catanese, S. Rollenske, Double Kodaira fibrations. J. Reine Angew. Math. **628**, 205–233 (2009)
8. The GAP Group, *GAP – Groups, Algorithms, and Programming, Version 4.11.0* (2020)
9. D.L. Gonçalves, J. Guaschi, On the structure of surface pure braid groups. J. Pure Appl. Algebra **186**, 187–218 (2004)
10. D. Gorenstein, *Finite Groups*, Reprinted edition by the AMS Chelsea Publishing (2007)
11. G.H. Hardy, E.M. Wright, *An Introduction to the Theory of Numbers*, 6th edn. (Oxford University Press, Oxford, 2008)
12. A. Kas, On deformations of a certain type of irregular algebraic surface. Amer. J. Math. **90**, 789–804 (1968)
13. R. Kirby (ed.), *Problems in Low-dimensional Topology*. AMS/IP Studies in Advanced Mathematics, 2.2, Geometric topology (Athens, GA, 1993) (American Mathematical Society, Providence, RI, 1997), pp. 35–473
14. K. Kodaira, A certain type of irregular, algebraic surfaces. J. Anal. Math. **19**, 207–215 (1967)
15. M. Isaacs, *Finite Groups Theory*, Graduate Studies in Mathematics, vol. 92 (American Mathematical Society, 2008)
16. J.A. Lee, M. Lönne, S. Rollenske, Double Kodaira fibrations with small signature. Int. J. Math. **31**(7), 2050052, 42 pp (2020)
17. W. Meyer, Die Signatur von Flächenbündeln. Math. Ann. **201**, 239–264 (1973)
18. F. Polizzi, Diagonal double Kodaira structures on finite groups, Current Trends in Analysis, its Applications and Computation, *Trends in Mathematics*, Birkäuser, Cham (2022), 111–128
19. F. Polizzi, P. Sabatino, Extra-special quotients of surface braid groups and double Kodaira fibrations with small signature. Geometriae Dedicata 216, article 65 (2022)
20. S. Rollenske, Compact moduli for certain Kodaira fibrations. Ann. Scuola Norm. Sup. Pisa Cl. Sci. **IX**(5), 851–874 (2010)
21. G.P. Scott, Braid groups and the groups of homeomorphisms of a surface. Proc. Camb. Phil. Soc. **68**, 605–617 (1970)
22. D.L. Winter, The automorphism group of an extra-special p-group. Rocky Mountain J. Math. **2**, 159–168 (1972)

Affine Cones over Fano–Mukai Fourfolds of Genus 10 are Flexible

Yuri Prokhorov and Mikhail Zaidenberg

To Ciro Ciliberto on the occasion of his 70th birthday.

Abstract We show that the affine cones over any Fano–Mukai fourfold of genus 10 are flexible in the sense of [1]. In particular, the automorphism group of such a cone acts highly transitively outside the vertex. Furthermore, any Fano–Mukai fourfold of genus 10, with one exception, admits a covering by open charts isomorphic to $\mathbb{A}^4$.

Keywords Flexible affine variety · Affine cone · Fano fourfold

1991 Mathematics Subject Classification Primary 14J35, 14J45; Secondary 14R10, 14R20

The research of the first author was partially supported by the HSE University Basic Research Program.

Y. Prokhorov (✉)
Steklov Mathematical Institute, Moscow, Russian Federation
e-mail: prokhoro@mi-ras.ru

National Research University Higher School of Economics, Moscow, Russian Federation

Department of Algebra, Moscow State Lomonosov University, Moscow, Russian Federation

M. Zaidenberg
Université Grenoble Alpes, CNRS, Institut Fourier, 38000 Grenoble, France
e-mail: Mikhail.Zaidenberg@univ-grenoble-alpes.fr

T. Dedieu et al. (eds.), *The Art of Doing Algebraic Geometry*, Trends in Mathematics,
https://doi.org/10.1007/978-3-031-11938-5_16

1 Introduction

1.1 Flexible Varieties

Our base field in this paper is the complex number field $\mathbb{C}$. Let X be an affine variety over $\mathbb{C}$. Consider the subgroup $\mathrm{SAut}(X)$ of the automorphism group $\mathrm{Aut}(X)$ generated by all the one-parameter unipotent subgroups of $\mathrm{Aut}(X)$. The variety X is called *flexible* if $\mathrm{SAut}(X)$ acts highly transitively on the smooth locus $\mathrm{reg}(X)$, that is, m-transitively for any natural number m [1]. There are many examples of flexible affine varieties, see, e.g., [1, 2, 4, 15]; studies on this subject are in an active phase. For a projective variety, the flexibility of the affine cones might depend on the choice of an ample polarization. If the affine cone over a smooth projective variety V with Picard number one is flexible, then V is a Fano variety. For a pluri-anticanonical polarization of a Fano variety V, the flexibility of the affine cone X over V is a nontrivial new invariant of V. Among the del Pezzo surfaces with their pluri-anticanonical polarizations, only the surfaces of degree ≥ 4 have flexible affine cones, see [17, 18]. Moreover, the group $\mathrm{SAut}(X)$ of the affine cone X over $(V, -mK_V)$, $m > 0$, is trivial for any del Pezzo surface V of degree at most 3, see [3, 12]. The affine cones over flag varieties of dimension ≥ 2 are flexible [2]. The secant varieties of the Segre–Veronese varieties provide another class of examples [15].

There are many examples of Fano threefolds with flexible affine cones, see, e.g., the survey article [4] and the references therein. However, we knew just two examples of such Fano fourfolds, and one of these is a special Fano–Mukai fourfold of genus 10, see [24, Theorem 14.3]. In the present paper, we extend the latter result to all the Fano–Mukai fourfolds of genus 10.

1.2 Main Results

Let V_{18} be a Fano–Mukai fourfold of genus 10 and degree 18. It admits a half-anticanonical embedding in $\mathbb{P}^{12}$. Recall ([11], [24, Remark 13.4]) that the moduli space of these fourfolds is one-dimensional. It contains two special members, namely, V_{18}^{s} with $\mathrm{Aut}^0(V) = \mathrm{GL}_2(\mathbb{C})$ and V_{18}^{a} with $\mathrm{Aut}^0(V_{18}^{\mathrm{a}}) = \mathbb{G}_{\mathrm{a}} \times \mathbb{G}_{\mathrm{m}}$ [24, Theorem 1.3.a,b]. The general member V_{18}^{g} is a member isomorphic to neither V_{18}^{a} nor V_{18}^{s}. One has $\mathrm{Aut}^0(V_{18}^{\mathrm{g}}) = \mathbb{G}_{\mathrm{m}}^2$ [24, Theorem 1.3.c]. There exists also a finer classification according to the discrete part of the automorphism group [25].

In the special case where $\mathrm{Aut}^0(V) = \mathrm{GL}_2(\mathbb{C})$ the flexibility of the affine cone with respect to the half-anticanonical polarization was established in [24, Theorem 14.3]. Now this is extended to any ample polarization and includes all the three cases above. The following theorem is our main result.

Theorem 1 *Let $V = V_{18}$ be a Fano–Mukai fourfold of genus* 10. *Then the affine cone over V is flexible for any ample polarization of V.*

The proof exploits the criteria of flexibility of affine cones borrowed from [18, Theorem 5] and [15, Theorem 1.4]; cf. also [19, Theorem 2.4]. These criteria are based on the existence of some special open coverings of the underlined projective variety, for instance, a covering by flexible affine charts, or by suitable toric affine varieties, or by the affine spaces, see Theorem 3.2. To apply these criteria, one needs to construct such a covering. To this end, we use Theorem 2 below.

Any Fano–Mukai fourfold $V = V_{18}$ can be represented, at least in two ways, as an $\mathrm{Aut}^0(V)$-equivariant compactification of the affine space $\mathbb{A}^4$ [24, Theorem 1.1]. More precisely, there exists at least two different $\mathrm{Aut}^0(V)$-invariant hyperplane sections A_1, A_2 of V such that $U_i := V \setminus A_i$ is isomorphic to $\mathbb{A}^4$, $i = 1, 2$. Statement (a) of the following theorem is proven in [24, Theorem 13.5(f)].

Theorem 2 *For the Fano–Mukai fourfolds $V = V_{18}$ of genus* 10 *the following hold.*

(a) If $\mathrm{Aut}^0(V) = \mathrm{GL}_2(\mathbb{C})$ then V admits a covering by a one-parameter family of Zariski open subsets U_t isomorphic to $\mathbb{A}^4$, where each U_t is the complement of a hyperplane section of V. Exactly two of these sets are $\mathrm{Aut}^0(V)$-invariant.

(b) If $\mathrm{Aut}^0(V) = \mathbb{G}_{\mathrm{m}}^2$ then V is covered by six $\mathrm{Aut}^0(V)$-invariant subsets $U_i \cong \mathbb{A}^4$ as in (a), $i = 1, \ldots, 6$.

(c) If $\mathrm{Aut}^0(V) = \mathbb{G}_{\mathrm{a}} \times \mathbb{G}_{\mathrm{m}}$ then there exist in V four $\mathrm{Aut}^0(V)$-invariant affine charts $U_i \cong \mathbb{A}^4$, $i = 1, \ldots, 4$ such that $V \setminus \bigcup_{i=1}^4 U_i$ is a projective line covered by a one-parameter family of polar $\mathbb{A}^2$-cylinders $U_t \cong \mathbb{A}^2 \times Z_t$, where Z_t is a smooth affine surface, $t \in \mathbb{P}^1$.

Some other families of Fano varieties demonstrate similar properties: the Fano threefolds of degree 22 and Picard number 1 [13, 21] and the Fano threefolds of degree 28 and Picard number 2 [26, Sect. 9]. Any member of the first family is a compactification of the affine 3-space. It is plausible that the same is true for the second family. It would be interesting to investigate the flexibility of the affine cones over these varieties.

The paper is organized as follows. In Sect. 2 we gather necessary preliminaries, in particular, some results from [11, 22, 24]; cf. also [23]. In Sect. 3 we provide a criterion of flexibility related to the existence of a special family of $\mathbb{A}^2$-cylinders on V. In Sect. 4 we prove Theorems 1 and 2 in the case where $\mathrm{Aut}^0(V) = \mathbb{G}_{\mathrm{m}}^2$. The proofs are based on studies of the action of the automorphism group and the existence of a covering of V by affine 4-spaces. In Sects. 5–7 we proceed with the proof of Theorem 2 in the case $\mathrm{Aut}^0(V) = \mathbb{G}_{\mathrm{a}} \times \mathbb{G}_{\mathrm{m}}$. In Sect. 5 we study the affine 4-spaces contained in V; they form an "almost covering" of V. In Sect. 6 a family of $\mathbb{A}^2$-cylinders is constructed for the smooth quadric fourfold Q^4 and for the del Pezzo quintic fourfold W_5. In Sect. 7 such a covering is transferred to any Fano–Mukai fourfold V of genus 10 via Sarkisov links. This enables us to complete the proofs of Theorems 1 and 2 in the remaining case where $\mathrm{Aut}^0(V) = \mathbb{G}_{\mathrm{a}} \times \mathbb{G}_{\mathrm{m}}$.

2 Cubic Scrolls in the Fano–Mukai Fourfolds V_{18}

In this section, we recall and extend some facts from [24] used in the sequel. Throughout the paper we let V be a Fano–Mukai fourfold $V = V_{18}$ of genus 10 half-anticanonically embedded in $\mathbb{P}^{12}$. Following [11] we call a *cubic scroll* both a smooth cubic surface scroll and a cone over a rational twisted cubic curve. The latter cones in $\mathbb{P}^4$ will be called *cubic cones* for short. A smooth cubic scroll $S \subset \mathbb{P}^4$ is isomorphic to the Hirzebruch surface $\pi : \mathbb{F}_1 = \mathbb{P}(\mathscr{O} \oplus \mathscr{O}(1)) \to \mathbb{P}^1$ embedded in such a way that the exceptional section and the fibers of π are lines in $\mathbb{P}^4$. A cubic cone is isomorphic to the surface obtained via the contraction of the negative section of the Hirzebruch surface $\mathbb{F}_3 = \mathbb{P}(\mathscr{O} \oplus \mathscr{O}(3))$.

Theorem 2.1 ([11, Propositions 1 and 2 and the proof of Proposition 4])

(a) Let $\Sigma(V)$ be the Hilbert scheme of lines in V. Then $\Sigma(V)$ is isomorphic to the variety $\mathrm{Fl}(\mathbb{P}^2)$ *of full flags on $\mathbb{P}^2$.*
(b) Let $\mathscr{S}(V)$ be the Hilbert scheme of cubic scrolls in V. Then $\mathscr{S}(V)$ is isomorphic to a disjoint union of two projective planes.

This theorem and the next lemma show that the cubic scrolls in V play the same role as do the planes in a smooth quadric fourfold, cf. [8, Lecture 22].

Lemma 2.2 ([24, Proposition 9.6]) *Let $\mathscr{S}_1$ and $\mathscr{S}_2$ be the connected components of $\mathscr{S}(V)$. Then for any $S_i \in \mathscr{S}_i$ and $S_j \in \mathscr{S}_j$, $i, j \in \{1, 2\}$ we have the following relation: $S_i \cdot S_j = \delta_{i,j}$ in $H^*(V, \mathbb{Z})$.*

Lemma 2.3 ([24, Lemma 9.2, Corollary 4.5.2])

(i) For any cubic scroll S on V there exists a unique hyperplane section A_S of V such that $\mathrm{Sing}(A_S) = S$. *This hyperplane section coincides with the union of lines on V meeting S, and any line contained in A_S meets S.*
(ii) For any point $P \in A_S \setminus S$ there is exactly one line $l \subset A_S$ passing through P and meeting S. Such a line meets S in a single point.
(iii) If $S \subset V$ is a cubic cone, then $V \setminus A_S \cong \mathbb{A}^4$.

Lemma 2.4 ([24, Proposition 8.2]) *There exists a divisor $\mathcal{B} \subset V$ such that the following hold:*

(i) for any point $v \in V \setminus \mathcal{B}$ there are exactly three lines passing through v;
(ii) for $v \in \mathcal{B}$ the number of lines passing through v either is infinite, or equals 2 *or* 1;
(iii) if the number of lines passing through v is infinite, then the union of these lines is a cubic cone S with vertex v, and $S \subset \mathcal{B}$.

Lemma 2.5 ([24, Lemmas 9.3, 9.4, 9.9(a), Corollaries 9.7.3, 9.7.4, 9.10.1])

(i) Any line on V can be contained in at most a finite number of cubic scrolls, and in at most two cubic cones;

(ii) *two cubic cones from different components of* $\mathscr{S}(V)$ *either are disjoint, or share a unique common ruling. Two cubic cones from the same component* $\mathscr{S}_i$ *meet transversally in a single point different from their vertices;*

(iii) *if two cubic cones* S *and* S' *are disjoint, then* $v_S \notin A_{S'}$*, where* v_S *stands for the vertex of* S*;*

(iv) *if two cubic cones* S *and* S' *share a common ruling* l*, then* $S \cup S'$ *coincides with the union of lines on* V *meeting* l*.*

Lemma 2.6 ([24, Lemma 9.7.2, Corollary 9.7.3(i)]) *In the notation of Theorem 2.1(a) let*

$$\mathrm{pr}_1 : \Sigma(V) = \mathrm{Fl}(\mathbb{P}^2) \longrightarrow \mathbb{P}^2 \quad \textit{and} \quad \mathrm{pr}_2 : \Sigma(V) = \mathrm{Fl}(\mathbb{P}^2) \longrightarrow (\mathbb{P}^2)^\vee \tag{1}$$

be the natural projections. Any line l *on* V *is the unique common ruling of exactly two cubic scrolls, say,* $S_1(l) \in \mathscr{S}_1$ *and* $S_2(l) \in \mathscr{S}_2$*. Hence, there are the well-defined morphisms*

$$\mathrm{pr}_i : \Sigma(V) \longrightarrow \mathscr{S}_i = \mathbb{P}^2, \quad l \mapsto S_i(l), \quad i \in \{1, 2\}$$

which coincide with the ones in (1). The fiber of pr_i *over* $S \in \mathscr{S}_i$ *is the line on* $\Sigma(V)$ *which parameterizes the rulings of* S*.*

Recall that the Fano–Mukai fourfolds $V = V_{18}$ are classified in three types according to the group $\mathrm{Aut}^0(V)$, which can be isomorphic to one of the following groups

$$\mathbb{G}_{\mathrm{m}}^2, \quad \mathbb{G}_{\mathrm{a}} \times \mathbb{G}_{\mathrm{m}}, \quad \mathrm{GL}_2(\mathbb{C}).$$

This classification reflects the geometry of V, namely, the number of cubic cones on V.

Lemma 2.7 *Let* $\mathscr{S}_i \cong \mathbb{P}^2$ *be a connected component of* $\mathscr{S}(V)$*. Then the following hold.*

(i) *If* $\mathrm{Aut}^0(V) = \mathbb{G}_{\mathrm{m}}^2$ *then* $\mathscr{S}_i$ *contains exactly* 3 *cubic cones for* $i = 1, 2$*. The six cubic cones in* V *form a cycle so that the neighbors have a common ruling and belong to different components of* $\mathscr{S}(V)$*, and the pairs of opposite vertices of the cycle correspond to the pairs of disjoint cubic cones.*

(ii) *If* $\mathrm{Aut}^0(V) = \mathbb{G}_{\mathrm{a}} \times \mathbb{G}_{\mathrm{m}}$ *then* $\mathscr{S}_i$ *contains exactly* 2 *cubic cones for* $i = 1, 2$*. The four cubic cones in* V *form a chain, that is, the neighbors have a common ruling and belong to different components of* $\mathscr{S}(V)$*, and the pair of extremal vertices corresponds to the unique pair of disjoint cones.*

(iii) *If* $\mathrm{Aut}^0(V) = \mathrm{GL}_2(\mathbb{C})$ *then the subfamily of cubic cones in* $\mathscr{S}_i \cong \mathbb{P}^2$ *consists of a projective line and an isolated point. The cubic cones* $S_i \in \mathscr{S}_i$*,* $i = 1, 2$*, represented by these isolated points are disjoint, and this is the only pair of* $\mathrm{Aut}^0(V)$*-invariant cubic cones on* V*.*

Proof By Lemma 2.6 the variety $\Lambda(S) \subset \Sigma(V)$ of rulings of a cubic scroll $S \subset V$ is the fiber of one of the projections pr_i, so it is a line under the Segre embedding

$\Sigma(V) \subset \mathbb{P}^2 \times (\mathbb{P}^2)^\vee \hookrightarrow \mathbb{P}^8$, and any line on $\Sigma(V)$ appears in this way. A line l on V is called a *splitting line* if the union of lines on V meeting l splits into a union of two cubic scrolls. Assuming $\mathrm{Aut}^0(V) \neq \mathrm{GL}_2(\mathbb{C})$ the subvariety $\Sigma_s(V) \subset \Sigma(V)$ of splitting lines is a del Pezzo sextic, which admits two birational contractions to $\mathbb{P}^2$ [24, Proposition 10.2]. The scroll S on V is a cubic cone exactly when the line $\Lambda(S)$ lies on $\Sigma_s(V)$ [24, Proposition 9.10]. The surface $\Sigma_s(V)$ is smooth in case (i), and has a unique node in case (ii) [24, Proposition 10.2]. It is well known that a smooth sextic del Pezzo surface contains exactly 6 lines, and these are arranged in a cycle. If $\Sigma_s(V)$ has a singularity of type A_1, then it contains exactly 4 lines, and these are arranged in a chain [5, Proposition 8.3], [6, Sects. 8.1.1, 8.4.2].

In both cases, a pair of intersecting lines on $\Sigma_s(V)$ corresponds to a pair of cubic cones on V sharing a common ruling. By Lemma 2.5(iv), such cones belong to distinct components of $\mathscr{S}(V)$. Thus, two neighbors of the same cubic cone belong to the same component $\mathscr{S}_i$ and meet transversally in a unique point, and two cubic cones separated by two others belong to distinct components of $\mathscr{S}(V)$ and are disjoint, see Lemma 2.5(ii). This gives (i) and (ii). See [24, Corollary 10.3.2] for (iii). □

Corollary 2.8 ([24, Lemmas 12.2 and 12.8.1]) *If* $\mathrm{Aut}^0(V) = \mathbb{G}_m^2$ *or* $\mathbb{G}_a \times \mathbb{G}_m$ *then any cubic cone in* V *is* $\mathrm{Aut}^0(V)$*-invariant. If* $\mathrm{Aut}^0(V) = \mathrm{GL}_2(\mathbb{C})$ *then* $\mathscr{S}_i$ *contains exactly one* $\mathrm{Aut}^0(V)$*-invariant cubic cone for* $i = 1, 2$.

Remark 2.9 By Lemma 2.10 the induced action of $\mathrm{Aut}^0(V)$ on $\mathscr{S}_i(V) \cong \mathbb{P}^2$, $i = 1, 2$ is effective. Since the 2-torus action on $\mathbb{P}^2$ can be diagonalized, in the case $\mathrm{Aut}^0(V) = \mathbb{G}_m^2$ this action gives the standard toric structure on $\mathbb{P}^2$. The six $\mathrm{Aut}^0(V)$-invariant cubic cones in V correspond to the $3 + 3$ toric fixed points in $\mathscr{S}(V)$.

Lemma 2.10 *The induced actions of* $\mathrm{Aut}(V)$ *on* $\Sigma(V)$, $\Sigma_s(V)$ *and* $\mathscr{S}(V)$ *are effective.*

Proof For $\Sigma(V)$ and $\Sigma_s(V)$ the statement is proven in [24, Lemma 11.2]. Suppose $\alpha \in \mathrm{Aut}(V)$ acts identically on $\mathscr{S}(V)$, that is, α leaves invariant any cubic scroll on V. By Lemma 2.6 any line l on V is the unique common ruling of a pair of cubic scrolls (S_1, S_2). Since both S_1 and S_2 are invariant under α, also l is. Hence α acts identically on $\Sigma(V)$, and then $\alpha = \mathrm{id}_V$ by the preceding. □

Lemma 2.11 *Let* S_1, $S_2 \subset V$ *be disjoint cubic cones and let* $\Gamma_1 = S_1 \cap A_{S_2}$, $\Gamma_2 = S_2 \cap A_{S_1}$ *where* A_S *has the same meaning as in Lemma 2.3(i). Then the following holds.*

(i) Γ_1 *and* Γ_2 *are rational twisted cubic curves;*
(ii) there exists a one-parameter family of lines l_t *on* V *joining* Γ_1 *and* Γ_2*;*
(iii) $D := \bigcup_{t \in \mathbb{P}^1} l_t$ *is a rational normal sextic scroll contained in* $A_{S_1} \cap A_{S_2}$*;*
(iv) If S_1 *and* S_2 *are* $\mathrm{Aut}^0(V)$*-invariant then the twisted cubic curves* Γ_i, $i = 1, 2$ *and the sextic scroll* D *are as well.*

Proof Since $S_1 \cap S_2 = \varnothing$ one has $v_i \notin A_{S_j}$ for $i, j = 1, 2, i \neq j$, see Lemma 2.5(iii). This yields (i). By Lemma 2.3(i)–(ii) for any $\gamma \in \Gamma_1$ there exists a unique line

$l_\gamma \subset A_{S_1} \cap A_{S_2}$ joining γ and S_2. This line l_γ meets $\Gamma_2 = S_2 \cap A_{S_1}$. This shows (ii) and the inclusion $D \subset A_{S_1} \cap A_{S_2}$, where D is as in (iii). For the first assertion in (iii) see, e.g., [8, Example 8.17]. The assertion of (iv) is immediate from the fact that the cubic cones S_i and the associated hyperplane sections A_{S_i}, $i = 1, 2$ are $\mathrm{Aut}^0(V)$-invariant. □

Lemma 2.12 *Let $\mathscr{S}_1 \subset \mathscr{S}(V)$ be a connected component. Then the set $\bigcap_{S \in \mathscr{S}_1} A_S$ coincides with the union of vertices of cubic cones in $\mathscr{S}_1$. In particular, $\bigcap_{S \in \mathscr{S}(V)} A_S = \varnothing$.*

Proof Let $v \in V$ be the vertex of a cubic cone $S_v \in \mathscr{S}_1$. By Lemma 2.2 one has $S \cap S_v \neq \varnothing$ for any $S \in \mathscr{S}_1$, and so, $v \in \bigcap_{S \in \mathscr{S}_1} A_S$ due to Lemma 2.3.

Conversely, let $v \in \bigcap_{S \in \mathscr{S}_1} A_S$. Assume to the contrary that the lines on V passing through v form a finite set, say, $\{l_1, \ldots, l_k\}$. By Lemmas 2.4(iii) and 2.5(i), the number of lines on V passing through a general point of l_i is finite. Hence, the family $\Sigma(V; l_1, \ldots, l_k)$ of lines in V meeting $\bigcup_{i=1}^k l_i$ is one-dimensional, and again by Lemma 2.5(i), any line in this family is contained in a finite number of cubic scrolls. It follows that the subfamily $\mathscr{S}_1(l_1, \ldots, l_k)$ of cubic scrolls from $\mathscr{S}_1$ meeting $\bigcup_{i=1}^k l_i$ is one-dimensional too. Since $\dim \mathscr{S}_1 = 2$, see Theorem 2.1(b), one has $\mathscr{S}_1(l_1, \ldots, l_k) \neq \mathscr{S}_1$, and the general cubic scroll $S \in \mathscr{S}_1$ does not meet $\bigcup_{i=1}^k l_i$. This implies $v \notin A_S$, a contradiction.

Thus, any point $v \in \bigcap_{S \in \mathscr{S}_1} A_S$ is the vertex of a cubic cone, say, S_v. Then one has $S_v \cap S \neq \varnothing$ for any $S \in \mathscr{S}_1$. Assuming $S_v \notin \mathscr{S}_1$ it follows from Lemma 2.2 that for any $S \in \mathscr{S}_1$ the intersection $S_v \cap S$ contains a curve. Then any line in S_v meets S, and so, $S_v \subset \bigcap_{S \in \mathscr{S}_1} A_S$. By the preceding, any point of S_v is a vertex of a cubic cone in V. This contradicts Lemma 2.5(i) and proves the first assertion. The second assertion follows from the first since two distinct cubic cones cannot share the same vertex, see Lemma 2.4(iii). □

For instance, in the case $\mathrm{Aut}^0(V) = \mathrm{GL}_2(\mathbb{C})$ the intersection $\bigcap_{S \in \mathscr{S}_i} A_S$ is the twisted cubic Γ_i, $i = 1, 2$ as in Lemma 2.11, see [24, Theorem 13.5(b)].

3 Criteria of Flexibility of Affine Cones

To formulate the flexibility criteria we need to recall the following notions.

Definition 3.1 ([18, Definitions 3–4]) An $\mathbb{A}^n$-*cylinder* over a variety Z is the product $Z \times \mathbb{A}^n$. An open covering $(U_i)_{i \in I}$ of a projective variety V by the $\mathbb{A}^1$-cylinders $U_i \cong \mathbb{A}^1 \times Z_i$ is called *transversal* if it does not admit any proper invariant subset. A subset $Y \subset V$ is *proper* if it is nonempty and different from V. It is called *invariant* with respect to this covering if for any cylinder $U_i \to Z_i$, $i \in I$, the intersection $Y \cap U_i$ is covered by the fibers of $U_i \to Z_i$.

Given an ample polarization H of V, a subset $U \subset V$ is called H-*polar* if $U = V \setminus \mathrm{supp}(D)$ for some effective divisor $D \in |dH|$ where $d > 0$. Clearly, any H-polar subset of V is open and affine. If $\operatorname{Pic}(V) = \mathbb{Z}$ then $U = V \setminus \operatorname{supp}(D)$ is H-polar for any ample polarization H on V and any effective divisor D.

Theorem 3.2 *Let (V, H) be a polarized smooth projective variety. Then the affine cone over (V, H) is flexible if one of the following holds:*

(i) ([18, Theorem 5]) V admits a transversal covering by a family of H-polar $\mathbb{A}^1$-cylinders $U_i = V \setminus \operatorname{Supp}(D_i) \cong \mathbb{A}^1 \times Z_i$ where Z_i is a smooth affine variety, $i \in I$.

(ii) ([15, Theorem 1.4]) V admits a covering by a family of flexible H-polar subsets $U_i = V \setminus \operatorname{Supp}(D_i)$, $i \in I$.

For instance, the affine cone over (V, H) is flexible provided one can find an open covering $\{U_i\}_{i\in I}$ of V by H-polar toric affine varieties $U_i = V \setminus \operatorname{Supp}(D_i)$ with no torus factor, that is, non-decomposable as a product $U_i = W_i \times (\mathbb{A}^1 \setminus \{0\})$. Indeed, any such variety U_i is flexible [2, Theorem 2.1]. In the simplest case where $\operatorname{Pic}(V) = \mathbb{Z}$ and V admits an open covering by the affine spaces, the affine cone X over (V, H) is flexible whatever is an ample polarization H of V. According to Theorem 2(a), (b) such a covering exists for any Fano–Mukai fourfold V_{18} with a reductive automorphism group. In the remaining case of $V = V_{18}^{\mathrm{a}}$ with $\operatorname{Aut}^0(V) = \mathbb{G}_{\mathrm{a}} \times \mathbb{G}_{\mathrm{m}}$ we ignore whether V admits a covering by flexible affine charts, cf. Proposition 5.1. Hence, we cannot apply the criterion of Theorem 3.2(ii) in this case. Instead, we will apply the following version, which mixes the two criteria of Theorem 3.2.

Proposition 3.3 *Let (V, H) be a smooth projective variety of dimension $n \geq 3$ with an ample polarization. Suppose V possesses a family of flexible polar $\mathbb{A}^1$-cylinders $U_i = V \setminus \operatorname{Supp}(A_i) \cong \mathbb{A}^1 \times Z_i$, $i \in I$, where Z_i is an affine variety of dimension $n - 1$. Then the affine cone over (V, H) is flexible provided the following hold.*

- *$D := \bigcap_{i\in I} A_i$ is a subvariety of V of dimension $m \leq n - 2$;*
- *through any point of V pass at most k components of D;*
- *any point $P \in D$ is contained either in a polar $\mathbb{A}^{m+1}$-cylinder $U_P \cong \mathbb{A}^{m+1} \times Z_P$, or in $k + 1$ polar $\mathbb{A}^m$-cylinders $U_{P,j} \cong \mathbb{A}^m \times Z_{P,j}$ in V, $j = 1, \ldots, k + 1$, where Z_P and $Z_{P,j}$ are affine varieties, and for any two cylinders $U_{P,j}$ and $U_{P,j'}$ with $j \neq j'$ the $\mathbb{A}^m$-fibers through P of the natural projections $U_{P,j} \to Z_{P,j}$ and $U_{P,j'} \to Z_{P,j'}$ meet properly, that is, the dimension of their intersection is smaller than m.*

Proof We use the criterion of Theorem 3.2(i), that is, we show the existence of a transversal covering of V by polar $\mathbb{A}^1$-cylinders. We take for such a covering the union of the collections $\{U_i\}$, $\{U_P\}$, and $\{U_{P,j}\}$, where each member is endowed with all possible structures of a polar $\mathbb{A}^1$-cylinder.

Let $Y \subset V$ be a nonempty subset invariant with respect to the above covering of V by polar $\mathbb{A}^1$-cylinders, see Definition 3.1. We claim that if $Y \cap U_i \neq \varnothing$ for some $i \in I$, then $Y \supset U_i$. Indeed, let $P \in Y \cap U_i$, and let l be the ruling of the $\mathbb{A}^1$-cylinder U_i passing through P. Then $l \subset Y$ because Y is invariant. Since U_i is flexible, for any point $P' \in U_i$ different from P one can find an automorphism $\alpha \in \mathrm{SAut}(U_i)$ such that $\alpha(P) = P$ and $\alpha^{-1}(P') \in l$. Then $\alpha(l)$ is a ruling of a new $\mathbb{A}^1$-cylinder structure on U_i. Since $P, P' \in \alpha(l)$ and $P \in Y$ where Y is invariant, then also $P' \in \alpha(l) \subset Y$. Hence $U_i \subset Y$, as claimed.

It follows that $Y \cap U_j \neq \varnothing$ for any $j \in I$. Thereby one has $Y \supset \bigcup_{i\in I} U_i = V \setminus D$. Due to our assumptions, for any point $P \in D$ one can choose either a polar $\mathbb{A}^{m+1}$-cylinder U_P, or a polar $\mathbb{A}^m$-cylinder $U_{P,j}$ such that the $\mathbb{A}^m$-fiber passing through P of the projection $U_{P,j} \to Z_{P,j}$ is not contained in D. Then for a suitable $\mathbb{A}^1$-cylinder structure on U_P or $U_{P,j}$, respectively, the ruling $l \cong \mathbb{A}^1$ passing through P is not contained in D, and so, meets $V \setminus D \subset Y$. Then one has $P \in l \subset Y$. Since this holds for any $P \in D$ one has $D \subset Y$, and so, $Y = V$, that is, Y is not a proper subset of V.

The latter argument shows as well that a nonempty invariant subset $Y \subset V$ cannot be contained in D. Thus, the criterion of Theorem 3.2(i) applies and yields the result. □

4 $\mathrm{Aut}^0(V)$-Action on the Fano–Mukai Fourfold V of Genus 10

In this section, we establish the facts on the geometry of the $\mathrm{Aut}(V)$-action on V used in the proofs of out main results. In the case where $\mathrm{Aut}^0(V) = \mathbb{G}_{\mathrm{m}}^2$ the proofs of Theorems 1 and 2 are short and provided in this section. The subsequent sections are devoted to the more subtle case where $\mathrm{Aut}^0(V) = \mathbb{G}_{\mathrm{a}} \times \mathbb{G}_{\mathrm{m}}$.

The Fixed Points of the Torus

Consider the simple complex algebraic group G_2 of rank 2 and of dimension 14. By the Mukai construction [16], see also [24, Theorem 7.1], any Fano–Mukai fourfold V of genus 10 is a hyperplane section of the homogeneous fivefold $\Omega = G_2/P \subset \mathbb{P}^{13}$, where $P \subset G_2$ is a parabolic subgroup of dimension 9 corresponding to a long root, and so, Ω is the corresponding adjoint variety. It is known [24, Theorem 1.2] that $\mathrm{Aut}^0(V)$ is the stabilizer of V in G_2 acting naturally on Ω.

Lemma 4.1 *Let $T \subset G_2$ be a maximal torus. Then T has exactly six fixed point in Ω.*

Proof The left coset $\omega = gP \in \Omega$, where $g \in G_2$, is a fixed point of T if and only if for any $t \in T$ there exists $p \in P$ with $tg = gp$, that is, $g^{-1}Tg \subset P$. Since the maximal tori in G_2 are conjugated and P contains the Borel subgroup of G_2, we may assume $T \subset P$. Consider the following subgroup of G_2:

$$\Gamma := \{g \in G_2 \mid g^{-1}Tg \subset P\}.$$

Clearly, $\Gamma \supset P$. Two T-fixed points $\omega_i = g_i P \in \Omega$, $i = 1, 2$, coincide if and only if $g_2 \in g_1 P$. So, the fixed points of T in Ω are in one-to-one correspondence with the elements of the left coset space Γ/P.

We have $N_{G_2}(T) \subset \Gamma$, where $N_G(H)$ stands for the normalizer subgroup of a subgroup $H \subset G$. The normalizer $N_{G_2}(T)$ acts on Γ/P via

$$N_{G_2}(T) \ni n : gP \mapsto ngP \quad \forall g \in \Gamma.$$

This action is transitive. Indeed, given $g \in \Gamma$, let $T' = g^{-1}Tg \subset P$. The maximal tori T and T' are conjugate in P, and so, $T' = p^{-1}Tp$ for some $p \in P$. Then $n := pg^{-1} \in N_{G_2}(T)$ verifies $nP = gP$. The stabilizer of the coset P in $N_{G_2}(T)$ under this action is $P \cap N_{G_2}(T) = N_P(T)$. Therefore, the fixed points of T in Ω are in one-to-one correspondence with the elements of the left coset space

$$N_{G_2}(T)/N_P(T) = \mathrm{W}(G_2)/\mathrm{W}(P) = \mathfrak{D}_6/\{\pm 1\} = \mathbb{Z}/6\mathbb{Z},$$

where $\mathrm{W}(G_2) = N_{G_2}(T)/T$ and $\mathrm{W}(P) = N_P(T)/T$ stand for the Weyl groups of G_2 and of P, respectively, and $\mathfrak{D}_n$ is the nth dihedral group. This yields the assertion. □

The case $\mathrm{Aut}^0(V) = \mathbb{G}_{\mathrm{m}}^2$

In this subsection we prove Theorems 1 and 2 in the case $\mathrm{Aut}(V) = \mathbb{G}_{\mathrm{m}}^2$, that is, for $V = V_{18}^{\mathrm{g}}$.

Recall that the Fano–Mukai fourfold V with $\mathrm{Aut}^0(V) = \mathbb{G}_{\mathrm{m}}^2$ contains exactly six cubic cones, see Lemma 2.7(i). Any cubic cone $S \subset V$ coincides with the union of lines in V passing through its vertex, see Lemma 2.4(iii). It is $\mathrm{Aut}^0(V)$-invariant and its vertex is fixed under the $\mathrm{Aut}^0(V)$-action. By Lemma 2.4(iii) the vertices of distinct cubic cones are distinct. Using Lemma 4.1 we deduce the following corollary.

Corollary 4.2 *Let* $\mathrm{Aut}^0(V) = \mathbb{G}_{\mathrm{m}}^2$. *Then the vertices* v_i *of the cubic cones* S_i, $i = 1, \ldots, 6$, *are the only fixed points of the torus* $T = \mathrm{Aut}^0(V)$ *acting on* V.

The next corollary yields Theorem 2(b).

Corollary 4.3 *One has* $\bigcap_{i=1}^{6} A_{S_i} = \varnothing$.

Proof Assume the contrary holds. Then by the Borel fixed point theorem, the intersection $\bigcap_{i=1}^{6} A_{S_i}$ contains a fixed point of the torus $\mathrm{Aut}^0(V) = \mathbb{G}_{\mathrm{m}}^2$. Due to Corollary 4.2 this point is the vertex of a cubic cone, say $S_1 \in \mathscr{S}_1$. By Lemma 2.7(i) one has $S_1 \cap S_4 = \varnothing$, and so, $v_{S_1} \notin A_{S_4}$ according to Lemma 2.5(iii). This gives a contradiction. □

Proof (*Proof of Theorem* 1 *in the case* $\mathrm{Aut}^0(V) = \mathbb{G}_{\mathrm{m}}^2$) Recall that the affine space $\mathbb{A}^4$ is a flexible variety, see, e.g., [10, Lemma 5.5]. By Lemma 2.3(iii) and Corollary 4.3, V is covered by the flexible Zariski open subsets $U_i = V \setminus A_{S_i} \cong \mathbb{A}^4$, $i = 1, \ldots, 6$. Thus, the criterion of Theorem 3.2(ii) applies and gives the result.

The case $\mathrm{Aut}^0(V) = \mathbb{G}_{\mathrm{a}} \times \mathbb{G}_{\mathrm{m}}$

In this subsection we let $V = V_{18}^{\mathrm{a}}$. Recall that by Lemma 2.10 the group $\mathrm{Aut}^0(V)$ acts effectively on $\mathscr{S}_i(V) = \mathbb{P}^2$ for $i = 1, 2$.

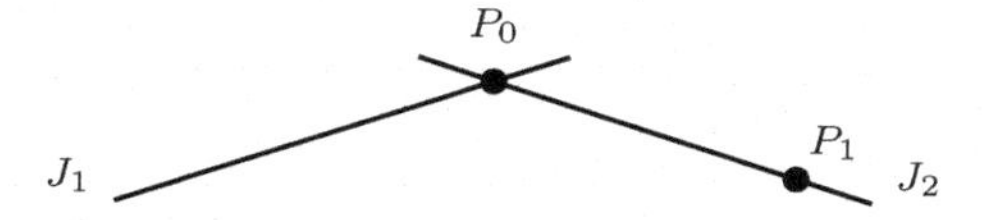

Fig. 1 Invariant lines and fixed points of a $\mathbb{G}_a \times \mathbb{G}_m$-action on $\mathbb{P}^2$

Lemma 4.4 *Any effective action of $\mathbb{G}_a \times \mathbb{G}_m$ on $\mathbb{P}^2$ can be given in suitable coordinates by the matrices*

$$\begin{pmatrix} \lambda & 0 & 0 \\ 0 & 1 & \mu \\ 0 & 0 & 1 \end{pmatrix} \quad \text{where} \quad \lambda \in \mathbb{C}^*, \ \ \mu \in \mathbb{C}. \tag{2}$$

This action has exactly two invariant lines J_1, $J_2 \subset \mathbb{P}^2$ and exactly two fixed points $P_0 \in J_1 \cap J_2$ and $P_1 \in J_2 \setminus J_1$, see Fig. 1.

Proof Since the group $\mathbb{G}_a \times \mathbb{G}_m$ is abelian and acts effectively on $\mathbb{P}^2$, (2) is the only possibility for the Jordan normal form of its elements modulo scalar matrices. □

Proposition 4.5 *For $V = V_{18}^{\mathrm{a}}$ the following assertions hold.*

(i) The action of $\mathrm{Aut}^0(V) = \mathbb{G}_a \times \mathbb{G}_m$ on each component $\mathscr{S}_i \cong \mathbb{P}^2$ of $\mathscr{S}(V)$, $i = 1, 2$ is given by matrices (2), and the action on $\Sigma(V) \cong \mathrm{Fl}(\mathbb{P}^2)$ is the induced one.

(ii) The subfamily $\mathscr{S}_1^{\mathrm{a}}$ of $\mathbb{G}_a$-invariant cubic scrolls corresponds to the line J_2 on $\mathscr{S}_1 = \mathbb{P}^2$, and the subfamily $\mathscr{S}_1^{\mathrm{m}}$ of $\mathbb{G}_m$-invariant cubic scrolls consists of the line J_1 and the point P_1, see Fig. 1.

(iii) There are exactly three $\mathrm{Aut}^0(V)$-invariant lines $l_{i,i+1}$, $i \in \{1, 2, 3\}$ and exactly four $\mathrm{Aut}^0(V)$-invariant cubic scrolls S_i, $i = 1, \ldots, 4$ on V. With a suitable enumeration, $l_{i,i+1}$ is the unique common ruling of S_i and S_{i+1}, while S_i and S_j have no common ruling if $j - i \neq \pm 1$. Furthermore, S_i and S_j belong to the same connected component of $\mathscr{S}(V)$ if and only if $j \equiv i \mod 2$.

(iv) Any cubic scroll S_i in (iii) is a cubic cone, and any cubic cone on V coincides with one of the S_i's.

(v) There are exactly two families of $\mathbb{G}_a$-invariant lines on V. These are the families of rulings of S_2 and S_3.

(vi) There are exactly three families of $\mathbb{G}_m$-invariant lines on V. These are the families of rulings of S_1, S_4 and of the smooth sextic scroll $D \subset V$, see Notation 4.6.

(vii) There are exactly four $\mathrm{Aut}^0(V)$-fixed points $v_1, \ldots, v_4$ on V, namely, the vertices of the cubic cones $S_1, \ldots, S_4$, where

$$v_2 = l_{1,2} \cap l_{2,3} = S_1 \cap S_3, \qquad v_3 = l_{2,3} \cap l_{3,4} = S_2 \cap S_4, \qquad S_1 \cap S_4 = \varnothing.$$

The three lines $l_{i,i+1}$ form a chain.

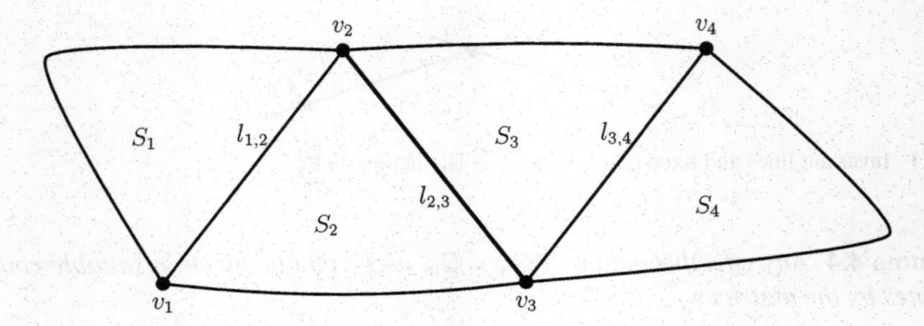

Fig. 2 Cubic cones on V_{18}^{a}

The proof is done below. Collecting the information from Proposition 4.5 we see that the configuration of the cones S_i looks like the one on Fig. 2.

Notation 4.6 There is exactly one pair of disjoint cubic cones on V, namely, (S_1, S_4). We let $D \subset V$ be the associated smooth sextic scroll as in Lemma 2.11(iii), that is, the union of lines on V joining the corresponding points of the rational twisted cubic curves $\Gamma_1 = S_1 \cap A_{S_4}$ and $\Gamma_4 = S_4 \cap A_{S_1}$. Notice that the curves Γ_1 and Γ_4, as well as the surface D are $\mathrm{Aut}^0(V)$-invariant.

Proof (*Proof of Proposition* 4.5) Assertions (i) and (ii) are immediate, see Lemma 4.4. Assertion (iii) follows from Lemmas 2.6 and 4.4.

(iv) By Corollary 2.8 any cubic cone on V is $\mathrm{Aut}^0(V)$-invariant. Hence, this is one of the $\mathrm{Aut}^0(V)$-invariant cubic scrolls $S_1, \ldots, S_4$ as in (iii). By Lemma 2.7(ii) there are exactly 4 cubic cones on V. So, the $\mathrm{Aut}^0(V)$-invariant cubic scrolls $S_1, \ldots, S_4$ are cubic cones.

(v) and (vi) By Theorem 2.1(a) we have $\Sigma(V) \cong \mathrm{Fl}(\mathbb{P}^2)$. By Lemma 4.4 the group $\mathrm{Aut}^0(V) \cong \mathbb{G}_{\mathrm{a}} \times \mathbb{G}_{\mathrm{m}}$ acts on $\Sigma(V) \cong \mathrm{Fl}(\mathbb{P}^2)$ via (2). Looking at Fig. 3.A one can select all the lines on V with stabilizers of positive dimension. There are exactly five such families of lines; they correspond to the following five families of flags on $\mathbb{P}^2$:

- (P_0, l), where l runs over the pencil of lines through P_0;
- (P_1, l), where l runs over the pencil of lines through P_1;
- (P, J_2), where P runs over J_2;
- (P, J_1), where P runs over J_1;
- (P, l), where P runs over J_1 and l passes through P_1.

The last family corresponds to the family of rulings of D, and the other four correspond to the families of rulings of the cubic cones $S_1, \ldots, S_4$.

(vii) Notice that the vertices of the cubic cones in V are fixed by $\mathrm{Aut}^0(V)$, because these cones are $\mathrm{Aut}^0(V)$-invariant. Let v be a fixed point of the $\mathrm{Aut}^0(V)$-action on V different from the vertices of cubic cones. According to Lemma 2.4(iii) the number of lines on V passing through v is finite. Hence each of these lines is $\mathrm{Aut}^0(V)$-invariant. By (iii) such a line coincides with the common ruling $l_{i,i+1}$ of a pair

(S_i, S_{i+1}), $i \in \{1, 2, 3\}$. However, by (v) and (vi) any line $l_{i,i+1}$ contains exactly two $\mathrm{Aut}^0(V)$-fixed points, namely, the vertices v_i and v_{i+1}. The remaining statements are immediate. □

Remark 4.7 Recall (Theorem 2.1(a)) that the Hilbert scheme $\Sigma(V)$ of lines in V is isomorphic to the variety $\mathrm{Fl}(\mathbb{P}^2)$ of full flags on $\mathbb{P}^2$ which can be realized as a divisor of type $(1, 1)$ on $\mathbb{P}^2 \times (\mathbb{P}^2)^\vee$ given by

$$x_0y_0 + x_1y_1 + x_2y_2 = 0.$$

Then the induced action of $\mathbb{G}_a \times \mathbb{G}_m$ on $\mathrm{Fl}(\mathbb{P}^2)$ can be given by

$$([x_0 : x_1 : x_2], [y_0 : y_1 : y_2]) \longmapsto ([\lambda x_0 : x_1 + \mu x_2 : x_2], [\lambda^{-1}y_0 : y_1 : y_2 - \mu y_1]).$$

In this realization of $\Sigma(V)$ the three $\mathrm{Aut}^0(V)$-invariant lines on V correspond to the following $\mathrm{Aut}^0(V)$–fixed points:

$$([0 : 1 : 0], [1 : 0 : 0]), \quad ([1 : 0 : 0], [0 : 0 : 1]) \quad \text{and} \quad ([0 : 1 : 0], [0 : 0 : 1]).$$

This provides an alternative way to verify the assertions of Proposition 4.5.

5 Affine 4-Spaces in V_{18}^{a} and Flexibility of Affine Cones

Affine 4-Spaces in V_{18}^{a}

In this subsection, we analyze affine charts isomorphic to $\mathbb{A}^4$ on the Fano–Mukai fourfold $V = V_{18}^{\mathrm{a}}$ of genus 10 with $\mathrm{Aut}^0(V) = \mathbb{G}_a \times \mathbb{G}_m$.

The following proposition proves the first part of Theorem 2(c); the second part will be proven in Proposition 7.3.

Proposition 5.1 *In the notation of Proposition 4.5 one has* $\bigcap_{j=1}^4 A_{S_j} = l_{2,3}$.

The proof is done below. We need the following auxiliary facts. Let $Y = \bigcap_{j=1}^4 A_{S_j}$. Clearly, Y is $\mathrm{Aut}^0(V)$-invariant.

Claim 5.2 One has $l_{2,3} \subset Y$.

Proof The line $l_{2,3}$ intersects all the S_i. Hence, $l_{2,3} \subset A_{S_i}$ for all i. □

Claim 5.3 $S_1 \cap Y = \{v_2\}$ and $S_4 \cap Y = \{v_3\}$.

Proof By Lemma 2.5(iii) we have $v_1 \notin A_{S_4}$ because $S_1 \cap S_4 = \varnothing$. Hence $S_1 \cap A_{S_4}$ is a smooth irreducible hyperplane section of the cone S_1. Since $v_1 \in A_{S_2}$, the intersection $S_1 \cap A_{S_2}$ is a singular hyperplane section of S_1. Therefore, the intersection $S_1 \cap Y$ is a finite set whose points are fixed by $\mathrm{Aut}^0(V)$. Then by Proposition 4.5(vii) and Claim 5.2 we have $S_1 \cap Y = \{v_2\}$. By symmetry, we have also $S_4 \cap Y = \{v_3\}$. □

In the sequel we let $\langle Y\rangle$ stand for the linear span of a subvariety $Y \subset \mathbb{P}^n$.

Claim 5.4 $S_2 \cap Y = S_3 \cap Y = l_{2,3}$.

Proof We have $v_2 \in S_2 \cap Y$ and $S_2 \not\subset Y$ because $l_{1,2} \not\subset Y$ by Claim 5.3. Since $S_2 \cap Y = S_2 \cap \langle Y\rangle$ the intersection $S_2 \cap Y$ consists of a finite number of rulings of S_2. These rulings are $\mathrm{Aut}^0(V)$-invariant. Since $l_{1,2} \not\subset Y$ by Claim 5.3 the only possibility is $S_2 \cap Y = l_{2,3}$, see Proposition 4.5(iii). □

Proof (*Proof of Proposition* 5.1) Assume there is a point $P \in Y \setminus l_{2,3}$. By Claims 5.3 and 5.4 $P \notin S_j$ for all j. By Lemma 2.3(ii) for any $j = 1, \ldots, 4$ through P passes a unique line $l_j \subset A_{S_j}$ meeting S_j. Since there are at most three lines on V passing through P, one has $l_i = l_j$ for some $i \neq j$. Set $l := l_i = l_j$. Thus, one has $P \in l \subset A_{S_i} \cap A_{S_j}$. By Lemma 2.4(iii), V contains no plane. Hence, the set of lines contained in the surface $A_{S_i} \cap A_{S_j}$ has dimension at most one, cf. [14, Lemma A.1.1]. Since $l \not\subset S_j$ for all j, the line l cannot be $\mathrm{Aut}^0(V)$-invariant, see Proposition 4.5(iii). Therefore, the stabilizer G of l under the $\mathrm{Aut}^0(V)$-action on $\Sigma(V)$ is one-dimensional. Since $l \not\subset S_j$ for all j, due to Proposition 4.5(v)–(vi), l is a ruling of D, and so, is $\mathbb{G}_m$-invariant. The rulings l and $l_{2,3}$ of D are disjoint, because $P \notin l_{2,3}$ by our choice. It follows from Claim 5.4 that $l \cap S_2 = l \cap S_3 = \varnothing$. So, we have $\{i, j\} = \{1, 4\}$, that is, l meets S_1 and S_4 in $\mathbb{G}_m$-fixed points. Besides, there exists a unique line $l_2 \neq l$ in V joining P and S_2. If the stabilizer of P in G is finite, then $l = \overline{G \cdot P} \subset Y$. In particular, the intersection point P_1 of l and S_1 lies on Y. By Claim 5.3 one has $P_1 = v_2$, contrary to the fact that $l \cap S_2 = \varnothing$. Hence, P is fixed by $\mathbb{G}_m$, and then the line l_2 is $\mathbb{G}_m$-invariant. Since l_2 is not a ruling of D or of one of the S_i, we get a contradiction with Proposition 4.5(vi). □

6 $\mathbb{A}^2$-Cylinders in Smooth Quadric Fourfolds and in the Del Pezzo Quintic Fourfold

The next lemma on the existence of an $\mathbb{A}^2$-cylinder will be used in the proof of Proposition 6.3 below.

Lemma 6.1 *Let $Q \subset \mathbb{P}^5$ be a smooth quadric, and let Q', $Q^\star$ be distinct hyperplane sections of Q. Let $Q_1, \ldots, Q_k$ be the members of the pencil $\langle Q', Q^\star\rangle$ generated by Q' and $Q^\star$ which have singularities outside $Q' \cap Q^\star$, and let P_i be the unique singular point of Q_i. Given a point $P \in Q \setminus (Q' \cup Q^\star \cup \{P_1, \ldots, P_k\})$ there exists a principal affine open subset $U = U_P$ inside the affine variety $Q \setminus (Q' \cup Q^\star)$ such that*

(i) $P \in U$;
(ii) $U \cong \mathbb{A}^2 \times Z$, *where Z is an affine surface.*

Proof Pick a general point $P^\bullet \in Q' \cap Q^\star$, and let $\mathbf{T}_{P^\bullet} Q \subset \mathbb{P}^5$ be the embedded tangent space to Q at $P^\bullet$. The projection with center $P^\bullet$ defines an isomorphism

$$Q \setminus Q^\bullet \cong \mathbb{P}^4 \setminus \mathbb{P}^3 \cong \mathbb{A}^4, \quad \text{where} \quad Q^\bullet := Q \cap \mathbf{T}_{P^\bullet} Q.$$

The quadric cone $Q^\bullet$ with vertex $P^\bullet$ coincides with the union of lines on Q passing through $P^\bullet$. If the quadric cone $\Delta_P(Q) = Q \cap \mathbf{T}_P Q$ with vertex $P \in Q \setminus (Q' \cap Q^\star)$ contains $Q' \cap Q^\star$, then $\Delta_P(Q)$ coincides with a member Q_i of the pencil $\langle Q', Q^\star \rangle$, which has the singular point $P = P_i$ for some $i \in \{1, \ldots, k\}$. However, the latter is excluded by our assumption. So, $Q' \cap Q^\star \not\subset \Delta_P(Q)$. Hence, for the general point $P^\bullet \in Q' \cap Q^\star$ the line joining P and $P^\bullet$ is not contained in Q. This implies $P \notin Q^\bullet$.

The images of $Q' \setminus Q^\bullet$ and $Q^\star \setminus Q^\bullet$ in $\mathbb{A}^4 = \mathbb{P}^4 \setminus \mathbb{P}^3$ under the projection with center $P^\bullet$ is a pair of affine hyperplanes with nonempty intersection. Thus, we obtain

$$P \in Q \setminus (Q^\bullet \cup Q' \cup Q^\star) \cong \mathbb{A}^2 \times (\mathbb{A}^1 \setminus \{\text{a point}\}) \times (\mathbb{A}^1 \setminus \{\text{a point}\}). \quad (3)$$

□

Recall that a smooth del Pezzo quintic fourfold $W = W_5 \subset \mathbb{P}^7$ is unique up to isomorphism [7]. This variety is quasihomogeneous, more precisely, the automorphism group $\mathrm{Aut}(W)$ has the open orbit $W \setminus R \cong \mathbb{A}^4$ in W, where R is the hyperplane section of W covered by the lines on W which meet the unique $\sigma_{2,2}$-plane $\Xi \subset W$, see [22, Sect. 4]. The planes on W different from Ξ form a one-parameter family, and their union coincides with R; we call them *Π-planes*.

Proposition 6.2 *Let $W = W_5 \subset \mathbb{P}^7$ be the del Pezzo quintic fourfold. Then the following hold.*

(i) *([22, Corollary 2.6]) The Hilbert scheme $\Sigma(W)$ of lines on W is smooth, irreducible of dimension* $\dim \Sigma(W) = 4$. *For any point $P \in W$ the Hilbert scheme $\Sigma(W; P) \subset \Sigma(W)$ of lines passing through P has pure dimension* 1.

(ii) *([22, Proposition 4.11.iv]) For any line $l \subset W$ there exists a unique hyperplane section B_l of W with* $\mathrm{Sing}(B_l) \supset l$. *This B_l is the union of lines meeting l.*

(iii) *Given a point $P \in W$, let Δ_P be the union of lines in W passing through P. If $P \in W \setminus R$, then Δ_P is a cubic cone. If $P \in R \setminus \Xi$, then Δ_P is the union of a plane Π_P passing through P and a quadric cone Δ'_P with vertex P.*

(iv) *Let $B \subset W$ be a hyperplane section whose singular locus is two-dimensional. Then $B = R$ and* $\mathrm{Sing}(B) = \Xi$.

(v) *Let $B \subset W$ be a hyperplane section whose singular locus is one-dimensional. Then $B = B_l$ for some line l.*

(vi) *Let $B \subset W$ be a hyperplane section, and let $C \subset B$ be an irreducible curve. Assume B contains a two-dimensional family of lines meeting C. Then one of the following holds:*

  (a) *C is contained in a plane on B;*

  (b) *$C = l$ is a line, and $B = B_l$;*

  (c) *$B = R$.*

Proof (iii) By (i) the universal family of lines $\mathfrak{L}(W) \subset \Sigma(W) \times W$ is smooth, and the natural projection $s : \mathfrak{L}(W) \to W$ is a flat morphism of relative dimension one. Its fiber $s^{-1}(P)$ is isomorphic to the base of the cone Δ_P. Let $P \in W \setminus R$. Since $W \setminus R$ is the open orbit of $\mathrm{Aut}(W)$, see, e.g., [24, (5.5.5)], the fiber $s^{-1}(P)$ is smooth in this case. Let H be a general hyperplane section of W passing through P. By Bertini's theorem, H is smooth, and by the adjunction formula, H is a del Pezzo threefold of degree 5. It is well known, see, e.g., [9, Chap. 2, Sect. 1.6] or [14, Corollary 5.1.5], that through a general point of H pass exactly three lines. This implies $\deg \Delta_P = 3$. On the other hand, one has $\Delta_P = \mathbf{T}_P W \cap W$. Since W is intersection of quadrics, Δ_P cannot be a cone over a plane cubic. It follows that Δ_P is a cubic cone.

Let further $P \in R \setminus \Xi$. By [20, Theorem 6.9], $\mathrm{Aut}(W)$ acts transitively on $R \setminus \Xi$. We have $\Delta_P \not\supset \Xi$, and Δ_P contains a Π-plane Π_P passing through P. Such a plane is unique because no two planes on W meet outside Ξ. By the flatness of s we have $\deg \Delta_P = 3$. There are lines on W which are not contained in R and meet $R \setminus \Xi$, and one of these lines passes through P. Therefore, one has $\Delta_P \neq 3\Pi_P$. Then the only possibility is $\Delta_P = \Pi_P + \Delta'_P$, where Δ'_P is a quadric cone.

(iv) Let Z_2 be an irreducible component of $\mathrm{Sing}(B)$ of dimension two. Choose general hyperplane sections H_1 and H_2 of W, and let $C = B \cap H_1 \cap H_2$. By Bertini's theorem, C is an irreducible curve with $\mathrm{Sing}(C) = Z_2 \cap H_1 \cap H_2$. By the adjunction formula one has $\mathrm{p}_a(C) = 1$. Hence $\deg Z_2 = 1$, i.e., Z_2 is a plane. Since B contains any line meeting Z_2 we have $B = R$, and then $\mathrm{Sing}(B) = \Xi$.

(v) Let Z_1 be the union of one-dimensional irreducible components of $\mathrm{Sing}(B)$. Since $Z_1 \subset \mathrm{Sing}(B)$ and B is a hyperplane section, B contains any line meeting Z_1. Hence, B is the union of lines meeting Z_1, see (i). By (iv) one has $B \neq R$. If Z_1 is a line, then $B = B_{Z_1}$ by (ii). Thus, we may assume Z_1 to be a curve of degree $d > 1$. Consider the general hyperplane section $H \subset B$. By Bertini's theorem, H has exactly d singular points, and these are the points of $H \cap Z_1$. By the adjunction formula, $-K_H$ is the class of a hyperplane section of H. Hence, H is a normal Gorenstein del Pezzo quintic surface.

By [5, Proposition 8.5] we have $d = 2$, that is, H has exactly two singular points, say, P_1 and P_2, and Z_1 is a conic. Again by [5, Proposition 8.5], H contains the line joining P_1 and P_2. Hence, B contains the linear span $\langle Z_1 \rangle$. Since R is the union of planes contained in W, we have $Z_1 \subset \langle Z_1 \rangle \subset R$. By (iv) one has $\langle Z_1 \rangle \neq \Xi$. So, $\langle Z_1 \rangle = \Pi$ is a Π-plane on R, where $\Pi \subset B$. Take a general line $l \subset \Pi$, and let $Z_1 \cap l = \{P_1, P_2\}$. Since Ξ meets $l \subset \mathrm{Sing}(B_l)$ one has $\Xi \subset B_l$. Likewise, since Ξ meets $Z_1 \subset \mathrm{Sing}(B)$ one has $\Xi \subset B$. Besides, the quadric cones Δ'_{P_1} and Δ'_{P_2} as in (iii) are contained in both B and B_l. Thus, we obtain

$$B \cap B_l \supset \Pi \cup \Xi \cup \Delta'_{P_1} \cup \Delta'_{P_2}.$$

Assuming $B \neq B_l$ the latter contradicts the fact that $\deg(B \cap B_l) = \deg W = 5$.

(vi) We may suppose that $B \neq R$ and the singular locus of B has dimension ≤ 1, see (iv). By assumption, $C \subset B$ is an irreducible curve, and there exists an irreducible two-dimensional family of lines $\Sigma(B, C) \subset \Sigma(B)$ on B meeting C. Let $r : \mathfrak{L}(C, B) \to \Sigma(C, B)$ be the universal family, and let $s : \mathfrak{L}(C, B) \to B$ be the natural projection. If $s(\mathfrak{L}(C, B)) \neq B$, then $s(\mathfrak{L}(C, B))$ is a plane, see, e.g., [14, Lemma A.1.1], and so, C is contained in a plane on B. Assume further $s(\mathfrak{L}(C, B)) = B$, and so, s is a generically étale morphism. We claim that the general line from $\Sigma(C, B)$ meets the singular locus of B. The argument below is well known, see, e.g., [9, Chap. 3, Prop. 1.3] or [14, Lemma 2.2.6], and we repeat it in brief for the sake of completeness. Suppose to the contrary that the general line l from $\Sigma(C, B)$ lies in the smooth locus of B. Using the fact that the restriction of s to $r^{-1}([l])$ is an isomorphism, we may identify l with $r^{-1}([l])$. For the normal bundles of l we have

$$\mathscr{N}_{l/\mathfrak{L}(C,B)} = \mathscr{O}_l \oplus \mathscr{O}_l, \qquad \mathscr{N}_{l/B} = \mathscr{O}_l(a) \oplus \mathscr{O}_l(-a), \quad a \geq 0. \tag{4}$$

Over the point $l \cap C$ the map s is not an isomorphism. Hence the differential

$$ds : \mathscr{N}_{l/\mathfrak{L}(C,B)} \longrightarrow \mathscr{N}_{l/B}$$

is not an isomorphism either. From (4) we see that ds degenerates along l. This means that s is not generically étale, a contradiction.

Thus, the general line from $\Sigma(C, B)$ meets $\operatorname{Sing}(B)$. Since B is not a cone, by our assumption we have $\dim \operatorname{Sing}(B) = 1$. Due to (v) there is a line l_0 on W such that $B = B_{l_0}$. If $l_0 = C$, then we are done. Otherwise, the lines on B passing through the general point $P \in C$ meet $l_0 = \operatorname{Sing}(B)$. The union of these lines is the cone with vertex P over l_0, that is, a plane. It follows that B is swept out by planes. Since any plane on W is contained in R, we conclude that $B = R$, contrary to our assumption. □

Proposition 6.3 *Let $B \subset W$ be a hyperplane section. For any point $P \in W \setminus B$ there exists a principal affine open subset U_P inside the affine variety $W \setminus B$ such that*

(i) $P \in U_P$;
(ii) $U_P \cong \mathbb{A}^2 \times Z_P$, where Z_P is an affine surface.

Proof If $B = R$ then $W \setminus B \cong \mathbb{A}^4$ [24, Corollary 2.2.2], and the assertion follows. So, we assume in the sequel $B \neq R$. We apply the following construction from [7] (see also [22, Proposition 4.11]). Fix a line $l \subset W$ not contained in R. There is the Sarkisov link

$$\begin{array}{ccc} & \tilde{W} & \\ {\scriptstyle \rho_W} \swarrow & & \searrow {\scriptstyle \varphi_Q} \\ W & \overset{\theta_l}{\dashrightarrow} & Q \end{array} \tag{5}$$

where $Q \subset \mathbb{P}^5$ is a smooth quadric, θ_l is induced by the linear projection with center l, ρ_W is the blowup of l, and φ_Q is the blowup of a smooth cubic scroll $\Lambda = \Lambda_3 \subset Q$. Furthermore, φ_Q sends the ρ_W-exceptional divisor $E_W \subset \tilde{W}$ onto the quadric cone $Q^\star = Q \cap \langle \Lambda \rangle$, while the φ_Q-exceptional divisor $\tilde{B}_l \subset \tilde{W}$ is the proper transform of $B_l \subset W$. Let $Q' := \theta_l(B)$.

Suppose our line $l \subset W$ satisfies the following conditions:

α) $l \subset B, l \not\subset R, P \notin B_l$, and
β) $\theta_l(P) \notin \{P_1, \dots, P_k\}$, where $P_1, \dots, P_k$ have the same meaning as in Lemma 6.1.

We have an isomorphism

$$W \setminus (B \cup B_l) \cong Q \setminus (Q' \cup Q^\star).$$

By Lemma 6.1 there exists a principal affine open subset $\tilde{U}_P$ inside the affine variety $Q \setminus (Q' \cup Q^\star)$ such that

(i) $\theta_l(P) \in \tilde{U}_P$;
(ii) $\tilde{U}_P \cong \mathbb{A}^2 \times Z_P$, where Z_P is an affine surface.

Then $U_P = \theta_l^{-1}(\tilde{U}_P) \subset W$ verifies (i)–(ii) of Proposition 6.3.

It remains to show the existence of a line $l \subset W$ satisfying α) and β).

Consider the union Δ_P of lines on W passing through P. By Proposition 6.2(iii), Δ_P is a (possibly reducible) cubic cone with vertex P on W of pure dimension two. Recall that $\Delta_P = (\mathbf{T}_P W \cap W)_{\text{red}}$, because W is an intersection of quadrics.

Let $C = \Delta_P \cap B$, and let $\Sigma(B, C)$ be the Hilbert scheme of lines in B which meet C, or, which is equivalent, which meet Δ_P. If dim $\Sigma(B, C) = 2$ then by Proposition 6.2 (vi), either the lines from $\Sigma(B, C)$ sweep out a plane, say, D on B, or $C = l$ is a line and $B = B_l$. However, the latter case is impossible. Indeed, B_l being singular along $l = C$, any ruling of the cone Δ_P meets $C \subset \text{Sing}(B_l)$, hence is contained in $B_l = B$. Then also $P \in B$, which is a contradiction.

Assume further dim $\Sigma(B, C) = 1$. Then the lines from $\Sigma(B, C)$ sweep out a surface scroll D on B. Thus, the latter holds whatever is the dimension of $\Sigma(B, C)$.

It follows from Proposition 6.2(i) that the threefold B is covered by lines on W. Since $B \neq R$ by our assumption, there is a point $P' \in B \setminus (D \cup R)$. Any line l through P' on B does not lie on $D \cup R$, and so, does not belong to $\Sigma(B, C)$. Hence, $\Delta_P \cap l = \varnothing$ for the general line l on B. Since B_l is the union of lines on W meeting l, see Proposition 6.2(ii), we deduce $P \notin B_l$. Thus, the general line l on B satisfies α).

To show β), we use the notation from the proof of (A). By the preceding, one has $\Delta_P \cap l = \varnothing$. Hence, the projection θ_l with center l is regular in a neighborhood of Δ_P. It follows that $\Delta_P^Q := \theta_l(\Delta_P)$ is a cone with vertex $P_Q := \theta_l(P)$. Suppose to the contrary that $P_Q = P_i$ for some $i \in \{1, \dots, k\}$. Then P_Q is the vertex of a quadric cone Q_i over $Q' \cap Q^\star$, see Lemma 6.1. Clearly, one has $\Delta_P^Q \subset Q_i$. From $l \not\subset D$ we deduce $C \not\subset B_l$.

On the other hand, we have $\theta_l(B_l) = \Lambda,\ Q^\star = Q \cap \langle\Lambda\rangle$, and

$$\theta_l(C) = \theta_l(\Delta_P \cap B) \subset \theta_l(\Delta_P) \cap \theta_l(B) = \Delta_P^Q \cap Q' \subset Q_i \cap Q' = Q^* \cap Q' \subset \langle\Lambda\rangle \cap Q.$$

From these inclusions we deduce $B_l = \theta_l^{-1}(\Lambda) \subset \theta_l^{-1}(\langle\Lambda\rangle \cap Q)$. The latter inclusion is an equality, since both sets are hyperplane section of W. Thus, one has $C \subset B_l$. This contradiction proves β).

7 $\mathbb{A}^2$-Cylinders in V_{18} and Flexibility of Affine Cones over V_{18}^{a}

In this section we finish the proofs of Theorems 1 and 2 in the case where $\mathrm{Aut}^0(V) = \mathbb{G}_{\mathrm{a}} \times \mathbb{G}_{\mathrm{m}}$. Using Propositions 3.3 and 5.1 we deduce such a flexibility criterion.

Corollary 7.1 *The affine cones over the Fano–Mukai fourfold V with* $\mathrm{Aut}^0(V) = \mathbb{G}_{\mathrm{a}} \times \mathbb{G}_{\mathrm{m}}$ *are flexible provided any point P of the common ruling $l_{2,3}$ of the cubic cones S_2 and S_3 on V is contained in a polar $\mathbb{A}^2$-cylinder U_P in V.*

In Proposition 7.3 we construct such a covering of $l_{2,3}$ in V by polar $\mathbb{A}^2$-cylinders. Combining with Corollary 7.1 this gives a proof of Theorem 1 in the remaining case $\mathrm{Aut}^0(V) = \mathbb{G}_{\mathrm{a}} \times \mathbb{G}_{\mathrm{m}}$. Indeed, the assertion of Theorem 1 follows immediately in this case by the flexibility criterion of Corollary 7.1 due to Proposition 7.3 below.

7.2 Let S be a cubic scroll on $V = V_{18}$. Then either S is smooth and isomorphic to the Hirzebruch surface $\mathbb{F}_1$, or S is a cubic cone. By virtue of [24, Proposition 3.1], in both cases the linear projection

$$\theta_S \colon V = V_{18} \subset \mathbb{P}^{12} \dashrightarrow \mathbb{P}^7$$

with center $\langle S \rangle = \mathbb{P}^4$ restricted to V yields a Sarkisov link

$$\begin{array}{ccccc} & & \tilde{V} & & \\ & \overset{\rho_V}{\swarrow} & & \overset{\varphi_W}{\searrow} & \\ V & & \overset{\theta_S}{\dashrightarrow} & & W \end{array} \tag{6}$$

where $W = W_5 \subset \mathbb{P}^7$ is the del Pezzo quintic fourfold, ρ_V is the blowup of S, and φ_W is the blowup of a smooth rational quintic scroll $F = F_5 \subset W$ isomorphic to $\mathbb{F}_1$. Furthermore, φ_W sends the ρ_V-exceptional divisor $E_V \subset \tilde{V}$ onto the hyperplane section $B_F := W \cap \langle F \rangle$ of W, while the φ_W-exceptional divisor $\tilde{A}_S \subset \tilde{V}$ is the proper transform of A_S. We have

$$V \setminus A_S \cong W \setminus B_F.$$

If S is a cubic cone then $B_F = R$ is the hyperplane section of W singular along the Ξ-plane and $V \setminus A_S \cong W \setminus R \cong \mathbb{A}^4$.

Proposition 7.3 *Let $V = V_{18}$ be a Fano–Mukai fourfold of genus* 10. *Then for any point $P \in V$ there exists a polar $\mathbb{A}^2$-cylinder U_P satisfying*

(i) $P \in U_P$;
(ii) $U_P \cong \mathbb{A}^2 \times Z_P$ where Z_P is an affine surface.

Proof Let $S \in \mathscr{S}(V)$ be a general cubic scroll on V and A_S be the hyperplane section of V with $\mathrm{Sing}(A_S) = S$, see Lemma 2.3. Then S is smooth, that is, not a cubic cone, and $P \notin A_S$ by Lemma 2.12. By Lemma 2.12 for the general $S \in \mathscr{S}(V)$

the map θ_S in (6) is well defined at P and $\theta_S(P) \notin B_F$. Now the assertion follows from Proposition 6.3. Indeed, by this proposition and in view of 7.2 we can find a principal open subset U_P inside $V \setminus A_S$ satisfying (i) and (ii). Then the complement $D_P := V \setminus U_P$ is an effective divisor. So, the $\mathbb{A}^2$-cylinder U_P is polar due to the fact that $\operatorname{Pic}(V) = \mathbb{Z}$. □

Acknowledgements The authors are grateful to Alexander Perepechko for a careful reading of the paper and valuable remarks. Our thanks are due as well to the anonymous referee for comments which allowed us to improve the presentation and to avoid certain inaccuracies slipped into the preliminary version. In particular, Remarks 2.9 and 4.7 were suggested by the referee.

References

1. I. Arzhantsev, H. Flenner, S. Kaliman, F. Kutzschebauch, M. Zaidenberg, Flexible varieties and automorphism groups. Duke Math. J. **162**(4), 767–823 (2013)
2. I.V. Arzhantsev, M.G. Zaidenberg, K.G. Kuyumzhiyan, Flag varieties, toric varieties, and suspensions: three examples of infinite transitivity. Mat. Sb. **203**(7), 3–30 (2012)
3. I. Cheltsov, J. Park, J. Won, Affine cones over smooth cubic surfaces. J. Eur. Math. Soc. (JEMS) **18**(7), 1537–1564 (2016)
4. I. Cheltsov, J. Park, Yu. Prokhorov, M. Zaidenberg. Cylinders in Fano varieties. EMS Surv. Math. Sci. **8**, 39–105 (2021)
5. D.F. Coray, M.A. Tsfasman, Arithmetic on singular del Pezzo surfaces. Proc. Lond. Math. Soc., III. Ser. **57**(1), 25–87 (1988)
6. I.V. Dolgachev, *Classical Algebraic Geometry* (Cambridge University Press, Cambridge, 2012)
7. T. Fujita, On the structure of polarized manifolds with total deficiency one. II. J. Math. Soc. Japan **33**(3), 415–434 (1981)
8. J. Harris, *Algebraic Geometry. A First Course*, *Graduate Texts in Mathematics*, vol. 133 (Springer, New York, 1992)
9. V.A. Iskovskikh, Anticanonical models of three-dimensional algebraic varieties. J. Sov. Math. **13**, 745–814 (1980)
10. Sh. Kaliman, M. Zaidenberg, Affine modifications and affine hypersurfaces with a very transitive automorphism group. Transform. Groups **4**(1), 53–95 (1999)
11. M. Kapustka, K. Ranestad, Vector bundles on Fano varieties of genus ten. Math. Ann. **356**(2), 439–467 (2013)
12. T. Kishimoto, Yu. Prokhorov, M. Zaidenberg, Group actions on affine cones. *Affine Algebraic Geometry: The Russell Festschrift.* CRM Proceedings and Lecture Notes, vol. 54 (Centre de Recherches Mathématiques, Montreal, 2011), pp. 23–163
13. A. Kuznetsov, Y. Prokhorov, Prime Fano threefolds of genus 12 with a $\mathbf{G}_m$-action. Épijournal de Géométrie Algébrique, 2(epiga:4560) (2018)
14. A. Kuznetsov, Y. Prokhorov, C. Shramov, Hilbert schemes of lines and conics and automorphism groups of Fano threefolds. Jpn. J. Math. **13**(1), 109–185 (2018)
15. M. Michałek, A. Perepechko, H. Süß, Flexible affine cones and flexible coverings. Math. Z. **290**(3–4), 1457–1478 (2018)
16. Sh. Mukai, Biregular classification of Fano 3-folds and Fano manifolds of coindex 3. Proc. Nat. Acad. Sci. U.S.A. **86**(9), 3000–3002 (1989)
17. J. Park, J. Won, Flexible affine cones over del Pezzo surfaces of degree 4. Eur. J. Math. **2**(1), 304–318 (2016)
18. A. Yu. Perepechko, Flexibility of affine cones over del Pezzo surfaces of degree 4 and 5. Funct. Anal. Appl. **47**(4), 284–289 (2013)

19. A. Perepechko, Affine cones over cubic surfaces are flexible in codimension one. Forum Math. **33**(2), 339–348 (2021)
20. J. Piontkowski, A. Van de Ven, The automorphism group of linear sections of the Grassmannians $G(1, N)$. Doc. Math. **4**, 623–664 (1999)
21. Yu. Prokhorov, Automorphism groups of Fano 3-folds. Russ. Math. Surv. **45**(3), 222–223 (1990)
22. Yu. Prokhorov, M. Zaidenberg, Examples of cylindrical Fano fourfolds. Eur. J. Math. **2**(1), 262–282 (2016)
23. Yu. Prokhorov, M. Zaidenberg, New examples of cylindrical Fano fourfolds, in *Proceedings of Kyoto Workshop "Algebraic Varieties and Automorphism Groups", July 7–11, 2014, Advance Studies in Pure Mathematics*, vol. 75 (Mathematical Society of Japan, Kinokuniya, Tokyo, 2017), pp. 443–463
24. Yu. Prokhorov, M. Zaidenberg, Fano-Mukai fourfolds of genus 10 as compactifications of $\mathbf{C}^4$. Eur. J. Math. **4**(3), 1197–1263 (2018)
25. Yu. Prokhorov, M. Zaidenberg, Fano-Mukai fourfolds of genus 10 and their automorphism groups. Eur. J. Math. **8**(2), 561–572 (2022)
26. V.V. Przyjalkowski, I.A. Cheltsov, K.A. Shramov, Fano threefolds with infinite automorphism groups. Izv. Math. **83**(4), 860–907 (2019)

Enriques Diagrams Under Pullback by a Double Cover

Joaquim Roé

To Ciro Ciliberto on the occasion of his 70th birthday.

Abstract Let $f : \hat{S} \to S$ be a ramified morphism between smooth surfaces, and $p \in \hat{S}$ a point where f has local degree 2. We describe an algorithm to determine the equisingularity type at p of the preimage by f of a reduced curve on S, in terms of Enriques diagrams.

Keywords Singularities of curve · Ramified covers · Smooth surfaces · Enriques diagrams

The question of describing the inverse and direct images of curve singularities under ramified morphisms of surfaces is of obvious importance. In [2], E. Casas-Alvero developed a theory to address this question in all generality. In this note, we apply the theory to the particular case of morphisms of local degree 2, which is the generic behavior for ramified maps and hence a case of particular interest.

Methods to determine the equisingularity type of the pullback of a curve under a double cover have existed, at least in principle, for long: for instance, splitting the curve into analytically irreducible components and computing square roots in the field of Puiseux series. Such a method, however, is not easily generalizable beyond the case of cyclic covers. In contrast, the methods of [2] are completely general; in the case of the double cover the situation is tame enough to obtain a simple algorithm. The method we propose is also better suited in problems where the *multiplicities* of the involved curves (including at infinitely near points) are relevant, for instance in applications to H-constants (see [5]) or more generally to study the effect of ramified

Partially supported by the Spanish MINECO grant MTM2016-75980-P.

J. Roé (✉)
Universitat Autònoma de Barcelona, 08193 Barcelona, Catalonia, Spain
e-mail: jroe@mat.uab.cat

T. Dedieu et al. (eds.), *The Art of Doing Algebraic Geometry*, Trends in Mathematics,
https://doi.org/10.1007/978-3-031-11938-5_17

covers on negativity. In fact this work grew out of conversations with T. Bauer, A. Küronya and S. Rollenske motivated by their work on the Bounded Negativity Conjecture. We are glad to acknowledge our debt to them for stimulating discussions and for raising the question in the first place.

1 Clusters and Enriques Diagrams

By a *surface* S, we mean a connected two-dimensional complex (analytic) manifold (so, smooth and irreducible). Unless otherwise stated we always work with the analytic topology.

Singularities of curves on a smooth surface S will be described in terms of their *clusters of singular points*, in the spirit of [1], i.e., taking into account the *infinitely near* singular points—those which have to be blown up in every embedded resolution. This description allows a convenient treatment of pullback curves. We begin by recalling the notions of infinitely near points and clusters.

Definition 1.1 ([1, 3.3]) Given a surface S and a point $p \in S$, denote $\pi_p : S_p \to S$ the blowing-up of S at p. Points on the exceptional curve $E_p = \pi^{-1}(p)$ are called *points in the first (infinitesimal) neighborhood of* p. Iteratively, a point q in the kth neighborhood of p is defined as a point in the first neighborhood of a point in the $(k-1)$th neighborhood of p. Note that in this case, the point in the $(k-1)$th neighborhood is uniquely determined; it is called *the immediate predecessor of* q. More generally, for every $1 \le i \le k-1$, q is in the ith neighborhood of a unique point in the $(k-i)$th neighborhood of p. The point p itself can be considered to be its 0th neighborhood.

On every blowing-up such as $\pi_p : S_p \to S$, it is convenient and natural to identify each point $q \in S$, $q \neq p$, with its unique preimage in S_p. To do such identifications consistently across different blowing-ups, and more generally across bimeromorphic models dominating S, we rely on *infinitely-near-ness*, a pre-order relation between points on such models. Points in the infinitesimal neighborhoods of $p \in S$ provide paradigmatic instances. The equivalence relation induced by the pre-order provides the desired identification of points, so the set of equivalence classes inherits a partial ordering by infinitely-near-ness.

A bimeromorphic map between surfaces is a proper holomorphic map $\pi : S_\pi \to S$ such that there exist proper analytic subsets $T \subset S$ and $T' \subset S_\pi$ such that π restricts to an isomorphism $S_\pi \setminus T' \to S \setminus T$.

Definition 1.2 If $\pi_1 : S_{\pi_1} \to S$, $\pi_2 : S_{\pi_2} \to S$ are bimeromorphic maps, $q_1 \in S_{\pi_1}$, and $q_2 \in S_{\pi_2}$, then q_2 is *infinitely near* to q_1 (we also write $q_2 \geq q_1$ and say q_1 precedes q_2) whenever there exist an open neighborhood $U \subset S_{\pi_2}$ of q_2 and a holomorphic map $\varpi : U \to S_{\pi_1}$, with $\varpi(q_2) = q_1$, such that the restriction $\pi_2|_U : U \to S$ factors as $\pi_2 = \pi_1 \circ \varpi$.

In particular, every point on a bimeromorphic model, $q \in S_\pi \to S$, is infinitely near to a unique point on S, namely $\pi(q)$. Obviously, if q is in the kth infinitesimal neighborhood of p for some $k \geq 1$, then $q \geq p$. Denote $\approx$ the equivalence relation induced by the pre-order, so that $q_1 \approx q_2$ if $q_1 \geq q_2$ and $q_2 \geq q_1$.

Proposition 1.3 ([5, 2.4]) *For every $p \in S$ and every $q \geq p$ there is a unique $n \geq 0$ and a unique point in the nth neighborhood of p equivalent to q.*

Definition 1.4 In the sequel, we shall identify equivalent points; thus for us a point infinitely near to p is by definition an equivalence class of points in bimeromorphic models of S mapping to p. Infinitely-near-ness is then a partial order on the set of points infinitely near to p. If q is a point in a bimeromorphic model of S, we denote the infinitely near point it determines by the same symbol q, recalling that equality of infinitely near points means equivalence of points in models of S.

Definition 1.5 *A cluster based at p* is a finite set of points K infinitely near to p such that, for every $q \in K$, if q' is a point infinitely near to p and q is infinitely near to q', then $q' \in K$. By assigning integral multiplicities $\nu = \{\nu_q\}_{q \in K}$ to the points of a cluster K one gets a *weighted cluster* (K, ν).

By Proposition 1.3 our notion of cluster agrees with the one in [1].

A curve C is said to go through the infinitely near point $q \in S_\pi \to S$ if its strict transform in S_π goes through q. The multiplicity of C at q, denoted $\operatorname{mult}_q C$, is defined as the multiplicity of its strict transform.

Let q, q' be two points equal or infinitely near to p. The point q is said to be *proximate* to q' if q is infinitely near to q' and the (strict transform of the) exceptional divisor of blowing up q' goes through q. A point $q \in S_\pi \to S$ in the kth neighborhood of $p = \pi(q)$ is said to be *free* if it is a smooth point of the reduced exceptional divisor $\pi^{-1}(p)$; otherwise it is called *satellite*. If q is free then it is proximate to exactly one point, namely its immediate predecessor in the $(k-1)$th neighborhood. If q is satellite then it belongs to exactly two components of the exceptional divisor, and thus it is proximate to exactly two points; one of them is its immediate prececessor q', and the other is a point q'' in the ℓth neighborhood for some $\ell < k-1$; moreover q' has to be proximate to q'' (otherwise the strict transforms of their exceptional divisors would not meet).

An infinitely near point $q \in S_\pi \xrightarrow{\pi} S$ is a singular point of C if C goes through q and the pullback divisor $\pi^*(C)$ is *not* a simple normal crossings divisor at q. Equivalently, q is a singular point of C if either $\operatorname{mult}_q C > 1$, q is a satellite point on C, or q precedes a satellite point on C.

Given a singular curve C in a smooth surface S, and a $p \in S$ a singular point of C, the set of all singular points of C infinitely near to p is a cluster $\operatorname{Sing}_p(C)$ [1, 3.7.1].

Sometimes a curve $B \subset S$ smooth at p is distinguished for some reason (for us, this will apply to the ramification/branch locus of a double cover). In that case, a point $q \in S_\pi \to S$ infinitely near to $p \in B$ is called B-satellite if the (reduced) divisor $\pi^{-1}(B)$ is singular at q; this happens if and only if $q \neq p$ belongs to B or is a satellite point. For a curve C through p, the cluster $\mathrm{Sing}_p(C+B)$ of singular points of the union of C and B consists exactly of the singular and B-satellite points of C (and p, if C is smooth at p).

Given a cluster K of points infinitely near to $p \in S$, one may *blow up all points in* K, as follows. First blow up S with center at p, then perform successive blowing-ups on the resulting surfaces, with centers which belong to K and are proper points of the surfaces obtained by previous blowing-ups. Subsequent centers may be chosen in any order compatible with the natural ordering by infinitely-near-ness (if q_1 precedes q_2, then q_1 must be blown up first); the final surface and bimeromorpic map obtained as the composition of all blowing-ups, which will be denoted $\pi_K : S_K \to S$, are independent on the order of these blowing-ups—up to unique S-biholomorphism [1, Proposition 4.3.2].

If $K = \mathrm{Sing}_p(C)$, then π_K is a minimal embedded resolution of the singularity of C at p. The multiplicities ν_q of C at the points of $\mathrm{Sing}_p(C)$ satisfy the *proximity inequalities*

$$\nu_q \geq \sum_{q' \text{ prox. to } q} \nu_{q'} \tag{1}$$

at every $q \in \mathrm{Sing}_p(C)$ (where the summation runs over all points proximate to q), see [1, Theorem 4.2.2], and the number $\rho_q = \nu_q - \sum_{q' \text{ prox. to } q} \nu_{q'}$, called the *excess multiplicity* at q, is the number of (transverse) intersections of the resolved curve $\tilde{C}$ with the exceptional component $\tilde{E}_q$.

In fact, every bimeromorphic map $\pi : S_\pi \to S$ whose exceptional locus is reduced to $p \in S$ is the blowing-up of all points in a convenient cluster: for every factorization of π as a finite sequence of point blowing-ups, the centers of the blowing-ups form a cluster K, and this cluster is independent of the factorization.

We denote by E_q (respectively, $\tilde{E}_q$) the pullback or total transform (respectively, the strict transform) in S_K of the exceptional divisor of the blowing-up centered at q. It is not hard to see that q_1 precedes q_2 if and only if $E_{q_2} - E_{q_1}$ is an effective divisor.

The set of points of a cluster K, equipped with the proximity relation, has an abstract combinatorial structure, which Enriques encoded in a convenient diagram, now called the *Enriques diagram* of the cluster (see [3, IV.I], [1]). It will be convenient for us to give a formal definition of Enriques diagrams along the lines of the one given by Kleiman and Piene in [4] (see also [7]).

A *tree* is a finite directed graph, without loops; it has a single initial vertex, or *root*, and every other vertex has a unique immediate predecessor. If p is the immediate predecessor of the vertex q, we say that q is a successor of p. An *Enriques diagram* is a tree with a binary relation between vertices, called *proximity*, which satisfies

1. The root is proximate to no vertex.
2. Every vertex except the root is proximate to its immediate predecessor.
3. No vertex is proximate to more than two vertices.
4. If a vertex q is proximate to two vertices then one of them is the immediate predecessor of q, and it is proximate to the other.

5. Given two vertices p, q with q proximate to p, there is at most one vertex proximate to both of them.

The vertices which are proximate to two points are called *satellite*, the other vertices are called *free*. We usually denote the set of vertices of an Enriques diagram $\mathbb{D}$ with the same letter $\mathbb{D}$. We shall consider two trees with the same number of vertices and satisfying the same proximity relations to be the same Enriques diagram.

To show graphically the proximity relation, Enriques diagrams are drawn according to the following rules:

1. If q is a free successor of q' then the edge going from q' to q is smooth and curved and if q' is not the root, it has at p the same tangent as the edge joining q' to its predecessor.
2. The sequence of edges connecting a maximal succession of vertices proximate to the same vertex q' are shaped into a line segment, orthogonal to the edge joining q' to the first vertex of the sequence.

The Enriques diagram of a cluster K is then the unique Enriques diagram for which there is a bijection from its set of vertices to K that preserves the proximity relation (see [1, p. 98]). If (K, ν) is a weighted cluster (for instance, the weights may be the multiplicities of some particular curve C at the points of K) this can be represented in the diagram by attaching the weight of each point to the corresponding vertex. Two curves are equisingular (or topologically equivalent) at p if and only if their clusters of singular points at p, weighted with their multiplicities, have the same Enriques diagram, see [1, Chap. 3] (the equisingularity class is also determined by the Enriques diagram—without the weights—by adding some nonsingular points, namely those which are not infinitely near to another nonsingular point [1, 3.9], but we don't use this method here.)

Example 1.6 We will use as running example a curve C with an E_{20} singularity, i.e., a curve topologically equivalent to $x^3 - y^{11} = 0$. Its embedded resolution requires blowing up 6 points; at the first three C has a triple point, at the fourth it has multiplicity 2 ($\tilde{C}$ has an ordinary cusp tangent to the preceding exceptional divisor), and the last two points are simple satellite points. See Fig. 1 for its weighted Enriques diagram, and also the possible Enriques diagrams with respect to a given smooth curve B (which will stand for a branch locus later on). The local intersection multiplicity of an E_{20} singularity with a smooth curve may be (a) 3, (b) 6, (c) 9, or (d) 11, depending on how many infinitely near points are shared by B and C.

1.1 The Algorithm

Input: Our algorithm to determine the equisingularity type of pullbacks under double covers takes as input the Enriques B-diagram $(\mathbb{D}, \nu)$ of $\mathrm{Sing}_{p'}(B + C)$ where C is a curve in the target surface S' of a double cover $f : S \to S'$, B is the branch curve

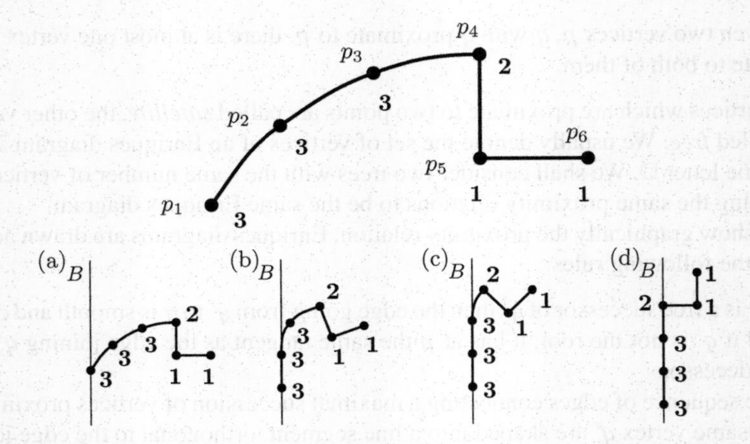

Fig. 1 Above, the Enriques diagram [1, 3.9] of the cluster of singular points of an E_{20} singularity as in Example 1.6. Each vertex in the diagram corresponds to one of the points, with each vertex joined to its immediate predecessor by an edge; edges are curved for free points, and straight segments for satellites, to represent the rigidity of their position. If there is a distinguished smooth curve B through p, the corresponding B-satellite points are also represented on a straight segment; the bottom part of the figure shows the four possibilities for an E_{20} singularity

$B \subset S'$, and ν_q is the multiplicity of C at $q \in \mathrm{Sing}_{p'}(B + C)$. This contains all the information on the proximities between points blown up in an embedded resolution of $C + B$, weighted with the multiplicities of the curve C, and the datum of which of these points belong to B.

Output: The output is given as the Enriques R-diagram of the singularity of the pullback $f^*(C)$ in the source surface S, with respect to the ramification curve R. Thus, we obtain the equisingularity type of $f^*(C)$ and also its relative position with respect to R.

Algorithm:

Step 1: **Coloring.** Some special vertices on the diagram are given the red and green colors (see Sect. 2 for the geometric meaning) as follows:

(a) The root is colored red;
(b) every B-free successor of a red vertex is colored red;
(c) a B-satellite successor of a red vertex is

i. colored red if it is proximate to a green vertex,
ii. colored green otherwise;

(d) every B-satellite successor of a green vertex is colored red;
(e) all other vertices are left black.

Step 2: **Completion.** For every red vertex which does *not* have a green successor, add one, as a B-satellite not proximate to a previous green vertex (there is always a unique way to do this). The new vertex is weighted with multiplicity 0.

Step 3: **Splitting.** Duplicate every black subdiagram (which always follows from a green vertex).

Step 4: **Mutations.** For every green vertex g which has no colored vertex after it, call r the (red) immediate predecessor and

(a) detach g from the straight segment containing r and g; if there are more vertices after g on that segment, keep them on the segment, the first of them being now joined directly to r;
(b) if r is the root or B-free, reattach g to r with a curved edge; otherwise, insert g in the straight segment emanating from r in the orthogonal direction to that previously joining g and r, immediately after r (if there is no such orthogonal segment, create it with g as its end);

For all other vertices infinitely near to r, the properties of being proximate to g and r have to be exchanged, which explicitly becomes

(c) detach every curved edge emanating from g (resp. from r) and reattach it to r (resp. to g);
(d) if there is a straight segment emanating from g and orthogonal to the edge joining r and g, and v is the first vertex after g on this segment, the two straight directions after v have to be exchanged, i.e., all edges after v prolonging the segment gv have to be moved to the orthogonal half line emanating from v, and conversely.
(e) change the weights, so that now r has multiplicity $\nu'_r = \nu_r + \nu_g$ and g has multiplicity $\nu'_g = \nu_r - \nu_g$;
(f) remove the coloring of r and g, and continue as long as there are colored vertices left.

Step 5: **Cleanup.** Remove any vertices of weight 0 and rename B as R (Fig. 2).

In Sect. 3, we shall prove that the resulting diagram corresponds to the equisingularity type of $f^*(C)$ (Figs. 3 and 4).

1.2 Infintely Near Points Verses Divisorial Valuations

Every point q infinitely near to p determines a valuation v_q on the local ring $\mathcal{O}_{S,p}$ (which is a ring of convergent power series in two variables). Let K_q be the cluster formed by all infinitely near points to which q is infinitely near and q itself (i.e., the minimal cluster containing q), and let $\pi_q : S_{K_q} \longrightarrow S$ be the surface obtained blowing up all points up to q. Then, for every $\phi \in \mathcal{O}_{S,p}$, $v_q(\phi)$ is the order of vanishing along $E_q \subset S_{K_q}$ of $\pi_q^*(\phi)$. The valuation v_q is *divisorial*, i.e., it is discrete and its residual field κ_q has transcendence degree 1 over $\mathbb{C}$.

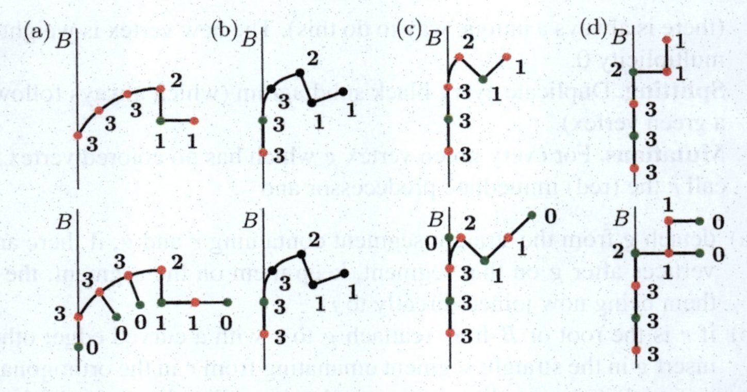

Fig. 2 The first two steps (coloring and completion) applied to the four possibilities for an E_{20} singularity

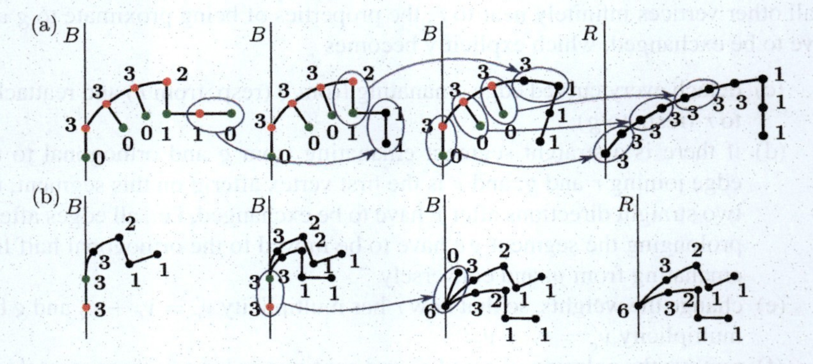

Fig. 3 Steps 3–5, applied to cases **a** and **b** of an E_{20} singularity. Mutations are indicated with arrows. If the intersection multiplicity of C with the branch curve is 3, the pullback has the Enriques diagram of an E_{42} singularity (equisingular to $x^3 - y^{22} = 0$), whose intersection multiplicity the ramification curve equals three. On the other hand, if the intersection multiplicity of C with the branch curve is 6, the pullback has two branches with distinct tangent directions, each equisingular to an E_{14} singularity (like $x^3 - y^8 = 0$)

The residual field κ_q can be identified with the field of rational functions of E_q. Indeed, the pullback to S_{K_q} of a meromorphic function with value zero restricts to E_q as a nonzero rational function, and two such functions agree modulo $\mathfrak{m}_{v_q}$ if and only if their difference has positive value, i.e., vanishes on E_q. Compatible generators of these fields (over $\mathbb{C}$) can be described as follows. Let q_1, q_2 be distinct free points in the first neighborhood E_q of q. By the results of [1, Chap. 4] there exist germs of curve C_1, C_2 in a neighborhood of p such that the strict transform $\tilde{C}_i$ on S_{K_q} of C_i intersects the exceptional divisor $\pi_q^{-1}(p)$ at q_i only and transversely. Let $\phi_1, \phi_2 \in \mathcal{O}_{S,p}$ be local equations at p of these germs. Being analytically irreducible, all proximity inequalities (1) are in fact equalities at each point in K_q different from q, hence C_1 and C_2 have the same multiplicity at every point of K_q, and the same value

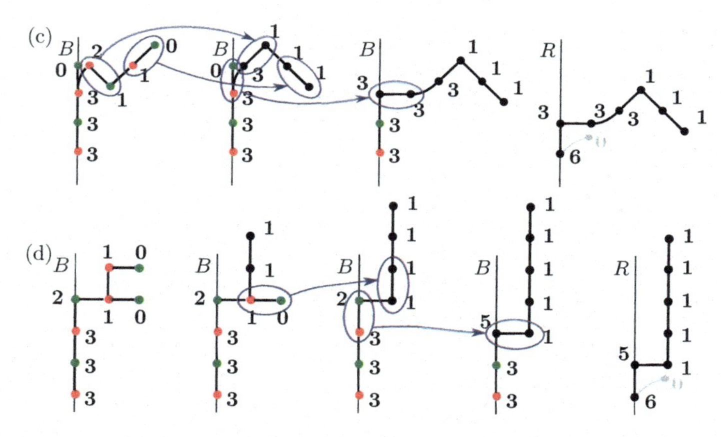

Fig. 4 Steps 3–5, applied to cases **c** and **d** of an E_{20} singularity. If the intersection multiplicity of C with the branch curve is 9, the pullback has the Enriques diagram of a singularity with two characteristic exponents, $1/2$ and $5/6$, tangent to the ramification curve. Finally, if the intersection multiplicity of C with the branch curve is 11, the pullback is equisingular to the curve $x^6 - y^{11} = 0$, tangent to the ramification curve as well

$v = v_q(\phi_1) = v_q(\phi_2)$. If $\tilde{x}$ is a local equation for E_q in a neighborhood of q_1, q_2, then $\pi_q^*(\phi_i)/\tilde{x}^v$ is a local equation of $\tilde{C}_i$, and therefore the restriction of $\pi_q^*(\phi_1/\phi_2)$ to E_q is a coordinate that generates $K(E_q)$ over $\mathbb{C}$. On the other hand, the class of ϕ_1/ϕ_2 generates κ_q over $\mathbb{C}$.

The classification of valuations of $\mathcal{O}_{S,p}$ (see [1, Chap. 8], or [8]) moreover shows that for every divisorial valuation v whose center is p (i.e., such that the functions $\phi \in \mathcal{O}_{S,p}$, with positive value $v(\phi) > 0$ are exactly those with $\phi(p) = 0$) there exists a point q infinitely near to p such that $v = v_q$.

2 Double Covers

Assume now that a morphism between surfaces $f : S \to S'$ is given, and $p \in S$ is a point where the local degree of f is 2 (i.e., every point in a small neighborhood of $p' = f(p)$ has two preimages near p). It is not hard to see that, in this case, there exist local (analytic) coordinates (x, y) at p and (u, v) at p' such that the morphism is given by $f(x, y) = (x^2, y)$. Restricting to suitable neighborhoods of p and p' we may assume (and we shall do so without further notice) without loss of generality that both S, S' are (isomorphic to) suitable open subsets of $\mathbb{C}^2$ where $f : S \to S'$ is the double cover given $f(x, y) = (x^2, y)$. The ramification divisor R is then given by the equation $x = 0$, and its image is the branch curve $B : u = 0$. Our goal is to describe the singularity at p of any curve $f^*(C)$ where $p' \in C \subset S'$.

All curves in the pencil defined locally at p by $\{C_\alpha : \alpha_1 x^2 + \alpha_2 y = 0\}$, $\alpha = \alpha_1/\alpha_2 \in \mathbb{C} \cup \{\infty\}$, formed by the pullbacks of the curves $\alpha_1 u + \alpha_2 v = 0$, go through

p and the point in its first neigbourhood in the direction of the curve $y = 0$ (which is transverse to R). These two points form the *cluster of base points* of f, denoted by $BP(f)$ [2]. Since they appear with multiplicity 1 in all germs of the pencil but $x^2 = 0$, these are the multiplicities in the weighted cluster of base points $\mathcal{BP}(f) = (BP(f), \{1, 1\})$.

Let $\pi_{p'} : S'_{p'} \to S'$ be the blowing-up centered at p'. Then $BP(f)$ is the cluster of points which need to be blown up to resolve the indeterminacy at p of the "meromorphic map" $\tilde{f} = \pi_{p'}^{-1} \circ f : S \dashrightarrow S_{p'}$ [5, Lemma 3.2].

Let $\mathcal{K} = (K, \mu)$ be an arbitrary weighted cluster of points infinitely near to the target point p' of f. Generalizing the cluster of base points of f, [5] defined a *pullback weighted cluster* $f^*(\mathcal{K}) = (f^*(K), f^*\mu)$ of points infinitely near to p in order to describe the singularities of pullbacks of general curves going through $\mathcal{K}$. Let $\pi_K : S'_K \to S'$ be the blowing up of all points of K, and let $D_{\mathcal{K}} = \sum_{q \in K} \mu_q E_q$ be the divisor on S'_K associated to the weights μ. The pullback cluster $f^*(K)$ is defined as the cluster of all points which need to be blown up to resolve the indeterminacy at p of the composition $\pi_{K'}^{-1} \circ f$, and its multiplicities $f^*\mu$ as determined by

$$\tilde{f}^*(D_{\mathcal{K}}) = \sum_{q \in f^*(K)} (f^*\mu)_q E_q.$$

It was proved in [5] that the number of points in $f^*(K)$ is bounded above by the number of points in K multiplied by the local degree of f. In our context this local degree is 2, so $f^*(K)$ has at most twice as many points as K, but in this particular case we are able to go further since our algorithm determines this number of points exactly.

2.1 Effect on Valuations

Let now v be a valuation of $\mathcal{O}_{S,p}$ centered at p. Composition with f^* determines a valuation on the image surface, denoted as $f(v)$:

$$f(v)(\psi) = v(f^*(\psi)) \quad \forall \psi \in \mathcal{O}_{S',p'}.$$

The center of $f(v)$ is p'; the value group $\Gamma_{v(f)}$ is a subgroup of the value group Γ_v of v, and the residual field $\kappa_{f(v)}$ is a subfield of κ_v. The index of $\Gamma_{v(f)}$ in Γ_v is called *ramification index* $\mathbf{e}_f(v)$ of f at v, and the degree of the extension $\kappa_{f(v)}/\kappa_v$ is called *inertia degree* $\mathbf{f}_f(v)$. By [6, 6.2, Theorem 3], we have for every valuation v' on $\mathcal{O}_{S',p'}$ an equality

$$2 = \deg_p(f) = \sum_{f(v)=v'} \mathbf{e}_f(v)\mathbf{f}_f(v). \tag{2}$$

In particular, the extension $\kappa_{f(v)}/\kappa_v$ is finite, and v is divisorial if and only if $f(v)$ is divisorial.

The field extension $\kappa_{f(v)}/\kappa_v$ has a geometric counterpart which we explain next. Let q be a point infinitely near to p and let v_q the associated divisorial valuation. Let q' be the point infinitely near to p' such that $f(v_q) = v_{q'}$. Consider the following commutative diagram, obtained blowing up the clusters $K_{q'}$ (the minimal cluster containing q') and its pullback.

$$\begin{array}{ccc} S_{f^*(K_{q'})} & \xrightarrow{\tilde{f}} & S'_{K_{q'}} \\ {\scriptstyle \pi_{f^*(K_{q'})}}\big\downarrow & & \big\downarrow{\scriptstyle \pi_q} \\ S & \xrightarrow[f]{} & S' \end{array} \tag{3}$$

The center of the valuation $v_{q'}$ on $S'_{K_{q'}}$ is the divisor $E_{q'}$; since $f(v_q) = v_{q'}$, necessarily the center of v_q on $S_{f^*(K_{q'})}$ must be a divisor, i.e., q belongs to the pullback cluster $f^*(K_{q'})$, and $\tilde{f}(\tilde{E}_q) = E_{q'}$. The restriction of $\tilde{f}$ to $\tilde{E}_q$ induces the field extension

$$\kappa_{q'} \cong K(E_{q'}) \overset{f^*}{\hookrightarrow} K(\tilde{E}_q) \cong \kappa_q .$$

The map between the first neighborhoods of q and q' was introduced and studied by Casas-Alvero in [2] (in all generality for germs of holomorphic maps between surfaces, not just at points of degree 2). The relationship with the construction given there is as follows. Consider a pencil of analytically irreducible germs of curve at S whose strict transforms have a variable tangent at q (i.e., they intersect transversely E_q at varying points); the images of these strict transforms by $\tilde{f}$ meet $E_{q'}$ at the points determined by $\tilde{f}|_{\tilde{E}_q}$. This shows that $K_{q'}$ is the cluster called *trunk of q* in [2] and the restriction of $\tilde{f}$ to the exceptional divisor of q is the *tangent map*. The *multiplicity of the trunk* in [2] equals the ramification index $\mathbf{e}_f(v_q)$.

Proposition 2.1 *Let $f : S \to S'$ be a morphism, let $p \in S$ a point where the local degree of f is 2, let $p' = f(p)$ and let q' be a point infinitely near to p'. Let as above $K_{q'}$ stand for the minimal cluster of points infinitely near to p' containing q', and keep the notations of diagram Eq. 3. Then exactly one of the following alternatives holds:*

1. *There are exactly two points, q_1, q_2 infinitely near to p, such that $f(v_{q_1}) = f(v_{q_2}) = v_{q'}$. In this case, the ramification index and inertia degree equal 1 for both preimages.*
2. *There is exactly one point, q infinitely near to p, such that $f(v_q) = v_{q'}$, $\mathbf{e}_f(v_q) = 2$ and $\mathbf{f}_f(v_q) = 1$.*
3. *There is exactly one point, q infinitely near to p, such that $f(v_q) = v_{q'}$, $\mathbf{e}_f(v_q) = 1$ and $\mathbf{f}_f(v_q) = 2$.*

Moreover, the tangent maps are isomorphisms onto $\tilde{E}_{q'}$ except when $\mathbf{f}_f(v_q) = 2$; in this latter case the tangent map $\tilde{E}_q \to E_{q'}$ is a double cover.

Depending on which of the alternatives above holds, we say that q' is *split*, *ramified* or *inert* under f, respectively. When q' is inert, one can also say that the exceptional component $\tilde{E}_q$ is *folded* over E'_q.

Proof Every valuation on $\mathcal{O}_{S',p'}$ has some extension to a valuation on the field of fractions, and hence to the field of fractions of $\mathcal{O}_{S,p}$, which is an algebraic extension of the former, by (2). Now, since the center of $v_{q'}$ is p', and p is an isolated preimage of p', every extension of $v_{q'}$ will have center equal to p, and since $v_{q'}$ is divisorial, every valuation v such that $f(v) = v'$ must be divisorial.

The three possible values of $\mathbf{e}_f(v)$, $\mathbf{f}_f(v)$ which give a sum equal to 2 in (2) lead to the three claimed cases.

The claim on the tangent maps follows from the connection, described above, with the field extension $\kappa_q/\kappa_{q'}$. □

2.2 The Main Trunk and Its Pullback

Let us now do some explicit computations on a degree 2 map, beginning with its cluster of base points $BP(f) = f^*(\{p'\})$. We assume without loss of generality that S, S' are open neighborhoods of the origin in $\mathbb{C}^2$, with $f(x, y) = (x^2, y)$. The cluster of base points is made up by the points infinitely near to p shared by all but finitely many of the curves $\lambda x^2 + \mu y = 0$, and these are exactly the point p and the point in its first neighborhood in the direction of $y = 0$, which we call q. Note that this is the "vertical" direction of the double cover, i.e., the kernel of $df : T_pS \to T_{p'}S'$.

Thus the morphism f lifts to $\tilde{f} : S_{BP(f)} \to S'_{p'}$ (see Fig. 2). We shall see in a moment (and it can also be checked directly using the expression of f in coordinates) that $\tilde{f}$ maps E_q isomorphically onto $E_{p'}$, whereas the strict transform $\tilde{E}_p$ of the first neighborhood of p contracts to the B-satellite point $q' = E_{p'} \cap \tilde{B}$ (Fig. 5).

Next we compute the image of v_p. Writing in coordinates as above $f(x, y) = (x^2, y)$ it is clear that for $\psi(u, v) = \sum a_{ij}u^iv^j$ one has

$$f(v_p)(\psi) = v_p(\psi \circ f) = v_p(\sum a_{ij}x^{2i}y^j) = \min\{2i + j \mid a_{ij} \neq 0\}.$$

This is the monomial valuation with $f(v_p)(u) = 2$ and $f(v_p)(v) = 1$, i.e., the divisorial valuation corresponding to the point q' in the first neighborhood of p' in the direction of the branch curve $B : u = 0$. So the trunk of f is the cluster $T = K_{q'} = (p', q')$ with multiplicities equal to 1.

This can be seen also with the algorithm of [2], which consists in computing the direct images of the curves $C_{ab} : ax + by = 0$, a pencil through the base point p. Each curve in the pencil can be parametrized as $t \mapsto (bt, -at)$, so its image by f, parametrized as $t \mapsto (b^2t^2, -at)$, has equation $a^2u - b^2v^2 = 0$, indeed a smooth curve through p' and q' (which has intersection multiplicity exactly 2 with $u = 0$ if $b \neq 0$). This moreover computes the tangent map $E_p \to E_{q'}$: the point with

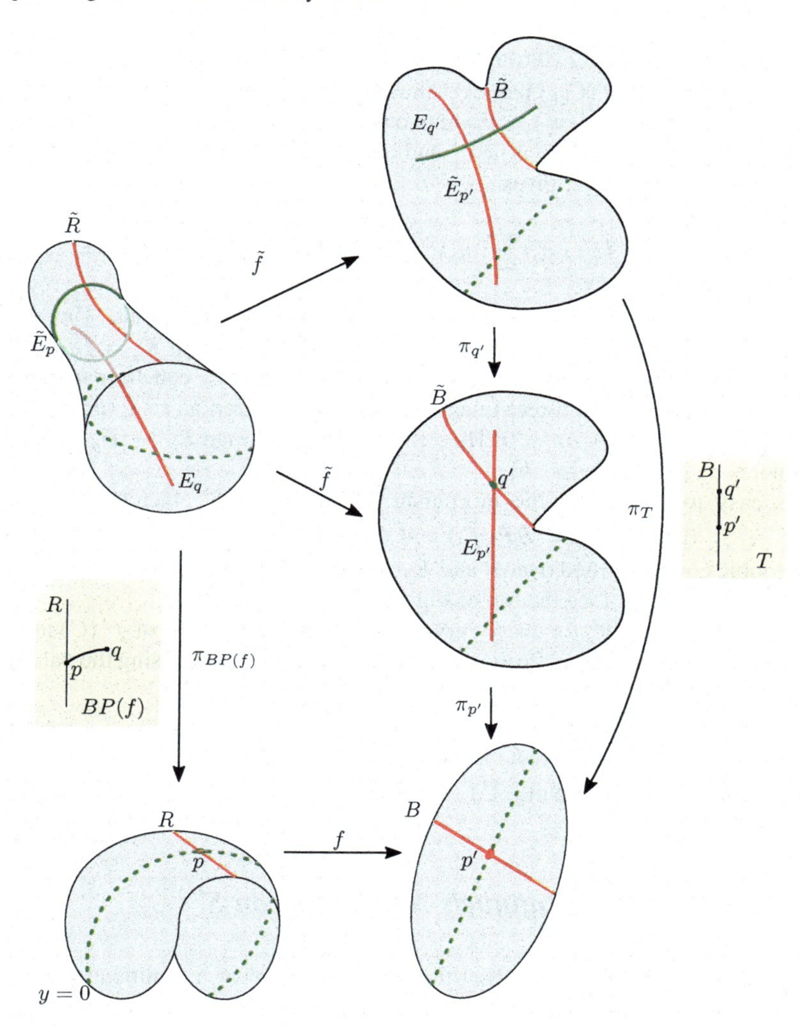

Fig. 5 The left part of the diagram depicts blowing up the cluster of base points $BP(f) = \{p, q\}$ of a morphism of local degree 2, which lifts to a morphism $\tilde{f}$ to the blowing up of the image point $p' = f(p)$ ($\tilde{f}$ contracts the exceptional component $\tilde{E}_p$ to q', the B-satellite point in the first neighborhood of p'.) The top right surface represents blowing up the trunk $T = \{p', q'\}$. Its pullback $f^*(T) = \{p, q\}$ coincides with the cluster of base points, and the double cover f lifts to a double cover $\tilde{f}$. Red lines represent the ramification/branch curves; green lines represent folded curves and their images

coordinates $[a : -b]$ in E_p (through which $\tilde{C}_{ab}$ passes) maps to the point $[a^2 : b^2]$ in $E_{q'}$ (through which $\tilde{f}(\tilde{C}_{ab})$ passes). Thus $\mathbf{f}_f(v_p) = 2$ and q' is an inert point.

We can also compute $f(v_q) = v_{p'}$ in two ways; the first method, directly computing the value of $\psi(u, v) = \sum a_{ij} u^i v^j$ and using that v_q is the monomial valuation with $v_q(x) = 1$, $v_q(y) = 2$, gives

$$f(v_q)(\psi) = v_q(\psi \circ f) = v_q(\sum a_{ij} x^{2i} y^j) = \min\{2i + 2j \mid a_{ij} \neq 0\} = 2v_{p'}(\psi)$$

which tells us also that $\mathbf{e}_f(v_q) = 2$, so p' is a ramification point. The second method, using the image of a pencil $C_{ab} : ax^2 + by = 0$ through $K_q = \{p, q\}$, confirms that the tangent map is an isomorphism. Indeed, C_{ab} can be parametrized as $t \mapsto (bt, -bat^2)$ so its direct image can be parametrized as $t \mapsto (b^2t^2, -bat^2)$ which has equation $au + bv = 0$. Therefore, the tangent map $E_q \to E_{p'}$ is given in coordinates by $[a : b] \mapsto [a : b]$.

It is easy to check that the morphism $\tilde{f} : S_{BP(f)} \to S'_{p'}$ factors through $\bar{f} : S_{BP(f)} \to S'_T$ (i.e., $f^*(T) = BP(f)$) and our computation shows that $\bar{f}$ is actually a double cover ramified over $\tilde{B}$ and $\tilde{E}_{p'}$, but not over the free points of $E_{q'}$ which is covered 2 : 1 by E_p (see the top of Fig. 2).

We can now compute, for every curve $C \subset S'$, the multiplicity of $f^*(C)$ at p and q, using [2, Theorem 4.1 and Proposition 13.3] (this can be done using the valuations v_p and v_q as well). Namely,

$$\begin{aligned} \mathrm{mult}_p(f^*(C)) &= \mathrm{mult}_{p'} C + \mathrm{mult}_{q'} C, \\ \mathrm{mult}_q(f^*(C)) &= 2\mathrm{mult}_{p'}(C) - \mathrm{mult}_p(f^*(C)) = \mathrm{mult}_{p'} C - \mathrm{mult}_{q'} C. \end{aligned} \tag{4}$$

2.3 *Classification of Infinitely Near Points on S'*

In the previous section we saw that the image point p' of a ramification point of degree 2 is always ramified, and that the B-satellite point in its first neighborhood is inert. We can now classify every point infinitely near to p' as ramified, inert or split, by induction, using the former statement as induction step, thanks to the fact that the lift $\bar{f}$ is again a double cover, so that every point on $\tilde{B}$ or $\tilde{E}_{p'}$ is ramified and every point of S'_T away from these curves is split.

Theorem 2.2 *Let $f : S \to S'$ be a morphism, let $p \in S$ a point where the local degree of f is 2, let $p' = f(p)$. Then, the behavior of points infinitely near to p' under f is as follows:*

1. *p' is ramified;*
2. *every B-free point in the first neighborhood of a ramified point is ramified;*
3. *a B-satellite point in the first neighborhood of a ramified point is*

   (a) *ramified if it is proximate to an inert point,*
   (b) *inert otherwise;*

4. *every B-free point in the first neighborhood of an inert point is split;*
5. *every B-satellite point in the first neighborhood of an inert point is ramified;*
6. *every point infinitely near to a split point is split.*

Proof Assume r' is a point infinitely near to p'; we will prove the claims by induction on the number $k = |K_{r'}|$ of points of the cluster $K_{r'}$ (the minimal cluster containing r'). If $r' = p'$ (i.e., $k = 1$) or $r' = q'$, the claim has already been proved. If r' is a B-free point in the first neighborhood of p', then r' is the image by $\bar{f}$ of a unique point r on E_q, and r is a point of $S_{BP(f)}$ where $\bar{f}$ has degree 2. So, applying the $k = 1$ case to $\overline{f}$ we see that r' is ramified, completing the proof for the $k = 2$ case. The same argument shows that the B-satellite points $\tilde{B} \cap E_{q'}$ and $\tilde{E}_{p'} \cap E_{q'}$ in the first neighborhood of the inert point q' are ramified. On the other hand, since $\bar{f}$ is an isomorphism locally near every point not on $\tilde{R}$ or E_q, it is clear that every free point in the first neighborhood of the (inert) point q' splits, and the same happens on the first neighborhood of every split point.

It only remains to prove the claim when r' is a point infinitely near to a ramified point s' on S'_T, which either belongs to $E_{p'}$ or equals the B-satellite point $\tilde{B} \cap E_{q'}$. In this case s' is the image by $\bar{f}$ of a unique point s on E_q or of the R-satellite point $\tilde{R} \cap \tilde{E}_p$, and s is a point of $S_{BP(f)}$ where $\bar{f}$ has degree 2. Moreover, the minimal cluster of points infinitely near to s' that contains r' is $K'_{r'} = K_{r'} \setminus \{p', q'\}$ so it has either $k - 2$ or $k - 1$ points (depending on whether q' belongs to $K_{r'}$ or not). Then the claims follow by the induction hypothesis applied to $\bar{f}$ (note that the third claim may be equivalently stated saying that the intersection of two ramification exceptional divisors is inert, whereas the intersection of a ramified and an inert divisor is ramified). □

Corollary 2.3 *The B-satellite point on the first neighborhood of a B-free ramified point is inert. Among the 2 B-satellite points on the first neighborhood of a B-satellite ramified point, one is ramified and the other is inert. In particular, there is exactly one inert point on the first neighborhood of each ramified point.*

Proof It follows immediately from the Theorem, observing that a B-satellite ramified point r is proximate to a ramified and an inert point. The corresponding satellites on the first neighborhood of r are inert and ramified, respectively. □

Corollary 2.4 *Every inert point q' is a satellite, and the unique preimage v_q of $v_{q'}$ is the valuation of a satellite point as well. Moreover, the tangent map $E_q \to E_{q'}$ is ramified exactly at the two satellite points on the first neighborhood of q, which map to the two satellite points on the first neighborhood of q'*

Remark 2.5 The coloring step of the algorithm described in Sect. 1.1 paints in red those vertices corresponding to ramified points, and in green those vertices corresponding to inert points.

3 Justification for the Algorithm

Let now C be a curve in S', going through p', and let $K = \mathrm{Sing}_{p'}(C + B)$ be the cluster of singular points of $C + B$. Consider the weighted cluster $\mathcal{K} = (K, \nu)$ where $\nu_{q'} = \mathrm{mult}_{q'}(C)$ for each $q' \in K$, and let $(\mathbb{D}, \nu)$ be its Enriques diagram. Since the blowing-up π_K is an embedded resolution of the singularities of $C + B$, the strict transform of C in S'_K is nonsingular *and transverse to the branch locus* of the lift $\tilde{f}$. Therefore its pullback in $S_{f^*(K)}$ is smooth and transverse to the ramification locus, so $\pi_{f^*(K)}$ is an embedded resolution of $f^*(C)$. So, we only need to show that the output of the algorithm is the diagram of the subcluster of $f^*(K)$ where $f^*(C)$ has positive multiplicity, weighted with this multiplicity.

If $K \subset \{p', q'\}$ (in the notation of the previous section), then the completion of the diagram of K consists exactly of one red and one green vertex, on B, the output of the algorithm is the diagram with two vertices the second of which is R-free, and formula (4) proves that the weights given by the algorithm coincide with the multiplicities of $f^*(C)$ at p and q. Thus, the algorithm is correct for diagrams with at most two vertices both on B, in particular it is correct for diagrams with one vertex. Moreover, formula (4) also shows that the multiplicities at p, q of $f^*(C)$ are the ones assigned by the algorithm to the two vertices coming from the red-green pair of vertices corresponding to p', q'.

Now we use induction on the number of points $|K|$. Let C' be the strict transform of C on the blowing-up S'_T of the trunk, described above. It may have several singular points p'_i on $E_{q'}$ or $\tilde{E}_{p'}$, and at each of them the cluster $K_i = \mathrm{Sing}_{p_i}(C')$ consists of strictly less than $|K|$ points infinitely near to p_i.

Each free point p_i on $E_{q'}$ has two distinct preimages on $S_{BP(f)}$, both of them free points on the first neighborhood of p at which $\bar{f}$ is a local isomorphism. The clusters (K_i, ν) have Enriques diagrams $\mathbb{D}_i$ which are subclusters of $\mathbb{D}$ following free vertices connected to the green vertex corresponding to q' and hence non-colored (black) in the algorithm, and by the local isomorphism appear twice as Enriques diagrams of $\bar{f}^*(C')$ at free points in the first neighborhood of p. Thus each such $\mathbb{D}_i$ appears twice as a subdiagram of $f^*(C)$, linked to the root by a curved edge, which is what the algorithm outputs after the splitting and mutation steps.

For points p_i on $\tilde{E}_{p'}$ or equal to the B-satellite point $E_{q'} \cap \tilde{B}$, each of them has a unique preimage on $S_{BP(f)}$, and $\bar{f}$ has local degree 2 there. So by the induction hypothesis, the singularity type of $\bar{f}^*(C')$ and its intersection with the ramification divisor are described by applying the algorithm to the subdiagrams $\mathbb{D}_i$, which are subdiagrams of the output of the algorithm applied to $\mathbb{D}$. For free points on $\tilde{E}_{p'}$ the output of the algorithm applied to the subdiagrams is simply a subdiagram of the output of the whole algorithm, linked to the second vertex (corresponding to q) by a curved edge (**Step 4**:c), which coincides with the corresponding subdiagram of $\mathrm{Sing}_p(f^*(C))$ as stated in the algorithm.

The only remaining detail to take care of are the singularities of C' at the B-satellite points on $E_{q'}$, which become singularities at the R-satellite points on $\tilde{E}_p$ by pullback. Here we notice that $\tilde{E}_p$ determines the "vertical" direction with respect to the double cover, i.e., the second point in the cluster of base points of $\tilde{f}$ at $E_q \cap \tilde{E}_p$ and $\tilde{R} \cap \tilde{E}_p$ belongs to the strict transform of $\tilde{E}_p$. Hence, the mutation of the subsequent green vertex in the algorithm is not a free vertex, but a satellite vertex. The exact proximities satisfied by this vertex are the ones described in **Step 4**:b of the algorithm because the exceptional divisors E_p, E_q are mapped to $E_{q'}$ and $E_{p'}$, respectively, i.e., the ordering is swapped, and this is encoded in exchanging orthogonal directions in the diagram; the same applies to further satellites, where the order exchange corresponds to the exchange of orthogonal directions in **Step 4**:d. This finishes the proof.

3.1 Additional Remarks

In applications, one might be interested in further information than the equisingularity type. For instance, the involution ι of S induced by the double cover f induces an involution on the set of divisorial valuations v_q which exchanges the preimages of split points. This involution respects proximities, and thus gives an involution on the Enriques diagram of the pullback of $\mathrm{Sing}_{p'}(C+B)$ for every $C \subset S'$. It is not hard to see that, in terms of the algorithm, this involution simply exchanges the duplicated black parts and leaves invariant all vertices coming (by mutation) from colored vertices, so one might want to omit the last step in the mutation part and keep the colors (maybe exchanging red and green to keep the meaning of ramified/inert) and make the involution 'visible'. These remarks allow to characterize the cases in which the pullback of a curve splits (locally) as two curves, $f^*(C) = C_1 + C_2$, as follows.

Proposition 3.1 *Let $f : S \to S'$ be a morphism, let $p \in S$ a point where the local degree of f is 2, let $p' = f(p)$, and let $C \subset S'$ be a curve through p', and let ι be the associated involution of S (defined in a neighborhood of p).*

Then, the curve C splits completely under pullback by f near p (i.e., $f^*(C) = C_1 + C_2$ where C_1, C_2 are curves in a neigborhood of p with no common components and $\iota(C_i) = C_j$) if and only if, after the coloring step in the algorithm above applied to the Enriques diagram of $K = \mathrm{Sing}_{p'}(B+C)$, no red vertex has positive excess.

Moreover, in that case the local intersection multiplicity of C_1 and C_2 is

$$I_p(C_1, C_2) = \sum_{q \in \mathbb{D} \text{ colored}} \frac{v_q^2}{2}.$$

Proof Since the involution fixes the ramification points, if C splits then its strict transform in S'_K does not intersect any branch divisor (which are those components $\tilde{E}_{q'}$ whose corresponding vertex is red), and hence all excesses at red vertices vanish. Conversely, if all excesses at red vertices vanish, $\tilde{C}$ intersects only exceptional components at points where the lifted double cover $\tilde{f} : S_{f^*(K)} \to S'_K$ is unramified, so each branch of C has two branches over it, exchanged by the involution.

In that case, the common points on C_1, C_2 infinitely near to p are exactly the points coming by mutation from colored vertices, and both C_i have the same multiplicity at each such point. Then the local intersection multiplicity can be computed with the aid of M. Noether's formula [1, Theorem 3.3.1] as

$$I_p(C_1, C_2) = \sum_{q \in f^*(K) \text{ colored}} \left(\frac{\nu'_q}{2}\right)^2 = \sum_{(r,g) \text{ pair } \subset \mathbb{D}} \left(\frac{\nu_r + \nu_g}{2}\right)^2 + \left(\frac{\nu_r - \nu_g}{2}\right)^2 =$$
$$\sum_{(r,g) \text{ pair } \subset \mathbb{D}} \frac{2\nu_r^2 + 2\nu_g^2}{4} = \sum_{q \in \mathbb{D} \text{ colored}} \frac{\nu_q^2}{2}.$$

□

References

1. E. Casas-Alvero, Singularities of Plane Curves, London Math. Soc. Lecture Notes Series, vol. 276 (Cambridge University Press, 2000)
2. E. Casas-Alvero, Local geometry of planar analytic morphisms. Asian J. Math. **11**(3), 373–426 (2007). https://doi.org/10.4310/AJM.2007.v11.n3.a3
3. F. Enriques, O. Chisini, *Lezioni sulla teoria geometrica delle equazioni e delle funzioni algebriche., Collana di Matematica [Mathematics Collection]*, vol. 5. Nicola Zanichelli Editore S.p.A. (Bologna 1985). Reprint of the 1915, 1918, 1924 and 1934 editions, in 2 volumes
4. S.L. Kleiman, R. Piene, Enumerating singular curves on surfaces, in *Proceedings of Conference on Algebraic Geometry: Hirzebruch 70 (Warsaw 1998)*, vol. 241 (A.M.S. Contemp. Math., 1999), pp. 209–238
5. P. Pokora, J. Roé, Harbourne constants, pull-back clusters and ramified morphisms. Results Math. **74**(3), Paper No. 109, 24 (2019). https://doi.org/10.1007/s00025-019-1031-x
6. P. Ribenboim, *The Theory of Classical Valuations*. Springer Monographs in Mathematics. (Springer, New York, 1999). https://doi.org/10.1007/978-1-4612-0551-7
7. J. Roé. Varieties of clusters and Enriques diagrams. Math. Proc. Camb. Philos. Soc. **137**(1), 69–94 (2004). http://dx.doi.org/10.1017/S0305004103007515
8. M. Spivakovsky, Valuations in function fields of surfaces. Am. J. Math. **112**(1), 107–156 (1990). https://doi.org/10.2307/2374856

The “Projective Spirit” in Segre’s Lectures on Differential Equations

Enrico Rogora

Abstract The “projective spirit” was undoubtedly a characteristic of Italian research in mathematics in the period 1860–1940, which pervades the successful researches of Luigi Cremona in projective geometry and Corrado Segre, Guido Castelnuovo, Federigo Enriques and Francesco Severi in algebraic geometry. However, its unconditional pursuit was also one of the reasons of the eclipse of Italian research in algebraic geometry in the period between the two world wars. This “projective spirit” was not limited to algebraic geometry but it can be traced, at various degree, in the background and sensibility of many Italian mathematicians of that period. It was an essential part of the mathematical culture disseminated at Italian Universities. The course given by Corrado Segre in the academic year 1920–1921 on the geometric theory of differential equations is, in some sense, paradigmatic of the pervasiveness of this spirit.

1 The “Projective Spirit” of Italian Mathematics, 1860–1940

Italian mathematicians believed that to mathematical research done Italy could be attached

> il requisito dell’italianità, vale a dire di quel quid che risulta dal connubio della serietà coll’agilità della parola e del pensiero, cioè dell’elaborazione artistica del materiale scientifico.[1]

Volterra made this “quid” a little more precise by saying that

> Il sentimento artistico, inteso nel suo significato più alto e comprensivo, ha avuto ed ha una gran parte nelle scoperte geometriche. Si comprende quindi come la matematica, la scienza

[1]These words are contained in a letter that Beltrami wrote to Cesàro and were quoted by Volterra in his address to the international congress of Mathematicians [56].

E. Rogora (✉)
Department of Mathematics, Sapienza Università di Roma, Rome, Italy
e-mail: enrico.rogora@uniroma1.it

T. Dedieu et al. (eds.), *The Art of Doing Algebraic Geometry*, Trends in Mathematics,
https://doi.org/10.1007/978-3-031-11938-5_18

Fig. 1 The Madonna painted by Ambrogio Lorenzetti in 1319 (left) and the one painted by Masaccio in 1426 (right) show the same scene from very different perspectives

> che non solo è la più pura e la più ideale, ma è la più schiettamente artistica delle scienze, abbia potuto trovare, sino dalle epoche lontane, un terreno favorevole per svilupparsi in Italia, ove il genio artistico è innato nelle genti, ben si comprende il carattere dell'opera matematica prodotta dagli ingegni italiani. Volterra, [56], p. 57.

In my opinion, one of the main aspects of what Volterra calls *elaborazione artistica del pensiero matematico* attributed to Italian mathematicians is their attitude towards space and spatial intuition, namely the idea that an important and often resolutive step to make when attempting to solve a mathematical problem is that of embedding the problem in the appropriate space in order to be able to use geometric intuition to understand it. When the space is projective and when the transformations which can be used to identify figures are projective I qualify this attitude as "*projective spirit*".

Projective spirit finds its root in Poncelet's pioneering work [41] and can be summarized by the phrase "looking at a problem from the correct prospective".

This attitude can be linked to that of the Renaissance painters which learned the rules of perspective in order to build the ideal space in which to place their subjects. Thus, one can try to pursue an analogy between the work of an artist who builds the pictorial space within which to place his work and that of a geometer who builds the space within which to place the object of his research. The artists of Renaissance discovered a new canon of ideal beauty expressible thanks to the language of perspective (Fig. 1).

Italian algebraic geometers were able to perceive new properties of algebraic varieties thanks to the language of higher dimensional projective spaces. For example, an algebraic curve can be immersed in the dual of the space of canonical divisors with canonical immersion, and hence it becomes possible to exercise a clear and powerful geometric intuition that was precluded or at least obscured in the primitive situation (Fig. 2).

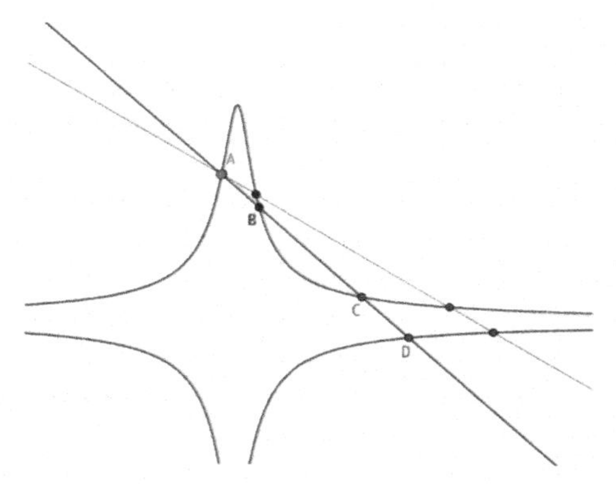

Fig. 2 Riemann—Roch theorem for algebraic curves gets a new geometric meaning on the canonical image of a curve, where the speciality of a divisor ($A + B + C + D$ in the figure) can be seen as the "defect" of dimension of its linear span (i.e. the number i of linearly independent hyperplanes containing the divisor) and the corresponding linear series has dimension $r = n + i - g$ where g is the genus and n is the degree of the divisor ($g = 2$ and $n = 4$ in the example)

In his address to the international congress of mathematicians, talking about the application of the "projective point of view" to the problems of the theory of algebraic curves, Castelnuovo wrote

> Era necessario che quei problemi li ritrovassimo noi stessi sotto una forma più adatta alla nostra mentalità. (...) [Segre] traduce le proprietà di una curva invarianti per trasformazioni birazionali in proprietà proiettive di un opportuno modello della curva, e trasporta così questioni per lui nuove nel terreno più familiare della geometria proiettiva iperspaziale. [6], p. 192.

The search for a geometric space in which to exercise geometric intuition is a characteristic attitude which is present in many of the Italian mathematicians of that period. In order to find or build the suitable space into which to place the mathematical objects to be studied, it is also necessary, as Klein pointed out in his famous program of Erlangen [26] and Corrado Segre constantly insisted on, to recognize the transformations that leave the nature of space unchanged, both to understand the nature of space and figures and to orient mathematical research towards the study of their invariant properties.

This "projective spirit" can be traced in all Corrado Segre's works and lectures, not only those on algebraic geometry but also those on the theory of continuous groups (1897-98, 1906–07, 1911–12), on the applications of the theory of invariants to geometry (1914–15), on hyperbolic geometry (1902-03), on the geometry of circles and spheres (1922–23) and, finally, even in his lectures on the geometry of differential equations (1920–21), which are considered in this work.

The "projective spirit" satisfied the aesthetic and "artistic" sense of many Italian mathematicians and characterized a great deal of the original contributions given by Italian mathematicians to algebraic geometry before the first world war. It deserves to be highlighted however also in works not strictly related to algebraic geometry, such as those of Vitali on the geometry of infinite dimensional spaces, of Amaldi on the theory of continuous groups, of Bompiani and Terracini on differential equations, of Bortolotti on projective connections and of Segre himself on the algebra of infinitesimals.

On the other hand, however, strict adherence to projective spirit severely blocked the diffusion of the new point of view of commutative algebra in Italian research in algebraic geometry. Abstract commutative algebra was developed in Germany by Emmy Noether and Wolfgang Krull in the period between the two world wars. They extended Hilbert's abstract approach contained in his landmark works on invariant theory. Commutative algebra does not aim at extending geometric intuition but at developing a new language and new methods. Italian geometers underestimated the importance of this new theory, believing that it limited itself to presenting well-known facts in new guises (see the excerpt of the letter of Fabio Conforto quoted in [38], p. 12). They did not realize that it was able to illuminate the outlines of areas of investigation where their geometric intuition was groping in the dark and to help finding a slow but safe path to get out of the swamps in which their research was bogged down, paving the way for the next revolution of sheaves and cohomology.

At the beginning of the twentieth century the idea that the algebraic point of view was necessary for an advancement of geometric knowledge was beginning to assert itself outside of Italy. The reduction of the importance and role of geometric intuition, testified for example by Kasner in his address [24], delivered before the section of Geometry of the international congress of art and science held in St Louis in 1904, of which we quote an excerpt, seems quite far from the spirit that still pervaded Italian researchers.

> The formulas move in advance of thought, while the intuition often lags behind ; in the oft-quoted words of d'Alembert, "L'algèbre est gènèreuse, elle donne souvent plus qu'on lui demande." As the field of research widens, as we proceed from the simple and definite to the more refined and general, we naturally cease to picture our processes and even our results. It is often necessary to close our eyes and go forward blindly if we wish to advance at all. [24], p. 285.

2 Segre's Notebooks

Thanks to the work of Livia Giacardi [18], Segre's notebooks, i.e. the manuscripts of Segre's lectures[2] are now on line.[3] Their importance for the knowledge of Italian geometry at the turn of the nineteenth and twentieth centuries is considerable. A critical edition of the manuscripts is very desirable and has begun with [52, 53] to which I refer for a biography of Segre, an introduction to the notebooks and further bibliography. I limit myself to quoting Gaetano Scorza's vivid description of the exceptional care that Segre put in the preparation of his lessons

---

[2] Part of the archival fond "Corrado Segre" of the Library "Peano" of the University of Torino.

[3] http://www.corradosegre.unito.it/quaderni.php.

Terminati gli esami della sezione estiva, egli fissava il tema del corso da tenere nel nuovo anno accademico; andando in villeggiatura, portava con sè, per le necessarie consultazioni, le memorie e le note che vi si riferivano; e, fermata fra agosto e settembre, la trama generale delle future lezioni, si poneva nell'ottobre a svolgerle per iscritto, sì da trovarsene redatte in modo definitivo un buon numero, quando, cessate le vacanze, tornava a Torino per la ripresa dell'attività didattica. Quivi l'esposizione orale delle lezioni già redatte e la stesura per iscritto delle successive procedevano di conserva con ritmo regolare, sì che mai gli avveniva di parlare di un argomento, di cui non avesse da tre o quattro settimane fissati fermamente i limiti e i modi della trattazione.

Così, con i 36 corsi di geometria superiore tenuti dal 1888 al 1924, mediante i quali egli venne a percorrere con ammirabile sicurezza di incesso tutte le regioni del vastissimo campo della geometria moderna, e quelle limitrofe dell'algebra e dell'analisi, di cui non mancava mai di raccomandare vivissimamente lo studio ai suoi allievi, egli venne a mettere assieme un'imponente collezione di manoscritti, di gran parte dei quali sarebbe altamente desiderabile che un qualche editore coraggioso intraprendesse la pubblicazione. G. Scorza, [46], pp. 133-4.

3 The Projective Point of View in the Theory of Differential Equations

The topic chosen by Corrado Segre for his 1920-21 course in Higher Geometry was the "Geometry of differential equations", as developed in the works of Clebsch [11], Darboux [12, 13], Klein and Lie [25] and Lie [29–32]. He already lectured on part of this topic in 1891–92 (General geometry), 1897-98 (Continuous groups of transformations) and 1911–12 (Continuous groups of transformations).

In the introduction of his 1920–21 notebook [51] he clearly highlights the approach he intends to follow and the purpose of his treatment.

*Il titolo del corso non stupisca. Due concezioni essenziali in Matematica: quella del numero e quella spaziale. Il metodo delle coordinate le collega. Quindi accanto ad un'*aritmetizzazione *si può porre una geometrizzazione.*

Una funzione di 1 o 2, ... *variabili equivale a una curva, una superficie, ...; e così le ordinarie equazioni: geometria delle equazioni algebriche*! [p.5]

[...]

Ora accanto a questi elementi geometrici: punti, rette, piani, ecc., se ne considerino altri, composti di un punto e una retta incidenti, o un punto e un piano incidenti, ecc. Aggregati di tali elementi sono gli enti geometrici che rappresentan le equazioni differenziali. [p.5]

[...]

la rappresentazione geometrica già accennata ci presenta enti che si capisce possono formare oggetto di studio, indipendentemente dalla possibilità di integrarle. [p.7]

For a geometric theory of differential equations, it is necessary to start by constructing a suitable space where embedding the objects of the theory. Segre, following a suggestion given by Clebsch in [11], defines a *differential element* (or simply an *element*) as a pair (x, l) where $x = (x_0, x_1, x_2) \in \mathbb{P}^2$ is a point, $l = (v_0, v_1, v_2) \in (P^2)^*$

is a line and $x \in l$, i.e. x and l are incident. The *duality* between lines and points in the plane is expressed analytically by the *incidence relation*, i.e. by the algebraic equation

$$x_0 v_0 + x_1 v_1 + x_2 v_2 = 0. \tag{1}$$

Equation (1) is an example of an algebraic equation $F(x_0, x_1, x_2, v_0, v_1, v_2) = 0$ which is separately homogeneous in the two series of variables x and v i.e., according to Clebsch's terminology, it is a *connex* [11]: an algebraic hypersurface in the algebraic variety $\mathbb{P}^2 \times (\mathbb{P}^*)^2$. If m is the degree of F w.r.t. the variables x and n is the degree of F w.r.t. the variables v, (m, n) is the *bidegree* of the connex $F = 0$.

A connex of bidegree (1,1) is called *bilinear*. For example, the incidence relation $I \subseteq \mathbb{P}^2 \times (\mathbb{P}^*)^2$ of Eq. (1) is bilinear and is called the *Identity connex*.

Each irreducible component of the intersection of two connexes is called a *coincidence*. If one of the two is the identical connex, the coincidence is said to be *principal*.

According to Clebsch's theory, the projective object which corresponds to an algebraic differential equation is a principal coincidence and the space in which is embedded is the incidence relation in the Segre's product $\mathbb{P}^2 \times (\mathbb{P}^*)^2$.

The coordinates v_1, v_2, v_3 of a line r joining two points x and x' can be computed by taking the minors of the matrix of the coordinates of x and x'. Each line through a point (x_1, x_2, x_3) joins this point to an "infinitely close" one, i.e. a point of form $(x_1 + dx_1, x_2 + dx_2, x_3 + dx_3)$. Therefore, the coordinates of a line through (x_1, x_2, x_3) can be chosen as the 2×2 minors of

$$\begin{vmatrix} x_1 & x_2 & x_3 \\ x_1 + dx_1 & x_2 + dx_2 & x_3 + dx_3 \end{vmatrix} = \begin{vmatrix} x_1 & x_2 & x_3 \\ dx_1 & dx_2 & dx_3 \end{vmatrix}$$

i.e.

$$(x_2 dx_3 - x_3 dx_2, x_3 dx_1 - x_1 dx_3, x_1 dx_2 - x_2 dx_1).$$

Then, the system

$$\begin{cases} F(x_0, x_1, x_2, v_0, v_1, v_2) = 0 \\ x_0 v_0 + x_1 v_1 + x_2 v_2 \quad = 0, \end{cases}$$

which defines a principal coincidence, can always be reduced to a single differential equation

$$F(x_1, x_2, x_3, x_2 dx_3 - x_3 dx_2, x_3 dx_1 - x_1 dx_3, x_1 dx_2 - x_2 dx_1) = 0. \tag{2}$$

In 1872, the idea of Clebsch to frame the theory of algebraic differential equations within that of connexes seemed quite promising. In Clebsch obituary [1], composed by Brill, Gordan, Klein, Lüroth, A. Mayer, Nöther and Von der Mühll and translated in Italian for *Annali di Matematica*, we read

> Con questo concetto Clebsch ha aperto alla ricerca geometrica un orizzonte interminato di nuove speculazioni, senza potere tuttavia segnare null'altro che l'indirizzo ch'egli inten-

deva di seguire nella trattazione dell' argomento. Egli voleva erigere il connesso a soggetto dell'investigazione geometrica, allo stesso titolo delle curve o delle superficie. Un connesso ha singolarità proprie, tra le quali hanno luogo equazioni analoghe a quelle date da Plucker per le singolarità delle curve plane. Si può far ricerca dei luoghi comuni a due, tre connessi, ecc. Tutte questo forme hanno un genere, relativamente ad ogni trasformazione univoca (estesa al doppio sistema ternario delle coordinate di punti e di rette); e così via. Ma ciò che sembra dotato di speciale importanza è il legame che sorge fra la teoria delle equazioni differenziali algebriche di prim'ordine (per ora fra due variabili assolute, ovvero fra tre variabili omogenee) e la teoria dei connessi. Ogni connesso trae con se un'equazione differenziale cosiffatta; le curve integrali sono definite da ciò che ciascun punto del piano e la tangente in esso soddisfanno all' equazione del connesso. Si ottiene in tal guisa la più generale equazione differenziale algebrica di prim' ordine, ma ad una data equazione differenziale corrispondono ancora infiniti connessi. In tal modo si viene a conseguire per queste equazione differenziale un punto di vista interamente nuovo, il quale quadra perfettamente colle nuove ricerche geometrico-algebriche. Si possono, per esempio, classificare queste equazioni differenziali in base ai loro generi, ecc. In ispecial modo poi si viene ad acquistare un algoritmo per la trattazione dei problemi differenziali, in armonia colle nuove vedute or dette. Questo algoritmo sembra destinato a completare le ricerche, di carattere piuttosto sintetico, iniziate in questi ultimi tempi sulle equazione differenziali, specialmente da Lie. [1], pp. 200–201.

Fifty years after Clebsch's work, when Segre wrote his lectures, the promises of the theory of connexes were not kept and I am not aware of any substantial progress since then. In my opinion, however, the inclusion in his lectures implies that Segre thought that Clebsch's theory was still worth to be explored within a projective approach, at a time when the applications of this approach to algebraic geometry were revealing the precariousness of its use in many fundamental questions.

The theory of connexes never became popular, despite its intriguing (and never pursued, as far as I know) connection with the theory of abelian integrals, pointed out by Klein in his American lectures [27], which must have appeared particularly promising to Segre.

This memoir[4] *implies an application, as it were, of the theory of Abelian functions to the theory of differential equations. It is well known that the central problem of the whole of modern mathematics is the study of transcendent functions defined by differential equations. Now Clebsch, led by the analogy of his theory of Abelian integrals, proceeds somewhat as follows. Let us consider, for example, an ordinary differential equation of the first order* $f(x, y, y') = 0$ *where* f *represents an algebraic function. Regarding* y' *as a third variable* z *we have the equation of an algebraic surface. Just as the Abelian integrals can be classified according to the properties of the fundamental curve that remain unchanged under a rational transformation, so Clebsch proposes to classify the transcendental functions defined by the differential equations according to the invariant properties of the corresponding surfaces* $f = 0$ *under rational one-to-one transformations.* [27], p. 8.

Clebsch projective theory of differential equations was soon outdated by Lie's work. Lie's geometric theory of differential equations is not cast in projective spaces but in spaces of jets which, in the simplest case, is just the 3-dimensional affine space (x, y, p) equipped with the contact form $dy - pdx = 0$. The connection with Clebsch theory is clear. If we introduce affine coordinates $x = x_1/x_0$,

[4] Alfred Clebsch, "Ueber ein neues Grundgebilde der analytischen Geometrie der Ebene", *Math Annalen*, **6** (1873), pp. 203-215.

$y = x_2/x_0$, $v = v_1/v_0$, $w = v_2/v_0$ in $\mathbb{P}^2 \times (\mathbb{P}^*)^2$, the affine equation of I becomes $xv + yw + 1 = 0$. Using this equation we can define the independent affine variable (x, y, p) for I by setting $p = -v/w$ and $v = p/(y - px)$, $w = 1/(px - y)$. There is a fundamental point, however, where Lie's approach is superior to Clebsch's: the introduction of the appropriate group of transformations. This, as already implicitly pointed out by Poncelet in [41] and made explicit and much further generalized by Klein in his Erlangen program [26], is the second fundamental pillar, (the first being a suitable redefinition of space) on which to build what I call a "projective approach" to a theory. The transformations with respect to which defining invariance in Lie's theory of differential equations, is the group of *contact transformations*,

> It is true that Jacobi used a particular contact transformation. So did Apollonius of Perga two hundred years before the present era. But we have no evidence that either Jacobi or Apollonius knew it, let alone the conception of the most general transformation of this kind; the idea of a contact transformation is Lie's. [37], p. 322.

Segre, in complete agreement with Lie's view, assigns ample space to contact transformations in his lectures, providing examples taken from projective geometry.

The complete list of topics treated by Segre in his lectures is contained in the notebook index

1. *Introduzione* (Introduction) [pp. 5–8].
2. *Equazioni differenziali ordinarie del primo ordine–Moltiplicatore* (First order ordinary differential equations—Integrating factor) [pp. 9–19].
3. *Sulle soluzioni singolari delle delle equazioni differenziali del primo ordine* (On singular solutions of first order ordinary differential equations) [pp. 20–31].
4. *Sui punti singolari delle equazioni differenziali del primo ordine dal punto di vista della realtà* (On singular points of first order ordinary differential equations from the standpoint of reality) [pp. 32–41].
5. *L'equazione differenziale del primo ordine di Jacobi e le curve W di Klein e Lie* (Jacobi first order differential equation and Klein and Lie W-curves) [pp. 42–53].
6. *Alcune proposizioni geometriche generali su le equazioni differenziali del primo ordine algebriche e sulle loro linee integrali* (Some general geometrical propositions about first order algebraic differential equations and their integral lines) [pp. 54–60].
7. *Estensione del concetto d'integrale. Trasformazioni di contatto del piano* (Extension of the concept of integral. Contact transformations ot the plane) [pp. 61–74].
8. *Cenni sui gruppi ∞^1 di trasformazioni puntuali. Trasformazioni infinitesime.* (Notes on one dimensional groups of point transformations. Infinitesimal transformations) [pp. 75–80].
9. *Applicazioni ai gruppi projettivi. Equazioni di Riccati* (Applications to projective groups. Riccati equations) [pp. 81–86].
10. *Equazioni differenziali di primo ordine con dati gruppi monomi di trasformazioni puntuali in sè* (First order ordinary differential equations with assigned monomial transformation groups) [pp. 87–94].

11. *Congruenze di linee e sistemi di equazioni differenziali del primo ordine fra* 3 *variabili* (Congruences of lines and systems of first order differential equations between 3 variables) [pp. 95–99].
12. *Equazioni di Monge e di Pfaff. Linee di un complesso* (Monge and Pfaff Equations. Lines of a complex) [pp.100–110].
13. *Equazioni alle derivate parziali del primo ordine* (First order partial differential equations) [pp. 111–126].
14. *Note e complementi* (Endnotes) [pp. 148–155].

4 A Taste of Segre's Lectures on Differential Equations

To give a taste of Segre's lectures on the Geometry of differential equations, I discuss the content of two chapters: "Jacobi differential equations and Lie-Klein theory of W curves" and "Congruences of curves and systems of ODE". Segre does not suggest a new approach to these topics, but he focus his attention on them because he thinks that further results can be obtained using the same geometric approach to projective hyperspaces which proved to be so successful in his dealing with algebraic geometry.

4.1 Jacobi Equations and W-curves

W-curves are curves in projective space which are invariant under a one parameter group of projective transformations. Conics, logarithmic spirals, powers, logarithms and helices are examples of W-curves.

The study of planar W-curves was unertaken by Klein and Lie in [25]. They also considered the classification of W-curves in space, but they decided to abandon it for the complexity of the details.

The classification of W-curves follows from that of projectivities. The classification of planar projectivities was well known to Klein and Lie, even if not set, according to Segre, in a completely rigorous way.[5] Segre provided a detailed classification of projectivities in $\mathbb{P}^n$ in [47]. His work could have easily allowed a detailed study of W-curves in space and hyperspaces but I am not aware that the problem has been addressed in the past century. A modern exposition of the classification of W-curves in space can be found in [14].

[5] According to Segre [47], the first complete geometrical classification of the projectivities in plane and space was due to Loria [36] in 1884, even if much work was done previously by Hirst, Sturm, Fiedler, Clebsch and Gordan, Reye and Battaglini. However the geometrical interpretation provided in [47] of Weierstrass' theorem on the classification of pairs of bilinear forms [57, 58] implies the complete classification of projectivities in $\mathbb{P}^n$. Independently of Weierstrass, also Jordan, in [23] provided a classification of linear substitutions in an arbitrary number of variables which gives, as a consequence, the classification of projectivities. A geometrical treatment of the classification of projectivities with exhaustive historical notes can be found in [16], vol. 2, pp. 658–686.

The topic of $W - curves$ is quite irrelevant nowadays,[6] but it is important from an historical point of view, since it is one of the first problem that have been studied with the help of continuous groups (see [20, 21, 42]).

Segre begins by considering *bilinear connexes*,

$$\sum a_{ij} x_i v_j = 0.$$

A *principal bilinear coincidence* is obtained by intersecting a bilinear connex with the identical connex (for which $\{a_{ij}\}$ is the identity matrix).

To a bilinear connex with matrix $A = \{a_{i,j}\}$ is associated the projectivity (or homography)

$$\mathcal{A}: \quad x \mapsto Ax$$

where x is the column vector of homogeneous coordinates. Note that A and $\lambda \cdot A$ define the same homography $\mathcal{A}$ and the same bilinear connex. Moreover, A and $\lambda \cdot A + \rho \cdot I$ define the same principal coincidence, i.e. the same first order ODE in the sense of Clebsch (see previous section).

Given the bilinear coincidence associated to a matrix A, it is easy to verify that the line $v(x)$ of the coincidence through a point x is the unique line joining x to $\mathcal{A}(x)$. Therefore, the principal coincidence is completely described by the homography $\mathcal{A}$.

When A is the matrix defining a bilinear connex, its associated differential equation (2) is

$$(a_{1,1}x_1 + a_{2,1}x_2 + a_{3,1}x_3)(x_2dx_3 - x_3dx_2) + (a_{1,2}x_1 + a_{2,2}x_2 + a_{3,2}x_3)(x_3dx_1 - x_1dx_3) + (a_{1,3}x_1 + a_{2,3}x_2 + a_{3,3}x_3)(x_1dx_2 - x_2dx_1) = 0 \tag{3}$$

The corresponding affine, non homogeneous differential equation obtained by setting $x_1 = 1, dx_1 = 0, x_2 = x, x_3 = y$ is

$$(a_{1,1} + a_{2,1}x + a_{3,1}y)(xdy - ydx) - (a_{1,2} + a_{2,2}x + a_{3,2}y)dy + (a_{1,3} + a_{2,3}x + a_{3,3}y)dx = 0 \tag{4}$$

This is the *Jacobi differential equation* (see [22], p. 22, where the analytic solution is given only when the matrix A of coefficients has three distinct eigenvalues, at p. 31). The Jacobi differential equation is a completely elementary projective object. Indeed, the direction field of a Jacobi differential equations, being defined by a projectivity, can be drawn with straightedge ruler only.[7]

---

[6] Aluffi and Faber considered the action of $PGl(3)$ on the projective space of all algebraic curve of given degree in [2, 3]. They gave a complete classification of the orbits of non maximal dimension (i.e. of dimension strictly less than 8) and computed their degree and other cohomological discrete characters. The orbits of the curves considered by Klein and Lie, have not maximal dimension since W-curves admit a one parameter group of projective endomorphisms. When they are algebraic, their orbits are therefore among those considered by Aluffi and Faber.

[7] With straightedge ruler only it is also possible to draw any finite number of points of the unique $W - curve$ through a given general point.

Segre points out that

L'integrazione fatta da Jacobi viene illuminata dalle considerazioni geometriche relative all'omografia. [51], p. 44

In order to pursue a geometric approach to the solution of Jacobi equation Segre begins with a geometric proof that planar projectivities can be split into five types, according to their fixed elements. The proof is equivalent to the reduction of 3×3 matrices to Jordan canonical forms.

Leaving aside the multiples of the identity, which give the null direction field, there are 5 types of projectivities corresponding to the 5 possible complex non scalar Jordan canonical forms. Their types, according to the convention used by Klein and Lie in [25], are.[8]

$$\begin{pmatrix} a & 0 & 0 \\ 0 & b & 0 \\ 0 & 0 & 1 \end{pmatrix} \quad \begin{pmatrix} a & 1 & 0 \\ 0 & a & 0 \\ 0 & 0 & 1 \end{pmatrix} \quad \begin{pmatrix} a & 1 & 0 \\ 0 & a & 1 \\ 0 & 0 & a \end{pmatrix} \quad \begin{pmatrix} a & 0 & 0 \\ 0 & a & 0 \\ 0 & 0 & 1 \end{pmatrix} \quad \begin{pmatrix} a & 1 & 0 \\ 0 & a & 0 \\ 0 & 0 & a \end{pmatrix}$$

Type I Type II Type III Type IV Type V

If in a Type I matrix, a an b are complex conjugates, the matrix can be transformed, with a linear substitution, into the form

$$\begin{pmatrix} \cos\theta & -\sin\theta & 0 \\ \sin\theta & \cos\theta & 0 \\ 0 & 0 & \rho \end{pmatrix}$$

Type Ib

Given a Jacobi differential equation, Segre discusses how to choose coordinates by suitably choosing the fixed elements of the associated homography. In the new coordinates the given Jacobi differential equation becomes very easy to solve and the solutions, which coincide with Klein and Lie W-curves can be written down explicitly. For example, for type I, Jacobi differential equation can be transformed by a projectivity into the homogeneous form

$$(a_2 - a_3)\frac{d\,x_1}{x_1} + (a_3 - a_1)\frac{d\,x_2}{x_2} + (a_1 - a_2)\frac{d\,x_3}{x_3}$$

with complex solutions[9]

$$x_1^{a_2 - a_3} \cdot x_2^{a_3 - a_1} \cdot x_3^{a_1 - a_2} = \text{const.}$$

[8] One can write each transformation in [25], p. 55 in homegeneous coordinates, i.e. in the form $\mathbf{x}' = A \cdot \mathbf{x}$, with A a 3×3 matrix. The matrix A has the same Jordan of the matrix of the same type shown here.

[9] Non algebraic, in general.

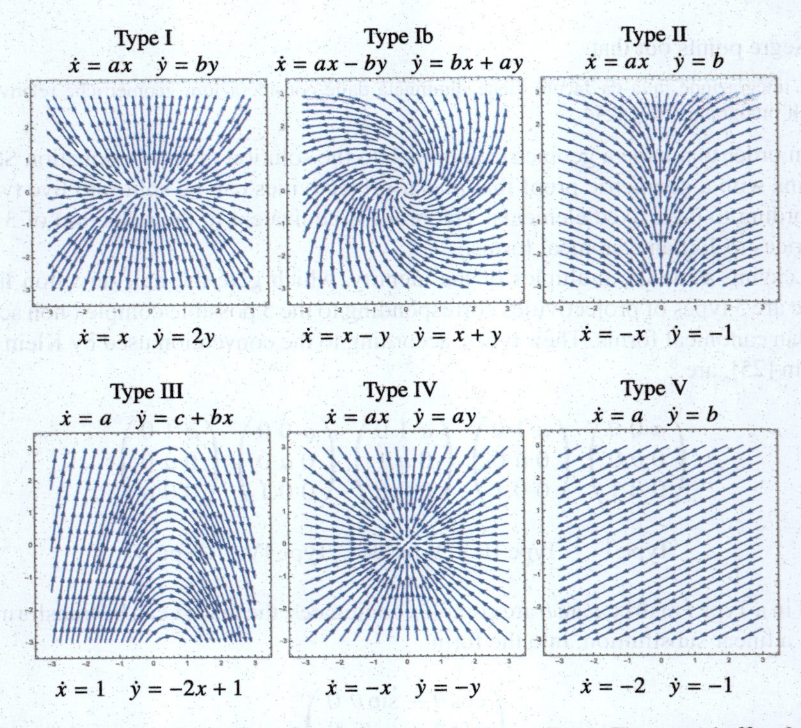

Fig. 3 Streamlines of Jacobi differential equations of different Types. The general affine form is given above the picture. The particular case whose streamlines are portrayed in the figure is given below the picture

According to Klein and Lie, W curves are orbits of a 1-parameter group. If $\mathcal{A}$ is the projectivity with which we build the direction field of the Jacobi equation, Klein and Lie extend "by continuity" the map $n \mapsto \mathcal{A}^n$ to a map $t \mapsto \mathcal{A}^t$ and consider the infinitesimal generator of this one parameter family of matrices, i.e. the derivative H of $t \mapsto \mathcal{A}^t$ at zero. Formally, one can introduce the logarithm and the exponential of a matrix and define $H = \log A$ and recover the one parameter group as $t \mapsto e^{tH}$. The orbits of this group are the W—curves which give the solutions of the Jacobi differential equation associated to A. This idea of infinitesimal transformation employed by Klein and Lie to analyze these curves turned out to be central in the development of the theory of continuous groups.

In Fig. 3 is a sketch the *streamlines* of the different types of Jacobi differential equations, written in affine form, as in [25].

In his lectures, Segre gives some further details about Klein Lie theory of W-curves. He begins with noticing that all curves of Type Ia contain two of the three fixed points of the homography but not the third and that the three lines through pairs of fixed points belong to the system. He also describes how some geometrical properties of the solutions of a differential equation of Type III can be explained by degenerating a family of differential equations of Type I. Since each of these differential equations is defined by a homography A, it is quite natural (at least after the work of Klein and Lie) to consider for each A the family of projectivities $\Omega(A)$ which commute with A and which transform the set W-curves into itself. For Type I and II, $\Omega(A)$ is bidimensional. Moreover, each pair of elements $\omega, \omega' \in \Omega(A)$ commutes. $\Omega(A)$ is a group which contains a subgroup $\Sigma(A) \subseteq \Omega(A)$ which sends each curve in itself. The group $\Omega(A)$ acts over the differential elements (in the sense of Lie) of the Jacobi differential equation determined by A.

In conclusion, the whole treatment is illuminated, as pointed out by Segre himself, by geometric considerations relating to homographies and is presented in a way which is easily generalizable, at least in principle, to a thorough study of W -curves in projective hyperspaces. In fact, the core of the discussione is based on the geometric classification of the homographies, that Segre himself obtained in general in [47].

4.2 Congruences of Curves and Systems of ODE

The second chapter that I want to discuss is dedicated to *congruences of space curves*. The main tools of investigation are *focal points* and *focal surfaces*, already considered in this context by Darboux in [13].[10] The study of the focal variety of an $n-1$ dimensional family of curves is a quite natural generalization of the of the case of $n-1$ dimensional families of lines of $\mathbb{P}^n$, to which Segre dedicated [48]. I consider therefore quite plausible that Segre intended to indicate a possible research topic with his choice to present Darboux theory of the focal surface of a congruence of curves, aimed at dealing geometrically with systems of ordinary differential equations of arbitrary dimension.

However, I am not aware that the study of the focal varieties of a family of curves in projective space of arbitrary dimension has been addressed before the nineties of the past century.

For Segre a *congruence of curves*[11] is a system of two algebraic equations

$$f(x, y, z, a, b) = 0 \qquad g(x, y, z, a, b) = 0 \tag{5}$$

depending on three variables x, y and z and two parameters a and b. Therefore, the differential elements $(x, y, y, dx : dy : dy)$ of the curves of the congruence verify the two equations

$$df = 0 \qquad dg = 0$$

Eliminating a and b from these, one gets two first order differential equations. The vice versa is more delicate and will be considered later.

[10] For a modern exposition, see [10].

[11] The name, according to [13], p. 2 is due to Plücker.

The general point of the space is contained in a finite number of curves of a congruence, but for some special points the curves may be infinite. For example: tangent lines to a surface which intersect a given curve (each point of the curve is contained in a one dimensional subfamily of the congruence); circles (in space) for two points (each of the two points are contained in all elements of the congruence).[12]

Given a congruence of curves, each function $b = \varphi(a)$ gives a *surface of the congruence*, of equations

$$f(x, y, z, a, \varphi(a)) = 0 \qquad g(x, y, z, a, \varphi(a)) = 0 \tag{6}$$

By taking the differential of (6) we one gets, for each surface of the congruence, the following differential constraints.

$$\begin{aligned} \tfrac{\partial f}{\partial x}dx + \tfrac{\partial f}{\partial y}dy + \tfrac{\partial f}{\partial z}dz + \tfrac{\partial f}{\partial a}da + \tfrac{\partial f}{\partial b}\varphi'(a)da &= 0 \\ \tfrac{\partial g}{\partial x}dx + \tfrac{\partial g}{\partial y}dy + \tfrac{\partial g}{\partial z}dz + \tfrac{\partial g}{\partial a}da + \tfrac{\partial g}{\partial b}\varphi'(a)da &= 0 \end{aligned} \tag{7}$$

Eliminating da from (7) we get

$$\frac{\tfrac{\partial f}{\partial x}dx + \tfrac{\partial f}{\partial y}dy + \tfrac{\partial f}{\partial z}dz}{\tfrac{\partial f}{\partial a} + \tfrac{\partial f}{\partial b}\varphi'(a)} = \frac{\tfrac{\partial g}{\partial x}dx + \tfrac{\partial g}{\partial y}dy + \tfrac{\partial g}{\partial z}dz}{\tfrac{\partial g}{\partial a} + \tfrac{\partial g}{\partial b}\varphi'(a)}$$

Given a point (x_0, y_0, z_0) of the surface, upon substituting dx, dy, dz with $(x - x_0)$, $(y - y_0)$, $(z - z_0)$, we get the equation of the tangent plane to the surface at that point.

Rewriting the equation in the form

$$\frac{\tfrac{\partial f}{\partial x}(X - x) + \cdots}{\tfrac{\partial g}{\partial x}(X - x) + \cdots} = \frac{\tfrac{\partial f}{\partial a} + \tfrac{\partial f}{\partial b}\varphi'(a)}{\tfrac{\partial g}{\partial a} + \tfrac{\partial f}{\partial b}\varphi'(a)} \tag{8}$$

Segre remarks that

> I piani tangenti in quel punto risultano formare un fascio attorno alla tangente alla linea; e dal secondo membro appare che quel fascio è proiettivo alla serie dei valori di $\varphi'(a)$. [51], p. 96.[13]

This remark immediately implies a result of Darboux: given 4 surfaces of the congruence containing the same curve γ of the congruence, the double ratio of the 4 tangent planes to the surfaces at a point $p \in \gamma$ does not depend on p.

---

[12] *Les deux exemples diffèrents que nous venons de signaler correspondent aux deux cas qui peuvent se prèsenter lorsqu'il y a indètermination. Dans le premier, les deux èquations auxquelles doivent satisfaire a et b se rèduisent à une seule; dans le second, elles sont, l'une et l'autre, identiquement vèrifièes.* [13] p.2.

[13] The general theory of projective varieties generated by linear spaces was developed by Segre in [50].

Morover, when

$$\frac{\partial f}{\partial a}\frac{\partial g}{\partial b} - \frac{\partial f}{\partial b}\frac{\partial g}{\partial a} = 0 \tag{9}$$

the tangent plane does not depend on $\varphi'(a)$ and all surfaces of the congruence containing a point satisfying (9) have the same tangent plane at that point. These special points are called *focal points* or *foci*.

If the polynomials in (5) have degree n and m in x, y, z, each curve of the congruence has degree $n \cdot m$ and the Eq. (9) has degree $n + m$. Hence Segre claims that the number of focal point on the general curve of the congruence is $m \cdot n(m + n)$. For example: a congruence of lines has 2 foci on the general line; a congruence of plane curves of degree n has $(n + 1)n$ foci on the general curve. He also claims that, in general, a fixed point of a bidimensional family of curves is a focal point of degree (at least) two and that, when the focal points describe a surface, the tangent plane to the focal surface at a given regular point is exactly the common tangent plane to all surfaces of the congruence which go through that point.

Following Darboux ([13], p. 6), Segre presents another way of defining focal points, starting from the equations for the envelope of a one dimensional family of curves. If a one dimensional family of plane curves has equation $F(x, y, a) = 0$, the equation of the envelope is obtained by eliminating a from

$$F(x, y, a) = 0, \qquad F_a(x, y, a) = 0.$$

Analogously, if a one dimensional family of space curves has equations

$$f(x, y, z, \alpha) = 0 \qquad g(x, y, z, \alpha) = 0$$

the equations of the envelope are

$$f = 0 \qquad g = 0 \qquad f_\alpha = 0 \qquad g_\alpha = 0. \tag{10}$$

provided that the system has ∞^1 solutions.

By setting $b = b(a)$ in (5) we get ∞^1 families of curves of the congruence. If one imposes the condition that such a one dimensional family of curves have an envelope, one obtains the same differential condition which describes focal points. Therefore, each point of a curve enveloped by a one dimensional family of curves of a congruence is a focal point of the congruence itself.

If the family of envelopes covers a surface Σ,[14] the surface Σ is a surface of *focal points* and the general curve of the congruence is tangent to the focal surface at its own focal points.

---

[14] This is not always true. In the congruence of lines through a point, for example, no one dimensional sub-family envelopes a curve.

Segre considers also a third possible characterization of a focal point as intersection of two "infinitesimally close" curves of the congruence. Given a curve

$$f(x, y, z, a, b) = 0 \quad g(x, y, z, a, b) = 0$$

the curves of the congruence which are "infinitesimally close to it" are, in the language of Segre, those of equations

$$f(x, y, z, a + \delta a, b + \delta b) = 0 \quad g(x, y, z, a + \delta a, b + \delta b) = 0$$

Expanding w.r.t. δ we get that the points of intersection of the two "infinitesimally close curves" are the solutions of the system

$$f = 0 \quad g = 0 \quad \delta a \cdot f_a + \delta b \cdot f_b = 0 \quad \delta a \cdot g_a + \delta b \cdot g_b = 0$$

which is equivalent to (9), hence a focal point is also definable as the intersection of a curve with another which is infinitesimally closed to the first.

Segre ends the chapter with some considerations about *singular solutions* of a first order ODE which are of interest for us. For a single first order differential equations $F(x, y, y') = 0$, the existence of *singular solutions*, was discovered by Brook Taylor in 1715 and by Alexis Clairaut in 1736, who considered the first order ordinary differential equation

$$y(x) = x\frac{dy}{dx} + f\left(\frac{dy}{dx}\right). \tag{11}$$

Taking the first derivative of (11) we get

$$\frac{dy}{dx} = \frac{dy}{dx} + x\frac{d^2y}{d^2x} + f'\left(\frac{dy}{dx}\right)\frac{d^2y}{d^2x},$$

so

$$\left[x + f'\left(\frac{dy}{dx}\right)\right]\frac{d^2y}{dx^2} = 0.$$

Hence, either

$$\frac{d^2y}{dx^2} = 0 \tag{12}$$

or

$$x + f'\left(\frac{dy}{dx}\right) = 0. \tag{13}$$

The system of Eqs. (11) and (12), gives the *general solution* i.e. the family of lines

$$y(x) = a \cdot x + f(a)$$

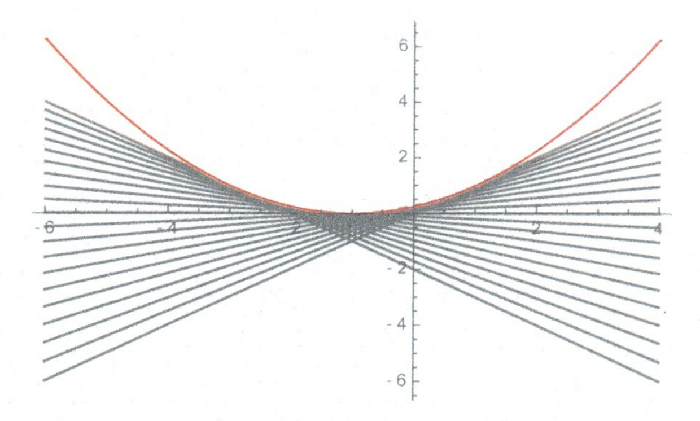

Fig. 4 In gray, the curves of the general solution. In red, their envelope, which gives the singular solution. A more rigorous description of the solutions can be found in [4], p. 113

The system of Eqs. (11) and (13) gives the *singular solution* which is the *envelope* of the general solutions (Fig. 4). For example, if $f(y') = y' - (y')^2$, the singular solution is

$$y(x) = \left(\frac{x+1}{2}\right)^2$$

If a one dimensional family of solutions of $F(x, y, y') = 0$ has an envelope, the envelope is still a solution. For a long period, this fact made mathematicians assume that, in general, first order differential equations have singular solutions and that the rule for finding them, at least when F is algebraic, was to eliminate y' from the system

$$F(x, y, y') = 0, \qquad \frac{\partial F}{\partial y'} = 0. \tag{14}$$

In a famous work of 1873 [12] Darboux proved that this assumption was wrong and instead

> *En gènèral, il n'y a pas de solution singulière, et l'èlimination de y' entre les èquations* (14) *conduit à l'èquation d'un courbe reprèsentant, non pas l'enveloppe des solutions gènèrales, mais le lieu de leurs points de rebroussement*

The way of proving Darboux's claim and to find the correct condition defining singular solutions is very interesting and surely appealed Segre's geometric sensibility who dedicates the entire chapter *Singular solutions of first order ODEs* to clarify this question following Darboux's idea to make use of the *principle of duality*.

> Une èquation diffèrentielle ordinaire prèsente un double caractère, sur lequel on n'insiste pas, en gènèral, quoiqu'il nous paraisse d'une grande importance. [12], p. 159.

The double character to which Darboux alluded is: the curve as a *locus of points*, specified by a condition on its tangent line; the curve as an *envelope of lines*, specified by a condition on its contact points.

> *Cette symètrie, introduite dans la thèorie qui nous occupe, est d'un grand secours dans une foule de questions, et elle va nous permettre, en particulier, d'èclaircir la question des solutions singulièrs.* [12], p. 160

Darboux proves that, eliminating y' from

$$F(x, y, y') = 0, \qquad \frac{\partial F}{\partial x} + \frac{\partial F}{\partial y} y' = 0. \tag{15}$$

one obtains an equation

le lieu [...] des points d'inflexion des courbes intègrales [12], p. 160.

By duality, the equation of *cuspidal points* is therefore obtained by eliminating y' from

$$F(x, y, y') = 0, \qquad \frac{\partial F}{\partial y} = 0. \tag{16}$$

The two results imply that

Pour qu'il y ait une solution singulière, il faut que les trois equations

$$F(x, y, y') = 0, \qquad \frac{\partial F}{\partial y} = 0 \qquad \frac{\partial F}{\partial x} + \frac{\partial F}{\partial y} y' = 0. \tag{17}$$

soient satisfaites en tous le points d'une courbe (A). [12], pp. 163–164

However, the application of the principle of duality, which helped Darboux to clarify the matter, was not enough for solving it completely. He needed to perform a more *elaborated* analytical analysis to make his last statement more precise. At the end of his analysis, he explains why mathematicians before him were convinced that in general a differential equation should admit singular solutions.

En terminant, nous devons nous demander quelle a ètè l'origine de l'erreur qui a durè si longtemps dans la thèorie des solutions singulières. Cette erreur tient à une confusion, que presque tous les gèomètres ont laissè s'ètablir dans toute cette question.
Comme un forme des èquations diffèrentielles par l'èlimination des constantes entre une èquation finie et ses dèrivèes, les auteurs ont supposè, a tort selon nous, qu'etant donnèe, par exemple, une èquation diffèrentielle du premier ordre, cette èquation admet toujours une intègrale du premier ordre, dèfinie par la formule

$$f(x, y, c) = 0,$$

où f est une fonction qui, dans toute l'ètendue du plan, possède les propriètès qu'on reconnaît gènèralement aux fonctions ètudièes dans l'Analyse. Cette fonction f ètait pour eux plus ou moins difficile à trouver; mais dans leur esprit elle existait toujours. Or c'est là prècisèment le point contestable, et les recherches nouvelles sur la thèorie des fonctions nous paraissent devoir changer cette manière de voir.

The topic of singular solutions should have appeared quite stimulating for Segre, for at least two reasons: the possibility to look for a generalization to systems of first order ODEs, and; the possibility to use more refined geometric methods for performing a finer analysis of relationships between singular solutions and other geometric loci attached to differential equations.

About the first point, Segre, at the end of the chapter "Congruences of lines and ODE systems", sums up Goursat's work [19] where the author extends Darboux's results to give a characterization of the singular solutions of a system of two first order ODEs. Goursat's approach is less geometric than Darboux's one and Segre could have thought the program to replace some of Gousat's analytic arguments with more geometric ones and to extend the whole geometric treatment to systems of first order ODEs in higher dimensional spaces worth to be considered for his students.

About the second point, even if Segre does not give explicit indications in this regard, I think that the topic of the relationships between focal varieties and other geometric loci attached to system of ODEs could have very well suggested to him the possibility of approaches based on geometric techniques that proved to be quite powerful in studying intersection of algebraic varieties, like that of blowing ups.

5 Conclusions

In many of his courses, Segre lectured on topics other than algebraic geometry, his main field of research, looking for new areas of application of the same "projective spirit" that achieved great success in the theory of algebraic curves and surfaces. In the twenties of last century, this approach began to show severe limitations in its applications to algebraic geometry. The need to unravel the intricacies of advanced research called for the development of new foundations and of an efficient algebraic calculus.

Betraying "projective spirit" however proved to be quite difficult for Italian geometers, initially because of their high appreciation of its intrinsic aesthetic beauty (see Sect. 1), and later for the stubborn determinacy of Francesco Severi, the leader of Italian mathematicians in the thirties of twentieth century, to frame the matter of geometric methods as an issue of nationalistic pride about Italian supremacy in the field of geometry (see [38]). The reactions to the crisis of the efficacy of these methods in algebraic geometry were manifold: Castelnuovo turned his interests elsewhere, Severi attempted to build a bald but ineffective geometric theory of equivalence of algebraic varieties; Enriques continued his fine analysis of important classes of examples to which it was possible apply geometric methods without compromises; Segre attempted to preserve the vitality of the approach by exploring new applications, like projective differential geometry and the geometric theory of differential equations.

I would like to conclude this work with some final thoughts about the seductive power of the projective spirit which characterized the research of Italian mathematicians but appears to be very limited in contemporary mathematics. In my opinion, there are some good reasons for not neglecting it in the teaching of mathematics to perspective teachers. (for a more detailed discussione, see [5]). It is a common feeling in the study of elementary geometry, that of not being satisfied with an analytical solution of a problem. It seems that analytical solution can be obtained just by brute force computation without illuminating the understanding of the problem and that synthetic solutions are more beautiful. The aesthetic appeal of elementary geometry appears to be overshadowed by analytical (and algebraic) methods. These methods may have as much aesthetic appeal in more advanced contexts, but geometric methods, in my opinion, make it easier to appreciate the beauty of mathematics in elementary contexts.

If we compare the study of elementary geometry to the exploration of an unknown territory we could use a metaphor. The projective spirit is like using a hang glider. You have to climb a mountain and throw yourself down. It is very thrilling and very aesthetic but it is not very practical. The analytic spirit is more like using a car: very practical but definitely less thrilling.

Or better, as Segre wrote.

> Spesso converrà alternare fra loro il metodo sintetico che appare più penetrante, più luminoso, e quello analitico che in molti casi è più potente, più generale, o più rigoroso. [49], p. 52.

At the end of this paper I would like to thank Ciro Ciliberto for having introduced me to the wonders and aesthetic pleasures of geometry.

References

1. Autori vari, Alfredo Clebsch e i suoi lavori scientifici. Annali di Matematica **8**, 153–207 (1873)
2. P. Aluffi, C. Faber, Plane curves with small linear orbits I. Ann. de l'Institut Fourier **50**, 151–196 (2000)
3. P. Aluffi, C. Faber, Plane curves with small linear orbits. II. Internat. J. Math. **11**(5), 591–608 (2000)
4. I. Vladimir, *Arnold Ordinary Differential Equations*, 3rd edn. (Springer, New York, 1992)
5. A. Brigaglia, *Maria Anna Raspanti, Enrico Rogora, "L'uso di un software di Geometria dinamica nella formazione dei futuri insegnanti"* (Cultura e Società. Rivista dell'Unione Matematica Italiana, to appear, Matematica, 2021)
6. G. Castelnuovo, La geometria algebrica e la scuola italiana, in *Atti del congresso internazionale dei Matematici, Bologna 3–10 Settembre 1928*, Zanichelli, Bologna, Tomo I, pp. 191–201 (1929)
7. A.-L. Cauchy, *Leçons de calcul diffèrentiel et de calcul intègral* (Bachelier, Paris, 1844)
8. A. Clebsch, *Leçons sur la gèometrie. Intègrales abèliennes et connexes, Tome 3* (Gauthier-Villars, Paris, 1883). https://mas.goetheanum.org/fileadmin/mas/downloads/Lou_de_Boer/Pathcurves.pdf
9. O. Chisini, Geometria numerativa. Seminario Mat. Fis. di Milano **XIX**, 1–16
10. C. Ciliberto, E. Sernesi, Singularities of the theta divisor and families of secant spaces to a canonical curve. J. Algebra **171**, 867–893 (1995)
11. A. Clebsch, *Leçons sur la gèomètrie, 3 voll* (Gauthier-Villars, Paris, 1879)
12. G. Darboux, Sur les solutions singulières des èquations aux dèrivèes ordinaires du premier ordre. Bulletin des sciences methèmatiques et astronomiques **4**, 158–176 (1873)
13. G. Darboux, *Leçons sur la thèorie gènèrale des surfaces*, vol. 2 (Gauthier Villars, Paris, 1889)
14. L. de Boer, Classification of real projective path curves. Mathematisch–Physikalische Korrispondenz **219**, 4–48
15. F. Enriques, *Conferenze di geometria: fondamenti di una geometria iperspaziale* (litogr.) (Tipografia L. Pongetti, Bologna, 1894–1895)
16. F. Enriques, *Lezioni sulla teoria geometrica delle equazioni e delle funzioni algebriche*, pubblicate per cura del dott. Oscar Chisini, Zanichelli, Bologna, vol. 1 (1915); vol. 2 (1918); vol. 3 (1930); vol. IV (1934)
17. Fouret, Mèmoire sur les systèmes gènèraux de courbes planes algèbriques ou transcendantes, dèfinis par deux caractèristiques, *Bulletin de la Sociètè Mathèmatique de France*, 2, Nota I, pp. 72–83; Nota II, pp. 96–100 (1873–1874)

18. Livia Giacardi, T. Varetto, Il Fondo Corrado Segre della Biblioteca "G. Peano" di Torino, in *Quaderni di storia dell'Università di Torino*, Torino, I, pp. 207–246 (1996)
19. E. Goursat, Sur les Solutions Singulièr des èquations Diffèrentielles Simultanèes. Am. J. Math. **11**(4), 329–372 (1889)
20. T. Hawkins, *Emergence of the Theory of Lie Groups* (Springer, New York, 2000)
21. T. Hawkins, Line geometry, differential equation and the Birth of ie's theory of groups, in *The History of Modern Mathematics*, vol. 1, ed. by J. McCleary, D. Rowe (Academic Press, New York, 1989), pp. 275–327
22. Edward Lindsay Ince, *Ordinary Differential Equations* (Dover, New York, 1956)
23. C. Jordan, *Traitè des Substitutions* (Gauthier Villars, Paris, 1870)
24. E. Kasner, The present problems of Geometry. Bull. Am. Math. Soc. **11**(6), 283–314 (1905)
25. F. Klein, S. Lie, Ueber diejenigen ebenen Curven.... Mathematische Annalen **4** (1871)
26. F. Klein, Vergleichende Betrachtungen über neuere geometrische Forschungen. Mathematische Annalen **43**, 63–100 (1893)
27. F. Klein, *Lectures on Mathematics* (Mc Millan, New York, 1894)
28. F. Klein (ed.), *Encyklopädie der mathematischen Wissenschaften mit EinschluSS ihrer Anwendungen, III-3* (Teubner, Leipzig, 1902)
29. S. Lie, With Friedrich Engel, Theorie der Transformationsgruppen, vols. 1–3 (1893) (Teubner, Leipzig, 1890)
30. S. Lie, *with Georg Sheffers* (Vorlesungen über Differentialgleichungen mit bekannter infinitesimalen Transformationen, Teubner, Leipzig, 1891)
31. S. Lie, *with Georg Sheffer* (Vorlesungen über continuierliche Gruppen mit geometrischen und anderen Awendungen, Teubner, Leipzig, 1893)
32. S. Lie, *with Georg Sheffers* (Geometrie der Berührungstransformationen, Teubner, Leipzig, 1896)
33. H. Liebmann, *Lehrbuch der Differentialgleichungen* (Verlag Von Veit, Leipzig, 1901)
34. H. Liebmann, Geometrische Theorie der Differentialgleichungen, in [28]
35. G. Loria, *Sulle corrispondenze projettive fra due piani e tra due spazii* (XXII, Giornale di Matematiche, 1884), pp. 1–16
36. G. Loria, *Spezielle algebraische und transzendente ebene Kurven* (Leipzig, B.G. Teubner, 1910–1911)
37. E.O. Lovett, Lie's geometry of contact transformations. Bull. Am. Math. Soc. **3**(9), 321–350 (1897)
38. P. Nastasi, E. Rogora, From internationalization to autarky: mathematics in Rome between the two world wars. Rendiconti di Matematica e delle sue applicazioni **41**(1), 1–50 (2020)
39. P. Painlevè, *Mèmoire sur les èquations diffèrentielles du premier ordre* (Gauthier-Villars et fils, Paris, 1892)
40. H. Poincarè, "Mèmoires sur les courbes dèfinès par une equatione differentielle"—*Journal de Mathèmatiques Pures et Appliquèes*: I. Mèmoire, (3)7, (1881), p. 375; II. Mèmoire, (3)8, (1882), p. 251; III. Mèmoire, (4)1, (1885), p. 167; IV. Mèmoire, (4)2, (1886), p. 151
41. J.V. Poncelet, *Traitè des proprièetes projectives des Figures* (Bacheliers, Paris, 1822)
42. D. Rowe, The early geometrical works of Felix Klein and Sophus Lie, in *The History of Modern Mathematics*, vol. 1, ed. by J. McCleary, D. Rowe (Academic Press, New York, 1989), pp. 209–273
43. G. Sansone, *Equazioni differenziali nel campo reale* (Parte I, Zanichelli, Bologna, 1941)
44. G. Sansone, *Equazioni differenziali nel campo reale* (Parte II, Zanichelli, Bologna, 1941)
45. G. Scheffers, Besondere Transcendente Kurven, in [?], pp. 185–268
46. G. Scorza La scuola geometrica italiana, in Gino Borgagli Petrucci, *L'Italia e la scienza*, Le Monnier, Firenze, pp. 117–145 (1932)
47. C. Segre, Sulla teoria e sulla classificazione delle omografie in uno spazio lineare ad un numero qualunque di dimensioni. Mem. R. Acc. Naz. Lincei **19**, 127–148 (1883–1884)
48. C. Segre, Un'osservazione sui sistemi di rette degli spazi superiori. Rend. Circolo Mat. Palermo **2**, 148–149 (1888)

49. C. Segre, Su alcuni indirizzi nelle investigazioni geometriche. Osservazioni dirette ai miei studenti. Rivista di Matematica **1**, 42–66 (1891)
50. C. Segre, Preliminari di una teoria delle varietà luoghi di spazi. Rendiconti del Circolo Matematico di Palermo **30**, 87–121 (1910)
51. C. Segre, Quaderno delle lezioni del corso di "Geometria delle equazioni differenziali", a.a. (1920–1921). http://www.corradosegre.unito.it/Quaderni/Quad34/1_34.php
52. C. Segre, Quaderno delle lezioni del corso di Geometria Superiore 1890–91. Introduzione alla Geometria sugli Enti Algebrici Semplicemente Infiniti, in G. Casnati et al., *From Classical to Modern Algebraic Geometry*, Birkhäuser, Cham, pp. 505–715 with an introduction by Alberto Conte, pp. 499–504 (2016)
53. C. Segre, *Lezioni inedite di due corsi universitari*, pubblicate nella collana *Lezioni e inediti di Maestri dell'Ateneo Torinese* a cura di Alberto Conte, Livia Giacardi, Maria Anna Raspanti per il Centro Studi di Storia dell'Università di Torino (2020)
54. Joseph Alfred Serret, *Cours d'algèbre supèrieure professè à la facultè des sciences de Paris* (Mallet-Bachelier, Paris, 1854)
55. A. Terracini, I quaderni di Corrado Segre, Atti del IV Congresso dell'Unione Mat. It., 1, Cremonese, Roma, pp. 252–262 (1953)
56. V. Volterra, Le matematiche in Italia nella seconda metà del secolo XIX, in *Atti del IV congresso internazionale dei Matematici*, vol. I, ed. by G. Castelnuovo (Tipografia della R. Accademia dei Lincei, Roma, 1909), pp. 55–65
57. K. Weierstrass, Weber ein die homogene Functionen zweiten Grades betreffendes Theorem. M'ber. Akad. Wiss. Berlin 233–246 (1858)
58. K. Weierstrass, Zur Theorie der quadratischen und bilinearen Formen. M'ber. Akad. Wiss. Berlin 311–338 (1868)

Printed by Printforce, United Kingdom